Ecological Studies, Vol. 221

Analysis and Synthesis

For further volumes see list at the end of the book and:
http://www.springer.com/series/86

Jörg Bendix • Erwin Beck • Achim Bräuning •
Franz Makeschin • Reinhard Mosandl •
Stefan Scheu • Wolfgang Wilcke
Editors

Ecosystem Services, Biodiversity and Environmental Change in a Tropical Mountain Ecosystem of South Ecuador

Springer

Editors
Jörg Bendix
Faculty of Geography
University of Marburg
Marburg, Germany

Erwin Beck
Department of Plant Physiology and Bayreuth Centre of Ecology and Environmental Research
University of Bayreuth
Bayreuth, Germany

Achim Bräuning
Institute of Geography
University of Erlangen-Nuremberg
Erlangen, Germany

Franz Makeschin
Institute of Soil Science and Site Ecology
Dresden University of Technology
Tharandt, Germany

Reinhard Mosandl
Institute of Silviculture
Technische Universität München
Freising, Germany

Stefan Scheu
J.F. Blumenbach Institute of Zoology and Anthropology
Georg August University Göttingen
Göttingen, Germany

Wolfgang Wilcke
Geographic Institute of the University of Bern (GIUB)
University of Bern
Bern, Switzerland

ISSN 0070-8356
ISBN 978-3-642-38136-2 ISBN 978-3-642-38137-9 (eBook)
DOI 10.1007/978-3-642-38137-9
Springer Heidelberg New York Dordrecht London

Library of Congress Control Number: 2013943580

Printed on acid-free paper

Springer is part of Springer Science+Business Media (www.springer.com)

Preface

Threat to biodiversity and ecosystem services by global change is meanwhile undisputed. Climate change, expansion of land use, atmospheric fertilization, and invasion by alien species have been identified as the main current and future drivers of ecosystem deterioration (Sala et al. 2000; Pereira et al. 2010). The Millennium Ecosystem Assessment (MEA 2005) connected for the first time the interdependence of ecosystem functioning and human interference with nature in a well-arranged, comprehensive manner by defining specific categories of services which the earth's ecosystems provide for their own stability and in particular for the benefits of their human inhabitants. While acknowledging those services as a major precondition for human well-being, the aim of the report was to assess the consequences of (predominantly) man's impact on ecosystems for human well-being and to provide the scientific basis for a responsible, sustainable use of ecosystems, including conservation. Notwithstanding the appreciation of the impressive conceptual work condensed in that report, its focus on the global dimension of ecosystem services inevitably generates scarcity of regional and local assessments. Thus, e.g. for the biodiversity hotspot of Ecuador information on the current or the predicted states is completely lacking (Fig. 6.1 in MEA 2005). At the same time, the report deplores insufficient knowledge, among others, on (1) long time series of local environmental data, (2) quantitative relationships between biodiversity and ecosystem services, particularly regarding regulative, cultural, and supporting services of specific ecosystems, which would allow predictions, and (3) the incapability to derive regional and local projections of the future development of ecosystem services (MEA 2005).

This book will contribute to fill such local gaps for one of the "hottest" biodiversity hotspots of the world, the south-eastern Andes of Ecuador. Assessment of the current and future state of biodiversity and ecosystem services in the valley of the Rio San Francisco is based on 15 years of comprehensive interdisciplinary ecosystem research, producing a wealth of data, and profound as well as far-reaching information on ecosystem structure and functioning, covering the biotic, abiotic, and socioeconomic spheres. A basis to this endeavor is the predecessor volume ("Gradients in a Tropical Mountain Ecosystem of Ecuador" in the same Series, Vol. 198, edited by Beck et al. 2008), which has been published five years ago. A special advantage of the selected study area is the direct spatial vicinity of the protected mountain rain forest as the natural ecosystem of the region on the one side of the valley and an anthropogenic agricultural replacement system on the opposite side. While the natural forest appears to be fairly resilient to climate changes, the agricultural systems, mostly pastures, turned out to be non-sustainable. The unique opportunity to conduct comparative field surveys and ecological experiments in both manifestations of the ecosystem allowed the authors to gather quantitative information on current ecosystem services which are subjected to the impacts of an ongoing climate and land-use change. With regard to ecosystem services, the book is based on an approach adapting the MEA (2005) service categories, as described in detail in Sect. 4.2.

Part II presents the current state of the different service categories. Naturally this part cannot claim to be exhaustive regarding the immense complexity of the tropical ecosystems. Thus, the authors have focused on services which are of major importance for the country, e.g., biodiversity as the main preserving but also cultural service, the regulation of climate, the water, carbon, and nutrient cycles, considering abiotic and biotic elements, the provision of water, the deposition of airborne nutrients, and various options of agricultural provisioning services (forestry and pasture management). The latter have been analyzed in a holistic way, ranging from ecological aspects to socioeconomic issues, in particular the sustainability of indigenous land-use systems.

Regarding prospective approaches, ecological intervention experiments on the one hand and numerical models calibrated and parameterized by a multitude of measured data on the other provide the basis for scenarios for the future development of the investigated ecosystems and ecosystem services. This is the concern of Part III. Special attention is given to derive a sustainable land-use portfolio from an ecologically adapted combination of suitable agricultural strategies and managements.

The main synthesis (Part IV) summarizes the accumulated comprehensive knowledge, culminating in a science-directed recommendation of sustainable land-use system for the hotspot area, which was the overarching aim of the past 6 years of research. Although the book reports projects of basic research, there is one major point which must not be overlooked. In the spirit of the Access and Benefit Sharing (ABS) principle publicized by the CBD (Convention on Biological Diversity), research in a developing country should address the needs of the local communities and should be conducted together with the local people, scientists, and stakeholders for the sake of building capacity. After 15 years of joint German–Ecuadorian research, a multitude of benefits have been achieved and are communicated in Part IV. This holds in particular for the academic scene of southern Ecuador. Furthermore, the compiled results and developed technologies of several projects are now ready for transfer into application to serve the local society. Consequently, the potential of the research results for knowledge transfer has been assessed here, too.

At this point, it should be stressed that the results of this book not only hold for the ecosystem of the Rio San Francisco Valley but *mutatis mutandis* show transferability to other forested tropical mountain areas of the Andes (and beyond), if located in a comparable altitudinal range of approximately 1,000–3,500 m a.s.l. The environmental background conditions of the study area are comparable to many other sites at the tropical eastern Andean ranges. The altitudinal level of the study area is subjected to the influence of a belt of high cloudiness and precipitation, the so-called Andes-Occurring System (AOS), ranging from Columbia to Peru (Bendix et al. 2006). As in the study area, the population pressure in the biodiversity hotspot of the entire tropical Andes is one of the highest in the world. This causes ongoing land-use changes, i.e., clearing of the natural forest to increase livelihood by exploiting provisioning services as revenues from agriculture. However, the needed conversion of natural forest into arable land at the same time deteriorates ecosystem services at other levels. As in the study area, the removal of forest for pastures is the current land-use practice everywhere in the tropical Andes (Mulligan et al. 2009). This type of land-use change is generally suspected to threaten cultural, supporting, regulating and provision services, and also knowledge which is associated with functional biodiversity.

However, many uncertainties of ecological, economic, and social nature remain with respect to the bouquet of ecosystem services from the natural and the man-made ecosystems in the research areas and beyond. The book takes up all these uncertainties and attempts to provide exemplarily transferable comments on the state of current ecosystem services and their management.

Last but not least, the endeavor of compiling an interdisciplinary book of this extent is a major challenge. This had not been possible without the extraordinary commitment of the 103 authors who contributed their excellent knowledge, creativity, and enthusiasm during the compilation of the manuscript. Many thanks go also to our editors for moderating partly controversial but fruitful discussions in order to match the individual chapters and to the publisher for supporting the publication of our results in the Ecological Studies series. Our assistant editor, Dr. Esther Schwarz-Weig (Mistelgau), deserves a special praise for her outstanding perseverance and patience in collecting, editing, and commenting on the chapters. Without her help, this book would certainly never have been realized. The authors would also like to thank the German Research Foundation (Deutsche Forschungsgemeinschaft DFG) for generously funding the research and the external board of advisors/reviewers for their help to refine the research program. For the achievement of the knowledge compiled in this book, the excellent cooperation with Ecuadorian colleagues and local people was instrumental, who became good friends over the time. The foundation Nature and Culture International (NCI) provided the facilities, in particular the very well-equipped research station ECSF (Estación Científica San Francisco) together with the surrounding research area. The effective running of this station by NCI in cooperation with the German scientific coordinators Dr. Felix Matt and Dipl. Geoecologist Jörg Zeilinger must be considered a stroke of luck for the entire enterprise "Ecosystem Studies in South Ecuador." The support of our counterparts from the Ecuadorian Universities, above all from the Universidad Técnica Particular de Loja (UTPL), the Universidad Nacional de Loja (UNL), and the University of Cuenca, also deserves special acknowledgment. The authorship and coauthorship of many Ecuadorian collaborators in this book witness the excellent cooperation. Last but not least, we thank the Ecuadorian governmental administration for enabling this exciting research and, on behalf of others, the Ecuadorian Ministry of Environment (MAE) for issuing the research permissions. As the space of this preface is limited, it is not possible to thank all people by names who have contributed to the success of our research and in turn to the realization of this book. Nevertheless, we are very grateful to them and their support of our venture is well appreciated.

Marburg, Germany Jörg Bendix
Bayreuth, Germany Erwin Beck

References

Beck E, Bendix J, Kottke I, Makeschin F, Mosandl R (eds) (2008) Gradients in a tropical mountain ecosystem of Ecuador. Ecological studies, vol 198. Springer, Berlin, pp 525

Bendix J, Rollenbeck R, Göttlicher D, Cermak J (2006) Cloud occurrence and cloud properties in Ecuador. Clim Res 30:133–147

Millennium Ecosystem Assessment (2005) Ecosystems and human well-being: biodiversity synthesis. World Resources Institute, Washington, DC

Mulligan M, Rubiano J, Hyman G, Leon JG, Saravia M, White D, Vargas V, Selvaraj J, Ball C, Farrow A, Marín JA, Pulido OL, Ramírez A, Gutierrez T, Cruz LA, Castro A, Andersson M (2009) The Andes Basin Focal Project. Final report to the CGIAR challenge program on water and food. CPWF Project Report PN63. http://mahider.ilri.org/bitstream/handle/10568/3872/PN63_KCL_Project%20Report_Dec09_final.pdf?sequence=1. Accessed 15 Sept 2012

Pereira HM, Leadley PW, Proença V, Alkemade R, Scharlemann JP, Fernandez-Manjarrés JF, Araújo MB, Balvanera P, Biggs R, Cheung WW, Chini L, Cooper HD, Gilman EL, Guénette S, Hurtt GC, Huntington HP, Mace GM, Oberdorff T, Revenga C, Rodrigues P, Scholes RJ, Sumaila UR, Walpole M (2010) Scenarios for global biodiversity in the 21st Century. Science 330:1496–1501

Sala OE, Chapin III FS, Armesto JJ, Berlow E, Bloomfield J, Dirzo R, Huber-Sannwald E, Huennecke LF, Jackson RB, Kinzig A, Leemans R, Lodge DM, Mooney HA, Oesterheld M, Poff NL, Sykes MT, Walker B, Walker M, Wall DH (2000) Global biodiversity scenarios for the year 2100. Science 287:1770–1774

Contents

Part I Introduction

1 **The Study Area** 3
Michael Richter, Erwin Beck, Rütger Rollenbeck, and Jörg Bendix

2 **Environmental Changes Affecting the Andes of Ecuador** 19
Thorsten Peters, Thomas Drobnik, Hanna Meyer, Melanie Rankl, Michael Richter, Rütger Rollenbeck, Boris Thies, and Jörg Bendix

3 **The Challenges of Sustainable Development in the Podocarpus-El Cóndor Biosphere Reserve** 31
Jan Barkmann, Sofia Eichhorn, Byron Maza, Frank v. Walter, and Roland Olschewski

4 **The Research Unit RU 816: Overall Approach in the Light of the Ecosystem Services Concept** 41
Jan Barkmann, Boris M. Hillmann, and Rainer Marggraf

Part II Current Situation of Biodiversity and Ecosystem Services

5 **Landscape History, Vegetation History, and Past Human Impacts** . . . 53
Fernando Rodriguez, Achim Bräuning, Andrés Gerique, Hermann Behling, and Franziska Volland

6 **Past Dynamics of Speciation in Andean Mountain Rainforests** 67
Konrad Fiedler and Patrick Strutzenberger

7 **Diversity in Soil Fungi, Protists, and Microarthropods** 81
Matthias C. Rillig, Tessa Camenzind, Julia Gawlik, Ingeborg Haug, Valentyna Krashevska, Mark Maraun, Dorothee Sandmann, and Stefan Scheu

8 Plant Diversity and Its Relevance for the Provision of Ecosystem Services 93
Jürgen Homeier, Florian A. Werner, Julia Gawlik, Thorsten Peters, Karl-Heinz J. Diertl, and Michael Richter

9 Supporting, Regulating, and Provisioning Hydrological Services . . . 107
Lutz Breuer, David Windhorst, Andreas Fries, and Wolfgang Wilcke

10 The Carbon Balance of Tropical Mountain Forests Along an Altitudinal Transect 117
Christoph Leuschner, Alexandra Zach, Gerald Moser, Jürgen Homeier, Sophie Graefe, Dietrich Hertel, Bärbel Wittich, Nathalie Soethe, Susanne Iost, Marina Röderstein, Viviana Horna, and Katrin Wolf

11 Current Regulating and Supporting Services: Nutrient Cycles 141
Wolfgang Wilcke, Jens Boy, Ute Hamer, Karin Potthast, Rütger Rollenbeck, and Carlos Valarezo

12 Natural Landslides Which Impact Current Regulating Services: Environmental Preconditions and Modeling 153
Jörg Bendix, Claudia Dislich, Andreas Huth, Bernd Huwe, Mareike Ließ, Boris Schröder, Boris Thies, Peter Vorpahl, Julia Wagemann, and Wolfgang Wilcke

13 Conservation, Management of Natural Forests and Reforestation of Pastures to Retain and Restore Current Provisioning Services . . . 171
Michael Weber, Bernd Stimm, Maria Fernanda López, Andrés Gerique, Perdita Pohle, Patrick Hildebrandt, Thomas Knoke, Ximena Palomeque, Baltazar Calvas, Sven Günter, Nikolai Aguirre, and Daniel Kübler

14 Mycorrhiza Networks Promote Biodiversity and Stabilize the Tropical Mountain Rain Forest Ecosystem: Perspectives for Understanding Complex Communities 187
Ingrid Kottke, Sabrina Setaro, Ingeborg Haug, Paulo Herrera, Dario Cruz, Andreas Fries, Julia Gawlik, Jürgen Homeier, Florian A. Werner, Andrés Gerique, and Juan Pablo Suárez

15 Current Provisioning Services: Pasture Development and Use, Weeds (Bracken) and Management 205
Kristin Roos, Jörg Bendix, Giulia F. Curatola, Julia Gawlik, Andrés Gerique, Ute Hamer, Patrick Hildebrandt, Thomas Knoke, Hanna Meyer, Perdita Pohle, Karin Potthast, Boris Thies, Alexander Tischer, and Erwin Beck

16 Current Provisioning Ecosystem Services for the Local Population: Landscape Transformation, Land Use, and Plant Use 219
Perdita Pohle, Andrés Gerique, Maria Fernanda López, and Regine Spohner

17 Sustainable Agriculture and Conservation Payments Are Key Factors in Mitigating Tropical Forest Loss 235
Baltazar Calvas, Thomas Knoke, Luz Maria Castro, Patrick Hildebrandt, Michael Weber, Bernd Stimm, Reinhard Mosandl, Sven Günter, and Nikolai Aguirre

Part III Future Environmental Changes and Their Impacts on Biodiversity and Ecosystem Services

18 Climate Change: Effects on Biodiversity and Ecosystem Functioning 247
Florian A. Werner, Nele Jantz, Valentyna Krashevska, Thorsten Peters, Hermann Behling, Mark Maraun, Stefan Scheu, and Gunnar Brehm

19 Global Climate Change Impacts on Local Climate and Hydrology . . . 265
Lutz Breuer, Jean-François Exbrayat, Ina Plesca, Wouter Buytaert, Theresa Ehmann, Thorsten Peters, Edison Timbe, Katja Trachte, and David Windhorst

20 Impacts of Local Land-Use Change on Climate and Hydrology . . . 275
David Windhorst, Brenner Silva, Thorsten Peters, Hanna Meyer, Boris Thies, Jörg Bendix, Hans-Georg Frede, and Lutz Breuer

21 Current and Future Variations of Nutrient Depositions and Influences on Tree Growth 287
Rütger Rollenbeck, Insa Otte, Peter Fabian, Wolfgang Wilcke, Darwin Pucha Cofrep, Achim Bräuning, and Jörg Bendix

22 Nutrient Additions Affecting Matter Turnover in Forest and Pasture Ecosystems 297
Ute Hamer, Karin Potthast, Wolfgang Wilcke, Hans Wullaert, Carlos Valarezo, Dorothee Sandmann, Mark Maraun, Stefan Scheu, and Jürgen Homeier

23 Effects of Nutrient Addition on the Productivity of Montane Forests and Implications for the Carbon Cycle 315
Jürgen Homeier, Christoph Leuschner, Achim Bräuning, Nixon L. Cumbicus, Dietrich Hertel, Guntars O. Martinson, Susanne Spannl, and Edzo Veldkamp

24 Climate Change and Its Impact on Current and Future Vegetation Dynamics and Carbon Cycling 331
Brenner Silva, Claudia Dislich, Ingo Voss, Kristin Roos, Renate Scheibe, Peter Vorpahl, Boris Schröder, Andreas Huth, Erwin Beck, and Jörg Bendix

25 Sustainable Use of Tropical Forests: A Plea for a Landscape View 343
Thomas Knoke, Baltazar Calvas, Patrick Hildebrandt, Michael Weber, Bernd Stimm, Sven Günter, Nikolai Aguirre, and Reinhard Mosandl

26 Future Provisioning Services: Repasturisation of Abandoned Pastures, Problems, and Pasture Management 355
Erwin Beck, Jörg Bendix, Brenner Silva, Rütger Rollenbeck, Lukas Lehnert, Ute Hamer, Karin Potthast, Alexander Tischer, and Kristin Roos

Part IV Synopsis

27 Synopsis: Towards a Sustainable Land Use Portfolio 373
Erwin Beck and Jörg Bendix

28 Knowledge Transfer for Conservation and Sustainable Management of Natural Resources: A Case Study from Southern Ecuador 395
Sven Günter, Baltazar Calvas, Thomas Lotz, Jörg Bendix, and Reinhard Mosandl

29 The Role of Biodiversity Research for the Local Scientific Community 411
Perdita Pohle, Maria Fernanda López, Erwin Beck, and Jörg Bendix

Index 429

Part I
Introduction

Chapter 1
The Study Area

Michael Richter, Erwin Beck, Rütger Rollenbeck, and Jörg Bendix

1.1 Why the Andes of Southern Ecuador?

The Andes of southern Ecuador are considered as one of the "hottest" global hotspots of vascular plant (Barthlott et al. 2007; Brummitt and Lughadha 2003; Jørgensen and Ulloa Ulloa 1994) and bird diversity (Orme et al. 2005). The major proportion of the biological diversity is found with the native mountain forest which as a rather stable ecosystem provides a multitude of services. However, this ecosystem is severely threatened by on-going deforestation which takes place at the highest annual deforestation rate (−1.7 %) of entire South America (Mosandl and Günter 2008). The forces driving forest decline in Ecuador are manifold (Rudel and Horowitz 1993; Pichón 1996; Mena et al. 2006). As main drivers in southern Ecuador the colonisation laws and land reforms, the population pressure and transmigration, the existence of state-owned land and unclear property regimes as well as the recently improving accessibility (road construction) could be identified (cf. Chap. 16).

More than 10 years of comprehensive interdisciplinary research in the San Francisco Valley in the eastern Andean Cordillera of southern Ecuador have proven the megadiverse character of the native mountain forest (Beck et al. 2008a). Apart from a high diversity of vascular plants (e.g. >280 tree and >337 orchid species) a variety of other organismic groups are represented with extraordinary species

M. Richter (✉)
Institute of Geography, Friedrich-Alexander University Erlangen-Nürnberg, Kochstr. 4/4, 91054 Erlangen, Germany
e-mail: sairecabur@web.de

E. Beck
Department of Plant Physiology and Bayreuth Centre of Ecology and Environmental Research (BAYCEER), University of Bayreuth, 95440 Bayreuth, Germany

R. Rollenbeck • J. Bendix
Laboratory of Climatology and Remote Sensing, Faculty of Geography, University of Marburg, Deutschhausstr. 10, 35032 Marburg, Germany

J. Bendix et al. (eds.), *Ecosystem Services, Biodiversity and Environmental Change in a Tropical Mountain Ecosystem of South Ecuador*, Ecological Studies 221,
DOI 10.1007/978-3-642-38137-9_1, © Springer-Verlag Berlin Heidelberg 2013

numbers (Liede-Schumann and Breckle 2007; Beck and Richter 2008). Records from the central part of the research area (natural forest area = 11.3 km^2, anthropogenic area = 1.91 km^2) revealed more than 500 species of bryophytes (mosses, liverworts, hornworts; up to 98 species on a single tree) and around 1,600 cormophytes. Even more diverse are insects, exemplified by a world record of 2,400 moth species. Birds and bats exhibit likewise extraordinary high diversity. A few organismic groups are nevertheless relatively poor in species, among them taxa of soil organisms, particularly the litter decomposers. Soil mites and earthworms are scarce or at least less abundant and diverse (Illig et al. 2005), probably due to nutrient limitations.

As in other parts of the country, the native forest has been and still is widely cleared in southern Ecuador for gaining pasture land. This particularly holds for well-accessible areas (e.g. close to roads) where there is no forest protection. Recurrent burning as the hitherto practised pasture management option is fostering the invasion and spreading of an aggressive weed, the Southern Bracken fern (*Pteridium arachnoideum* (Kaulf.) Maxon and *Pt. caudatum* (L.) Maxon) which outcompetes the pasture grasses, finally leading to the abandonment of the spoiled land (Roos et al. 2010; Hartig and Beck 2003). As expected, biodiversity of this highly disturbed sites ("Llashipa" = fern) is generally lower compared to the native forest, but this does not hold for all organismic groups (e.g. Haug et al. 2010).

The status of the mountain ecosystems encompassing native forests and pastures as the anthropogenic replacement systems shows that the current mode of exploiting the provisioning services in the Andean ecosystems of southern Ecuador is not sustainable. Hence there is serious concern that essential ecosystem services at all levels are affected particularly by the land use practices not refraining from further clearing of the natural forest. If no effective counter-measures are taken, biodiversity and ecosystem services are supposedly further declining, as they are additionally challenged by climate change for which evidence in the area will be presented (Bendix et al. 2010, Chap. 2). Based on the results of a comprehensive research effort in the Andes of southern Ecuador, this book aims to provide deeper insights into current and future ecosystem functioning and services in a mega-biodiverse hotspot area.

1.2 The Location of the Study Site

The entire research area comprises the Andean part of south-eastern Ecuador between the provincial capital Loja in the west, the provincial capital Zamora in the east and the town Vilcabamba in the South (Fig. 1.1). The main connection road between the two capitals runs along the valley of the Rio San Francisco (RSF). The core area of the research activities expands around the research station ECSF (Estación Científica San Francisco, lat. 3°58′18″ S, long. 79°4′45″ W, 1,860 m a.s.l.) and the focal experimental site in the natural forest RBSF (Reserva Biológica San Francisco) which ranges from 1,600 to 3,140 m a.s.l. The lower parts of the

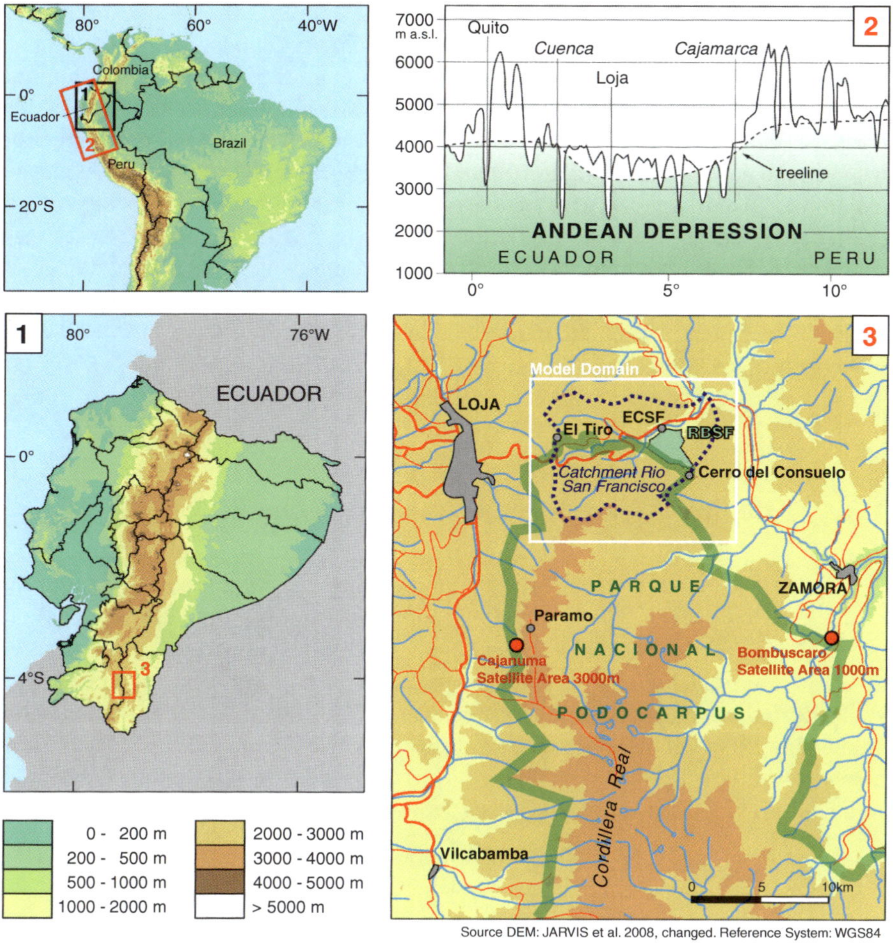

Fig. 1.1 Location of the research area. The image on *top* (*left*) shows the area of the map 1 and the site of the cross section of map 2 highlighting the Andean depression. Graph 3 shows the location of the research area (Reserva Biologica San Francisco, RBSF), the catchment of the Rio San Francisco and the two mountains El Tiro and Cerro del Consuelo at the borders of the research area. The satellite areas at Bombucaro and at Cajanuma were used to study altitudinal gradients. The digital elevation model (DEM) is based on Jarvis et al. (2008)

south-facing slopes opposite the forest are covered by active but also abandoned pastures. The border between the provinces Loja and Zamora-Chinchipe runs along the crest of the Cordillera Real (eastern cordillera range) which is the core of the Podocarpus National Park (PNP). For a complete altitudinal gradient (Beck et al. 2008a) over the natural forest, two experimental satellite areas (Bombuscaro at 1,000 m a.s.l. and Cajanuma at 3,000 m a.s.l., see Fig. 1.1) have been established near the western and eastern border of the PNP, respectively.

The elevation of the Cordillera Real does not exceed 3,900 m a.s.l. With this relatively low vertical extension, the Andes of southern Ecuador are the lowest part of the tropical Andes. As stressed, e.g. in Richter et al. (2009), this depression (the Amotape-Huancabamba Depression) forms a transition zone between the higher northern and the central tropical Andes in the South and in east–west direction between the moist Amazon rainforest and the dry Sechura desert. As such, it plays a major role in the development of the extremely high biological diversity of the ecoregion (see Sect. 1.4.3). Furthermore, the SE-Ecuadorian Andes are characterised by a high topographic fragmentation which results in numerous isolated basins and ridges, thus offering a great variety of habitats which might foster development and maintenance of a high organismic diversity, as well as of endemism (c.f. Oesker et al. 2008).

1.3 Ecological Measurements and Experiments

The scale concept of the research programme correlates with the clustering of the ecological experiments. The satellite research sites at 1,000 and 3,000 m elevation, respectively, allow comparative surveys and ecological experiments along a 2,000 m altitudinal gradient. The experimental sites have been equipped with sophisticated instrumentation and tailor-made ecological treatments produced the data and results which are presented in this book. The location of the research sites are shown in Fig. 1.2.

The following measurements and experiments were carried out:

1. The still on-going Nutrient Manipulation Experiment (NUMEX) investigates effects of nutrient input into the megadiverse natural forest, simulating the natural atmospheric deposition, however in selected constant amounts. Research objectives are the effects on biomass allocation of trees, species composition (biodiversity) and the biogeochemical cycle. The consortium operates main experimental sites at three elevations and several additional plots in the matrix area, to consider also the influence of the topographic variability (see Chaps. 22, 23 and 26).
2. On the pasture site, the corresponding FERPAST experiment investigates the effects of nutrient addition on pasture yields in relation to the biology and chemistry of the soils (Chaps. 22 and 26).
3. In the forest area, several long-term experiments monitor temporal oscillations of key components of the biogeochemical cycle (long term ecosystem study) or investigate the outcome and ecological impact of a moderate promotion of valuable timber species in the natural forest (natural forest experiment, Chap. 13).
4. On the abandoned land, experiments were performed (1) to understand the competition of bracken fern with the common pasture grass *Setaria sphacelata*, (2) to elaborate optimal pasture management protocols and (3) to study the

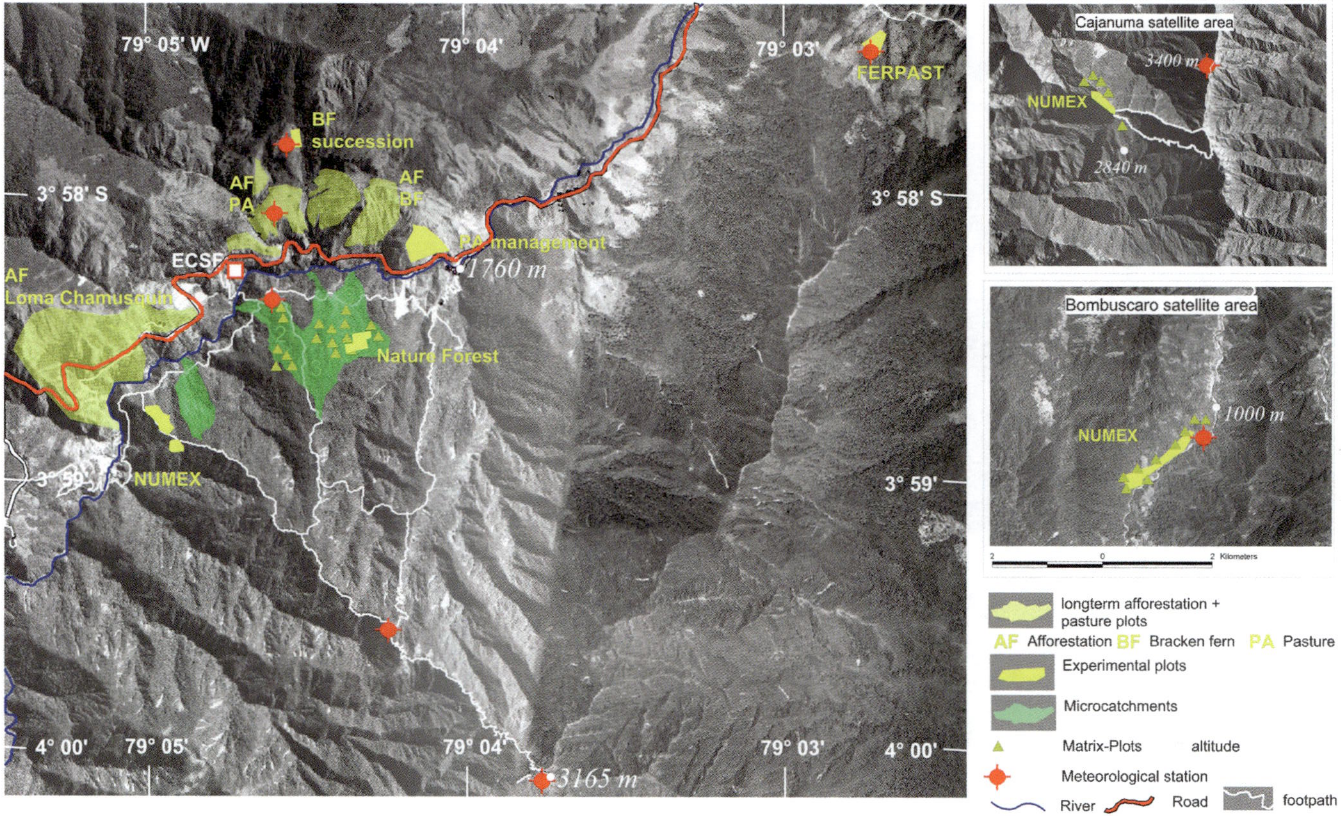

Fig. 1.2 Location of experimental sites and the research infrastructure. Further information about measurements, the Nutrient Manipulation Experiment (NUMEX, description in Chap. 23), the Fertilization of Pastures experiment (FERPAST, description in Chap. 22) and the Bracken Fern succession experiment (description in Chap. 26) is provided in the text. *ECSF* Estación Científica San Francisco

success of afforestation with indigenous tree species on active and abandoned pastures as well as under the shelter of exotic trees (see Chaps. 15 and 26).
5. The experimental setup is completed by basic research infrastructure as, e.g. a network of meteorological and hydrological stations.
6. In order to upscale the experimental findings to the landscape scale, numerical models have been developed and/or parameterized on the scale of the (1) experimental sites, (2) the RBSF (model domain 1) or (3) the total catchment of the Rio San Francisco (Fig. 1.1, framed in blue) and beyond. The adapted models are used to understand ecosystem processes and to generate future scenarios as presented in Part III of the book.

1.4 Selected General Features of the Study Area

The general abiotic and biotic zonation, as well as the geology, topography, the soils and the entire biogeography of the area including population features of southern Ecuador has already been described in Beck et al. (2008a) and Beck and Richter (2008). However, selected new findings which are completing the overall knowledge of the study area deserve special attention with regard to the following aspects: (1) the complex dynamics of rainfall formation in the study area and other climate peculiarities, (2) the natural disturbances as an internal trigger of biological megadiversity in the tropical mountain forest system and (3) the reasons for the megabiodiversity of the area.

1.4.1 Climate Peculiarities

The climate of the study area is generally characterised by extreme horizontal gradients. Over a very short distance of ~25 km, annual rainfall increases from less than 500 mm in the semi-arid inter-Andean basin of Catamayo west of Loja to more than 6,000 mm (including cloud water deposition) at the top region of the Cordillera Real (Emck 2007; Richter 2003), where the crest of the cordillera constitutes a clear weather divide with extremely wet conditions on the eastern escarpment (see Fig. 1.3). At the same time, the seasonality changes from two rainfall maxima in austral spring and autumn at Loja to one rainfall peak in austral winter (JJA: June to August) for the Rio San Francisco Valley.

In austral winter quasi-permanent easterly winds (>70 %) lead to rainfall maxima and a high diurnal persistence of the rain from relatively shallow cap clouds (cloud top between 4 and 5 km; Bendix et al. 2006b) resulting from forced convection and condensation of moist Amazonian air masses impinging on the eastern slopes of the Andes (Fig. 1.3). From October to January (austral summer), dry periods but also strong convective, thermally induced rainfall events are mostly combined with (north-) westerly weather conditions (WWCs) and with

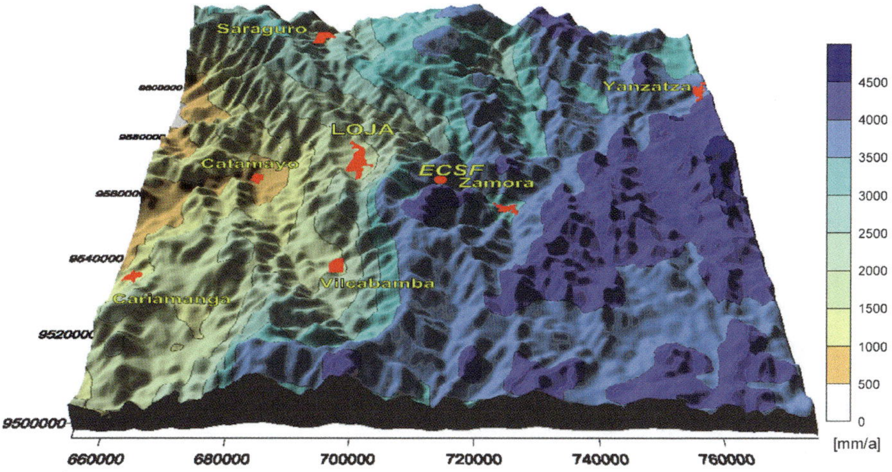

Fig. 1.3 Map of average annual rainfall (1998–2010) derived from rain radar data and ground measurements projected on a digital elevation model of the region around Loja (see also Rollenbeck and Bendix 2011)

VdN-situations (Veranillo del Niño) with a maximum incidence in November (Emck 2007; Bendix et al. 2008a). During this drier period convective showers induced by the valley-breeze system are also frequent in the afternoons of the San Francisco Valley. With regard to the vertical cloud moisture gradient in the valley, a clear increase of cloud frequency and thus, cloud water deposition is observed above 2,600 m (Rollenbeck et al. 2010). During the early morning hours around sunrise, however, cloud frequency increases almost linearly upwards from the valley bottom, starting with higher values also in its lowermost range (Bendix et al. 2008b). Radiation fog (visibility < 1 km) at the valley bottom is scarce (Fig. 1.3) and relates to nocturnal radiation losses during dry periods, while at higher altitudes particularly in austral winter cloud fog and cloud water deposition is extremely prominent, due to the cap clouds touching the upper slopes (cloud fog).

A novel finding is the occurrence of two daily maxima of precipitation in the Rio San Francisco Valley. Apart from the afternoon showers during drier periods the other maximum can be attributed to cap clouds in austral winter. These early morning rains are also related to highland–lowland interactions and mesoscale atmospheric dynamics. The confluence of nocturnal katabatic flows from the escarpment—caused by the specific concave shaping of the eastern Andes in the East of the research area—produces local cold fronts in the warm-moist Amazon air during the night. In combination with an Andean-parallel low level jet (LLJ) above 1,500 m a.s.l. they cause strong atmospheric instability leading to the formation of mesoscale convective systems (MCS) over the eastern foothills. In their mature phase around sunrise (MSC_{ms} in Fig. 1.4), these systems are forced towards the eastern escarpment by the easterlies, fostering rainfall over the valley itself (details in Bendix et al. 2009; Trachte et al. 2010).

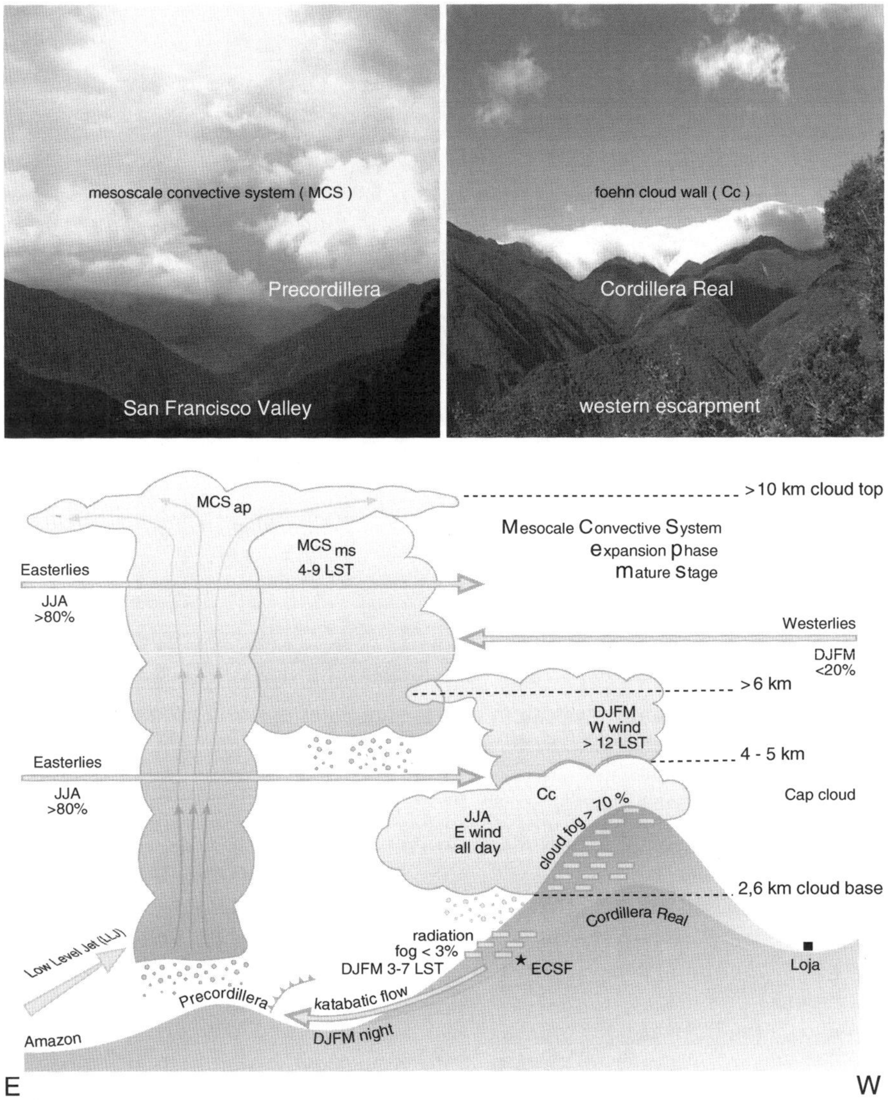

Fig. 1.4 Summary of cloud and rain formation processes affecting the study area. *Note*: The graph contains climate features of various seasons. JJA = austral winter (June–July–August), DJFM = austral summer (December–January–February–March). *ECSF* Estación Científica San Francisco, *LST* local standard time, *MCS* mesocale convective system. Photos: Michael Richter (*left*) and Paul Emck (*right*)

The great number of rainy and cloudy days results in low amplitudes of the persisting high relative humidity during the peak season of rainfall in austral winter (JJA). In contrast, in October–November pronounced fluctuations occur during VdNs, when abrupt drops of 80 % can happen within few hours. The most dramatic case was recorded at night (!) of 17/18 October 1998, when the summit area was stressed by a sudden decrease from relative humidity (rH) >95 % down to <15 %, initiated by an enhanced downswing of air masses towards a strong low at the eastern foothills of the Andes.

Extraordinarily high values of global irradiance of up to 1,832 W m^{-2} were measured in the paramó above the tree line during such irregular meteorological events. Incidence of "superirradiance" of over 170 % of the potential "clear sky" irradiance occur predominantly under easterly weather conditions (EWC) when global radiation is enhanced by reflections and diffractions of transparent clouds screening the sun disk and through lensing effects by water droplets (Emck and Richter 2008). Ultraviolet radiation with its mutagenic potential must be considered to reach record levels as well, which might enhance the genetic dynamic of the organisms.

1.4.2 Landforms, Erosion and Mass Movement

Typically, landform characteristics in high mountains result from past as well as recent geomorphological processes. Among the latter, gelifluction and/or glaciation traces of the ice ages are obvious in the Cordillera Real, but not down in the valley. Glacial cirques and moraines are absent as are obvious signs of frost debris remnants. Both phenomena are, however, present in the upper regions of the neighbouring Rio Sabanilla Valley. Hence, apart from the quasi-continuous uplift since the Tertiary, land-forming processes are to a large extent restricted to fluvial erosion and slope denudation (Fig. 1.5).

Deeply incised V-shaped valleys are a typical landform in the entire eastern range of the wet tropical Andes. Side crests and offsets are mostly staggered and graded on concordant levels showing relics of formerly interrelated and meanwhile dissected rock terraces. They are considered residuals of ancient valley floors which developed during phases of slow uplift of the Andes. Narrowly delimited flat sections on side crests are generally fixed on hard parent rock such as quartz, which builds resistant strata between the phyllites or fine sand stones. In the RBSF terrain prominent terrace levels are at 2,200 m a.s.l. and around 2,580 m a.s.l. (pale yellow areas in Fig. 1.6, left). The steep appearance of the research area becomes apparent by the fact that around 55 % of its surface belongs to the inclination class between 25 and 40°, and slopes >40° are represented by almost 20 %.

Due to such inclinations under a perhumid climate and an on-going lowering of the erosion base level, slope stability is weak and landslides are a frequent natural phenomenon. The most striking morphological processes of the San Francisco Valley are numerous translation slides in the rainforest area (Fig. 1.6, right), which are a characteristic feature of destructive mass movements in the Rio San Francisco Valley (see Chap. 12). Vanacker et al. (2003) stated that under forests, slope steepness affects slope stability. Additionally, in case of the forested research area slide activity results from the destabilisation of thick, water-soaked organic and humus layers and in particular of the weight of dense tree stands (Richter 2009,

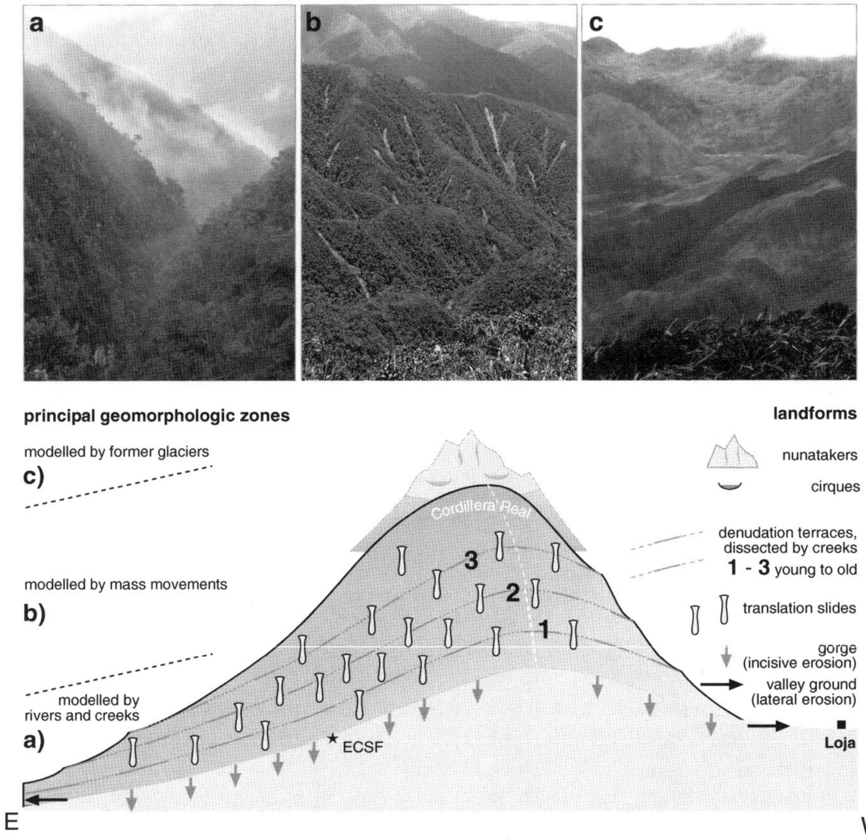

Fig. 1.5 Geomorphologic zonation and landform distribution in the RSF Valley including the summit region few kilometres further south in the Cordillera Real. (**a**) V-shaped RSF Valley near the research station at 1,800 m a.s.l., (**b**) slide section around 2,500 m a.s.l. in the RBSF, (**c**) view towards Lagunas de los Compadres, 3,100–3,550 m a.s.l. Photos: Michael Richter

see also Chap. 12). After deforestation, however, slumps, debris and mud flows on pasture land become rare. Such areas are devoid of the thick organic layer and the humic topsoil is very compact due to trampling by the cattle. If any, bulk density and soil wetness gain some importance as the above ground weight of the vegetation is negligible and the compact root systems of the grasses reach deep.

The importance of landslides for ecological processes in the research area is obvious from the numbers in Table 1.1.

Denudation rates by sheet erosion and splash are present throughout, although of scarce visibility. At around 2,000 m a.s.l., continuous measurements on soil erosion test plots on $40 \pm 5°$ inclined slopes show denudation rates between 200 and 1,100 kg ha^{-1} a^{-1} in the rainforests of RBSF, depending on the density of the understory, while the values on active or abandoned pastures vary between 200 and 300 kg ha^{-1} a^{-1}. Interestingly, these results of removal by sheet erosion resemble

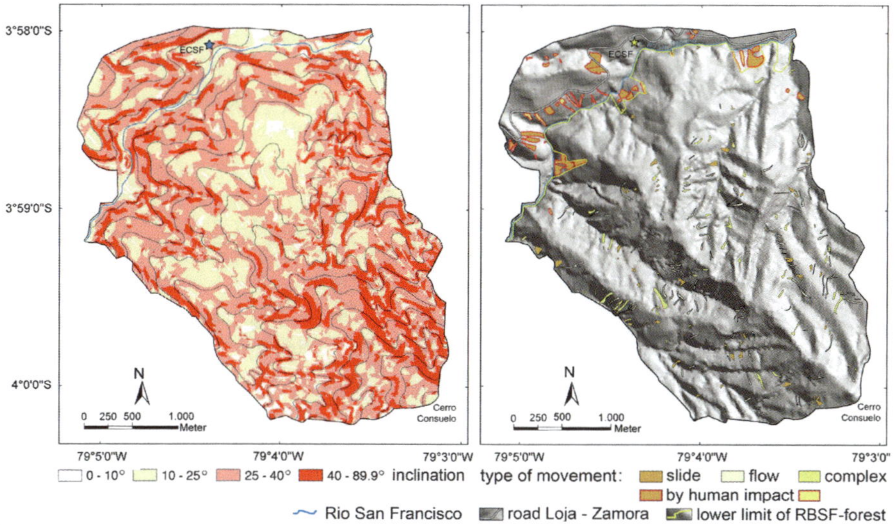

Fig. 1.6 Inclination (*left*) and landslides (*right*; each slide clearly visible in 1998 is indicated, after Münchow et al. 2012) mapped in RBSF including parts of Llashipa

Table 1.1 Landslides in the research area (relating to all incidents between 1963 and 2000; Münchow et al. 2012)

	Natural forest	Roadside and pasture land s.l.
Number of landslides	691	138
Landslide density (km^2)	12	14
Mean landslide size (m^2)	558	1,070
Proportion of slided terrain surface (%)	3.41	7.73

Note: By far, most of the mass movements in pasture and abandoned land are caused by roadside slides

the transport amounts produced by splash, which vary between 250 and 850 kg ha^{-1} a^{-1} in forest stands and around 80–150 kg ha^{-1} a^{-1} on pastures. The natural denudation rates caused by sheet wash and splash depend considerably on the structure of the mountain rainforest, on the soil structure and on the quantity of rainfall. At 1,000 m a.s.l., splash and sheet erosion rates in the rainforests of Bombuscaro are up to three times higher compared to RBSF, as rain intensities and tree size in the lower site are much higher and such are the energies of the droplets (Hagedorn 2001).

1.4.3 *Causes for the Outstanding Biodiversity*

Available reports of recent botanical expeditions and own observations document a high probability to discover unknown plant species in the so-called Amotape-Huancabamba floristic zone (Weigend 2002). With respect to the uplift of the

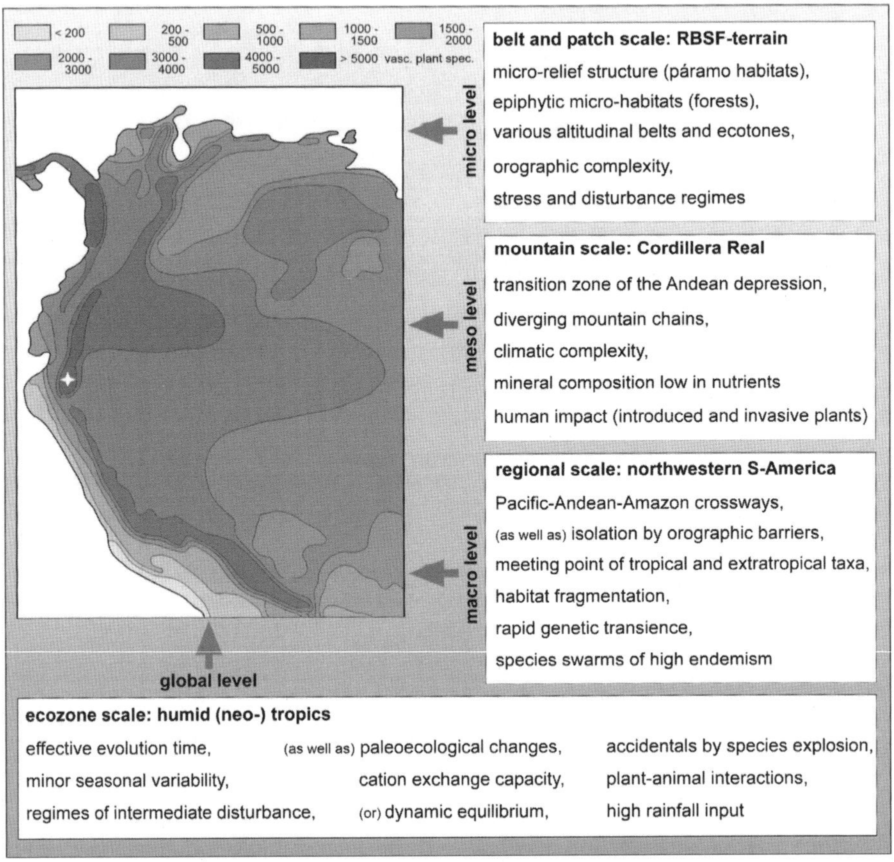

Fig. 1.7 Scale dependent factors contributing to plant diversity in the study area in southern Ecuador (Richter et al. 2009). Four levels from global to microscale are expressed in separate *boxes*. The map of vascular plant species (vasc. Plant spec.) richness is based on a world map by Barthlott et al. (2007). *Star* indicates the position of the RBSF research area

mountain system, this zone represents a geologically young part of the tropical Andes and since the uplift is apparently the crucial event for speciation, a high dynamics of the biological diversity can be expected. On the family level, far above average endemism has been shown for Orchidaceae (55 % of the occurring species), Bromeliaceae (50 %), Asteraceae (37 %) and Piperaceae (37 %). The ample supply of digestible plant organs, flowers and fruits provides a broad spectrum of food for generalists as well as for specified consumers and in turn affects faunal diversity.

Several factors may be considered as driving forces for the development of the high diversity of vascular plants (see the box "ecozone scale" in Fig. 1.7). Examples are the long effective evolution time in combination with a low seasonal variability, the reasonably fair cation exchange capacities of strongly weathered tropical subsoils, high rainfall and the multiplicity of plant–animal interactions (for details see Richter et al. 2009). On a regional and mountain scale, corridors for species migration but also barrier effects from the orographic heterogeneity between ridges and valleys give rise to a variety of microclimates producing a multitude of ecological niches. In addition habitat fragmentation triggered by the high relief energy is of importance. On the plot scale slight terrain differences in particular in the páramo ecosystems or bark peculiarities favouring a variety of epiphytic populations in the RBSF forests are decisive variables for the coexistence of ecological microniches.

Disturbances take place at all scales. On the plot scale, gaps originate by falling branches or tree veterans, while landslides produce vertically oriented gaps of several hundred metres length, each triggering the onset of a succession sequence and thus contributes to the maintenance of the biological diversity of the hotspot. Big disturbances are man-made agricultural areas, mostly pastures, a significant portion of which are not sustainably managed and thus are abandoned after some years of cattle farming (see Chap. 17). After clearing the natural forest by slash and burn, pasture grasses, predominantly exotic species are planted, because grasses are rare (except bamboo) in the mountain forest. The non-natural species composition of the pastures is not resistant to the invasion by fast growing and propagating weeds, as e.g. the Southern Bracken Fern (*Pteridium arachnoideum* and *Pt. caudatum*) or shrubs like *Ageratina dendroides*, *Baccharis latifolia, Brachyotum spec.* and *Monochaetum lineatum* which are accompanied by a specific fauna. They contribute to collateral non-native elements, most of them of ubiquitous occurrence in the tropics and commonly considered as undeserved organisms. Species numbers of most of the native taxonomic groups decline upon human impact (e.g. Rubiaceae, Lauraceae, Araceae among vascular plants and Geometridae among moths) while others increase considerably (e.g. Asteraceae, Poaceae, Melastomataceae and Arctiidae, respectively; Peters et al. 2010; Brehm et al. 2005). Biodiversity is replaced by high abundances of a few species. These changes are accompanied by an expansion of the daily temperature ranges, a decrease in relative humidity, resulting from the conversion of a dense humid forest into open pastures, i.e. into brightly illuminated, drier and warmer environments (Fries et al. 2009, 2012).

References

Barthlott W, Hostert A, Kier G, Küper W, Kreft H, Mutke J, Rafiqpoor D, Sommer JH (2007) Geographic patterns of vascular plant diversity at continental to global scales. Erdkunde 61:305–315

Beck E, Richter M (2008) Ecological aspects of a biodiversity hotspot in the Andes of South Ecuador. In: Gradstein SR, Homeier J, Gansert D (eds) Biodiversity and ecology series 2: the tropical mountain forest – patterns and processes in a biodiversity hotspot, pp 195–217

Beck E, Bendix J, Kottke I, Makeschin F, Mosandl R (eds) (2008a) Gradients in a tropical mountain ecosystem of Ecuador. Ecological studies, vol 198. Springer, Berlin, 525 pp

Bendix J, Rollenbeck R, Göttlicher D, Cermak J (2006b) Cloud occurrence and cloud properties in Ecuador. Clim Res 30:133–147

Bendix J, Rollenbeck R, Fabian P, Emck P, Richter M, Beck E (2008a) Climate variability. In: Beck E, Bendix J, Kottke I, Makeschin F, Mosandl R (eds) Gradients in a tropical mountain ecosystem of Ecuador. Ecological studies, vol 198. Springer, Berlin, pp 280–291

Bendix J, Rollenbeck R, Göttlicher D, Nauß T, Fabian P (2008b) Seasonality and diurnal pattern of very low clouds in a deeply incised valley of the eastern tropical Andes (South Ecuador) as observed by a cost-effective WebCam system. Meteorol Appl 15:281–291

Bendix J, Trachte K, Cermak J, Rollenbeck R, Nauß T (2009) Formation of convective clouds at the foothills of the tropical eastern Andes (South Ecuador). J Appl Meteor Climatol 48:1682–1695. doi:10.1175/2009JAMC2078

Bendix J, Behling H, Peters T, Richter M, Beck E (2010) Functional biodiversity and climate change along an altitudinal gradient in a tropical mountain rainforest. In: Tscharnke T, Leuschner C, Veldkamp E, Faust H, Guhardja E, Bidin A (eds) Tropical rainforests and agroforests under global change. Environmental science series. Springer, Berlin, pp 293–308

Brehm G, Pitkin LM, Hilt N, Fiedler K (2005) Montane Andean rain forests are a global diversity hotspot of geometrid moths. J Biogeogr 31:1621–1627

Brummitt N, Lughadha EN (2003) Biodiversity: where's hot and where's not. Conserv Biol 17:1442–1448

Emck P (2007) A climatology of South Ecuador – with special focus on the major Andean ridge as Atlantic-Pacific climate divide. PhD Thesis. Institute of Geographie, FAU Erlangen, 275 p

Emck P, Richter M (2008) An upper threshold of enhanced global shortwave irradiance in the troposphere derived from field measurements in tropical mountains. J Appl Meteor Climatol 47:2828–2845

Fries A, Rollenbeck R, Göttlicher D, Nauss T, Homeier J, Peters T, Bendix J (2009) Thermal structure of a megadiverse Andean mountain ecosystem in southern Ecuador and its regionalization. Erdkunde 59:321–335

Fries A, Rollenbeck R, Nauss T, Peters T, Bendix J (2012) Near surface air humidity in a megadiverse Andean mountain ecosystem of southern Ecuador and its regionalization. Agric For Meteorol 152(1):17–30

Hagedorn A (2001) Extent and significance of soil erosion in southern Ecuador. Die Erde 132:75–92

Hartig K, Beck E (2003) The bracken fern (Pteridium arachnoideum (Kaulf.) Maxon.) dilemma in the Andes of southern Ecuador. Ecotropica 9:3–13

Haug I, Wubet T, Weiß M, Aguirre N, Weber M, Günter S, Kottke I (2010) Species-rich but distinct arbuscular mycorrhizal communities in reforestation plots on degraded pastures and in neighboring pristine tropical mountain rain forest. Trop Ecol 51:125–148

Illig J, Langel R, Norton RA, Scheu A, Maraun M (2005) Where are the decomposers? Uncovering the soil food web of a tropical montane rain forest in southern Ecuador using stable isotopes. J Trop Ecol 21:589–593

Jarvis A, Reuter HI, Nelson A, Guevara E (2008) Hole-filled seamless SRTM data V4. International Centre for Tropical Agriculture (CIAT). Available from http://srtm.csi.cgiar.org. Accessed 13 Oct 2011

Jørgensen PM, Ulloa Ulloa C (1994) Seed plants of the high Andes of Ecuador – a checklist. Aarhus University (AAU) Reports 34

Liede-Schumann S, Breckle SW (eds) (2007) Provisional checklists of fauna and flora of the San Francisco valley and its surroundings. Ecotropical monographs 4. 256 pp

Mena C, Barbieri A, Walsh S, Erlien C, Holt F, Bilsborrow R (2006) Pressure on the Cuyabeno wildlife reserve: development and land use/cover change in the northern Ecuadorian Amazon. World Dev 34(10):1831–1849

Mosandl R, Günter S (2008) Sustainable management of tropical mountain forests in Ecuador. In: Gradstein SR, Homeier J, Gansert D (eds) Biodiversity and ecology series 2, pp 177–193

Münchow J, Brenning A, Richter M (2012) Geomorphic process rates of landslides along a humidity gradient in the tropical Andes. Geomorphology 139–140:271–284

Oesker M, Daliz H, Günter S, Homeier J, Matezki S (2008) Spatial heterogeneity patterns – a comparison between Gorges and Ridges in the upper part of an evergreen lower montane forest. In: Beck E, Bendix J, Kottke I, Makeschin F, Mosandl R (eds) Gradients in a tropical mountain ecosystem of Ecuador. Ecological studies, vol 198. Springer, Berlin, pp 267–274

Orme CDL, Davies RG, Burgess M, Eigenbrod F, Pickup N, Olson VA, Webster AJ, Ding T-S, Rasmussen PC, Ridgely RS, Stattersfield AJ, Bennett PM, Blackburn TM, Gaston KJ, Owens IPF (2005) Global hotspots of species richness are not congruent with endemism or threat. Nature 436:1016–1019. doi:10.1038/nature03850

Peters T, Diertl K, Gawlik J, Rankl M, Richter M (2010) Vascular plant diversity in natural and anthropogenic ecosystems in the Andes of Southern Ecuador – studies from the Rio San Francisco valley. Mt Res Dev 30:344–352

Pichón FJ (1996) Settler agriculture and the dynamics of resource allocation in frontier environments. Hum Ecol 24(3):341–371

Richter M (2003) Using epiphytes and soil temperatures for eco-climatic interpretations in Southern Ecuador. Erdkunde 57:161–181

Richter M (2009) To what extent do natural disturbances contribute to Andean plant diversity? A theoretical outline from the wettest and driest parts of the tropical Andes. Adv Geosci 22:95–105. doi:10.5194/adgeo-22-95-2009

Richter M, Diertl KH, Emck P, Peters T, Beck E (2009) Reasons for an outstanding plant diversity in the tropical Andes of southern Ecuador. Landsc Online 12:1–35. doi:10.3097/LO.200912

Rollenbeck R, Bendix J (2011) Rainfall distribution in the Andes of southern Ecuador derived from blending weather radar data and meteorological field observations. Atmos Res 99:277–289

Rollenbeck R, Bendix J, Fabian P (2010) Spatial and temporal dynamics of atmospheric water inputs in tropical mountain forests of South Ecuador. Special Issue Hydrometeorology of tropical montane cloud forests. Hydrol Process 25(3):344–352

Roos K, Rollenbeck R, Peters T, Bendix J, Beck E (2010) Growth of tropical bracken (Pteridium arachnoideum): response to weather variations and burning. Inv Plant Sci Manage 3:402–411. doi:10.1614/IPSM-D-09-00031.1

Rudel T, Horowitz B (1993) Tropical deforestation. Small farmers and land clearing in the Ecuadorian Amazon. Columbia University Press, New York, 234 pp

Trachte K, Rollenbeck R, Bendix J (2010) Nocturnal convective cloud formation under clear-sky conditions at the eastern Andes of South Ecuador. J Geophys Res 115, D24203. doi:10.1029/2010JD014146

Vanacker V, Vanderschaeghea M, Govers G, Willems E, Poesen J, Deckers J, De Bievre B (2003) Linking hydrological, infinite slope stability and land-use change models through GIS for assessing the impact of deforestation on slope stability in high Andean watersheds. Geomorphology 52:299–315

Weigend M (2002) Observations on the biogeography of the Amotape-Huancabamba zone in northern Peru. Bot Rev 68:38–54

Chapter 2
Environmental Changes Affecting the Andes of Ecuador

Thorsten Peters, Thomas Drobnik, Hanna Meyer, Melanie Rankl, Michael Richter, Rütger Rollenbeck, Boris Thies, and Jörg Bendix

2.1 Introduction

It is indisputable since the announcement of the Millennium Ecosystem Assessment (2005) that global environmental change, especially land use and climate change, are threatening biodiversity. Although it is widely supposed that climate change will lead to the extinction of many species in the future (Colwell et al. 2008; Williams et al. 2007), human land use is currently the most important threat to biodiversity (Pimm and Raven 2000; Köster et al. 2009; de Koning et al. 1998; Southgate and Whitaker 1994; Bebbington 1993). Sala et al. (2000) have pointed out in this regard that global terrestrial biodiversity will be most severely affected by expanding agriculture by the year 2100, with climate change and nitrogen deposition being the next most important factors. Tropical forests have recently undergone great changes, due mainly to land use activities that annihilate ecological niche diversity and lead to the extinction of species (Sala et al. 2000). In this context it must be emphasised that the tropical Andes contain about one-sixth of all known plant species in a space of $<1\ \%$ of the world's terrestrial area (Mittermeier et al. 1997).

The area of our research—southern Ecuador—comprises dry and humid mountain biomes as well as lowland tropical rainforests. A great variety of ecosystems are found in this area, ranging from high altitude habitats harbouring only a few species to complex, extremely species-rich habitats on the eastern escarpment of the Andes (Richter et al. 2009). Williams et al. (2007) argued that the climate

T. Peters (✉) • T. Drobnik • M. Rankl • M. Richter
Institute of Geography, University of Erlangen-Nuremberg, Kochstr. 4/4, 91054 Erlangen, Germany
e-mail: tpeters@geographie.uni-erlangen.de

H. Meyer • R. Rollenbeck • B. Thies • J. Bendix
Laboratory for Climatology and Remote Sensing, Faculty of Geography, University of Marburg, Deutschhausstr. 10, 35032 Marburg, Germany

J. Bendix et al. (eds.), *Ecosystem Services, Biodiversity and Environmental Change in a Tropical Mountain Ecosystem of South Ecuador*, Ecological Studies 221,
DOI 10.1007/978-3-642-38137-9_2, © Springer-Verlag Berlin Heidelberg 2013

conditions that favoured this biodiversity hotspot during the twentieth century may disappear entirely during the twenty-first century. Consequently, the extraordinarily high biodiversity of this region appears as strongly endangered (Myers et al. 2000; Brummitt and Lughadha 2003; Southgate and Whitaker 1992; Harden 1993; Myers 1988, 1993; Bendix et al. 2010). Scientific research is therefore essential for developing and fostering conservation strategies.

In this chapter current knowledge of the recent development of the three major ecosystem threats of land use dynamics, climate change and nitrogen deposition (Sala et al. 2000) will be discussed for the study area of southern Ecuador.

2.2 Land Use Dynamics

Andean environments have undergone modifications by human activities for at least 7000 years (Bruhns 1994; Jokisch and Lair 2002; Sarmiento and Frolich 2002), but the intensity of land use has accelerated considerably during the past century (Ellenberg 1979; Luteyn 1992; Peters et al. 2010). This especially holds true for Ecuador, which exhibits the highest deforestation rate in South America (FAO 2005; Mosandl et al. 2008). Figure 2.1 shows the decisive role of road construction for land reclamation in this country. In 1938 only few roads existed within the coastal plain and the Andes of Ecuador, while the eastern regions of the country were still untouched. At least 75 % of the western part of the country was forested at that time, and as of 1969 primary forest covered still ca. 63 % (Dodson and Gentry 1991). Only a few new roads were constructed in western and central Ecuador during the interim, while the eastern part of the country remained almost unexploited. During the period up to the year 2000 various factors initiated a rapid expansion of road construction, which also encroached the eastern lowlands (Fig. 2.1). The population increased from less than 4 to more than 10 million people during the same period, and land reform programmes effectively promoted inner colonisation of government-owned forested lands. Large sums were invested in road construction to provide communication between and transport to new cities and transfer sites (Dodson and Gentry 1991). Petroleum became the most important export commodity, and the cultivation of cash crops contributed to the derogation of natural environments. Today an extensive network of primary and secondary roads has opened up most of western and central Ecuador, while parts of the Oriente have been converted into protected areas and safeguarded to certain extents.

Facilitated access has had a devastating effect on the mountain rainforest of the Rio San Francisco (RSF) valley. Between 1960 and 1980 approximately 0.25 % of the south Ecuadorian Andean forests were cut by slash-and-burn annually (Keating 1997; Marquette 2006). While most of the north-facing slopes of the Reserva Biológica San Francisco (RBSF, see Fig. 1.1, map 3) are still covered by primary mountain rainforest today, much of the forest on the south-facing slopes has been converted into pastures. The construction of the road from Loja to Zamora led to a rapid establishment of settlements in the region subsequent to 1957 (Pohle et al. 2009),

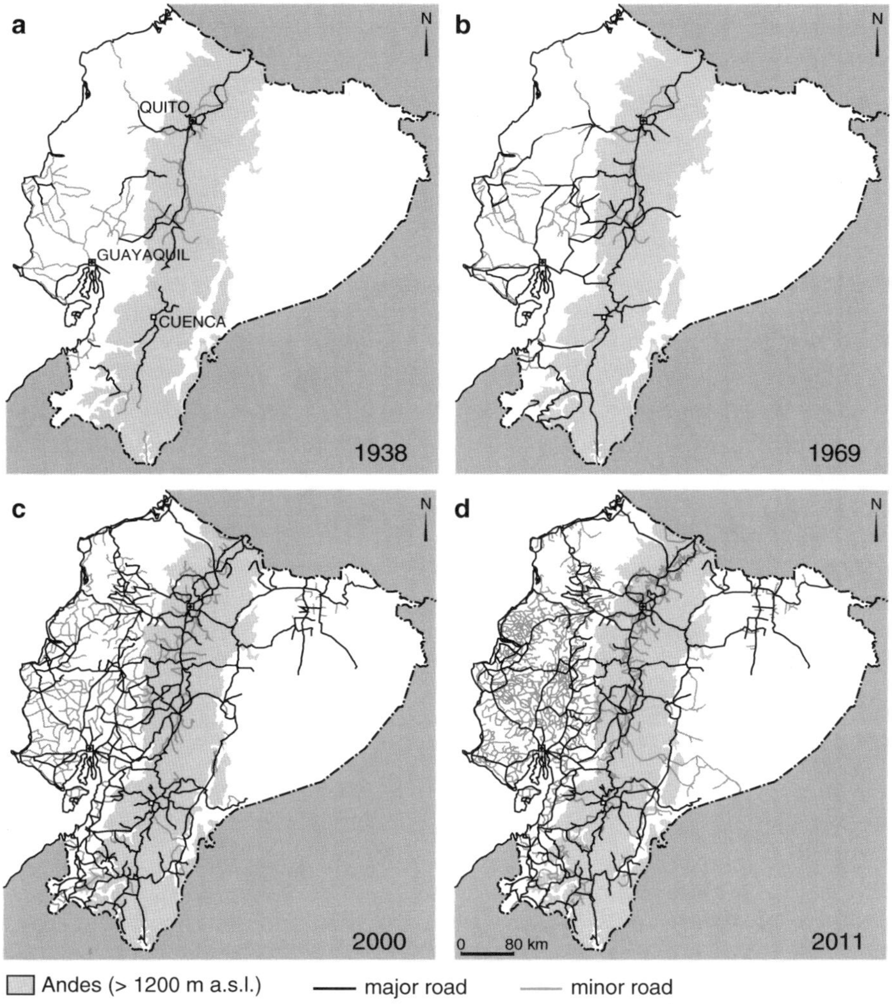

Fig. 2.1 Road networks of Ecuador in 1938, 1969, 2000 and 2011. The maps are based on the following sources: the American Geographical Society of New York (**a**), the Head Office of Geodesy and Cartography, German Democratic Republic, Berlin (**b**), the Instituto Geográfico Militar, Quito, Ecuador (**c**) and own inquiries (**d**), respectively

and pastoral land use increased rapidly between 1962 and 1989. With the foundation of the Podocarpus National Park (PNP) in 1982 (Pohle and Gerique 2008) and the RBSF in 1997, major parts of the Cordillera Real were declared as protected areas. Land use was intensified on the northern slopes of the RSF valley and on unprotected areas near Sabanilla (Fig. 2.2).

In order to quantify land use changes within the wider area between the two cities of Loja and Zamora (cf. model domain, Fig. 1.1), two orthorectified Landsat TM/ETM+ scenes from 1987 and 2001 were pre-processed and classified.

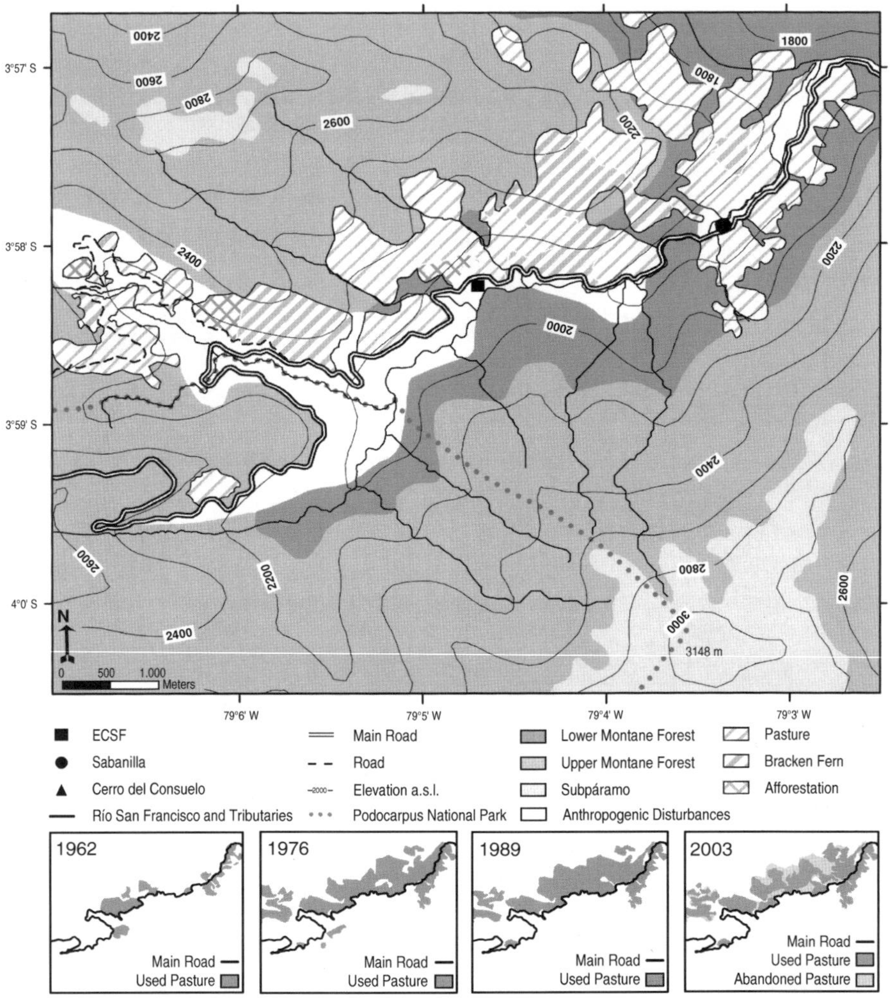

Fig. 2.2 Principal plant formations in the study area of the upper Rio San Francisco valley. The *large upper map* dates from July 2009. The *four smaller maps* illustrate the land use forms in the vicinity of the road between Loja and Zamora over a 40-year period. *ECSF* research station Estación Científica San Francisco

A post-classification change detection analysis revealed the changes that took place within an area of 4,800 km^2 during the 14-year period. Analysis was conducted by a pixel per pixel comparison based on univariate image differencing (post-classification intercomparison technique; Singh 1989), whereby cloud-covered areas in both scenes were excluded from the analysis (for more details on the classification technique see Göttlicher et al. 2009). The results (Fig. 2.3) demonstrate that the greater part of the region is still covered by forests, of which large parts can be found within the protected area of the PNP. Grassland is frequently encountered,

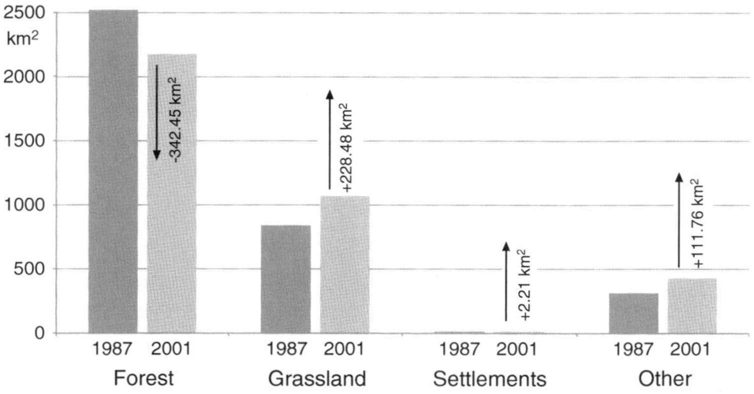

Fig. 2.3 Land use changes between 1987 and 2001 within the 4,800 km^2 study region encompassing Loja and Zamora. The *black arrows* indicate the quantitative changes in km^2. Details as to the methods are provided in the text

particularly in the valleys. Urbanised areas occur especially in and around Loja and Zamora. Forests suffered the greatest quantitative losses (Fig. 2.3). A total of 424.6 km^2 were cleared between 1987 and 2001, and even after taking the reforestation of 82.1 km^2 into consideration, the net loss of forest was 342.45 km^2. The expansion of urban land amounted to 2.2 km^2 and became especially apparent in the growth of Loja. The detected deforestation rate of 13.61 % for the study area corresponds to an annual deforestation rate of 0.97 %. Goerner et al. (2007) reported a deforestation rate of 0.9 % per year for the same study area.

2.3 Climate Change and Its Effects

Using a regional climate model based on the prospective IPCC (Intergovernmental Panel on Climate Change) climate scenarios A2 and B2 (Meehl et al. 2007), grid cell maps presented by Urrutia and Vuille (2009) show that several regions of the tropical Andes may sustain dramatic temperature and precipitation changes as a result of currently progressing climate change. For the grid cell of southern Ecuador a slight increase in rainfall (+8 %) and cloud cover (+4 %) is expected, together with a marked increase of air temperature of +3°K relative to the average of 1980–1999 (Meehl et al. 2007). Climate change may thus severely affect a floristic region which harbours one of the global diversity hotspots for vascular plant species (Barthlott et al. 2007; Richter et al. 2009). Unfortunately, long-term series of meteorological measurements for SE-Ecuador are available to an only very limited extent. Only the INAMHI (Instituto Nacional de Meterología e Hidrología) station “Loja” (2,160 m a.s.l.) that is located in the inter-Andean basin west of the main Cordillera has provided continuous data since 1964. The main station of the Oriente in Zamora (970 m a.s.l.) was closed in the1990s. In the San Francisco Valley itself,

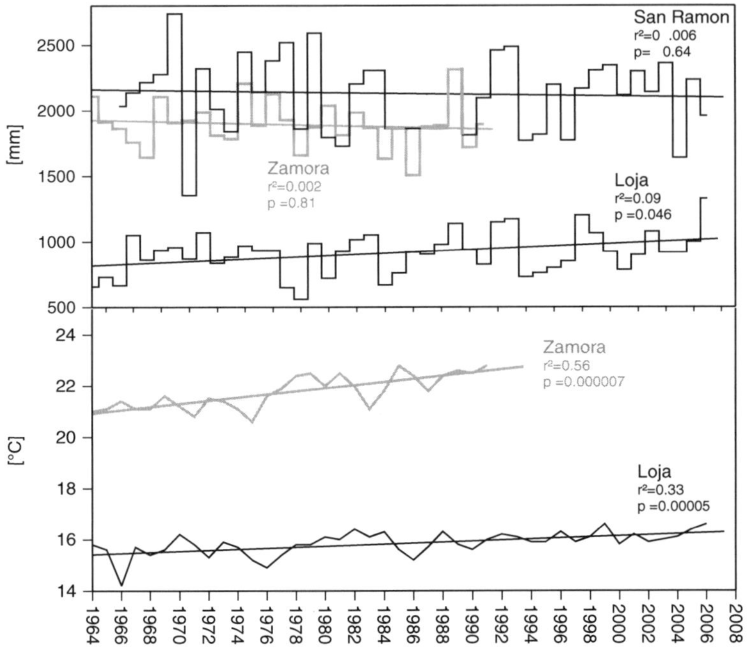

Fig. 2.4 Climate trends in the study area in southern Ecuador over a 40-year period. *Top*: annual rainfall [mm]. *Bottom*: average annual air temperature [°C] at selected locations. Data taken from the INAMHI (Instituto Nacional de Meterología e Hidrología) and EERSSA (Empresa eléctrica regional del Sur S.A.) climate stations

the regional electricity company EERSSA (Empresa Electrica Regional de Sur S.A.) has operated a rain gauge since 1966 at a location close to ECSF's main meteorological research station (1,950 m a.s.l.) that has been recording meteorological data since 1998. The insufficient supply of data is a general problem with regard to the fact that the regional climate of SE-Ecuador is locally highly variable and exhibits pronounced seasonal changes. Along a W–E distance of only 40 km it ranges from semiarid conditions and a relative dry season in MJJ at Catamayo in the west of the main Cordillera to a perhumid climate east of the main Cordillera (at the Cerro met station of the research programme, see Fig. 1.2), with peak rainfall occurring in the same period (Richter 2003; Emck 2007; Bendix et al. 2008a). However, some general climatic trends can be observed in spite of the poor data basis (Fig. 2.4). Both the western inter-Andean basin of Loja and the eastern Andean escarpment at Zamora reveal a significant warming trend. The air temperature at the station "Loja" about 30 km to the west of the ECSF meteorological station evidences a warming of ~0.6 °C over the 45 years of 1961–2008 (0.13 °C per decade). The station "Zamora" (San Ramon) situated in the eastern Andean foothills was characterised by an even stronger warming trend up to 1990.

The climate data in conjunction with model calculations thus point to a clear warming trend in the study area. The ecological importance of such a thermal shift is obvious. Assuming a stationary average annual lapse rate of -0.61 °C 100 m^{-1}

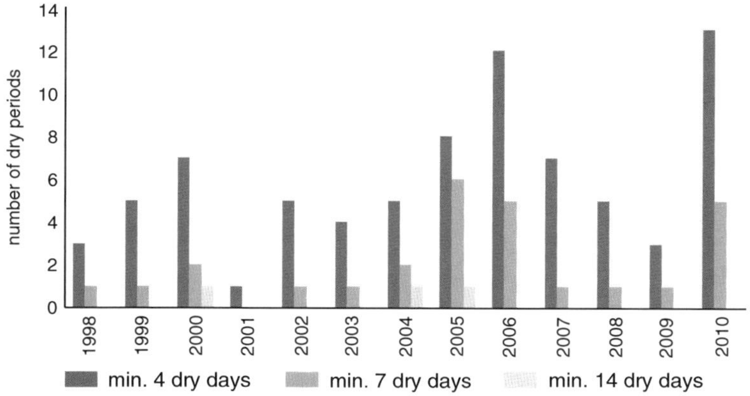

Fig. 2.5 Number of periods at the ECSF meteorological station during which no rain fell and that were continuously dry for at least 4, 7 or 14 days

(Bendix et al. 2008b) until 2100, the increase of temperature must result in an altitudinal shift of ecothermal belts in the study area (Bendix et al. 2010). An increase in temperature of 3 °C during the course of the twenty-first century would result in climatic conditions currently prevailing at a particular altitude being found at significantly lower altitudes. For example, the average air temperature of 15.5 °C now recorded by the ECSF meteorological station at an altitude of 1,860 m a.s.l. would be found at an altitude of 2,300 m a.s.l. at the end of the current century. This would lead to an upslope migration of thermophilous species, for which, however, suitable habitat corridors to higher areas are a prerequisite (Colwell et al. 2008). The numerous valleys of the Precordillerean and Amazon forelands might represent such corridors in the case of the Cordillera Real. Many lowland rain forest species have outposts in western Amazonia (Miles et al. 2004), from where they would be able to spread into the valleys and Precordillerean ranges. Drought- and heat-tolerant species are accordingly most likely to migrate into new terrains, and taxonomic input of invasive species from anthropogenic habitats can play a significant role in re-shuffling communities. Sources of invasive species are pastures, abandoned former cultivated land, roadsides and exotic tree plantations close to the RBSF and further downstream in the valley.

The situation of rainfall is more complex (Fig. 2.4). There has been a weak but significant trend towards an increase in rainfall in Loja over the observation period of 1964–2006. The slight decrease in rainfall in the eastern Andean foothills at Zamora is, however, not significant. The station San Ramon between Loja and Zamora also features an almost unnoticeable negative trend over the observation period. The areas west of the main Cordillera are subject to a slight increase in the amount of rainfall, while the humid eastern regions received a little less. The ECSF climate station has documented an obviously accelerating decrease in precipitation during the last 10 years of observations and particularly after 2005 (for more details see Chap. 19). This negative trend can mainly be attributed to an increase in the number of dry days and a more frequent occurrence of longer lasting dry phases in 2005, 2006 and 2010 (Fig. 2.5). However, it should be kept in mind that the high

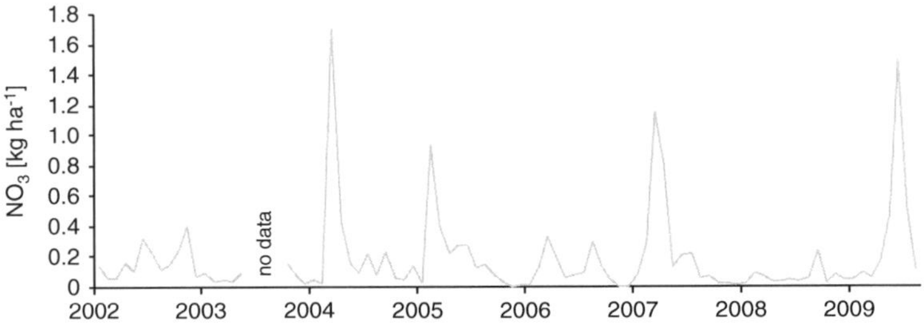

Fig. 2.6 NO_3 depositions at the meteorological station ECSF between 2002 and 2009

variability of annual precipitation and the limited length of the data series do not yet permit a distinct trend analysis.

2.4 Atmospheric Nutrient Deposition

The spatial distribution and temporal dynamics of precipitation and wind also regulate the deposition of water-dissolved matter into mountain ecosystems. These depositions can contain relatively high concentrations of plant nutrients (particularly N, S, P, and K) and acids (Boy et al. 2008), that are transported over long distances in the atmosphere. Especially Amazonia and even northern Africa are source regions for the nutrients that are transported to the mountain forests of Ecuador. Several authors showed that biomass burning (as, e.g. takes place in the Amazon) has a fertilising effect on ecosystems that thereby receive an input of N and P (Da Rocha et al. 2005; Fabian et al. 2005; Boy et al. 2008). To identify processes involved in these mineral depositions, rain water samples are being analysed for principal cations and anions, pH-value and electrical conductivity. The sampled water generally has very low concentrations of ions, and conductivity rarely exceeds values of 20 $\mu S\ cm^{-1}$. The values are generally lower during the rainy season due to dilution effects (Rollenbeck et al. 2006). The results shown in Fig. 2.6 suggest seasonal differences in element concentrations present in rainfall, probably resulting from biomass burning in Amazonia. For example, depositions of nitrate (NO_3) were higher during "fire episodes" in 2004 and 2007 than during "no fire" periods, and these peaks of nutritional input could in the long run affect the floristic composition of the mountain forests (Rollenbeck 2010). A thorough analysis of nutrient inputs into the mountain forest and its temporal development is presented in Chaps. 11 and 21.

2.5 Aspects of Future Threats

The expansion of agricultural land use in conjunction with an augmentation of population pressure is a major threat to the biodiversity hotspot on the eastern escarpment of the Ecuadorian Andes and in the adjacent lowlands of the Oriente.

They may well result in irretrievable losses of natural forests. Further fragmentation of already rare animal and plant population refuges will proceed along with the expansion of the road network. The imminence of this threat to plant and animal life is highly topical, since substantial mineral deposits have been detected in the Andes and their immediate forelands, even though they have not yet been exploited on a large scale. Especially in the south-eastern part of the country, in the surroundings of the Cordillera del Condor, gold and copper beds are assumed to exist. The Zamora-Loja connection road passing the RBSF will suffer from a dramatic rise of traffic volume if these resources are exploited. Increasing traffic and industrial activities in regions as remote as the booming megacities of Brazil will contribute to increasing inputs of aerosols and airborne pollutants. Not all of these are harmful, though, and some might even contribute to plant nutrient deposition. Acid fallouts, however, will exacerbate an acidification that is already burdening a majority of the natural ecosystems. According to Urrutia and Vuille (2009), warming is predicted to be moderate in western Ecuador, while the eastern part will suffer from increasing heat. Changes in precipitation are expected to be spatially much less cohesive, with manifold increases and decreases in rainfall throughout the Andes (refer to Chap. 19). More important yet are changes expected for the austral summer, when weakened mid- and upper tropospheric easterlies and strengthened westerlies may result in longer and more frequent dry spells. These will extend from October to January and could promote the migration of thermophilous organisms towards higher reaches. More serious appears, however, an increase of fire incidence.

Acknowledgements The authors would like to thank Paul Ziegler and Erwin Beck for linguistic proofreading this work.

References

American Geographical Society of New York (1938) South America 1:1.000.000. New York

Barthlott W, Hostert A, Kier G, Küper W, Kreft H, Mutke J, Rafiqpoor D, Sommer JH (2007) Geographic patterns of vascular plant diversity at continental to global scales. Erdkunde 61:305–315

Bebbington A (1993) Sustainable livelihood development in the Andes: local institutions and regional resource use in Ecuador. Dev Policy Rev 11:5–30

Bendix J, Rollenbeck R, Richter M, Fabian P, Emck P (2008a) Climate. In: Beck E, Bendix J, Kottke I, Makeschin F, Mosandl R (eds) Gradients in a tropical mountain ecosystem of Ecuador, vol 198, Ecological studies. Springer, Berlin, pp 63–73

Bendix J, Rollenbeck R, Fabian P, Emck P, Richter M, Beck E (2008b) Climate variability. In: Beck E, Bendix J, Kottke I, Makeschin F, Mosandl R (eds) Gradients in a tropical mountain ecosystem of Ecuador, vol 198, Ecological studies. Springer, Berlin, pp 280–291

Bendix J, Behling H, Peters T, Richter M, Beck E (2010) Functional biodiversity and climate change along an altitudinal gradient in a tropical mountain rainforest. In: Tscharntke T, Leuschner C, Veldkamp E, Faust H, Guhardja E, Bidin A (eds) Tropical rainforests and agroforests under global change, Environmental Science and Engineering. Subseries Environmental Science. Springer, Berlin, pp 239–268

Boy J, Rollenbeck R, Valarezo C, Wilcke W (2008) Amazonian biomass burning-derived acid and nutrient deposition in the north Andean montane forest of Ecuador. Global Biogeochem Cycles 22, GB4011

Bruhns KO (1994) Ancient south America. Cambridge University Press, Cambridge

Brummitt N, Lughadha EN (2003) Biodiversity: where's hot and where's not. Conserv Biol 17:1442–1448

Colwell RK, Brehm G, Cardelius CL, Gilman AC, Longino JT (2008) Global warming, elevational range shifts, and lowland biotic attrition in the wet tropics. Science 322:258–261

Da Rocha GO, Allen AG, Cardoso AA (2005) Influence of agricultural biomass burning on aerosol size distribution and dry deposition in southeastern Brazil. Environ Sci Technol 39 (14):5293–5301

De Koning GHJ, Veldkamp A, Fresco LO (1998) Land use in Ecuador: a statistical analysis at different aggregation levels. Agric Ecosyst Environ 70:231–247

Dodson CH, Gentry AH (1991) Biological extinction in western Ecuador. Ann Mo Bot Gard 78:273–295

Ellenberg H (1979) Man's influence on tropical mountain ecosystems in south America. J Ecol 67 (2):401–416

Emck P (2007) A climatology of South Ecuador. With special focus on the major Andean Ridge as Atlantic-Pacific Climate Divide. PhD thesis, University of Erlangen http://www.opus.ub.uni-erlangen.de/opus/frontdoor.php?source_opus=656 Cited 1.12.2011

Fabian P, Kohlpaintner M, Rollenbeck R (2005) Biomass burning in the Amazon-fertilizer for the mountainous rain forest in Ecuador. Environ Sci Pollut Res Int 12(5):290–296

FAO (2005) State of the world's forests 2005. Food and Agriculture Organization of the United Nations, Rome

Goerner A, Gloaguen R, Makeschin F (2007) Monitoring of the Ecuadorian mountain rainforest with remote sensing. J Appl Remote Sens 1:013527

Göttlicher D, Obregón A, Homeier J, Rollenbeck R, Nauss T, Bendix J (2009) Land cover classification in the Andes of southern Ecuador using Landsat ETM+ data as a basis for SVAT modelling. Int J Remote Sens 30:1867–1886

Harden C (1993) Land use, soil erosion, and reservoir sedimentation in an Andean drainage basin in Ecuador. Mt Res Dev 13:177–184

Head office of Geodesy and Cartography (1969) World Map 1:2.500.000. German Democratic Republic, Berlin

Instituto Geográfico Militar (2000) Mapa del Ecuador 1:1.000.000. Quito, Ecuador

Jokisch BD, Lair BM (2002) One last stand? Forests and change on Ecuador's eastern Cordillera. Geogr Rev 92(2):235–256

Keating PL (1997) Mapping vegetation and anthropogenic disturbances in southern Ecuador with remote sensing techniques: implications for park management. Yearb Conf Latin Am Geogr 23:77–90

Köster N, Friedrich K, Nieder J, Barthlott W (2009) Conservation of epiphyte diversity in an Andean landscape transformed by human land use. Conserv Biol 23(4):911–919

Luteyn JL (1992) Páramos: why study them? In: Balslev H, Luteyn JL (eds) Páramo: an Andean ecosystem under human influence. Academic, London, pp 151–170

Marquette CM (2006) Settler welfare on tropical forest frontiers in Latin America. Popul Environ 27(5–6):397–444

Meehl GA, Stocker TF, Collins WD, Friedlingstein P, Gaye AT, Gregory JM, Kitoh A, Knutt R, Murphy JM, Noda A, Raper SCB, Watterson IG, Weaver AJ, Zhao Z-C (2007) Global climate projections. In: Solomon S, Qin D, Manning M, Chen Z, Marquis M, Avery KB, Tignor M, Miller HL (eds) Global climate projections – climate change 2007: the physical science basis. Contribution of working group I to the fourth assessment report of the intergovernmental panel on climate change. Cambridge University Press, Cambridge, 996 p

Miles L, Grainger A, Philips O (2004) The impact of global climate change on tropical forest biodiversity in Amazonia. Glob Ecol Biogeogr 13:553–565

Millennium Ecosystem Assessment (2005) Ecosystem and human well-being: biodiversity synthesis. World Resources Institute, Washington, DC
Mittermeier RA, Robles Gil P, Mittermeier CG (1997) Megadiversity: earth's biologically wealthiest nations. Monterrey, Mexico
Mosandl R, Günter S, Stimm B, Weber M (2008) Ecuador suffers the highest deforestation rate in South America. In: Beck E, Bendix J, Kottke I, Makeschin F, Mosandl R (eds) Gradients in a tropical mountain ecosystem of Ecuador, Ecological studies. Springer, Berlin, pp 37–40
Myers N (1988) Threatened biotas: 'hot spots' in tropical forests. Environmentalist 8:187–208
Myers N (1993) Tropical forests: the main deforestation fronts. Environ Conserv 20:9–16
Myers N, Mittermeier RA, Mittermeier CG, da Fonseca GAB, Kent J (2000) Biodiversity hotspots for conservation priorities. Nature 403:853–858
Peters T, Diertl K, Gawlik J, Rankl M, Richter M (2010) Vascular plant diversity in natural and anthropogenic ecosystems in the Andes of southern Ecuador – studies from the Rio San Francisco valley. Mt Res Dev 30:344–352
Pimm R, Raven P (2000) Biodiversity – extinction by numbers. Nature 403:843–845
Pohle P, Gerique A (2008) Sustainable and non sustainable use of natural resources by indigenous and local communities. In: Beck E, Bendix J, Kottke I, Makeschin F, Mosandl R (eds) Gradients in a tropical mountain ecosystem of Ecuador, Ecological studies. Springer, Berlin, pp 331–346
Pohle P, Gerique A, Park M, Lopez Sandoval MF (2009) Human ecological dimensions in sustainable utilization and conservation of tropical mountain forests under global change in southern Ecuador. In: Tscharnke T, Leuschner C, Veldkamp E, Faust H, Guhardja E, Bidin A (eds) Tropical rainforests and agroforests under global change, Environmental Science and Engineering. Subseries Environmental Science. Springer, Berlin, pp 477–509
Richter M (2003) Using plant functional types and soil temperatures for eco-climatic interpretation in southern Ecuador. Erdkunde 57:161–181
Richter M, Diertl KH, Emck P, Peters T, Beck E (2009) Reasons for an outstanding plant diversity in the tropical Andes of Southern Ecuador. Landscape Online 12/2009
Rollenbeck R (2010) Global sources – local impacts: natural and anthropogenic matter deposition in the Andes of Ecuador. Geoöko 1–2:1–25
Rollenbeck R, Bendix J, Fabian P (2006) Spatial and temporal dynamics of atmospheric water- and nutrient inputs in tropical mountain forests of southern Ecuador. In: Bruijnzeel LA et al (eds) Mountains in the Mist: science for conserving and managing tropical montane cloud forests. University of Hawaii, Honolulu
Sala OE, Chapin FS III, Armesto JJ, Berlow R, Bloomfield J, Dirzo R, Huber-Sanwald E, Huenneke LF, Jackson RB, Kinzig A, Leemans R, Lodge D, Mooney HA, Oesterheld M, Poff NL, Sykes MT, Walker BH, Walker M, Wall DH (2000) Global biodiversity scenarios for the year 2100. Science 287:1770–1774
Sarmiento FO, Frolich LM (2002) Andean cloud forest tree lines. Naturalness, agriculture and the human dimension. Mt Res Dev 22(3):278–287
Singh A (1989) Digital change detection techniques using remotely-sensed data. Int J Remote Sens 10:989–1003
Southgate D, Whitaker M (1992) Promoting resource degradation in Latin America: tropical deforestation, soil erosion, and coastal ecosystem disturbance in Ecuador. Econ Dev Cult Change 40:787–807
Southgate D, Whitaker M (1994) Economic progress and the environment: one developing country's policy crisis. Oxford University Press, New York, p 150
Urrutia R, Vuille M (2009) Climate change projections for the tropical Andes using a regional climate model: temperature and precipitation simulations for the end of the 21th century. J Geophys Res 114, D02108
Williams JW, Jackson ST, Kutzbach JE (2007) Projected distributions of novel and disappearing climates by 2100 AD. Proc Natl Acad Sci USA 104:5738–5742

Chapter 3
The Challenges of Sustainable Development in the Podocarpus-El Cóndor Biosphere Reserve

Jan Barkmann, Sofia Eichhorn, Byron Maza, Frank v. Walter, and Roland Olschewski

3.1 Introduction

Since the publication of the Brundtland Report in 1987 (WCED 1987), sustainable development has been the prime guiding principle for discussing the challenges of human development in view of limited natural resources. Numerous activists and scholars have taken issue during the past two and a half decades with the anthropocentric core of sustainable development, and with its apparently poor capacity to prescribe specific courses of development. Still, the fundamental proposition is as relevant today as ever: any sustainable development must focus on the needs of the current generation without compromising the ability of future generations to fulfil their needs. Science will only be able to contribute directly to this task if it meets the challenges of interdisciplinary collaboration.

From a sustainable development perspective, science has to analyse (1) on which ecosystem states, processes, or structures the ecological services depend that human

Frank v. Walter died in March 2012. Although he could not directly contribute to writing this chapter, it does, nevertheless, include some of his contributions to Research Unit RU 816.

J. Barkmann (✉) • S. Eichhorn • F.v. Walter
Faculty of Agricultural Sciences, Department of Agricultural Economics and Rural Development, Environmental and Resource Economics, Georg-August-Universität Göttingen, Platz der Göttinger Sieben 5, 37073 Göttingen, Germany
e-mail: jbarkma@uni-goettingen.de

B. Maza
Faculty of Agricultural Sciences, Department of Agricultural Economics and Rural Development, Environmental and Resource Economics, Georg-August-Universität Göttingen, Platz der Göttinger Sieben 5, 37073 Göttingen, Germany

Universidad Nacional de Loja, Ciudad Universitaria Guillermo Falconí Espinosa, La Argelia PBX 072547252, Casilla Letra "S", Loja, Ecuador

R. Olschewski
WSL Swiss Federal Institute for Forest, Snow and Landscape Research, Environmental and Resource Economics, Zürcherstrasse 111, 8903 Birmensdorf, Switzerland

J. Bendix et al. (eds.), *Ecosystem Services, Biodiversity and Environmental Change in a Tropical Mountain Ecosystem of South Ecuador*, Ecological Studies 221,
DOI 10.1007/978-3-642-38137-9_3,

society utilises, (2) which pressures threaten the respective ecosystems and their services, and (3) how the long-term capacity of these ecosystems to provide ecosystem services can be protected and developed (Barkmann et al. 2008). The first point is the main topic of several of the following chapters including Chap. 4, which provides an overview on the ecosystem services investigated by Research Unit RU 816. With regard to pressures (point 2), this book focuses on the conversion of near natural forests to pasture ecosystems (cf. Chap. 15). External atmospheric nutrient inputs and climate change are also considered (Chaps. 11, 23, and 24). Concerning the last point, the research unit addresses several protection options: the conservation of biological diversity by protecting the remaining forests, reforestation and rehabilitation of abandoned pastures, and the general improvement of farm management in order to reduce forest conversion caused by poverty (see Chaps. 22, 26, and Sect. 3.3.2).

After discussing the sustainable development concept (Sect. 3.2), implications for scientific decision-making support at tropicap biodiversity hotspots are derived with reference to the Ecosystem Approach of the Convention on Biological Diversity (CBD) and the Aichi Biodiversity Targets (Sect. 3.3). The chapter closes with two examples of those conservation-development trade-offs that research needs to address to meet the challenges of sustainable development.

3.2 Sustainable Development

In the year 1713 the Chief Mining Officer of Saxony, v. Carlowitz, called for the "greatest art, science and industry" in order to "conserve and produce wood in a way that there be a continuing, stable and sustaining use" (Carlowitz 1713, p. 105; translation jb). But it became more and more apparent to the general public only in the 1960s that many natural resources were being exploited at rates impossible to sustain indefinitely (e.g. Meadows et al. 1972). The well-being or even the existence of humankind appeared to be threatened. By the late 1970s it had become well established that the strain on global natural resources resulted from the combined impacts of a poor majority struggling for a livelihood and an affluent minority consuming a disproportionately large share of these resources (IUCN/WWF/UNEP 1980). Within this strand of the global environment and development discourse, the term sustainable development was coined.

In 1983 the General Assembly of the United Nations adopted the term sustainable development. The UN created a "Special Commission", later known as the World Commission on Environment and Development (WCED) or, simply, the Brundtland Commission (WCED 1987, p. ix; United Nations 2012). It was tasked with firmly establishing the search for a global "environmental perspective" on the international political agenda. The most widely quoted definition of sustainable development is found in the report which WCED published 4 years later (WCED 1987, p. 43):

> "Sustainable development is development that meets the needs of the present without compromising the ability of future generations to meet their own needs."

Following the Brundtland approach, three essentials of sustainable development should be emphasised (Barkmann 2002):

- orientation towards (basic) human needs,
- environmental equity and justice (intergenerational and international justice), and
- retinity (from Latin *retis*: the net); retinity refers to the complex interaction of the ecological, economic, and social spheres required for responsible decision making (SRU 1994).

The Brundtland approach does recognise absolute limits of development set by the global life support system (WCED 1987, p. 45). Still, WCED points out that there is no single point beyond which ecological disaster is to be expected. Different resources have different limits, and technology and knowledge can extent the limits of the resource base at least at times. Without doubt, the Brundtland definition is a socio-economic and anthropocentric approach—a fact for which it is frequently criticised. There is, however, little debate that prime attention to basic human needs and to environmental justice forms a minimum requirement for any consistent sustainable development paradigm (e.g. Birnbacher 1980; Krebs 1997).

Sustainable development in the WCED sense is often depicted by three overlapping circles (the ecological, economic, and social spheres). This graphical representation suggests equal importance of these three spheres (or "pillars"). However, all goals of sustainable development originate from the social sphere. The environmental as well as the economic limitations and preconditions of human production and consumption are of "only" instrumental value with respect to the satisfaction of basic human needs and the demands of environmental justice.

In order to stress the interrelatedness of the three spheres, the German Advisory Council on the Environment coined the term "retinity" ("Retinität"; SRU 1994). Retinity demands that all conservation and development decisions take the interrelatedness of three spheres into account. The WCED highlights the retinity norm regarding ecological and economic aspects in a subsection on *Merging Environment and Economics in Decision Making* (WCED 1987, p. 62ff.).

Technically speaking, systematic decision making within a sustainable development framework means that multi-dimensional bundles of ultimate and proximate objectives are to be taken into account. The complexity of issues involved in properly assessing alternative courses of action precludes the existence of a simple "blueprint" for sustainable development (WCED 1987, p. 40). In the event of conflict, there is no ecological meta-guideline according to which conflicting objectives, for example biodiversity conservation and utilisation of the economically productive resource base, can be prioritised. The ecosystem service concept as popularised by the Millennium Ecosystem Assessment (MA 2005; see also Chap. 4) provides an analytical basis for *investigating* these conflicts, not for solving them. In this respect, the carrying capacity/ecocapacity approach is too deeply entangled in a mesh of unresolved social and normative questions as to be a generally applicable, normatively justified guideline (Barkmann et al. 2008).

Its restrictions do not make the sustainable development concept useless. In fact, several authors have pointed out that sustainable development should be interpreted as a *regulative idea* of the discourse on environment and development (cf. Jörissen et al. 1999; Hirsch-Hadorn 1999). Like other regulative ideas such as freedom or justice, sustainable development has a clear normative dimension (basic needs orientation, intergenerational and international justice), and guides the search for human courses of action. If interpreted as a regulative idea, it is evident why a reference to the sustainable development paradigm itself cannot select the "best" option for development or conservation: this must be done by the "regulated" discourse on environment and development.

In combination with the retinity demands of sustainable development, the regulative character of the concept establishes certain information requirements that science needs to respond to. With respect to the situation in the project area of RU 816—i.e. the northern part of the UNESCO Biosphere Reserve Podocarpus-El Cóndor, we turn to an investigation of two exemplary information needs in the second part of the next section.

3.3 Implications for Research in Tropical Biodiversity Hotspots

3.3.1 CBD Ecosystem Approach and Aichi Biodiversity Targets

Accepting the proposition that sustainable development is a regulative idea (see above) has consequences for the way in which applied sciences have to interpret their role. Most fundamentally, applied ecological and socio-economic sciences need to generate and present knowledge in a way that diverse groups of stakeholders can make good use of in decision-making processes. In consequence, applied sciences have to make their expertise "discourse-able" (Barkmann 2001).

In the project area, various land users compete for access to limited natural resources. During the first phase of RU 816 in 2007, the Podocarpus–El Cóndor region was officially recognised by UNESCO as a Biosphere Reserve (BR; Bendix et al. 2010). A BR is a protected area specifically dedicated to the principles of sustainable development. In accordance with UNESCO's Seville Strategy (2012), the *core zone* of the BR is formed by strictly protected Podocarpus National Park. Around this core, several *buffer zones* are located. One of the buffer zones is the protected forest area "Bosque Protector Corazón de Oro" (Fig. 3.1). Between this buffer zone and the national park, a less protected *transition zone* is wedged along the main road from Loja to Zamora.

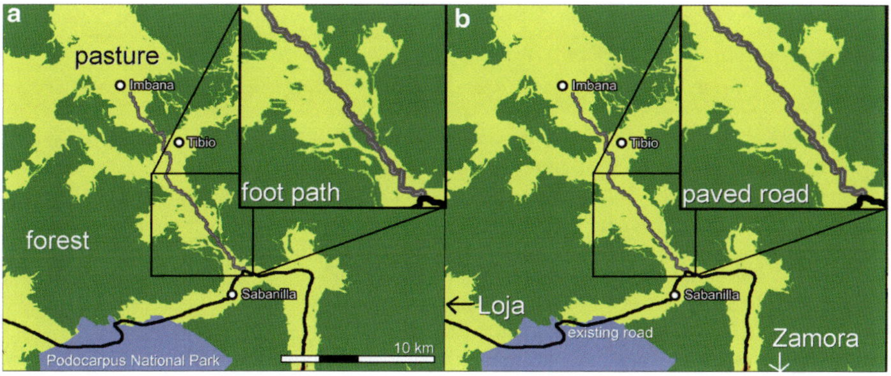

Fig. 3.1 Predicted extent of forest habitat loss if the footpath between Imbana and Sabanilla is replaced by a paved road; the *inset* highlights an example of substantially reduced habitat connectivity between the Sabanilla/Podocarpus (South) and the Yacuambi (Northeast) habitats of the Andean bear (*Tremarctos ornatus*)

Local farmers are interested in the continued or even extended availability of productive pastures (see Chaps. 15 and 26). The urban public is interested in the quality of drinking water (see Chap. 4). National and international conservationists focus on the conservation of biological diversity. Finally, tourists and recreational users are interested in the amenity value of the South Ecuadorian landscape. Principally, all of these land-use interests can be traced to some socially legitimate objective. Thus, references to sustainable development cannot be directly drawn upon to prioritise these interests. Nevertheless, sustainable development as a regulative idea requires that decision making must respect the sustainable development essentials mentioned in Sect. 3.2.

The CBD does not directly refer to sustainable development as a regulative idea. Yet, the twelve Malawi Principles (Hartje et al. 2002; see Table 3.1), on which the CBD Ecosystem Approach is based, provide important guidelines how the sustainable development of tropical biodiversity hotspots should be organised. It is the basic idea of the CBD Ecosystem Approach (CBD 2012a) that biological diversity can only be conserved successfully if the multitude of its interactions with humans is taken into consideration in a balanced manner–as required by the retinity norm.

The Aichi Biodiversity Targets were approved within the legal framework of the CBD in 2010. They combine some of the most important procedural principles of the CBD Ecosystem Approach with quantitative targets for biodiversity protection. For example, Principle 3 demanding management decisions within an "economic context" is now represented by Target 3

> "By 2020, at the latest, incentives, including subsidies, harmful to biodiversity are eliminated, phased out or reformed in order to minimise or avoid negative impacts, and positive incentives for the conservation and sustainable use of biodiversity are developed and applied, consistent and in harmony with the Convention and other relevant international obligations, taking into account national socio economic conditions" (CBD 2012b).

Table 3.1 Selected principles of the CBD Ecosystem Approach

Number	Principle of the CBD Ecosystem Approach
1	The objectives of management of land, water, and living resources are a matter of social choice
2	Management should be decentralised to the lowest appropriate level—hoping to achieve greater efficiency, effectiveness, and equity
4	There is a need to understand and manage the ecosystem in an economic context because biological diversity is threatened by ecosystem conversion to more intensive land-use systems. Any such ecosystem-management program should (1) reduce market distortions that adversely affect biological diversity, (2) align incentives to promote diversity conservation and sustainable use, and (3) internalise costs and benefits in the given ecosystem to the extent feasible.
5	Ecosystem structure and functioning should be a priority target of the CBD Ecosystem Approach in order to maintain ecosystem services
10	The CBD Ecosystem Approach should seek for an appropriate balance between and strive for an integration of conservation and use of biological diversity
11	The approach should consider all forms of relevant information, including scientific, indigenous and local knowledge, as well as innovations and practices

The principles are partly abbreviated and slightly modified

The fundamental match of a discursive framework for sustainable development and the CBD Ecosystem Approach is exemplified by the first principle of the approach: *The objectives of management of land, water, and living resources are a matter of social choice*. Although more detailed than the essentials of sustainable development, already the CBD Ecosystem Approach's first principle indicates that the fundamental ambiguities of social decision making will remain when it comes to local application. Even though the quantitative nature of the Aichi Biodiversity Targets may suggest otherwise, the described ambiguities also remain unsolved there.

3.3.2 Examples from the Project Area

How should a binding management plan for the BR Podocarpus-El Cóndor be developed? Some of the most severe land-use conflicts in the project area of RU 816 can be expected in the protected forest area Corazón de Oro. According to the Seville Strategy, only "activities compatible with the conservation objectives" may take place here (e.g. environmental education, ecotourism, research) (UNESCO 1996, p. 17). Even in the transition zone, agricultural activities and settlements are only allowed under certain restrictions. In fact, however, a few hundred households have been settling in the Corazón de Oro region for several decades. The inhabitants are predominantly pastoralists. They also have small arable fields or home gardens, and at times extract valuable timber trees from remote parts of the forest (Maza 2011). Thus, a realistic management and development plan for the area cannot consider conservation issues only; it must also account for the existing villages and their socio-economic development. Any other approach would clearly be against the spirit and the letter of the fundamental documents of sustainable development and their implementation through the CBD.

For example, the inhabitants of the indigenous villages of El Tibio Alto and El Tibio Bajo have access to the main road between Loja and Zamora only via a long footpath. Both villages belong administratively to the province of Zamora-Chinchipe. In consequence, the inhabitants can reach their own provincial capital Zamora faster if they first travel to the capital of adjacent Loja Province than if they go directly to Zamora. Given this obstacle to accessing essential administrative services, the construction of a road from Imbana to Sabanilla, which better connects these remote villages to the network of main roads, has been suggested. However, analyses by project C3.2 of RU 816 suggest that the construction of the road will have substantial negative effects on biodiversity (Eichhorn et al. 2010).

Making information "discourse-able" in the context of this potential sustainable development conflict means that the prospective impacts of road construction have to be analysed with respect to the ecosystem services that main stakeholders care about. Local farming households with a shortage of accessible land or with poor access to markets and public and private services may be interested in knowing how much faster or less expensive they can access land, markets or services. In contrast, conservationists want to know how much forest will be lost and how Andean bear (*Tremarctos ornatus*) populations are affected that inhabit the forests of the region.

In addition to the direct loss of forests, a further expansion of current pasture area is likely as improved road access facilitates the establishment of new pastures, i.e. of additional farms (Eichhorn 2009; Eichhorn et al. 2010). In turn, the road and the new pastures will reduce connectivity between Andean bear habitats along the Yacuambi–Podocarpus–Sabanilla biological corridor, which forms a regional habitat bottleneck (inset in Fig. 3.1). A first quantitative assessment of the trade-offs of road construction concludes that about 600–700 ha of forest may be lost, mainly by conversion to pastures. This equals roughly the habitat size of two adult female bears, i.e. of two *effective individuals*. Also, habitat connectivity will suffer substantially. The smallest distance between disconnected but adjacent main habitat patches of the Andean bear will increase from less than 150 m to more than a critical value of about 1,000 m. On the other hand, the additional pastures may provide livelihood for 30–40 average farm households. Without additional sources of livelihood and at current production technology, income would be close to absolute poverty levels, though (Eichhorn et al. 2010).

The bulk of the pasture area shown in Fig. 3.1 is legally located in the protected "forest" area, and in the—no-agriculture—buffer zone of the BR. From a conservationist point of view, massive reforestation in the buffer zone could be justified. At least, further forest conversion should be stopped. However, the majority of the affected rural population consists of poor smallholders. Any policy measure of the BR management plan that affects their livelihood should be investigated and considered carefully. We explore some of the most important impacts that a ban on further forest conversion in the Corazón de Oro region is likely to engender in the following paragraphs. An economic analysis of reforestation options is found in Chap. 25.

Specifically, we assume that losses of future farming income by the deforestation ban are offset by Payments for Ecosystem Services (PES). Payments that provide incentives for smallholders to stop deforestation are potentially available from the *Socio-Bosque* program of the Ecuadorian government (Ministerio del Ambiente 2012). Socio-Bosque aims at combining forest conservation with poverty alleviation. Using a typical Socio-Bosque budget, Maza et al. (2011) found that ~36 % of the threatened forest in the research area could be covered by a PES scheme that exactly compensates lost pasture incomes. At the resulting compensation rates, positive effects on poverty alleviation are unlikely in the long run (cf. also Olschewski et al. 2005; Olschewski and Benítez 2005). Only if payments (1) are targeted at those farm households most in need and (2) substantially overcompensating income losses, the income of the poor will rise, and existing economic inequalities be reduced. However, pronounced "pro-poor" PES scheme come at a considerable cost: At a fixed, increased compensation of US \$300 ha^{-1} $year^{-1}$, which would substantially reduce local rural poverty, only 10 % of the threatened forest can be covered by the program.

Ideally, a public debate on the pros and cons of different development and conservation options is initiated using this type of trade-off information (Olschewski et al. 2010). Because of heavily conflicting land-use interests, it is far from sure that a consensus on the implementation of a certain option will be achieved—or its final inclusion in the management plan of the BR will be accomplished. International income transfers are potentially available to facilitate consensus for mega-diverse "hotspots" of biological diversity (Hillmann and Barkmann 2009, Chap. 4). Although purely selfish arguments can and will be put forward in public debate, arguments that can refer to basic needs, or to intergenerational or international justice should be given prime consideration. The same holds for arguments that take all available information into account instead of focussing exclusively on ecological or socio-economic criteria.

For all practical purposes, providing information on conservation-development trade-offs is one of the most useful ways of responding to the information requirements of concerned stakeholders. Gearing the generation of knowledge towards the elucidation of such trade-offs is one way for the ecological as well as the socio-economic sciences to meet their own sustainable development challenges.

Acknowledgments We would like to thank our field assistants, respondents and interview partners for support and sharing their knowledge with us. Furthermore, we gratefully acknowledge financial support for co-author B. Maza by the German Academic Exchange Service (DAAD), and final publishing support by the editorial team and by P. Ziegler.

References

Barkmann J (2001) Angewandte Ökosystemforschung zwischen Biodiversitäts-, Landschafts- und Ressourcenschutz. Petermanns Geogr Mitt 145:16–23

Barkmann J (2002) Modellierung und Indikation nachhaltiger Landschaftsentwicklung – Beiträge zu den Grundlagen angewandter Ökosystemforschung. EcoSys – Beiträge zur Ökosystemforschung 9:1–368

Barkmann J, Blume H-P, Irmler U, Kluge W, Kutsch WL, Reck H, Reiche EW, Trepel M, Windhorst W, Dierßen K (2008) Ecosystem research and sustainable land use management. In: Fränzle O, Kappen L, Blume H-P, Dierßen K (eds) Ecosystem organization of a complex landscape, vol 202, Ecological studies. Springer, Berlin, pp 313–344

Bendix J, Paladines B, Ribadeneira-Sarmiento M, Romero LM, Valarezo C, Beck E (2010) Benefit sharing by research, education and knowledge transfer – a success story of biodiversity research in southern Ecuador. https://new.cbd.int/abs/side-events/abs-9/id1694-dfg-article.pdf. Accessed 11 Mar 2012

Birnbacher D (ed) (1980) Ökologie und Ethik. Reclam, Stuttgart

CBD [Convention on Biological Diversity] (2012a) Ecosystem approach. http://www.cbd.int/ecosystem/. Accessed 16 Aug 2012

CBD [Convention on Biological Diversity] (2012b) Aichi biodiversity targets. http://www.cbd.int/sp/targets/. Accessed 16 Aug 2012

Eichhorn S (2009) Effekte eines Straßenbaus im Biosphärenreservat Podocarpus-El Cóndor auf Landnutzung und Biodiversität. Unpublished diploma thesis, Albrecht-von-Haller-Institute for Plant Sciences. Georg-August-Universität Göttingen, Göttingen

Eichhorn S, von Walter F, Barkmann J (2010) Two for the road? Andenbär (*Tremarctos ornatus*) und Straßenbau in Ecuador. In: 40th Anniversary conference, Gesellschaft für Ökologie. Book of Abstracts, p. 420

Hartje V, Klaphake A, Schliep R (2002) Consideration of the ecosystem approach on the conservation on biological diversity in Germany. BfN-Skripten 69, Bonn

Hillmann B, Barkmann J (2009) Conservation: a small price for long-term economic well-being. Nature 461:37

Hirsch-Hadorn G (1999) Natur und ihr Wert - Nachhaltige Entwicklung und der Wert der Natur. GAIA – Ökologische Perspektiven für Wissenschaft und Gesellschaft 8:269–274

IUCN/WWF/UNEP (1980) World Conservation Strategy. Living resource conservation for sustainable development. IUCN/WWF/UNEP, Gland, Switzerland

Jörissen J, Kopfmüller V, Brandl V, Paetau M (1999) Ein Integratives Konzept nachhaltiger Entwicklung. Forschungszentrum Karlsruhe, Germany

Krebs A (1997) Naturethik im Überblick. In: Krebs A (ed) Naturethik. Grundtexte der gegenwärtigen tier- und ökoethischen Diskussion. Suhrkamp, Frankfurt

MA [Millenium Ecosystem Assessment] (2005) Millennium ecosystem assessment ecosystems and human well-being: synthesis. Island Press, Washington, DC

Maza Rojas BV (2011) Heterogeneity of peasant land use decisions as an effect of differences in financial and personal capitals in the area of the Biosphere Reserve "Podocarpus-El Cóndor", Ecuador Dissertation, Faculty for Agricultural Sciences, Georg-August-Universität Göttingen. http://webdoc.sub.gwdg.de/diss/2011/maza_rojas/maza_rojas.pdf. Accessed 15 Aug 2012

Maza Rojas BV, Barkmann J, von Walter F, Marggraf R (2011) Efficiency and distributional impacts of protected area planning using PES schemes in the Biosphere Reserve "Podocarpus-El Cóndor", Ecuador. In: Maza RBV, pp 114–146

Meadows D, Meadows D, Zahn E, Milling P (1972) Die Grenzen des Wachstums. DVA, Stuttgart, 180 p

Ministerio del Ambiente (2012) SocioBosque – Programa de protectión de bosques. http://sociobosque.ambiente.gob.ec/. Accessed 16 Aug 2012

Olschewski R, Benítez P (2005) Secondary forests as temporary carbon sinks? The economic impact of accounting methods on reforestation projects in the tropics. Ecol Econ 55 (3):380–394

Olschewski R, Benítez P, de Koning GHJ, Schlichter T (2005) How attractive are forest carbon sinks? Economic insights into supply and demand of certified emission reductions. J For Econ 11:77–94

Olschewski R, Klein AM, Tscharntke T (2010) Economic trade-offs between carbon sequestration, timber production and crop pollination in tropical forested landscapes. Ecol Complex 7 (3):314–319

SRU [Rat von Sachverständigen für Umweltfragen] (1994) Umweltgutachten 1994 - Für eine dauerhaft-umweltgerechte Entwicklung. Der Rat von Sachverständigen für Umweltfragen (SRU). Metzler-Poeschel, Stuttgart

UNESCO [United Nations Educational, Scientific and Cultural Organization] (1996) Biosphere reserves: the Seville strategy and the statutory framework of the world network. UNESCO, Paris

United Nations (2012) Process of preparation of the environmental perspective to the year 2000 and beyond. General assembly document A/RES/38/161. http://www.un.org/documents/ga/res/38/a38r161.htm. Accessed 16 Aug 2012

von Carlowiz HC (1713) Sylvicultura oeconomica. Anweisung zur wilden Baum-Zucht. Johann Friedrich Braun, Leipzig

WCED [The World Commission on Environment and Development] (1987) Our common future (brundtland report). Oxford University Press, Oxford

Chapter 4
The Research Unit RU 816: Overall Approach in the Light of the Ecosystem Services Concept

Jan Barkmann, Boris M. Hillmann, and Rainer Marggraf

4.1 Introduction

Over the last decade, "ecosystem service" has become one of the most influential but also most controversial scientific concepts at the interface of biodiversity science and environmental policy. The most widely accepted definition of ecosystem service is provided by the Millennium Ecosystem Assessment (MEA, 2003), which highlights the role of ecological systems for the provisioning of benefits for human society. The establishment of IPBES (Intergovernmental Science-Policy Platform for Biodiversity and Ecosystem Services) has recently underlined the steadily growing importance of the ecosystem service concept. Concurrently, RU 816 "Biodiversity and Sustainable Management of a Megadiverse Mountain Ecosystem in South Ecuador" strove to identify the characteristics of science-based sustainable land use management systems that should inform conservation decisions in the biodiversity hotspot of the South Ecuadorian Andes. Such management system(s) should at the same time

1. preserve biodiversity and the ecosystem processes underlying its evolution and conservation,
2. rehabilitate degraded biodiversity and lost land usability, and
3. improve livelihoods for the local population.

This triple reseach challenge poses the mein research question of RU816–a question directly related to the most urgent gaps of current knowledge requiring integrative research on the ecological/socio-economic science interface (Ehrlich and Mooney 1983; Daily 1997; Carpenter and Turner 2000; Farber et al. 2002; Heal et al. 2005). The required research is much facilitated if ecosystems with differing anthropogenic

J. Barkmann (✉) • B.M. Hillmann • R. Marggraf
Department of Agricultural Economics and Rural Development, Environmental and Resource Economics, Georg-August-Universität Göttingen, Platz der Göttinger Sieben 5, 37073 Göttingen, Germany
e-mail: jbarkma@uni-goettingen.de

J. Bendix et al. (eds.), *Ecosystem Services, Biodiversity and Environmental Change in a Tropical Mountain Ecosystem of South Ecuador*, Ecological Studies 221,
DOI 10.1007/978-3-642-38137-9_4,

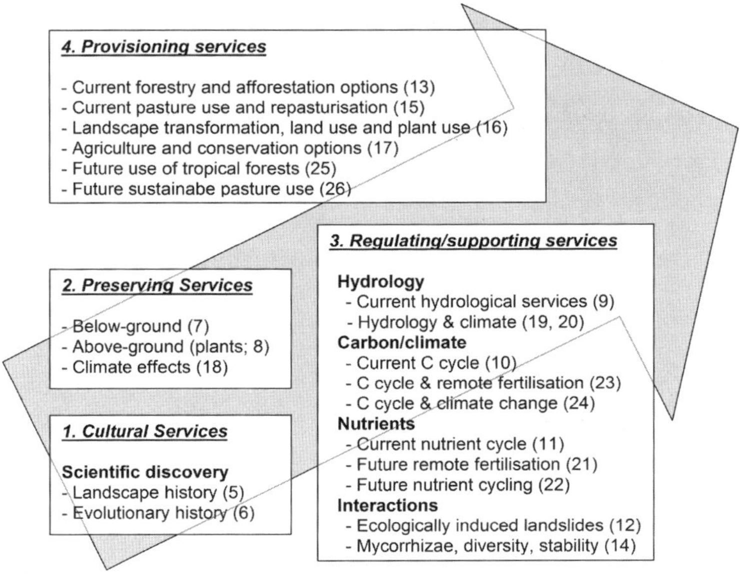

Fig. 4.1 The chapters of the book can roughly be arranged in four groups of ecosystem service categories that increase (*grey arrow* with respect to direct applicability to landscape management in the project area)

impact can be compared in direct spatial proximity. Comparative research can be based on field studies as well as on ecological experiments (refer to preface). Fortunately, the research area presents a dichotomy of protected natural forest and adjacent pasture ecosystems for which the forest had been cleared by slash and burn. Consequently, differences in the capacity of these two ecosystem manifestations to provide benefits to humans, i.e. ecosystem services, are at the analytic focus of this book.

When translating the main research question of RU 816 into a suitable research approach, it became apparent that all MEA (2003) categories of ecosystem services are covered to some degree by the investigations presented in this book. While the ecosystem service typology of the MEA provides the conceptual frame, the specific foci of the single projects suggest to place the chapters of this book into four groups that differ slightly from MEA ecosystem service categories: (1) cultural, (2) preserving, (3) regulating/supporting and (4) provisioning services (see Fig. 4.1). These four groups roughly indicate the differing relation of the chapters of this book to direct application in sustainable landscape management on the one hand and more basic research on the other.

In order to use an ecosystem services approach for the science-based design of sustainable land use management systems, ecosystem services need to be quantified in a suitable way. Many of the underlying ecological processes can be directly quantified such as water discharge and element leaching. With respect to benefits more directly appropriated by humans, ecosystem contributions—e.g. by soil quality or microclimate—to crop yields and farming income can also be quantified readily. For other ecosystem services, quantitative assessment is more difficult. For example, the usefulness of a preserved species for science or the more comprehensive importance of nutrient cycling or of a dampening of climate extremes is difficult to relate directly to human benefits. Contrary to a widespread belief, cultural ecosystem services including those based on aesthetical and ethical motivations pose no specific challenge to quantification in economic benefit terms (see e.g. Cerda et al. 2012; Barkmann et al. 2010).

The utilisation of an ecosystem service approach does not prescribe a specific valuation and/or decision-making method. Quantified data on the provisioning of ecosystem services can be input into multi-criteria methods, which do not monetise ecosystem services, as well as into cost-effectiveness or cost–benefit analysis. In cost–benefit analysis, for example, data on the effect of vegetation on the frequency of landslides is translated into a monetary estimate of avoided damages. Likewise, the contribution of a fascinating landscape to tourism profits can be assessed. Comparison of two or more differing manifestations of an ecosystem facilitates such valuation tasks considerably because the natural ecosystem can be used as a reference.

4.2 The RU 816 Approach in the Light of the Slightly Modified MEA Ecosystem Services Concept

Cultural services refer to the fact that ecosystems possess cultural value. Among other aspects, ecosystems serve as an object for research—including ecological research on ecosystem compartments, components and functionality. Understanding of landscape history and an evolutionary approach to speciation are further aspects of ecosystem research that tap cultural ecosystem services. With the generated information, ecosystem traits can be assessed and the conservation value of an ecosystem addressed. Plant, animal and fungal communities of the natural forest have been investigated during the past one and a half decades. The studies document that the research area is one of the "hottest" biodiversity hotspots worldwide (Liede-Schumann and Breckle 2008).

Preserving services are a specific manifestation of cultural ecosystem services, and refer to the maintenance of biological diversity irrespective of any specific ecological function. Preserving services provide benefits because many people assign a high value to the existence of species and to high biological diversity. Consequently, the research area deserves a high degree of conservation and protection from an existence and bequest value perspective (see Sect. 4.3). The existence value of the natural forests—as well as of the high altitude subpáramo vegetation—is underscored by the uniqueness of these ecosystems and by the irreversibility of ecosystem degradation: The systems will not restore themselves once converted to agricultural land, and then abandoned. The forests have, however, the potential for slow recovery after moderate disturbance that removes or destroys part of the trees

(Martinez et al. 2008). This knowledge might be transferable to other tropical forests. Thus, desiderata listed under (1) and (2) of the main research question can be materialised by the conservation of the natural forest manifested by its status as a national park.

Supporting services play a major role for desiderata (1)–(3) of the main research question because they are the fundament for the generation of ecosystem services from all other categories. Supporting ecosystem services range, e.g., from primary production via seed set and dispersal to soil formation, nutrient cycling and regional hydrology. As such the processes constituting supporting services can be quantified, but the importance of these quantitative figures for sustainable land management is difficult to assess.

Regulating services are necessary for the stabilisation of other services and, thus, often provide insurance services (Rajmis et al. 2010). Consequently, they are also inherently related to (1–3) of the main research question, addressing the capacity of the ecosystem to mitigate hazardous events by regulating climate extremes, erosion as well as water flow and nutrient cycles. Of particular importance is the question how changes of the environment affect the regulating capacity of the ecosystem, or how resilient the ecosystem is against changes of climate and/or of land use. A change of land use, e.g. conversion of the forest to pastures, may affect the single regulating services differently although the whole ecosystem changes.

Regulating and supporting services of an ecosystem are closely intertwined ecologically. Thus, the joint investigation of topics relating to *regulating and supporting ecosystem services* constitute the main focus of RU 816 research. The relevant book chapters are weaving around four themes essential for the valuation of the ecological basis (supporting services) and ecosystem stabilisation (regulating services) of provisioning services. The selected topics are highly relevant for Ecuadorian society. The four themes mainly address:

- Hydrological services, where the influence of the ecosystem and its changes on the water cycle with special reference to the soil–vegetation–atmosphere interface are investigated. This is a precondition for assessing the provision services potable water supply and water supply for hydropower generation.
- Nutrient services, where the current situation, regulation and stability (towards environmental change) of nutrient cycling are analysed, primarily supporting not only the provisioning service of agricultural production but also growth of the natural forest.
- Carbon/climate regulation services, whose investigation generates urgently needed knowledge on how to mitigate global climate change by carbon sequestration and how to prevent hazardous effects of climate extremes in the local climate due to land use change. Obviously, these services equally affect all other service categories.
- Interaction services that provide stability due to ecosystem interactions permeating the biotic and the abiotic spheres. One prominent example is the occurrence of landslides in the natural forest due to biotic–abiotic interactions, which at the same time enhances sediment transport in and fosters biodiversity of the natural system.

Provisioning services are a service category directly related to human needs (food and fibres, timber/non-timber forest products, potable water, hydropower, etc.), and thus to the livelihood of the local population. While provisioning services are obviously related to (3) of the main research question, the utilisation of provisioning services has feedbacks to (1) and is a precondition for achieving (2). The investigations of direct provisioning services in this book is related to current (non-sustainable) and potential future (sustainable) land use portfolios and thus, mainly related to agricultural production options including (a) forestry, (b) non-timber products, (c) pasture management and (d) indigenous home gardens. To answer the main research question of developing sustainable land use systems, benefits due to agriculture are investigated alongside other ecosystem services. One example is the design of Payment for Ecosystem Service schemes (PES) that foster afforestation of abandoned pasture land.

At this point it should be stressed that the perspective of investigating ecosystem services in a science-directed manner must not necessarily be in accordance with the priorities of the local population. However, reaching (1)–(3) without the acceptance and thus, the support of the local population is neither possible nor worthwhile. Therefore, the study presented in the next section was conducted to assess the preferences of the local population regarding selected ecosystem services.

4.3 Stated Preferences for Selected Ecosystem Services

Hillmann, Barkmann and Marggraf conducted a social science survey in order to estimate the economic preferences that the population of the project region holds with respect to a subset of the most important non-marketed ecosystem services. As the conservation of biological diversity in the Andean biodiversity hotspot is of global importance, we also estimated preferences of a German sample of respondents representing global stakeholders focusing on the conservation of "charismatic species" and of valuable habitats (see Fig. 4.2). While global stakeholders are expected to have only preferences with respect to the existence value of biological diversity of the research region—which relates to cultural ecosystem services-, local and regional respondents are expected to be interested economically also in other benefit classes. We report here briefly on results calculated form the two Ecuadorian samples using the Choice Experiment method (CE; Hensher et al. 2005). The design and wording of the CE instrument for these samples uses an ecosystem service approach specifically developed for the economic valuation of functional aspects of biological diversity (Barkmann et al. 2008).

During a qualitative pre-study, we tested several CE attributes, of which three were finally included in the main study. The number of moth species occurring in the project area was among the tested attributes. In the pre-study interviews, respondents did not indicate any economic preferences for the moths. The proximity

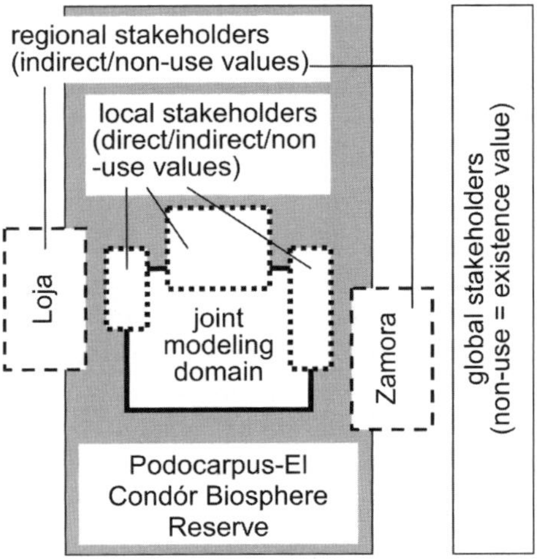

Fig. 4.2 Conceptual sketch of the spatial distribution of local, regional and global stakeholders valuing ecosystem services and biodiversity (for further explanations, see text)

of the forest to the home of respondents was initially included as a proxi for the provisioning services of regional forests. Averaged over interviewed local and regional stakeholders, preferences were very low in the pilot study, and the attribute subsequently dropped from the main study. Also, an attribute on the number of landslides (regulating service) was included in the qualitative interviews but not in the final valuation interviews as only low "demand" was recorded. In result, the main study addressed respondent preferences for changes (1) to endemic plant biodiversity at the species level, (2) to the quality of potable water supply, and (3) to offsets in average individual C emissions. We operationalised the carbon emissions attribute via an inquiry into the personal responsibility to mitigate the own C emissions of respondents by regional afforestation measures (cf. Rajmis et al. 2009). Current average annual C emissions stand at about 2 t C per capita in Ecuador.

For the Ecuadorian CE study, 310 of ~400 "local" rural households and 401 "regional" households from Loja and Zamora cities were sampled. Regional respondents were systematically selected from a stratified random sample of urban neighbourhoods. The local sample includes all households in the selected villages of the study area willing to participate (~78 %).

Positive preferences for the conservation of endemic plant species could clearly be detected ($p < 0.001$; see Fig. 4.3). However, this is the relatively least important CE attribute from the perspective of the Ecuadorian respondents. If scaled to a prevention of a loss of 20 endemic plant species, stated willingness to pay (WTP) is between ~1 US\$ year^{-1} and ~2 US\$ year^{-1}. WTP between 17 and 24 US\$ year^{-1} was estimated for a substantial reduction in infections brought about by hygienically cleaner potable water. Economic preferences for compensating 1 t carbon (C) emissions by regional afforestation ranged from 20 US\$ year^{-1} to 45 US\$ year^{-1}. In each case, the lower values are the mean of rural residents. As *willingness* to pay is restricted by *ability* to pay, lower economic preferences by rural respondents commanding lower average incomes were to be expected.

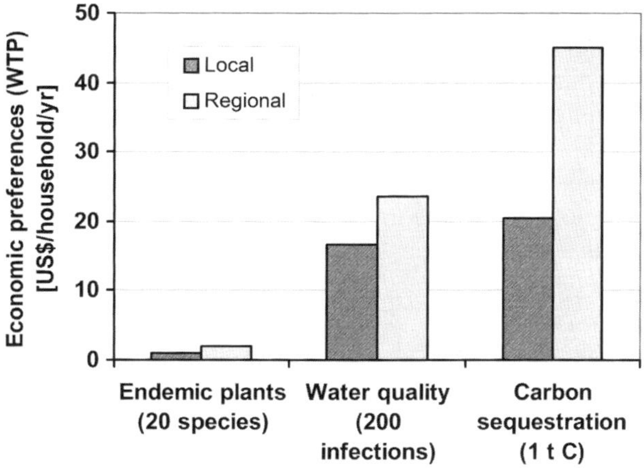

Fig. 4.3 Results of the local and regional Choice Experiment studies; WTP (willingness to pay) values are for changes in preserving, regulating and provisioning ecosystem relative to the status quo

Although direct benefits from offsetting individual C emissions by afforestation are low for any single individual, stated WTP is very substantial. Thus, respondents assign high importance to safeguarding global regulating services. High importance is also assigned to the direct "provisioning" benefits of cleaner potable water.

4.4 Conclusion

The ecosystem service concept was used as a tool to structure research at the interface of the natural and the social sciences. Thus, assigning a certain chapter to a specific part of the book should not be interpreted as a claim to the direct relevance of the ecological processes or functions investigated in terms of environmental assessment or decision making. Some of the proximity of the presented results in relation to their direct application in sustainable land management is captured by the used ecosystem service categories, though (Fig. 4.1). Within each category, however, there is substantial heterogeneity. This heterogeneity stems from the overall approach of RU 816 to advance the understanding of ecosystem structure and functioning at applied but also at more fundamental levels.

The social science results presented in Sect. 4.3 indicate that project area respondents clearly differentiate among different ecosystem services in terms of perceived benefits. For example, many respondents accepted a responsibility for offsetting their own contribution to climate change by local afforestation measures. Likewise, measures to improve the hygienic quality of potable water were strongly demanded. Without the demonstration of additional tangible benefits of biodiversity conservation—or without sufficient financial incentives-, local farming households are unlikely to forgo income opportunities even if income generation threatens plant species diversity.

The conservation of the exceptional biological diversity of the research area is a matter of potentially global concern. Results from the German sample indicate that international stakeholder preferences in the order of 130 € ha^{-1} $year^{-1}$ for the better protection of biodiversity hotspot areas such as the project area exist. Thus, there is substantial global demand for the protection of the "cultural" ecosystem services provided by biological diversity in southern Ecuador. Overall sustainable development of the project area and of the wider Andean biodiversity hotspot will require that synergies and trade-offs between potential development and conservation options be carefully examined—and that global WTP for biodiversity protection can actually be channelled into economically and ecologically sound conservation action.

References

Barkmann J, Glenk K, Keil A, Leemhuis C, Dietrich N, Gerold G, Marggraf R (2008) Confronting unfamiliarity with ecosystem functions: the case for an ecosystem service approach to environmental valuation with stated preference methods. Ecol Econ 65:48–62

Barkmann J, Yan J, Zschiegner A-K, Marggraf R (2010) The Dao of the sceptic and the spiritual: attitudinal and cultural influences on preferences for sustainable tourism services in the domestic Chinese tourism market. Int J Serv Sci Tech Manag 13(3/4):281–304

Carpenter SR, Turner M (2000) Opening the black boxes: ecosystem science and economic valuation. Ecosystems 3:1–3

Cerda C, Barkmann J, Marggraf R (2012) Application of choice experiments to quantify the existence value of an endemic moss: a case study in Chile. Environ Dev Econ 18(2):207–224

Daily G (ed) (1997) Nature's services: societal dependence on natural ecosystems. Island Press, Washington, DC

Ehrlich PR, Mooney HA (1983) Extinction, substitution, and ecosystem services. Bioscience 33:248–254

Farber SC, Costanza R, Wilson MA (2002) Economic and ecological concepts for valuing ecosystem services. Ecol Econ 41:375–392

Heal GM, Barbier EB, Boyle KJ, Covich AP, Gloss SP, Hershner CH, Hoehn JP, Pringle CM, Polasky S, Segerson K, Shrader-Frechette K (2005) Valuing ecosystem services: toward better environmental decision-making. National Academies Press, Washington, DC, 277 p

Hensher DA, Rose JM, Greene WH (2005) Applied choice analysis: a primer. Cambridge University Press, Cambridge

Liede-Schumann S, Breckle S-W (2008) Provisional checklist of flora and fauna of the San Francisco Valley and its surroundings. Ecotrop Monogr 4:256

Martinez A, Mahecha MD, Lischeid G, Beck E (2008) Succession stages of vegetation regeneration: secondary tropical mountain forests. In: Beck E, Bendix J, Kottke I, Makeschin F, Mosandl R (eds) Gradients in a tropical mountain ecosystem of Ecuador, vol 198, Ecological studies. Springer, Berlin, pp 425–432

MEA [Millennium Ecosystem Assessment] (2003) Ecosystems and human well-being: a framework for assessment. Island Press, Washington, DC

Rajmis S, Barkmann J, Marggraf R (2009) User community preferences for climate change mitigation and adaptation measures around hainich national park, Germany. Clim Res 40 (1):61–73

Rajmis S, Barkmann J, Marggraf R (2010) Pythias Rache: Zum Tauschwert ökologischer Versicherung vor unbekannten Risiken. GAIA – Ökologische Perspektiven für Wissenschaft und Gesellschaft 19(2):114–121

Part II
Current Situation of Biodiversity and Ecosystem Services

Chapter 5
Landscape History, Vegetation History, and Past Human Impacts

Fernando Rodriguez, Achim Bräuning, Andrés Gerique, Hermann Behling, and Franziska Volland

5.1 Introduction

The ecosystem of the Podocarpus National Park (PNP, see Fig. 5.1) in the Southern Ecuadorian Andes is well recognized worldwide as a "hotspot" of biodiversity (Richter et al. 2009; see Chap. 8). The altitudinal range has allowed the development of highly diverse habitats which contribute to the extraordinarily high degree of biodiversity (Richter et al. 2009). Other ecosystems present in the area are evergreen tropical mountain forests (montane rainforests) and elfin forests in the transition zone to the páramo ecosystem (see Fig. 1.6 in Beck et al. 2008). However, natural and anthropogenic impact has influenced the transformation of ecosystem areas concomitant with the loss of original ecological integrity. The present landscape of the PNP is the result of human impact and climate change that have been taking place for thousands of years. Some valleys that were formed during the cold Pleistocene (from around 2.6 million years ago until 10000 B.P.) are witness to ancient glaciations in the Podocarpus plateau. The landscape has been transformed during recent centuries upon human colonization to satisfy the population needs. Deforestation caused by timber, agriculture, and grazing activities has changed the original ecosystem vegetation to the point where it has been replaced by other opportunistic and pioneer plant communities that were able to develop in the transformed areas. Today, only small patches of natural forests remain due to the long history of occupation and, in particular, due to increasing human impact during recent decades (Dodson and Gentry 1991; Pohle 2008).

F. Rodriguez (✉) • H. Behling
Department of Palynology and Climate Dynamics, Albrecht-von-Haller-Institute for Plant Sciences, University of Göttingen, Untere Karspüle 2, 37073 Göttingen, Germany
e-mail: frodriguez@ecociencia.org

A. Bräuning • A. Gerique • F. Volland
Institute of Geography, University of Erlangen-Nuremberg, Kochstrasse 4/4, 91054 Erlangen, Germany

J. Bendix et al. (eds.), *Ecosystem Services, Biodiversity and Environmental Change in a Tropical Mountain Ecosystem of South Ecuador*, Ecological Studies 221,
DOI 10.1007/978-3-642-38137-9_5, © Springer-Verlag Berlin Heidelberg 2013

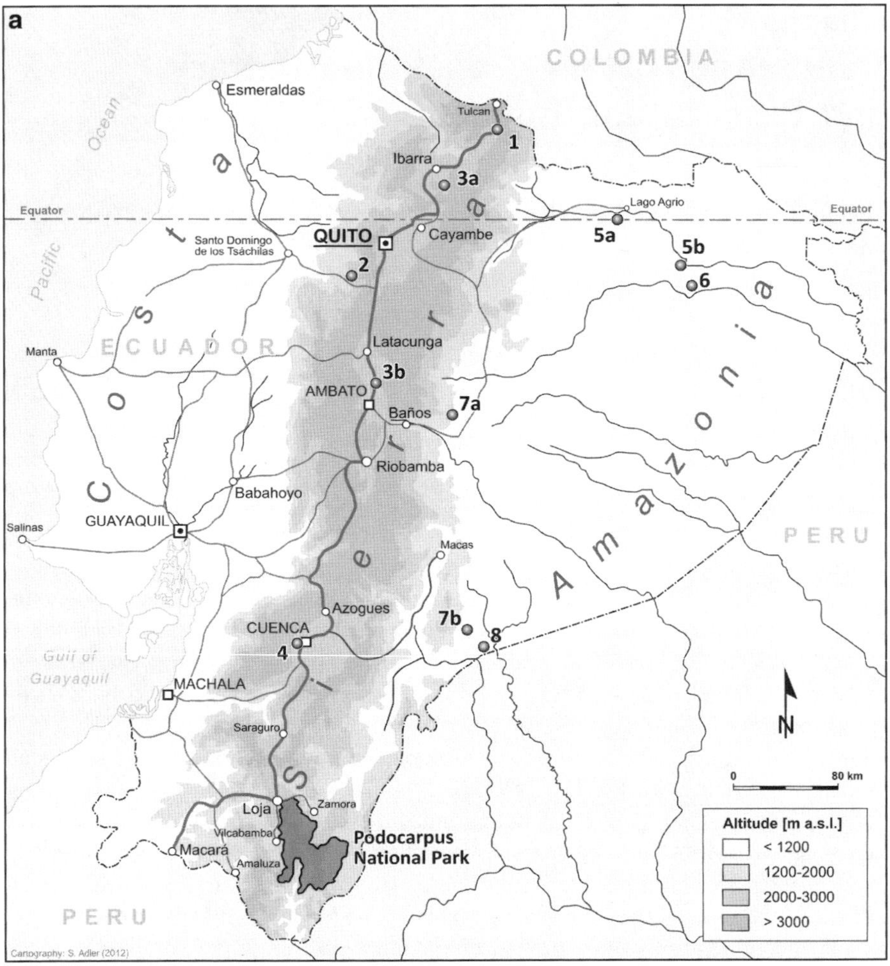

Fig. 5.1 (continued)

The reconstruction and understanding of past vegetation and climate changes as well as of human impact, can enhance our knowledge as to how past processes have influenced ecosystem dynamics and led to the development of the outstanding diversity in the study region. This will further improve our understanding of the reactions of the tropical mountain ecosystems to environmental changes on local and regional scales. Such information is crucial for developing management and conservation strategies within the context of changing climatic conditions (see Chap. 27). In this chapter we consider the landscape dynamics having taken place during the last 16,000 years (Late Quaternary) in the wider area of the PNP region including changes in vegetation, biodiversity, climate, tree growth, fire, and land use.

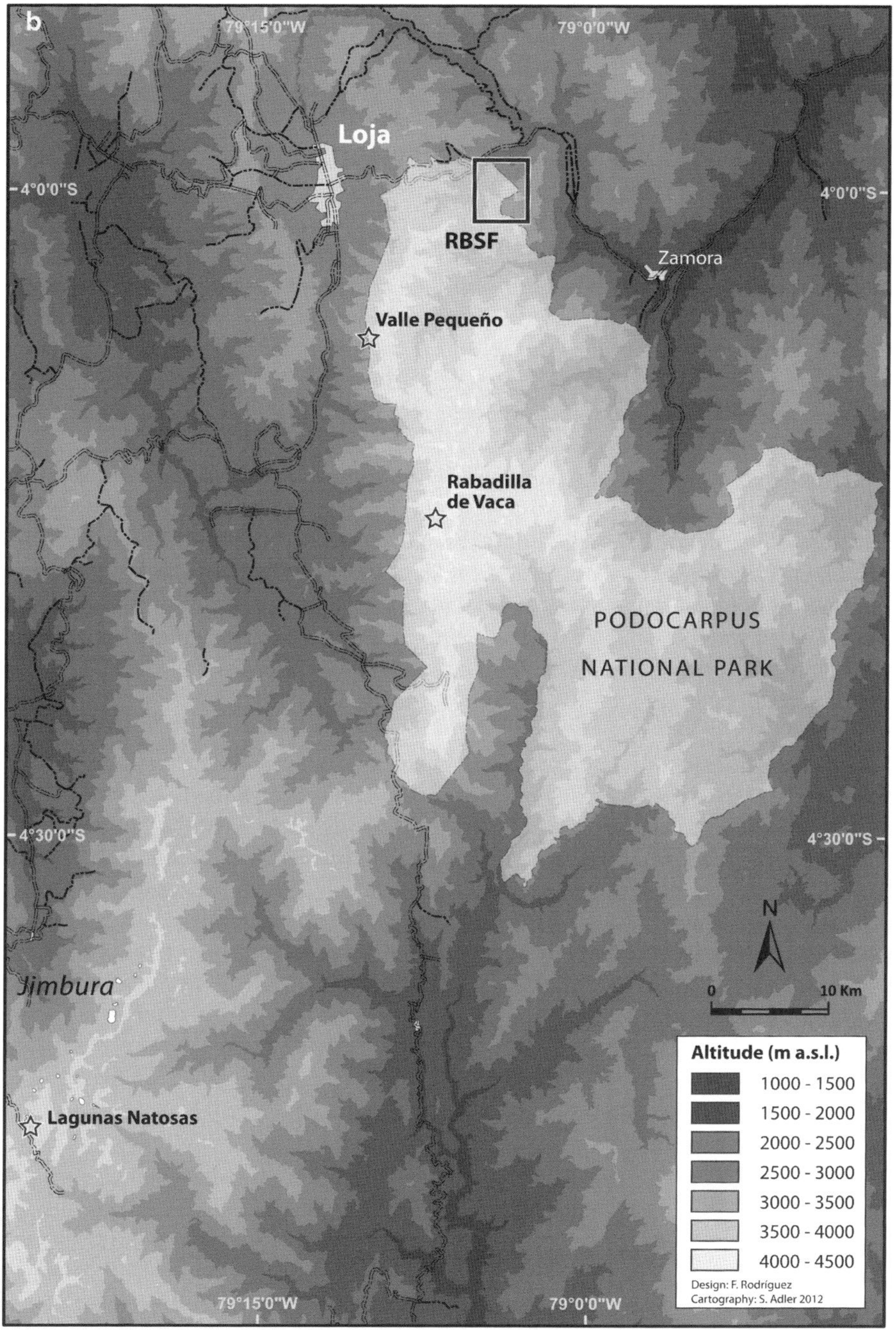

Fig. 5.1 (**a**) The map marks the locations of palynological study sites in Ecuador (*gray*, *numbered circles* and smaller *white circles*) in relation to the Podocarpus National Park (PNP). Numbers refer to the following references: (1) Bakker et al. (2008)/Di Pasquale et al. (2008); (2) Wille et al. (2002); (3a/b) Colinvaux et al. (1988b); (4) Colinvaux et al. (1997)/Hansen et al. (2003); (5a/b) Colinvaux et al. (1988a); (6) Weng et al. (2002); (7a) Bush et al. (1990)/Liu and Colinvaux (1985);

5.2 Methods and Analyses

Pollen analysis was carried out on three soil cores taken in the transition zone between the forest and the páramo ecosystems (Fig. 5.1). Two cores originate from north of the PNP (Ravadilla de Vaca, RVM, and Valle Pequeño, VP), the third being taken in a glacial valley (Lagunas Natosas Forest, LNB) close to the Peruvian border, 12 km to the south of the town of Jimbura. Radiocarbon ages were determined for each sample using CalPal (Weninger et al. 2004). For pollen and charcoal analyses, subsamples (0.25 cm^3) were taken at 4 cm intervals along the cores. Standard analytical methods were used to process the samples (Fægri and Iversen 1989). Known amounts of exotic *Lycopodium* spores were added to each sample before treatment as internal controls of charcoal concentrations (grains or particles cm^{-3}) and influx rates (grains or particles cm^{-2} $year^{-1}$). A minimum of 300 pollen grains were counted for each sample. Tree, shrub, and herb pollen grains were included in the pollen sum, whereas spores and aquatic taxa were excluded. Pollen types were grouped according to the ecosystem or habitat type for which they are most characteristic. Four main vegetation types were defined: Lower Mountain Rainforest (LMF), Upper Mountain Rainforest (UMF), Subpáramo, and Páramo. Pollen and spore data are presented as percentage diagrams based on the pollen sum using the programs TILIA and TILIAGRAPH for calculation, illustration, and zonation (Grimm 1987). The zonation of the pollen record is based on important changes in the pollen assemblages' cluster analysis by CONISS (Grimm 1987).

Tree-ring widths of 70 increment cores taken from 32 *Cedrela montana* (Meliaceae) individuals from the Reserva Biologica San Francisco (RBSF; Fig. 5.1) were measured at a precision of 0.01 mm. To remove the biological age trend inherent in ring-width series, the measured raw values were standardized with a cubic smoothing spline removing 50 % of the tree-growth variance at 2/3 of the original series length (Cook and Kairiukstis 1990). Climate-growth relationships were calculated as correlation functions with local climate station data and variations of the Southern Oscillation Index (SOI) taken from the 'Global Change Master Directory' (GCMD 2012).

Fig. 5.1 (continued) (7b) Bush et al. (1990); (8) Bush and Colinvaux (1988). (**b**) The map marks the locations of the study sites Valle Pequeño (VP), Rabadilla de Vaca (RVM), and Lagunas Natosas Forest bog (LNB) in and near the Podocarpus National Park in more detail by means of stars, as well as indicating the location of the Reserva Biológica San Francisco (RBSF) by a *square*

5.3 Results and Discussion

5.3.1 Vegetation of the Southern Ecuador

The páramo areas of southern Ecuador are characterized by irregular topography (Maldonado 2002). The vegetation of the herbaceous páramo (Paramo herbáceo) found at about 3,100–3,400 m a.s.l. is rich in *Neurolepis nana*, *Calamagrostis macrophylla*, and *Niphogeton dissecta*. This vegetation unit is 0.2–1 m in height and covers flat slopes and concave terrains, as well as ridge areas. The shrub paramo (Páramo arbustivo bajo) occurring at about 2,900–3,400 m a.s.l. is rich in *Weinmannia rollottii*, *Oxalis spiralis*, and *Ilex andicola*. This páramo type is found mainly on steep slopes (Lozano et al. 2003) and exhibits shrubs and herbs of from 0.5 to 1.2 m height. The subpáramo (approx. 2,800–3,100 m a.s.l.) is characterized by *Puya nitida*, *Brachyotum rotundifolium*, and *Oritrophium peruvianum*. Shrubs and herbs grow up to 1 m in height, while individual shrubs can be 2–3 m tall. The upper montane rainforest (UMF) is present between about 2,100–2,750 m a.s.l. and is represented by a low, single-tree stratum of between 5 and 10 m in height, occasionally up to 15 m. Characteristic trees are *Morella pubescens* (Myricaceae) and *Myrsine andina* (Myrsinaceae) (Bussmann 2005; Lozano et al. 2003).

5.3.2 Past Climate, Vegetation, and Fire Dynamics in the PNP

A few palynological records for the Ecuadorian Andes (Colinvaux et al. 1997; Hansen et al. 2003; Fig. 5.1) and some more recent studies on upper forest line (UFL) changes in northern Ecuador (Bakker et al. 2008; Di Pasquale et al. 2008; Wille et al. 2002; Fig. 5.1) have contributed to the environmental history of the northern Ecuadorian Andes and thus to our understanding of the present vegetation distribution in the PNP area. However, palaeoecological research in Ecuador has concentrated mainly on the Amazon basin (e.g. Bush and Colinvaux 1988; Bush et al. 1990; Colinvaux et al. 1988a; Liu and Colinvaux 1985; Weng et al. 2002), and only one study is available as to the inter-Andean Plateau (Colinvaux et al. 1988b; Fig. 5.1).

5.3.2.1 The Late Glacial Environment

The area's outstanding biodiversity has evolved from the interplay of several regional and local scale environmental factors (e.g., climatic, geologic, orographic and edaphic conditions, as well as natural and anthropogenic disturbance regimes; see Chap. 1) and the environmental history that has figured importantly in the development of the current plant diversity pattern (Broecker 1997; Churchill et al. 1995; Richter et al. 2009). Since the onset of the Pleistocene around 2.6 million years ago, recurrent climate alterations between cold and warm periods have resulted in changes in the number and distribution of ecosystems in the area.

During the Last Glacial Maximum (LGM, 18,000 years ago), most Andean glaciers advanced and reached their lowermost positions at about 3,000 m a.s.l. in the eastern Andes of Colombia, Ecuador, and northern Peru (Clapperton 1993; Rodbell 1992; Heine 2011). Temperatures in the area gradually increased at the end of the LGM. As the retreating glaciers shrunk, they left behind them moraines and small glacial lakes (Heine 2011; Mark et al. 2004). Data for the Lagunas Natosas forest bog at 3,495 m a.s.l (Fig. 5.2) respective of the late Pleistocene and Holocene periods (the last approximately 10,000 years) show that páramo vegetation (Poaceae) was dominant during the period of 15930–11660 B.P. (zone LNB-I, Fig. 5.2). The pronounced abundance of *Plantago rigida* and pteridophytes indicates that cold and humid conditions prevailed locally during the Late Glacial period and formed cushion mires in depressions. The limited number of mountain forest taxa reflects the prevalence of low temperatures during the Last Glacial period. The presence of clayey sediment poor in organic matter may also reflect the lack of forest vegetation near the coring site. In this context, it is assumed that the forest line was at a markedly lower elevation than it is at present.

The Bølling–Allerød interstadial during the period of from 14700 to 12700 B.P. was characterized by relatively warm and moist conditions, which favored an abundant presence of *Hedyosmum* and pteridophytes. Small populations of *Polylepis* also became established during the Last Glacial period in the region. Moraceae/Urticaceae can easily be transported by wind over longer distances, and their ample representation may thus be related to this factor (Bush et al. 1990; Niemann and Behling 2009).

The Younger Dryas period was a cold climate period at the end of the Pleistocene between 12,800 and 11,500 years ago. Rapid climatic fluctuations during the last deglaciation phase of this period are documented by glacier advances in the Ecuadorian Andes before 12500 B.P. and between 10,500 and 9,000 years ago (Heine 2011). The local absence of *P. rigida* in the study area at approximately 12,600 years ago, whereas it had previously represented the most abundant taxon, suggests a drying of the climate at this time.

The incidence of fires during this period can be disregarded due to the wet and cold environmental conditions. This notion is supported by the almost complete absence of carbonized particles, which suggests that anthropogenic influences played no role during this period.

5.3.2.2 Early to Mid-Holocene Environment

The LNB core points a marked decrease of páramo vegetation between 11660 and 4280 B.P. (Zone LNB III, Fig. 5.2). The expansion of woody dwarf shrub vegetation (Melastomataceae and Asteraceae) is mostly associated with the subpáramo ecosystem. Dwarf shrubs can survive and reproduce well above the forest line. Furthermore, an indication of upper forest line (UFL) shift is less evident in

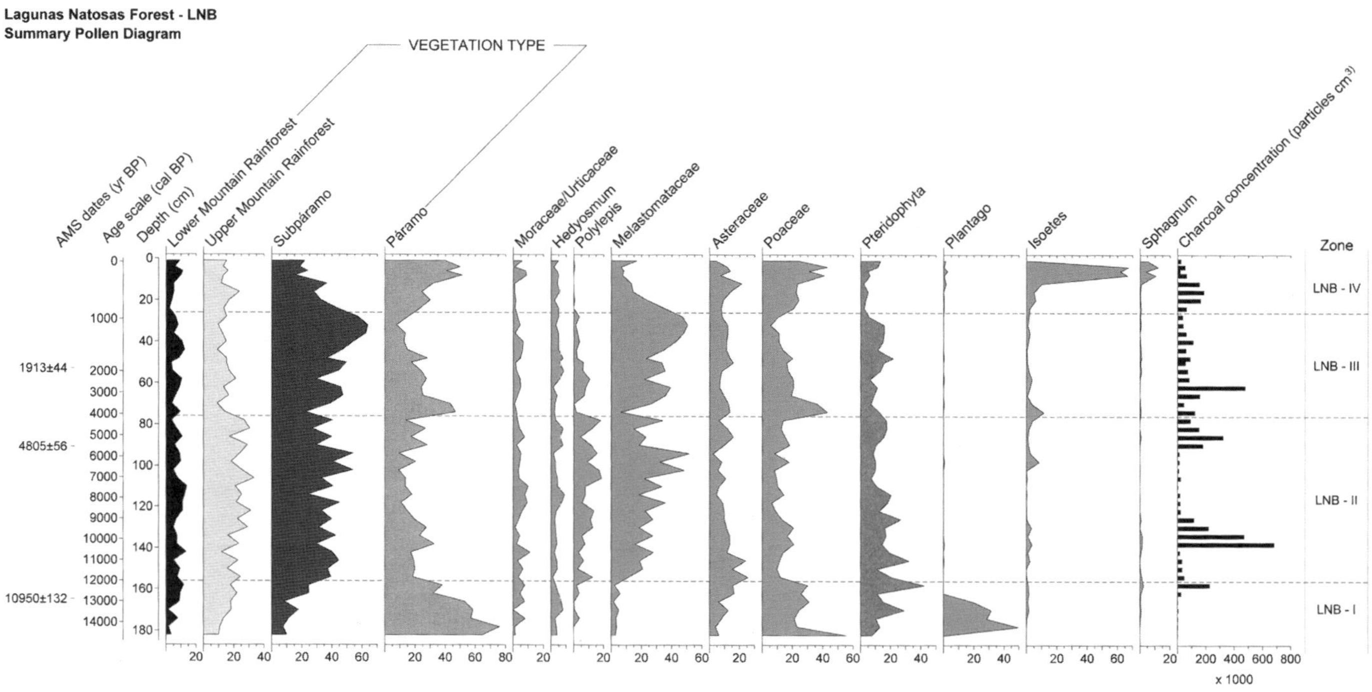

Fig. 5.2 Analysis of the Lagunas Natosas Forest core (3,495 m a.s.l.) showing radiocarbon dates, age scale, ecological groups, values of charcoal concentration, and zones

southern Ecuador than on other sites where the typical treeline genus *Polylepis* dominates the UFL. *Polylepis* was well represented and most likely widespread in southern Ecuador, as evidenced by marked changes in not only vegetation structure but also ecosystem composition. Warmer and drier conditions characterized the transition from the Pleistocene to the Holocene, allowing an upslope shift of the UFL. According to Thompson et al. (2000), the early Holocene climate was 1.5–2 °C warmer than that today, favoring the development of forest vegetation. The period was marked by a widespread recession of glaciers in most parts of the Andes, but a return to cooler and more humid conditions apparently occurred after about 5000 B.P. This phenomenon is evident in soil core data for approximately 4200 B.P. in the LNB area and corresponds to an increase in páramo vegetation.

5.3.2.3 Late Holocene Environment and Human Settlement History

According to Guffroy (2004), signs of human activity in the Loja region are evident for times as early as about 4,500 years ago. However, human activities may have occurred much earlier in the dry inter-Andean valley. At Paso El Tiro, located 10 km east of Loja, Niemann and Behling (2008) found evidence of increased fire intensity during the wetter late Holocene. This suggests the occurrence of frequent fires of anthropogenic origin used for hunting purposes and slash-and-burn activities in the drier lower valleys. The palynological data show a dynamic transition between subpáramo and páramo from 4,280 years ago until the present (Zone LNB III and IV). Poaceae were dominant around 4200 B.P., but suffered a critical decline in the vicinity of 3,100 years ago. An important increase in the abundance of Melastomataceae and Asteraceae at the same time has been registered. The results of the RVM record (Fig. 5.3) from 125 km north of the LNB (Zone RVM-I) show that the UMF was most abundant from about 2100–1720 B.P. as evidenced by the greater abundance of *Hedyosmum*, while the subpáramo was more strongly represented as indicated by an increased abundance of Melastomataceae. Pollen data in the RVM core show that Moraceae/Urticaceae became more abundant after 1200 B.P. This period also shows a greater abundance of *Weinmannia* in the UMF compared to the previous and subsequent periods. Melastomataceae were abundant at the LNB around 1,200 years ago, while Poaceae declined radically.

This vegetation composition reveals that somewhat warmer and drier environmental conditions most likely prevailed during this period. In the VP record no marked vegetation changes can be detected (Fig. 5.4).

After the fifteenth century the area around the upper range of the Cordillera Oriental close to the present city of Loja was the Eastern border of the Inca Empire and the Shuar territories (cf. Hocquenghem et al. 2009). The Shuar had triumphantly defended themselves against the Inca armies of Túpac Yupanqui, who finally considered the Shuar to be indomitable (González-Suarez 1890). After 60 years of coexistence the Shuar successfully revolted in 1599 against the Spaniards (Conde 1988), who then had to abandon the gold mines and settlements on the eastern part of the Cordillera. The region was probably sparsely inhabited

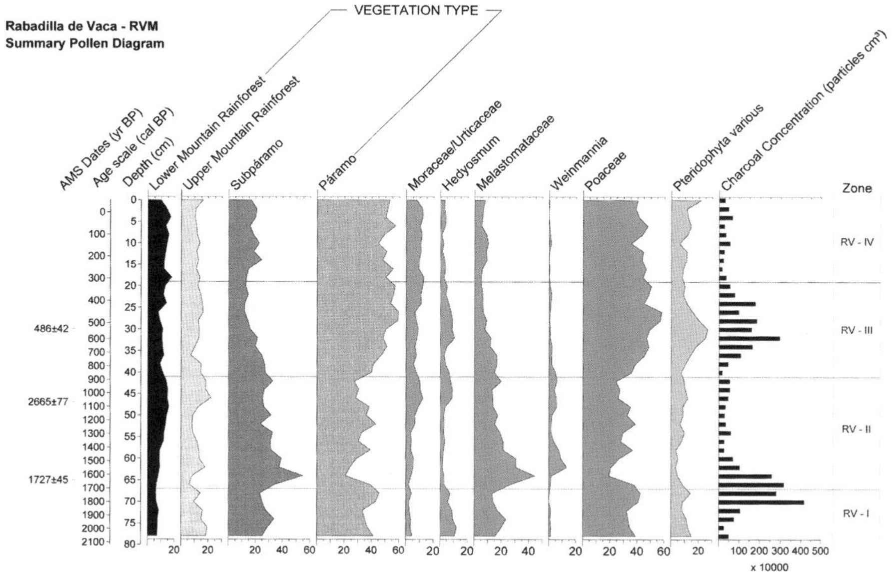

Fig. 5.3 Analysis of the Rabadilla de Vaca Mire core (3,200 m a.s.l.) showing radiocarbon dates, sums of ecological groups, records of charcoal concentration, and zones

between the fifteenth century and the end of the nineteenth century. It was not until the end of the nineteenth century that missionaries, soldiers, and gold miners successfully installed permanent settlements in the mountain area, initiating the current processes of forest exploitation and agricultural expansion (Gerique 2010; Chap. 2). The evidence of a very strong increase in páramo taxa (Poaceae and Cyperaceae), *Isoëtes*, and *Sphagnum* during the last 500 years in the LNB core shows the area to be a moister ecosystem than it was during the previous period. This is corroborated by the fact that the oldest trees of the PNP as dated by radiocarbon are *Prumnopitys montana* (Podocarpaceae) growing in a valley north of the RVM pollen site at an elevation of about 2,100 m a.s.l. at the lower limit of the UMF. After calibration with the Southern Hemisphere Calibration dataset (McCormac et al. 2004), ^{14}C dates collected from the inner parts of a *P. montana* stem (lab code Erl-12870) showed the tree to be 418 $\pm$ 35 years old, corresponding to A.D. 1453–1504. Since there are even taller trees than the tested one in the forest, it can be inferred that the lower (dry) altitudinal limit of the UMF has not decreased during the past ca. 500 years.

The strong decline of Melastomataceae up to the present day suggests that the UFL has shifted slightly downslope. This situation is similar to that inferred from the RVM and VP cores, which show that the proportions of the LMF and the UMF remained relatively stable during the period between 880 and 310 B.P. In contrast, the páramo vegetation expanded continuously due to the spread of Poaceae, while the subpáramo vegetation declined as indicated by the progressively more infrequent occurrence of Melastomataceae. The proportions of the LMF and the UMF have remained relatively stable in the RVM and VP areas during the last

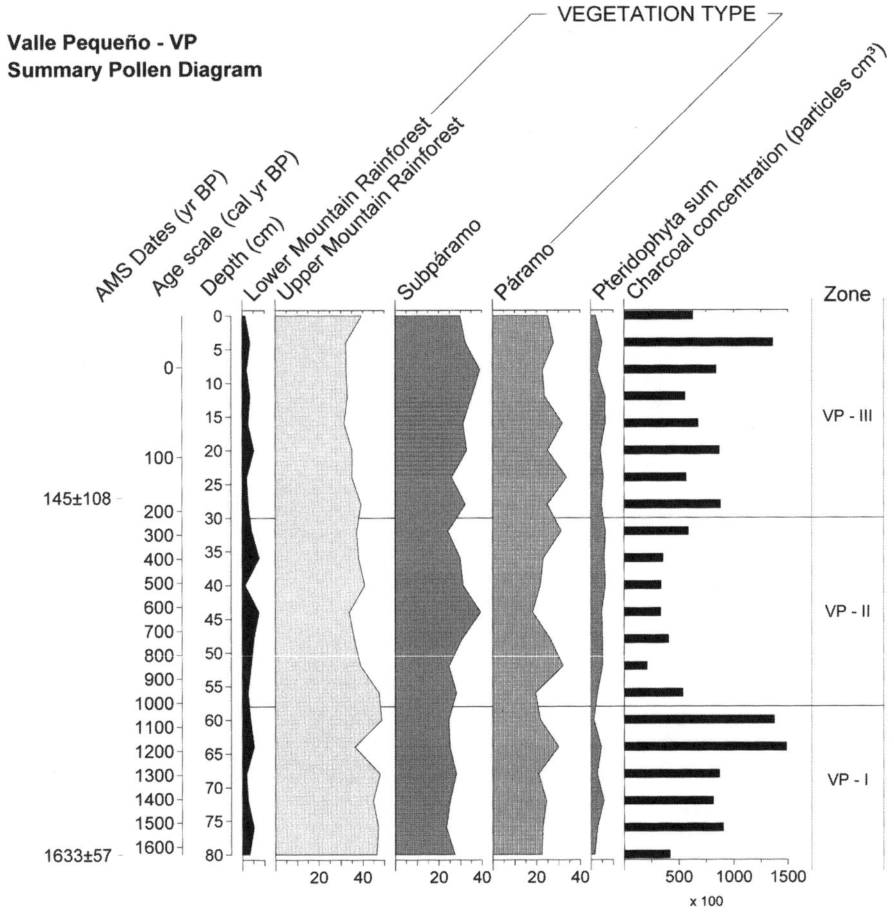

Fig. 5.4 Analysis of the Valle Pequeño bog core (3,244 m a.s.l.) showing radiocarbon dates, sums of ecological groups, records of charcoal concentration, and zones

approximately 300 years. One important difference between these two areas as shown by the pollen records is that *Alnus*, which used to be rare, has become quite frequent in the RVM area. The páramo ecosystem was dominant in the RVM and LNB areas, as shown by the frequent occurrence of Poaceae. The different vegetation types remained relatively stable in the VP area during this period.

The carbon particle concentration has increased in the LNB during the last 900 years. In spite of the very wet local conditions, human activities to promote hunting or grazing could have led to frequent fires. Other activities such as deforestation to provide for a supply of wood, grazing, and cultivation may have taken place during this period. Charcoal records indicate that fires occurred at the RVM and VP sites throughout the late Holocene. Two periods of enhanced fire frequency (during 1800–1600 and 600–400 B.P.) are evident from the RVM core.

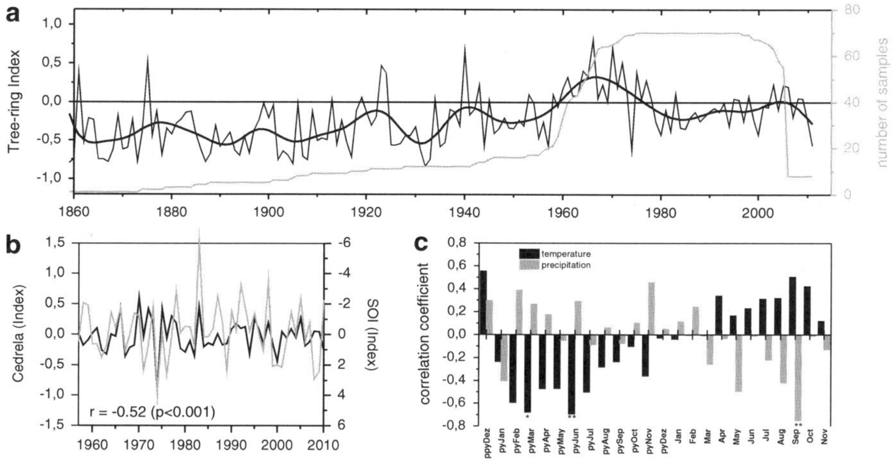

Fig. 5.5 (**a**) Ring-width index chronology of *Cedrela montana* (*thin black line*) with 5-year smoothing (*thick black line*) and number of samples included in the chronology (*gray line*); (**b**) comparison between the *Cedrela* ring-width index (*black*) and variations of the November–February Southern Oscillation Index (SOI; *gray*); (**c**) correlation function diagram between the *Cedrela* ring-width index and monthly means of temperature and monthly sums of precipitation of local climate station data at the Reserva Biologica San Francisco (RBSF; 1998–2011). Correlations significant at $p < 0.05$ and $p < 0.01$ (t-test) are marked with one and two *asterisks*, respectively. py and ppy correspond to respective months in the year prior to growth or 2 years prior to growth

The oldest tree-ring record presently available for the PNP dates back to 1840 A.D. It originates from the deciduous species *Cedrela montana* (Meliaceae), the wood of which forms clearly visible annual rings and exhibits cambial activity mainly between January and April (Bräuning et al. 2009). However, *Cedrela* ceases growth activity during dry periods and is thus sensitive to drought (Bräuning 2009). Although the oldest individuals having been found are up to 140 years of age, most trees are not older than 60 years. Since the average correlation between all individual ring-width series averaged in the final chronology is rather low (mean $r = 0.19$), the present dating of *Cedrela* has attained a sufficient level of confidence only for growth having commenced after 1950, when more than 20 series became available for the chronology analysis. The growth of *Cedrela* at the local level is promoted by lower temperatures having occurred during previous year's growing season. This relationship is reversed during April–September of the current growth year (Fig. 5.5c). Since local climate data from the RBSF are only available for the short time period since 1998, we also tested correlations between the tree-ring chronology and regional climatic data and found significantly negative correlations ($r = -0.52$) with the Southern Oscillation Index (SOI, Fig. 5.5b). Decadal ring-width variations (Fig. 5.5a) may accordingly be related to large-scale atmospheric pressure variations in the tropical Pacific Ocean. A larger quantity of older tree-ring data is nevertheless required to improve the chronology quality for time periods having occurred prior to the twentieth century.

5.4 Conclusions

During the Late Pleistocene (15930–11660 B.P.) grass páramo vegetation dominated the PNP and its surroundings, and the presence of *Plantago rigida* suggests that the area was cold and wet. The minor warming of the climate during the Bølling--Allerød interstadial allowed the expansion of forest taxa (*Hedyosmum* and Moraceae/Urticaceae), and *Polylepis* appeared in the south of Ecuador. A short cold period comprised by the Younger Dryas event (12800–11500 B.P.) exercised constraints on the *P. rigida* population and promoted the expansion of grass páramo (Poaceae). It is assumed that the UFL was lower at that time than it is nowadays. The climate was warmer and drier than it is today during the early and mid-Holocene (11,660–4280 B.P.), promoting the dominance of forest taxa (Melastomataceae and Asteraceae) and the decline of páramo vegetation, the latter being adversely influenced by frequent fires. An upslope shift in the UFL is suggested by an increase in the proportion of *Polylepis*. The late Holocene was marked by the expansion of subpáramo vegetation which could have led to the UFL upslope shifting. *Polylepis* became rare and almost absent during the Late Holocene due to wetter conditions and frequent fires. The vegetation and fire records of the investigation sites make it evident that fires favored the expansion of grass páramo vegetation and curtailed the expansion of subpáramo vegetation and the upper montane forest. Because fire probably played an important role in determining the upper forest limit during the Late Holocene, it is difficult to conclude whether climate change played an important role in changes in the UFL taking place during this period. The fire events occurred diachronously between sites, suggesting strong anthropogenic influence. Although regional vegetation is influenced by large-scale climate variability, local conditions of temperature, precipitation, wind regimes, radiation, geomorphological features and human land-use history also have a strong impact on vegetation patterns and are important in determining the structure, heterogeneity, and distribution of ecosystems.

References

Bakker C, Moscol Olivera M, Hooghiemstra H (2008) Holocene environmental change at the upper timberline in northern Ecuador. Holocene 18:1–17

Beck E, Makeschin F, Haubrich F, Bendix J, Richter M, Valerezo C (2008) The ecosystem (Reserva Biológica San Francisco). In: Beck E, Bendix J, Kottke I, Makeschin F, Mosandl R (eds) Gradients in a tropical mountain ecosystem of Ecuador. Ecological studies, vol 198. Springer, Berlin, pp 1–13

Bräuning A (2009) Climate variability of the tropical Andes since the Pleistocene. Progress Geosci 22:13–25

Bräuning A, Volland-Voigt F, Burchardt I, Ganzhi O, Nauss T, Peters T (2009) Climatic control of radial growth of *Cedrela montana* in a humid mountain rain forest in southern Ecuador. Erdkunde 63(4):337–345

Broecker W (1997) Future directions of paleoecological research. J Quaternary Sci 16:821–825

Bush MB, Colinvaux PA (1988) A 7000-year pollen record from the Amazon lowlands, Ecuador. Vegetatio 76:141–154

Bush MB, Colinvaux PA, Wiemann MC, Piperno DR, Liu KB (1990) Late Pleistocene temperature depression and vegetation change in Ecuadorian Amazonia. Quaternary Res 34:330–345

Bussmann RW (2005) Bosques andinos del sur de Ecuador, clasificación, regeneración y uso. Rev Peru Biol 12:203–216

Churchill SP, Balslev H, Forero E, Luteyn JL (eds) (1995) Biodiversity and conservation of neotropical montane forests. The New York Botanical Garden, Bronx, NY

Clapperton CW (1993) Quaternary geology and geomorphology of South America. Elsevier, Amsterdam, 779 pp

Colinvaux PA, Bush MB, Steinitz-Kannan M, Miller MC (1997) Glacial and postglacial pollen records from the Ecuadorian Andes and Amazon. Quat Res 48:69–78

Colinvaux PA, Frost M, Frost I, Liu K-B, Steinitz-Kannan M (1988a) Three pollen diagrams of forest disturbance in the western Amazon basin. Rev Palaeobot Palynol 55:73–81

Colinvaux PA, Olson K, Liu K-B (1988b) Late-Glacial and Holocene pollen diagrams from two endorheic lakes of the inter-Andean Plateau of Ecuador. Rev Palaeobot Palynol 55:83–99

Conde PFrT (1988) Los Yaguarzongos. Historia de los Shuar de Zamora. Abya-Yala, Quito. Reprint of the publication "Conde PFrT (1931) Los Yaguarsongos. Historia de los Jívaros de Zamora. Loja"

Cook E, Kairiukstis LA (1990) Methods of dendrochronology. Applications in the environmental sciences. Springer

Di Pasquale G, Maziano M, Impagliazzo S, Lubritto C, De Natale A, Bader M (2008) The Holocene treeline in the northern Andes (Ecuador): first evidence from soil charcoal. Palaeogeogr Palaeoclimatol Palaeoecol 259:17–34

Dodson CH, Gentry AH (1991) Biological extinction in Western Ecuador. Ann MO Bot Gard 78:273–295

Fægri K, Iversen J (1989) Textbook of pollen analysis (4th edn by Fægri K, Kaland PE, Krzwywinski K). Wiley, Chichester

Gerique A (2010) Biodiversity as a resource: plant use and land use systems among the Shuar, Saraguros, and Mestizos in tropical rainforest areas of southern Ecuador. PhD Thesis, Friedrich-Alexander-Universität Erlangen-Nürnberg

Global Change Master Directory (GCMD) (2012) NASA. http://gcmd.nasa.gov. Accessed 28 Dec 2011

González-Suarez F (1890) Historia General de la República del Ecuador. Tomo I, Quito. Cited in Jaramillo Alvarado P (2002) Historia de Loja. Municipio de Loja-Senefelder. Loja

Grimm EC (1987) CONISS: a Fortran 77 program for stratigraphically constrained cluster analysis by the method of the incremental sum of squares. Comput Geosci 13:13–35

Guffroy J (2004) Catamayo precolombino. Investigaciones arqueológicas en la provincia de Loja (Ecuador). IRD Editions, Paris

Hansen BCS, Rodbell DT, Seltzer GO, Leon B, Young KR, Abbott M (2003) Late-glacial and Holocene vegetational history from two sites in the western Cordillera of southwestern Ecuador. Palaeogeogr Palaeoclimatol Palaeoecol 194:79–108

Heine K (2011) Late Quaternary glaciations of Ecuador. In: Ehlers J, Gibbard PL (eds) Developments in quaternary science, vol 15. Elsevier, Amsterdam, pp 803–813

Hocquenghem AM, Poma J, Salcedo L (2009) La Red Víal Incaica en la Región Sur del Ecuador. Centro de Investigación y Apoyo al Desarrollo Local-Regional, Universidad Nacional de Loja, Loja

Liu K-B, Colinvaux PA (1985) Forest changes in the Amazon Basin during the last glacial maximum. Nature 318:556–557

Lozano P, Delgado T, Aguirre Z (2003) Estado actual de la flora ende´mica exclusiva y su distribucio´n en el occidente del Parque Nacional Podocarpus. Fundacio´n Funbota´nica, Herbario y Jardı´n Bota´nico "Reinaldo Espinosa". Loja, Ecuador

Maldonado N (2002) Clima y vegetación de la región sur del Ecuador. In: Aguirre Z, Madsen JE, Cotton E, Balslev H (eds) Botánica Austroecuatoriana. Estudios sobre los recursos vegetales en las provincias de El Oro, Loja y Zamora Chinchipe. Abya Yala, pp 1–28

Mark BG, Seltzer GO, Rodbell DT (2004) Late Quaternary glaciations of Ecuador, Peru and Bolivia. In: Ehlers J, Gibbard PL (eds) Quaternary glaciations – extent and chronology, Part III: South America, Asia, Africa, Australasia, Antarctica: Developments in quaternary science, vol 2. Elsevier, Amsterdam, pp 151–163

McCormac FG, Hogg AG, Blackwell PG, Buck CE, Higham TFG, Reimer PJ (2004) SHCAL04 Southern Hemisphere Calibration, 0–11.0 CAL KYR BP. Radiocarbon 46:1087–1092

Niemann H, Behling H (2008) Late Quaternary vegetation, climate and fire dynamics inferred from the El Tiro record in the southeastern Ecuadorian Andes. J Quaternary Sci 23:203–212

Niemann H, Behling H (2009) Late Pleistocene and Holocene environmental change inferred from the Cocha Caranga sediment and soil records in the southeastern Ecuadorian Andes. Palaeogeogr Palaeoclimatol Palaeoecol 276:1–14

Pohle P (2008) The people settled around Podocarpus National Park. In: Beck E, Bendix J, Kottke I, Makeschin F, Mosandl R (eds) Gradients in a tropical mountain ecosystem of Ecuador. Ecological studies, vol 198. Springer Verlag, Berlin, pp 25–36

Richter M, Diertl KH, Emck P, Peters T, Beck E (2009) Reasons for an outstanding plant diversity in the tropical Andes of Southern Ecuador. Landsc Online 12:1–35

Rodbell DT (1992) Late Pleistocene equilibrium-line reconstructions in the Northern Peruvian Andes. Boreas 21:43–52

Thompson LG, Mosley-Thompson E, Henderson KA (2000) Ice core palaeoclimate records in tropical South America since the last glacial maximum. J Quaternary Sci 15:377–394

Weng C, Bush MB, Athens JS (2002) Holocene climate change and hydrarch succession in lowland Amazonian Ecuador. Rev Palaeobot Palynol 120:73–90

Weninger B, Jöris O, Danzeglocke U (2004) Calpal – the Cologne radiocarbon CALibration and PALaeoclimate research package. http://www.calpal.de. Accessed Aug 2010

Wille M, Hooghiemstra H, Hofstede R, Fehse J, Sevink J (2002) Upper forest line reconstruction in a deforested area in northern Ecuador based on pollen and vegetation analysis. J Trop Ecol 18:409–440

Chapter 6
Past Dynamics of Speciation in Andean Mountain Rainforests

Konrad Fiedler and Patrick Strutzenberger

6.1 Introduction

The staggering species richness of tropical biota, already recognized by Darwin, Wallace, and their nineteenth century contemporaries, continues to stimulate the development and testing of ecological theory. This concerns the mechanisms that allow species to coexist in dense packing, be it under niche (Schluter 2000) or neutralistic concepts (Hubbell 2001), but also the role of biodiversity in ecosystem functioning (Hooper et al. 2005), or the challenges of tropical species richness for conservation biology (Barlow et al. 2007). Especially for megadiverse regions like the focal area of the RU816 in South Ecuador, the conditions that allow species to coexist in extremely dense packing are still far from being understood (see Chap. 1). Another important dimension of biodiversity research, however, is primarily devoted to unraveling the evolutionary origin of exceptionally high tropical species richness. The tropics have alternately been viewed as cradles, or museums, of biological diversity (Arita and Vázquez-Domínguez 2008). This debate can be simplified to the provocative question whether high tropical species richness is ancient or recent in origin. "Recent origin" in this context means that evolutionary processes during the Quaternary [i.e., the Pleistocene (2.588–0.0116 Ma) and Holocene (0.0116 Ma to present)] have played a key role in generating the high tropical biodiversity we can observe today. "Ancient origin," in contrast, is here viewed as those speciation processes that have taken place earlier, viz. in the Neogene and before.

K. Fiedler (✉)
Department of Tropical Ecology and Animal Biodiversity, University of Vienna, Rennweg 14, 1030 Vienna, Austria
e-mail: konrad.fiedler@univie.ac.at

P. Strutzenberger
Senckenberg Natural History Collections, Königsbrücker Landstraße 159, 01109 Dresden, Germany

J. Bendix et al. (eds.), *Ecosystem Services, Biodiversity and Environmental Change in a Tropical Mountain Ecosystem of South Ecuador*, Ecological Studies 221,
DOI 10.1007/978-3-642-38137-9_6,

The Quaternary period is characterized by recurrent and ongoing cycles of global climatic cooling (glacial) and warming (interglacial) during the past 2.5 Ma. Undoubtedly these ice ages and their interruptions have left an immense legacy in current biodiversity (Hewitt 2004). In tropical realms, fragmentation of tropical moist forests during phases of drought has for a long time been thought to be the main mechanism by which glacial periods affect biodiversity. According to this concept, tropical moist lowland forest persisted only in Pleistocene refuges, allowing for allopatric speciation processes to start in (potentially) every individual forest fragment (Haffer 1969). Repeated cycles of forest fragmentation and expansion, then, could explain the generation of multiple species from one single ancestral form, by means of recurrent allopatric speciation processes. The spatial concordance of distributions of endemic species, across widely divergent groups of organisms, in apparent centers of endemism, has for decades been accepted as support for this theory.

As a corollary for tropical mountain environments, Quaternary climatic oscillations have pushed the tree line down during glacial periods, whereas warmer climate in interglacials allowed a re-expansion of forest into higher elevations. Paleo-climatological and paleo-ecological studies (for example by analyzing pollen records in sediment deposits) confirmed that Pleistocene climate oscillations have shifted forest ecosystems down and up again in the tropical Andes (Brunschön and Behling 2010; see Chap. 5). Such oscillations would provide scope for accelerated allopatric speciation, for example in multiple forest fragments that persisted in isolated valleys or in otherwise restricted landscape patches as a function of stark topographic heterogeneity. Alternatively, climate oscillations might have forced forest species, or even entire forest species assemblages, to move down and up to keep track with their climatic niche requirements (at least if niche conservatism were to prevail instead of rapid niche evolution; Wiens and Graham 2005). Both these ideas—increased generation of new species through repeated cycles of habitat fragmentation (in analogy to the refuge hypothesis) or the tracking of climate niches by vertical migrations of individual species or inter-connected communities, but without concomitant increase in speciation rates—are plausible against the background of ecological and evolutionary theory. The challenge, therefore, remains to empirically test which of these scenarios holds true to explain the observed megadiversity of tropical Andean mountain rainforests.

6.2 Pleistocene Climate Cycles and Forest Refuges

Most tests of the Pleistocene refuge hypothesis have thus far dealt with Amazonian lowland rainforest organisms. Biogeographers rather quickly adopted this concept (e.g., Mayr and O'Hara 1986) and concluded, from apparent concordance of endemism centers, that such refuges provide a plausible explanation of current patterns of species distributions and densities. Yet, at approximately the same time critical voices emerged against this hypothesis (Endler 1982). It turned out that part of the biogeographical evidence was spurious, due to spatial heterogeneity of sampling coverage (Nelson et al. 1990). There also has been an intensive debate as to how pronounced the contraction and fragmentation of Amazonian lowland forests was as a consequence of drought during the glacial periods (Colinvaux 1996; Bush and de Oliveira 2006). Current perception based on more comprehensive evaluation of pollen records and climate models is that much of Amazonia had persistent forest cover during the Pleistocene; especially at the foothills of the Andes moist forest was continuously present. Molecular phylogeny and phylogeography also have refuted the refuge theory to an increasing extent (Moritz et al. 2000; Knapp and Mallet 2003; Bridle et al. 2004).

Even though biogeographical, paleo-ecological, and molecular evidence has undermined the refuge hypothesis as a general explanation for Amazonian speciation, the possibility remains that in the more heterogeneous relief of the Andes, the undisputed cyclic changes in the elevational extent of forest might have set the stage for Pleistocene speciation events to occur more frequently, such that climate oscillations and concomitant ecosystem dynamics could have served as a motor of a speciation pump (see Chap. 1). In the absence of adequate fossil evidence for most species-rich groups of terrestrial organisms in the Andean region, and also in view of the extremely limited knowledge about distributional ranges of most of the arthropod species that make up the megadiverse Andean faunas, the only way to decide between the "ancient" and "recent" origin hypothesis is by means of data that can be obtained from extant organismal samples. The most promising venue relies on the application of modified "molecular clock" approaches to phylogenetic hypotheses derived from gene sequences.

6.3 Calibrated Phylogenies as a Tool to Assess Temporal Speciation Patterns

Time calibrated phylogenies or "timetrees" are becoming more and more customary to answer questions related to the origin of biodiversity. The establishment of a common repository for temporal data (http://www.timetree.org) together with the publication of an accompanying volume (Hedges and Kumar 2009) is ample testament to the rapidly increasing number of time-calibrated phylogenies. Peterson et al. (2007) coined the name "molecular paleobiology" for using molecular genetic data to answer essentially paleontological questions. Until recently, dating efforts were largely focused on deep divergences (e.g., among animal phyla; Wray et al. 1996). With the advent of ever cheaper sequencing opportunities and ever higher computing power, the focus of molecular dating studies has shifted to include investigations on lower taxonomic levels with dense sampling on species level (e.g., Santos et al. 2009; Ramirez et al. 2010; Strutzenberger and Fiedler 2011). Those kinds of studies do not only provide age estimates but also contain valuable information on relative evolutionary rates and their change over time. Importantly, molecular paleobiology provides opportunities to gain insight into the history of taxa devoid of any fossil record. An increasing number of analytical tools are available for the investigation of temporal diversity patterns. Arguably the simplest and most commonly used one is the lineage through time plot (LTT). The log-transformed cumulative number of lineages within a clade is plotted against the age of each node. LTTs are useful to visualize overall temporal patterns, but are not sensitive to changes in rates of diversification. Other measures have to be applied in order to investigate changes in evolutionary rates. A commonly used metric here is the γ statistic (Pybus and Harvey 2000) to test whether observed rates of diversification deviate from a rate constant model. More complex patterns of rate changes can be analyzed by model fitting. Rabosky (2006) developed algorithms to fit several common models to a set of branching times derived form calibrated phylogenies.

The application of timetree approaches to species-rich tropical taxa is faced with a number of challenges. Tropical arthropod communities often have high species numbers including many rare species (Brehm et al. 2008; Coddington et al. 2009). For the calculation of evolutionary rates and inferences based upon them, it is vital to either have a near-complete sampling of species for the target taxon, or be able to estimate the number of missing species. Monte Carlo approaches to estimate the effect of missing species on evolutionary rates are now available (Rabosky 2006). Animals in tropical forest ecosystems have a low probability of fossilization due to the lack of sedimentation and rapid biological degradation in tropical forests (Benton and Harper 2009). This problem is exacerbated for terrestrial arthropods whose propensity to fossilize is much lower to begin with, leading to a profound lack of fossil calibration points for tropical terrestrial arthropods. All published molecular dating studies on Neotropical arthropods were therefore forced to use calibration points outside the target taxon. In comparison to studies on groups with good fossil records, this augments confidence intervals for time estimates of arthropods. The use of calibration points far back in time also increases the likelihood to introduce a substantial bias into age estimates.

6.4 Material and Methods

Eois is a genus of geometrid moths and forms a substantial fraction of megadiverse moth communities in Andean montane forests (Brehm et al. 2005; Hilt et al. 2006; examples are displayed in Fig. 6.1). *Eois* has been the focus of investigations on host plant specialization (Bodner et al. 2010a, 2012; Rodríguez-Castañeda et al. 2010) and host–parasitoid interactions (Connahs et al. 2009). Assessment of local *Eois* diversity in a DNA barcode assisted integrative taxonomy approach revealed that at least 166 species of *Eois* occur in the Reserva Biológica San Francisco (RBSF) and surrounding areas in southern Ecuador alone (Strutzenberger et al. 2011; see Fig. 1.1), whereas total Neotropical *Eois* species richness (Brehm et al. 2011) may be as high as 733–1,710 species, as opposed to less than 250 described

Fig. 6.1 Eight representative members of the diverse moth genus *Eois* from southern Ecuador. Within each row, a pair of closely related species, with their assignment to species groups sensu Strutzenberger et al. (2010), is depicted. None of the eight species has been taxonomically described and named thus far. The datings (Ma, million years ago) given on the right of each pair are extracted from a calibrated molecular phylogeny (Strutzenberger and Fiedler 2011). In all four taxon pairs, individual species can hardly be told apart by means of external morphological characters, but are clearly differentiated in terms of sequence markers. Moreover, molecular dating shows these divergences to be rather old, clearly predating the period of Pleistocene climatic oscillations. Photographs: L. Lehner (*upper right*), G. Brehm (all others)

ones. Neotropical *Eois* form a monophyletic clade (Strutzenberger et al. 2010). Strutzenberger and Fiedler (2011) performed a molecular dating study on *Eois* based on a phylogeny calculated from nuclear *Ef1α* and mitochondrial *COI* sequences. Representatives of 102 *Eois* species sampled in the RBSF and its surroundings were sequenced by standard methods (Sanger sequencing). Sequence data were used to infer the phylogeny of *Eois* (Strutzenberger et al. 2010) and subsequently to date divergences in a relaxed molecular clock approach. We used the BEAST algorithm (Drummond and Rambaut 2007) to estimate divergence times. Absolute ages were calibrated with two calibration points, one derived from a larger scale dating study (Yamamoto and Sota 2007) and the second from fossil evidence of a member of the same moth subfamily Larentiinae (Cockerell 1922). As there is no known fossil record for *Eois,* both calibration points are outside of *Eois*. To compare the fit of several rate constant and rate variable models of diversification to the observed pattern, we used the "R" package *laser* (Rabosky 2006), after estimating the γ statistic corrected for the number of unsampled *Eois* species (Pybus and Harvey 2000).

6.5 Results and Discussion

6.5.1 The Megadiverse Moth Genus Eois *as a Case Study*

The origin of Neotropical *Eois* was inferred in the mid-Oligocene at 30.96 Ma (+4.94 Ma; −5.17 Ma). The LTT plot (Fig. 6.2) shows that diversification started at a high rate and was constantly declining towards the present. The rate of divergence dropped abruptly around the Miocene–Pliocene boundary, leveling off into a plateau phase. Only three splits were detected in the Pleistocene, all of them among taxa with uncertain species status. This observed plateau in the LTT plot starting in the late Pliocene could be explained either by increased extinction rates or by stasis. It is conceivable that Pleistocene climate changes have disrupted the established microhabitat structure in Andean montane forests and therefore put a halt to the diversification of *Eois* and its host plants. It is equally possible that changes in climate and the resulting shifts of elevational zones may have lead to increased extinction rates. Using sophisticated statistical models, a strong and statistically significant slow down of the diversification rate towards the present was detected. This could still be recovered when correcting for unsampled species with the lower estimate of 733 species, but not with the maximum richness estimate of 1,710 *Eois* species to exist in South America (Brehm et al. 2011). However, even this "failure" to reject a constant rate of evolution with the highest species richness estimate does not necessarily mean that diversification rates were actually constant. Fitting of temporal models of evolutionary rate changes identified a density dependent logistic model (DDL) as the best explanation for the observed pattern ($\Delta AIC_{DDL,\ yule3rate} = 7.4$; $\omega = 0.97$). This model assumes that the rate of speciation

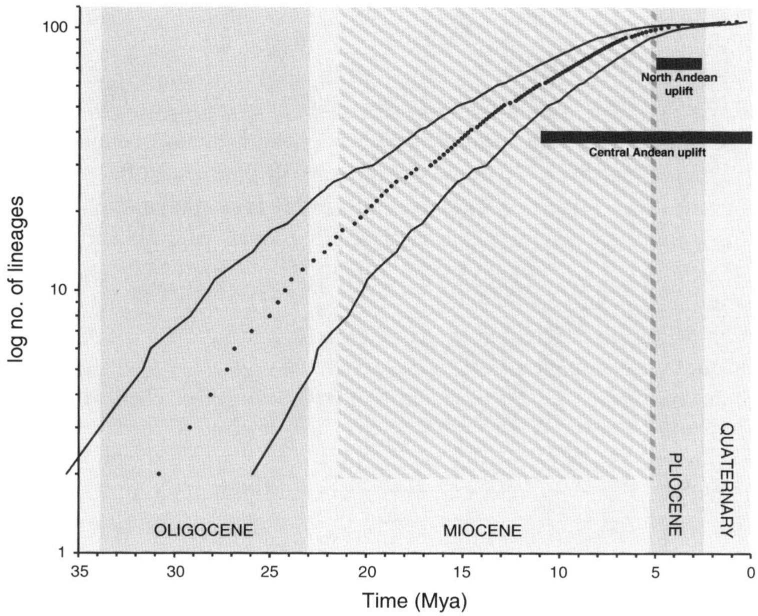

Fig. 6.2 Lineage through time (LTT) plot of Neotropical *Eois* moths (modified after Strutzenberger and Fiedler 2011). *Dotted line*: cumulative number of lineages [when the log of the number of lineages (*y*-axis) is plotted through time (*x*-axis) a *straight line* would indicate that diversification rate has been constant over time]. *Solid lines* represent 95 % credibility intervals. Hatched area indicates period of massive species level diversification of *Piper* (according to Smith et al. 2008), the most important host plant genus of *Eois* moths. Periods of increased uplift for the north and central Andes are indicated with *black bars*. See Strutzenberger and Fiedler (2011) for further details

is inversely proportional to the number of species present at any given time. The second best model (yule3rate) is clearly an inadequate representation of the data given the high Akaike weight of 0.97 for the DDL model. This timing of divergences within *Eois* reveals two major events that are likely to have at least influenced, if not directly facilitated, the diversification of *Eois* in parallel.

1. Diversification of *Eois* overlaps with the diversification of its host plants. According to a calibrated phylogeny for the Piperacean genera *Piper* and *Peperomia*, species level diversification of *Piper* was dated as mid-late Miocene to earliest Pliocene (Smith et al. 2008). The majority of known host plant affiliations of *Eois* are with either *Piper* or *Peperomia*. Density-dependent speciation points to processes of niche filling and saturation (Ricklefs 2009) and is considered a hallmark of adaptive radiation (Schluter 2000). In the case of *Eois* it is plausible that these moths may have diversified to occupy available niches (i.e., host plant species), and diversification has slowed down in concordance with a slowdown of diversification in *Piper* and *Peperomia*.

2. Much of the diversification of *Eois* took place during a period of increased Andean uplift. Twenty million years ago the Andes were only 25–50 % of their present day elevation. Rates of uplift in the central Andes increased about 12 Ma while the northern Andes underwent rapid uplift during the Pliocene period (Gregory-Wodzicki 2000). Both episodes of Andean uplift could potentially have affected the diversification of *Eois*. Due to the lack of data on the paleo-distribution or even the present day distribution of *Eois*, it cannot be determined to what extent the different temporal scales of the central and north Andean uplift affected *Eois*. Owing to higher average temperatures during the Miocene, the latitudinal as well as elevational range of *Eois* might well have been larger than it is today. The flora and fauna in the RBSF region is partitioned on a very fine spatial scale (Homeier et al. 2010; Günter et al. 2008; see Chap. 8). Plant communities in ravine forests are strikingly different from ridge forest, and the same is true for moth communities. It is therefore obvious that the constant formation and modification of these small-scale habitat structures happening during an active orogeny provided ample opportunities for speciation to occur (see Chap. 1).

The two factors described above may have acted on *Eois* not separately, but rather in tight interaction with each other. Andean uplift is likely to have driven the speciation of *Eois* directly, but also speciation in its host plants. The diversification of *Eois* might also have influenced the evolution of Piperaceae plants, even though larval densities of *Eois* on *Piper* shrubs are usually too low for these herbivores to exert a massive selection pressure on their host plants (Bodner et al. 2010b, 2012). A more detailed picture of the exact nature of present day interactions between *Eois* and Piperaceae plants will be necessary to assess the potential coevolutionary importance of these interactions with more confidence. On a global scale the most dramatic event during the diversification of *Eois* was the global cooling that started in the Oligocene and eventually lead to the Pleistocene glacial cycles (Zachos et al. 2001). This could have affected Andean taxa by restricting their latitudinal and elevational ranges, which in the case of herbivorous *Eois* might have acted either directly or indirectly via their host plants. Currently, it is impossible to disentangle these biotic effects from those caused by Andean uplift with the available data on Andean climate and species diversity over time.

6.5.2 What Do Other Calibrated Phylogenies Tell About Andean Speciation?

Calibrated phylogenies for tropical taxa are still few (see Sect. 6.3), but the advent of molecular dating analyses for a growing number of Neotropical taxa challenged traditional concepts. In a meta-analysis Rull (2008) found that various Neotropical plant and animal taxa diversified at a constant rate starting in the Eocene [about 56–34 Ma] and obtained no evidence for a concentration of speciation events in the

Pleistocene. Insects turned out to be one of the few taxa where a glacial origin of species seemed to predominate. This can be attributed to the gross under-sampling of insects in regard to calibrated phylogenies at the time this meta-analysis was conducted.

Thus far, most calibrated phylogenies of Neotropical taxa have a focus on lowland Amazonian organisms. Nevertheless, most of these studies revealed the respective taxa diversified during the period of Andean uplift, starting ~12 Ma (Antonelli et al. 2010). There are several ways in which the growing mountain range did influence the Amazonian lowland. The uplift of the Andes had major impact on Amazonian habitats by changes in climate, especially in the distribution of rainfall (Sepulchre et al. 2010). The growing mountain range also acted as a barrier to dispersal. It has been proposed that the Andes might have served as a species pump for the Amazonian lowland (Fjeldsa and Lovett 1997). This notion proposes speciation in the Andes with subsequent migration into the Amazonian lowland. This model was at least partially supported in a number of studies (Wahlberg and Freitas 2007; Santos et al. 2009; Sedano and Burns 2010) and might well be applicable to *Eois*, the majority of which have been found in the Andes (Brehm et al. 2011). More complex biogeographical patterns were recently uncovered in a butterfly group associated with Poaceae as larval food (the *Taygetis*-clade of Euptychiina: Matos-Maraví et al. 2013), but also in their study only few speciation processes involving high-elevation specialists occurred during the Pleistocene.

Antonelli et al. (2010) reviewed the literature available on Neotropical tetrapods and found that the overwhelming majority of Amazonian tetrapods diversified in the Miocene and Pliocene periods. Studies with a focus on Andean organisms, however, are still rare. Few studies on primarily Andean Lepidoptera have recently seen the light of the day. Besides *Eois*, we have started to unravel the phylogeny of another species-rich geometrid moth genus, *Psaliodes*, in an analogous approach. Though data are not yet complete, available sequences for 37 species from southern Ecuador give no indication of massive Pleistocene radiation (P. Strutzenberger and K. Fiedler, unpublished results). Rather, patterns appear to be broadly similar to those observed in *Eois*, even though *Psaliodes* larvae have totally different feeding habits (viz. they are specialized on ferns; Bodner et al. 2010a). Elias et al. (2009) investigated temporal patterns of diversification in the Ithomiinae butterfly genera *Ithomia* and *Napeogenes*. These genera, comprising narrow specialists with larval host plants in the Solanaceae, diversified in the second half of the Miocene and in the Pliocene. Pleistocene diversification was detected in only two species pairs. Casner and Pyrcz (2010) investigated the Andean Satyrinae butterfly genus *Lymanopoda* that mostly occurs in páramo and subpáramo habitats, where glacial divergences could be expected with highest likelihood. Again, however, diversification within *Lymanopoda* took largely place in the late Miocene and Pliocene, whereas the contribution of Pleistocene divergences was small. The authors concluded that the elevational zonation of *Lymanopoda* species to be observed today was established prior to the Pleistocene. No study on Andean insects was thus far able to detect a massive signal of rapid diversification during the Pleistocene.

We therefore suggest that an alternative hypothesis should be considered in future studies, i.e., that Pleistocene glacial cycles and the concomitant habitat disruptions dramatically slowed down diversification of herbivorous insects and may even have lead to increased extinction rates.

6.6 Conclusions

Available evidence from calibrated phylogenies suggests that much of Andean biodiversity is not "recent": rather, most extant species originated earlier than during the Pleistocene. This is especially true for forest-dwelling organisms like *Eois* moths which are bound to forested habitats through their larval food plants. Rather, in such clades speciation rates increased during and after the formation of the Andes. Coevolution with host plants, over long periods of mutual association, emerged as the prevailing mode of speciation in *Eois*, and it is likely that similar patterns will be uncovered in many other groups of organisms that are integrated in tight biotic interaction networks. This might indicate that these organisms, herbivorous insects and their host plants alike, had to track their climatic and biotic niches by moving vertically down and up in the Andes, yet these range movements did not necessarily translate into cladogenesis. This does not preclude, of course, that also in Andean mountain forests some very young species should exist. From a biogeographical as well as an ecological perspective, however, instances of young speciation are expected to be more prevalent in the vegetation units above the tree line (such as the páramos). Páramos are nowadays island-like habitats between which gene flow can be severely limited by forests, high mountain ridges, and low valleys that serve as barriers (as is the case with alpine plants and animals elsewhere in high mountains of the world). Hence, sequences of habitat contraction and expansion are more likely to interrupt gene flow to a degree that facilitates allopatric speciation (Hewitt 2004). For Andean forest organisms, in contrast, there is presently no evidence that the contraction of the elevational extent of forest habitats during glacial maxima really fostered speciation in a general manner. This leads us to conclude that the continuous existence of Andean forest ecosystems from the time of the Andean uplift onwards was the prime requisite for the tremendous extant species richness of mountain forests to accumulate, though not in a constant manner.

Under a conservation perspective it follows from these considerations that maintaining sufficiently large stretches of near-natural forest is essential to preserve this species richness (Laurance et al. 2012; see Chaps. 8 and 13). However, the challenge for the decades to come is whether individual species and entire sets of interacting species will be able to track the rapid climate changes that may shift climatic niches of forest species upwards (Colwell et al. 2008, see also Part III).

Acknowledgments We thank Florian Bodner, Manuela Zimmermann, Lisamarie Lehner, Ivonne Daniela Vásquez Quezada, Brigitte Gottsberger, and Christine Truxa for assistance with field sampling, laboratory work, access to literature, and/or comments on the manuscript. Further thanks go to the collaborators of the Research Unit in Ecuador, and especially to the staff and inhabitants of the Estación Científica San Francisco. The Ministerio del Medio Ambiente del Ecuador issued research permits, and the foundation Nature and Culture International (Loja, Ecuador) as well as the Universidad Técnica Particular de Loja gave logistic support and allowed access to the study area and their facilities.

References

Antonelli A, Quijada-Mascarenas A, Crawford AJ, Bates JM, Velazco PM, Wüster W (2010) Molecular studies and phylogeography of Amazonian tetrapods and their relation to geological and climatic models. In: Hoorn C, Weeslingh F (eds) Amazonia – Landscape and species evolution: a look into the past. Wiley-Blackwell, Chichester, pp 387–404

Arita HT, Vázquez-Domínguez E (2008) The tropics: cradle, museum or casino? A dynamic null model for latitudinal gradients of species diversity. Ecol Lett 11:653–663

Barlow J, Gardner TA, Araujo IS, Avila-Pires TC, Bonaldo AB, Costa JE, Esposito MC, Ferreira LV, Hawes J, Hernandez MI, Hoogmoed MS, Leite RN, Lo-Man-Hung NF, Malcolm JR, Martins MB, Mestre LA, Miranda-Santos R, Nunes-Gutjahr AL, Overal WL, Parry L, Peters SL, Ribeiro-Junior MA, da Silva MN, da Silva MC, Peres CA (2007) Quantifying the biodiversity value of tropical primary, secondary, and plantation forests. Proc Natl Acad Sci USA 104:18555–18560

Benton MJ, Harper DAT (2009) Introduction to paleobiology and the fossil record. Wiley-Blackwell, Chichester

Bodner F, Brehm G, Homeier J, Strutzenberger P, Fiedler K (2010a) Caterpillars and host plant records for 59 species of Geometridae (Lepidoptera) from a montane rainforest in southern Ecuador. J Insect Sci 10:67

Bodner F, Mahal S, Reuter M, Fiedler K (2010b) Feasibility of a combined sampling approach for studying caterpillar assemblages – a case study from shrubs in the Andean montane forest zone. J Res Lepid 43:27–35

Bodner F, Strutzenberger P, Brehm G, Fiedler K (2012) Species richness and host specificity among caterpillar ensembles on shrubs in the Andes of southern Ecuador. Neotrop Ent 41:375–385

Brehm G, Pitkin LM, Hilt N, Fiedler K (2005) Montane Andean rain forests are a global diversity hotspot of geometrid moths. J Biogeogr 32:1621–1627

Brehm G, Fiedler K, Häuser CL, Dalitz H (2008) Methodological challenges of a megadiverse ecosystem. In: Beck E, Bendix J, Kottke I, Makeschin F, Mosandl R (eds) Gradients in a tropical mountain ecosystem of Ecuador, vol 198, Ecological Studies. Springer, Berlin, pp 41–47

Brehm G, Bodner F, Strutzenberger P, Hünefeld F, Fiedler K (2011) Neotropical *Eois* (Lepidoptera: Geometridae): checklist, biogeography, diversity, and description patterns. Ann Entomol Soc Am 104:1091–1107

Bridle JR, Pedro PM, Butlin RK (2004) Habitat fragmentation and biodiversity: testing for the evolutionary effects of refugia. Evolution 58:1394–1396

Brunschön C, Behling H (2010) Reconstruction and visualization of upper forest line and vegetation changes in the Andean depression region of southeastern Ecuador since the last glacial maximum – a multi-site synthesis. Rev Palaeobot Palynol 163:139–152

Bush MB, de Oliveira PE (2006) The rise and fall of the refugial hypothesis of Amazonian speciation: a paleoecological perspective. Biota Neotrop 6:1. http://www.biotaneotropica.org.br/v6n1/en/abstract?point-of-view+bn00106012006

Casner KL, Pyrcz TW (2010) Patterns and timing of diversification in a tropical montane butterfly genus, *Lymanopoda* (Nymphalidae, Satyrinae). Ecography 33:251–259
Cockerell TDA (1922) A fossil moth from Florissant, Colorado. Am Mus Novit 34:1–2
Coddington JA, Agnarsson I, Miller JA, Kuntner M, Hormiga G (2009) Undersampling bias: the null hypothesis for singleton species in tropical arthropod surveys. J Anim Ecol 78:573–584
Colinvaux PA (1996) Quaternary environmental history and forest diversity in the Neotropics. In: Jackson JBC, Buff AF, Coates AG (eds) Evolution and environment in tropical America. University of Chicago Press, Chicago, pp 359–405
Colwell RK, Brehm G, Cardelús CL, Gilman AC, Longino JT (2008) Global warming, elevational range shifts, and lowland biotic attrition in the wet tropics. Science 322:258–261
Connahs H, Rodríguez-Castañeda G, Walters T, Walla T, Dyer L (2009) Geographic variation in host-specificity and parasitoid pressure of an herbivore (Geometridae) associated with the tropical genus *Piper* (Piperaceae). J Insect Sci 9:28
Drummond AJ, Rambaut A (2007) BEAST: bayesian evolutionary analysis by sampling trees. BMC Evol Biol 7:214
Elias M, Joron M, Willmott K, Silva-Brandão KL, Kaiser V, Arias CF, Gomez Piñerez LM, Uribe S, Brower AVZ, Freitas AVL, Jiggins CD (2009) Out of the Andes: patterns of diversification in clearwing butterflies. Mol Ecol 18:1716–1729
Endler JA (1982) Pleistocene forest refuges: fact or fancy? In: Prance GT (ed) Biological diversification in the tropics. Columbia University Press, New York, pp 641–657
Fjeldsa J, Lovett JC (1997) Geographical patterns of old and young species in African forest biota: the significance of specific montane areas as evolutionary centres. Biodivers Conserv 6:325–346
Gregory-Wodzicki KM (2000) Uplift history of the Central and Northern Andes: a review. Geol Soc Am Bull 112:1091–1105
Günter S, Cabrera O, Weber M, Stimm B, Zimmermann M, Fiedler K, Knuth J, Boy J, Wilcke W, Iost S, Makeschin F, Werner F, Gradstein R, Mosandl R (2008) Natural forest management in Neotropical mountain rain forests – an ecological experiment. In: Beck E, Bendix J, Kottke I, Makeschin F, Mosandl R (eds) Gradients in a tropical mountain ecosystem of Ecuador, vol 198, Ecological Studies. Springer, Berlin, pp 347–359
Haffer J (1969) Speciation in Amazonian forest birds. Science 165:131–137
Hedges BS, Kumar S (eds) (2009) The timetree of life. Oxford University Press, New York
Hewitt GM (2004) The structure of biodiversity – insights from molecular phylogeography. Front Zool 1:4
Hilt N, Brehm G, Fiedler K (2006) Diversity and ensemble composition of geometrid moths along a successional gradient in the Ecuadorian Andes. J Trop Ecol 22:155–166
Homeier J, Breckle SW, Günter S, Rollenbeck RT, Leuschner C (2010) Tree diversity, forest structure and productivity along altitudinal and topographical gradients in a species-rich Ecuadorian montane rain forest. Biotropica 42:140–148
Hooper DU, Chapin FS, Ewel JJ, Hector A, Inchausti P, Lavorel S, Lawton JH, Lodge DM, Loreau M, Naeem S, Schmid B, Setälä H, Symstad AJ, Vandermeer J, Wardle DA (2005) Effects of biodiversity on ecosystem functioning: a consensus of current knowledge. Ecol Monogr 75:3–35
Hubbell SP (2001) The unified neutral theory of biodiversity and biogeography. Princeton University Press, Princeton
Knapp S, Mallet J (2003) Refuting refugia? Science 300:71–72
Laurance WL, Useche DC et al (2012) Averting biodiversity collapse in tropical forest protected areas. Nature 489:290–294
Matos-Maraví PF, Peña C, Willmott KR, Freitas AVL, Wahlberg N (2013) Systematics and evolutionary history of butterflies in the "*Taygetis* clade" (Nymphalidae: Satyrinae: Euptychiina): towards a better understanding of Neotropical biogeography. Mol Phyl Evol 66:54–68

Mayr E, O'Hara RJ (1986) The biogeographic evidence supporting the Pleistocene forest refuge hypothesis. Evolution 40:55–67

Moritz C, Patton JL, Schneider CJ, Smith TB (2000) Diversification of rainforest faunas: an integrated molecular approach. Annu Rev Ecol Syst 31:533–563

Nelson BW, Ferreira CAC, DaSilva MF, Kawasaki ML (1990) Endemism centres, refugia and botanical collection density in Brazilian Amazonia. Nature 345:714–716

Peterson KJ, Summons RE, Donoghue PCJ (2007) Molecular palaeobiology. Paleontology 50:775–809

Pybus OG, Harvey PH (2000) Testing macro-evolutionary models using incomplete molecular phylogenies. Proc R Soc Lond B 267:2267–2272

Rabosky DL (2006) Likelihood methods for detecting temporal shifts in diversification rates. Evolution 60:1152–1164

Ramirez SR, Roubik DW, Skov C, Pierce NE (2010) Phylogeny, diversification and historical biogeography of euglossine orchid bees (Hymenoptera: Apidae). Biol J Linn Soc 100:552–572

Ricklefs RE (2009) Speciation, extinction and diversity. In: Butlin R, Bridle J, Schluter D (eds) Speciation and patterns of diversity. Cambridge University Press, New York, pp 257–277

Rodríguez-Castañeda G, Dyer LA, Brehm G, Connahs H, Forkner RE, Walla TR (2010) Tropical forests are not flat: how mountains affect herbivore diversity. Ecol Lett 13:1348–1357

Rull V (2008) Speciation timing and neotropical biodiversity: the tertiary-quaternary debate in the light of molecular phylogenetic evidence. Mol Ecol 17:2722–2729

Santos JC, Coloma LA, Summers K, Caldwell JP, Ree R, Cannatella DC (2009) Amazonian amphibian diversity is primarily derived from late Miocene Andean lineages. PLoS Biol 7: e1000056

Schluter D (2000) The ecology of adaptive radiation. Oxford University Press, Oxford

Sedano RE, Burns KJ (2010) Are the Northern Andes a species pump for neotropical birds? phylogenetics and biogeography of a clade of neotropical tanagers (Aves: Thraupini). J Biogeogr 37:325–343

Sepulchre P, Sloan LC, Fluteau F (2010) Modelling the response of Amazonian climate to the uplift of the Andean mountain range. In: Hoorn C, Weeslingh F (eds) Amazonia – Landscape and species evolution: a look into the past. Wiley-Blackwell, Chichester, pp 211–222

Smith JF, Stevens AC, Tepe EJ, Davidson C (2008) Placing the origin of two species-rich genera in the late Cretaceous with later species divergence in the Tertiary: a phylogenetic, biogeographic and molecular dating analysis of *Piper* and *Peperomia* (Piperaceae). Plant Syst Evol 275:9–30

Strutzenberger P, Fiedler K (2011) Temporal patterns of diversification in Andean *Eois*, a species-rich clade of moths (Lepidoptera, Geometridae). J Evol Biol 24:919–925

Strutzenberger P, Brehm G, Bodner F, Fiedler K (2010) Molecular phylogeny of *Eois* (Lepidoptera, Geometridae): evolution of wing patterns and host plant use in a species-rich group of Neotropical moths. Zool Scr 39:603–620

Strutzenberger P, Brehm G, Fiedler K (2011) DNA barcoding-based species delimitation increases species count of *Eois* (Geometridae) moths in a well-studied tropical mountain forest by up to 50%. Insect Sci 18:349–362

Wahlberg N, Freitas AVL (2007) Colonization of and radiation in South America by butterflies in the subtribe Phyciodina (Lepidoptera: Nymphalidae). Mol Phylogenet Evol 44:1257–1272

Wiens JJ, Graham CH (2005) Niche conservatism: Integrating evolution, ecology, and conservation biology. Annu Rev Ecol Evol Syst 36:519–539

Wray GA, Levinton JS, Shapiro LH (1996) Molecular evidence for deep Precambrian divergences among Metazoan phyla. Science 274:568–573

Yamamoto S, Sota T (2007) Phylogeny of the Geometridae and the evolution of winter moths inferred from simultaneous analysis of mitochondrial and nuclear genes. Mol Phylogenet Evol 44:711–723

Zachos J, Pagani M, Sloan L, Thomas E, Billups L (2001) Trends, rhythms, and aberrations in global climate 65Ma to present. Science 292:686–693

Chapter 7
Diversity in Soil Fungi, Protists, and Microarthropods

Matthias C. Rillig, Tessa Camenzind, Julia Gawlik, Ingeborg Haug, Valentyna Krashevska, Mark Maraun, Dorothee Sandmann, and Stefan Scheu

7.1 Introduction and Background

This chapter is one of the three that are concerned with current preserving services of a biodiversity hotspot of the South-Ecuadorian Andes. While the other two chapters (Chaps. 6 and 8) deal with plants and aboveground animals, here we cover selected soil biota groups. These are the arbuscular mycorrhizal (AM) fungi, forming a key mutualism with plant roots in tropical forests, soil microarthropods and testate amoebae, which are crucial components of the soil food web. While separated into chapters, it is clear that these components are interdependent and interact in various ways (e.g., Wardle et al. 2002). For example, plant diversity can have a strong influence on AM fungal diversity and vice versa (e.g., van der Heijden et al. 1998), at least as known from temperate ecosystems (mostly grasslands). Other microorganisms including bacteria and saprobic fungi, as well as decomposer invertebrates and their interactions, affect plant performance and thereby ecosystem processes (Scheu et al. 1999, 2005; Wardle 2002; Scheu 2003). Thus, it is important to consider these components together.

Preserving Services and Soil Biota. Preserving services refer to the maintenance of genetic or species diversity irrespective of any function. Soil biota generally play a key role in preserving services, since the reservoir of organismal diversity in soil is

M.C. Rillig (✉) • T. Camenzind
Institut für Biologie, Freie Universität Berlin, Berlin, Germany
e-mail: matthias.rillig@fu-berlin.de

J. Gawlik
Georgraphic Institute, University of Erlangen, Erlangen, Germany

I. Haug
Eberhard Karls-Universität Tübingen, Tübingen, Germany

V. Krashevska • M. Maraun • D. Sandmann • S. Scheu
J.F. Blumenbach Institute of Zoology and Anthropology, Georg August University Göttingen, Göttingen, Germany

J. Bendix et al. (eds.), *Ecosystem Services, Biodiversity and Environmental Change in a Tropical Mountain Ecosystem of South Ecuador*, Ecological Studies 221,
DOI 10.1007/978-3-642-38137-9_7, © Springer-Verlag Berlin Heidelberg 2013

typically vast, greatly exceeding the aboveground diversity of biota in terrestrial ecosystems (Wardle 2002). The strategy taken in research summarized in this chapter was not to tackle belowground diversity exhaustively, but to instead focus on a number of key groups. These are, for soil microbes, the group of AM fungi, for soil protists, testate amoebae, and for soil microarthropods, oribatid mites.

7.2 Methods

Terrestrial vascular plants of the natural forest were compared with strongly modified vegetation types (pastures, pine plantations, and bracken fern areas). The methods are described in greater detail in Chap. 8.

In order to assess molecular diversity of root-colonizing AM fungi we conducted a survey of the small subunit (SSU), comparing pristine forest with reforestation sites. The mycobionts were examined by molecular tools as given in Haug et al. (2010), additionally the primers AML1 and AML2 (Lee et al. 2008) were used. Operational Taxonomic Units (OTUs) were defined as surrogates for species on the basis of sequence similarity by use of OPTSIL (Göker et al. 2009) at <1 % divergence. In addition, we also carried out a morphological survey of root colonization at $\times 200$ magnification.

Testate amoebae are unicellular protists and represent major consumers of bacteria in many terrestrial ecosystems in particular acidic forests such as the studied montane tropical rainforests (Krashevska et al. 2010, 2012a, b). They were extracted by washing soil and litter samples over filters, inspected under the microscope, and live specimens and empty shells were determined to species (see Chap. 18). Oribatid mites represent major soil mesofauna taxa reaching maximum density in acidic forests (Maraun and Scheu 2000) including the studied system (Maraun et al. 2008). They predominantly feed on fungi but are trophically diverse with some species functioning as not only primary decomposers but also predators (Schneider et al. 2004; Heidemann et al. 2011). They were extracted from soil cores of a diameter of 5 cm by heat (Kempson et al. 1963).

7.3 Results

7.3.1 Changes in Plant Diversity as a Driver of Other Diversities

The highly diverse natural forests in the San Francisco valley are exposed to a strong land use pressure. As a consequence large areas have been deforested and converted into pastures (see Chap. 2). This land use change causes a strong impact on plant diversity which is clearly shown by a differing vegetation composition and

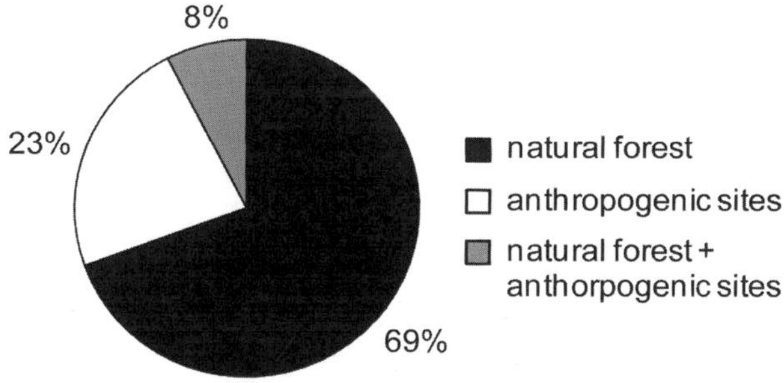

Fig. 7.1 Distribution of plant species in the natural forest, on anthropogenic sites, and in both systems (data from Peters et al. 2010)

the loss of the vast majority of natural forest species (Peters et al. 2010; Chap. 8). With the conversion of the multi-storied natural forest ecosystem into a single or double layered anthropogenic replacement system about 90 % of the forest species (without epiphytes) are lost. Furthermore, species number average per 100 m^2 decline from 114 forest species to 39 on the pasture sites. Exchange ratio between both sites is very low and just 8 % of 880 plant species can be found on both sites (Fig. 7.1). The vast majority of species growing on human influenced sites, except the introduced fodder grass species, is native and has their distribution center in the Andes.

The data reported here are for aboveground diversity surveys; however, they will translate to the belowground as well, in terms of carbon input from living roots (rhizodeposition), from dead roots (root litter), and aboveground litter input.

7.3.2 Soil Fungi: Arbuscular Mycorrhizal Fungi

AM fungi are considered keystone mutualists in many terrestrial ecosystems; they are clearly the dominant mycorrhizal association type in tropical mountain forests in Southern Ecuador, with 112 out of 115 examined forest tree species forming this association (Kottke and Haug 2004).

Given the vast difference between the natural forest and the anthropogenic replacement system, the pasture, in terms of the respective plant community type and species composition, a strong effect on this group of plant root symbionts is expected. Additional factors contributing to any community divergence are higher frequency of disturbance through fire, altered microclimate, as well as soil properties (see Chaps. 1 and 2; Beck et al. 2008a, b).

In tropical ecosystems, disturbance by natural phenomena (e.g., treefall gaps and landslides) as well as by slash and burn or other agricultural practices may affect AM fungal diversity and abundance (Allen et al. 1998; Cuenca et al. 1998; Bastias et al. 2006). Already Janos (1980) proposed that a transformation of primary forests towards managed systems will impact the root-associated mycorrhizal fungi, even though he rather focused on the state of mycotrophy of the newly arising plant community.

Previous studies on the AM status of forests and pastures do not clearly support this view. The abundance/inoculum potential of AM fungi in soil and roots tended to not be reduced (Fischer et al. 1994; Maldonado et al. 2000; Guadarrama et al. 2008; Stürmer and Siqueira 2011), although methodologies applied may alter the outcome of a study (Muthukumar et al. 2003). Changes concerning the AM fungal community composition are complex. Picone (2000) compared tropical lowland forests in Costa Rica with converted pastures, finding a similar species composition and no decline in species diversity, though gamma-diversity was significantly reduced. Johnson and Wedin (1997) reported the same pattern in Costa Rican dry forests. Conversely, Gavito et al. (2008) found higher species richness in primary forests and a change in species composition in a tropical dry ecosystem in Mexico, whereas Stürmer and Siqueira (2011) report the lowest species richness in pristine forests in the Brazilian Amazon. These data were collected by analyzing spore communities, which generally raises problems because sporulation patterns can be affected by biotic and abiotic factors (Bever et al. 1996; Clapp et al. 1995). PCR-based methods provide a novel tool to directly assess fungal diversity in the roots itself. Few studies in the tropics applied molecular tools on this topic (e.g., Aldrich-Wolfe 2007). Basing on the analysis of only one tree species growing in tropical premontane forests in Costa Rica, seedlings planted into adjacent pastures formed a distinct mycorrhizal community. In the data described in the present chapter, the AM fungal community is likewise surveyed using a molecular approach.

A morphological comparison of AM fungal root colonization revealed no obvious divergence between forests and pastures, regarding both the abundance of AM fungal structures and the morphological traits. The percentage of AM fungal root colonization in the pristine forest has been assessed for various root samples collected at the NUMEX plots (2,020 and 2,120 m a.s.l., 3°59′S, 79°05′W; see Fig. 1.2; Chap. 23). Percentage of root colonization is on average 40–50 % (T. Camenzind, unpublished data). Roots sampled on the near reforestation plots and pastures were studied as well and the AM fungal colonization was examined visually after staining. These samples were well colonized throughout with no apparent morphological differences to the forest.

A small subunit (SSU)-based molecular survey of root-colonizing AM fungi comparing pristine forest with reforestation sites was conducted. In the pristine forest mycorrhizal roots were sampled from trees, planted (Kuptz et al. 2010) and naturally regenerated (seedlings). On reforestation sites of different successional stages, mycorrhizal roots were taken from tree seedlings of four planted native species (Günter et al. 2009) and surrounding trees, shrubs, and grasses. Members of Archaeosporales, Diversisporales, and Glomerales were found on all sites, with the Glomerales dominating. Omitting the singleton-OTUs, a total of 56 OTUs were found: 35 OTUs on the reforestation plots, 42 OTUs in the pristine forest, of these 21 OTUs occurring in both systems (Fig. 7.2a). No plant specific OTUs were recognizable. Few OTUs were found in high numbers, many in low numbers (Fig. 7.2b). Most of the frequent OTUs occurred in both systems. The results not only indicate a slight decrease in AM fungal richness and changes in species

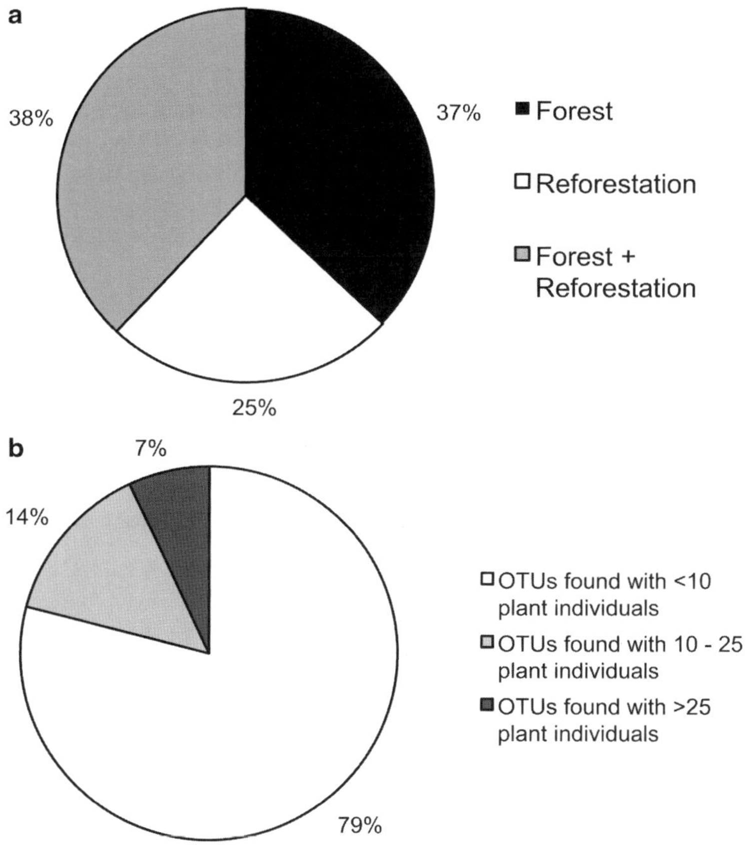

Fig. 7.2 (**a**) Distribution of the occurrence of AM fungal Operational Taxonomic Units (OTUs) in the ecosystems of pristine forest, reforestation sites, and both systems. (**b**) Percentage of AM fungal OTUs found on different numbers of plant individuals

composition on the reforestation sites but also show a high fraction of AM fungi that are robust to the system change.

7.3.3 Soil Protists and Microarthropods: Testate Amoebae and Oribatid Mites

Micro- and mesofauna form main components of the soil food web and essentially affect decomposition of organic matter and thereby carbon and nitrogen cycling (Bradford et al. 2002; Hackl et al. 2005; Wilkinson and Mitchell 2010), but this remains poorly investigated in particular in tropical ecosystems (Coleman et al. 2004; Scheu et al. 2005).

Soil bacteria and fungi are imbedded in a complex food web in soil. Major consumers of bacteria include unicellular organisms such as testate amoebae, and major consumers of fungi include microarthropods such as oribatid mites. Both are considered to play an important role in regulating prey abundance and thereby decomposition processes in virtually all ecosystems on earth. This is also likely to be true in tropical montane rain forests of southern Ecuador, in particular since macrofauna decomposers are relatively rare in these ecosystems (Maraun et al. 2008; Scheu et al. 2008). Over the last 6 years we investigated the response of microorganisms, testate amoebae, and microarthropods to altitudinal changes (Krashevska et al. 2007; Scheu et al. 2008). Further, we explored the limitation of microorganisms and testate amoebae by carbon and nutrients (Krashevska et al. 2010), their dependency on litter quality (Illig et al. 2008; Krashevska et al. 2012a) and response to the exclusion of precipitation (Krashevska et al. 2012b). Here we summarize these findings and show how sensitively these components of tropical montane rainforests respond to environmental changes and to changes of altitude. As opposed to AM fungi, no comparable data are available for the comparison of forest and replacement system, so this aspect cannot be discussed.

7.3.3.1 Testate Amoebae

The density of testate amoebae live specimens was at maximum at 2,000 m, lower at 3,000 m and lowest at 1,000 m a.s.l. (Fig. 7.3a). Parallel to the density of live specimens, total density of testate amoebae peaked at 2,000 m, was similar at 3,000 m and lower at 1,000 m (Krashevska et al. 2007). Species number of live specimens did not change with altitude (Fig. 7.3b). In contrast, total species number decreased significantly in the order 2,000 $\leq$ 3,000 $<$ 1,000 m. Krashevska et al. (2007) identified a total of 135 taxa and only few more species are to be expected, as suggested by rarefaction analysis (Fig. 7.4a). Indeed, in the last few years we found only few more species. Generally, more than 160 taxa of testate amoebae were identified in more than 200 samples. In comparison with plant diversity (about 1,200 species; Homeier and Werner 2007) protist diversity in the studied tropical montane rainforests is low, however, compared to testate amoebae diversity in a rain forest of Puerto Rico (104 species), in a temperate forest in Germany (62 species) and in acidic soils in Chile (94 species) the number of species of testate amoebae in the studied tropical montane rain forests of Ecuador is high. High species diversity is particularly remarkable as with a maximum of about 3,000 live specimens and about 13,000 total ind./g dry weight the density of testate amoebae is low compared to the above mentioned ecosystems with densities of 23,000–370,000 ind./g dry weight (Bonnet 1966; Wanner 1989; Bamforth 2007). The results suggest that the diversity but not the density of testate amoebae in tropical mountain forests is high exceeding that in other tropical forests and temperate forests. Additionally, the majority of the taxa found are geographically widespread, less than 10 % are endemic and we judge that less than 5 % are new to science. However, more detailed studies including molecular investigations are

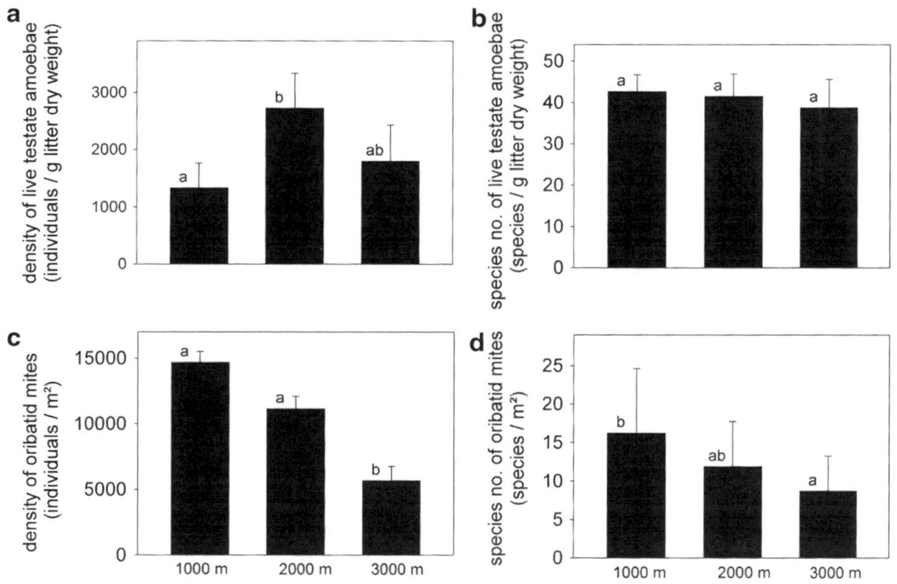

Fig. 7.3 Density and species richness of live testate amoebae (**a** and **b**, respectively) and oribatid mites (**c** and **d**, respectively) in the litter/fermentation layer at three altitudes (1,000, 2,000, and 3,000 m a.s.l.) of the studied tropical montane rainforests. Means with SE; bars with different letters vary significantly (Tukey's HSD test, $\alpha < 0.05$)

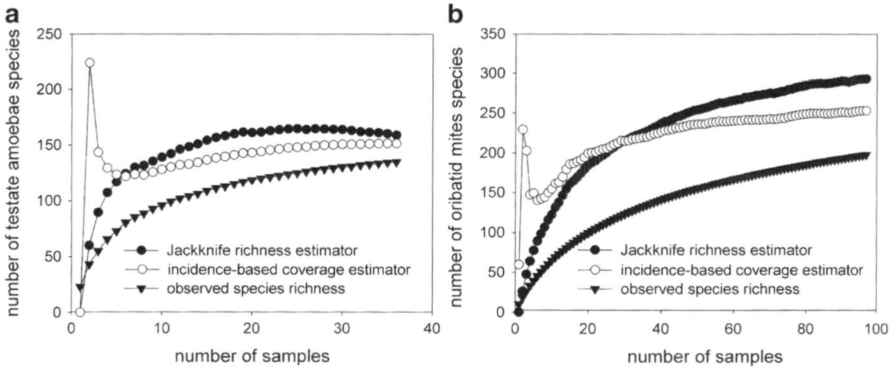

Fig. 7.4 Sample-based rarefaction curves and corresponding estimators for testate amoebae (**a**) and oribatid mites (**b**)

needed to confirm these findings. As abiotic conditions at the study site are rather beneficial for soil invertebrates such as protists, low density of testate amoebae suggests low availability of resources with nitrogen presumably being most important (Iost et al. 2008; Krashevska et al. 2010).

7.3.3.2 Oribatid Mites

Total density of oribatid mites significantly decreased with increasing altitude in order 1,000 $\leq$ 2,000 < 3,000 m with 13,452 $\pm$ 740, 10,942 $\pm$ 946 and 5,597 $\pm$ 1,090 ind. m^{-2}, respectively (Fig. 7.3c). These densities resemble those of other tropical forests (rainforest in French Guiana ca. 25,000 ind. m^{-2}, Kounda-Kiki et al. 2004; rainforest in Australia ca. 43,000 ind. m^{-2}, Plowman 1981) but are much lower compared to acidic deciduous forests of the temperate zone and boreal forests where oribatid mites reach densities of >200,000 ind. m^{-2} (Maraun and Scheu 2000). Surprisingly and in contrast to adult oribatid mites, the density of juveniles significantly increased with increasing altitude (data not shown). The higher proportion of juveniles at higher altitudes at least in part is likely due to longer developmental time of juveniles resulting from lower temperature. Corresponding to the density the average number of species found in one sample strongly decreased with increasing altitude (Fig. 7.3d). The total number of identified species also decreased with altitude, in particular at the highest altitude, with 169, 178, and 73 taxa at 1,000, 2,000, and 3,000 m, respectively. The number of species which exclusively occurred on one altitude decreased with increasing altitude with 60 %, 52 %, and 27 % at 1,000, 2,000, and 3,000 m, respectively, suggesting that at higher altitude the number of ubiquitous species increases.

In more than 350 samples investigated, a total of 335 taxa of oribatid mites were identified along the three study sides at 1,000, 2,000, and 3,000 m. The number of species corresponded with those estimated by rarefaction analysis (Fig. 7.4b). Although very preliminary, we estimate the number of new species to be high; about 40 % of the identified species could not be named with certainty using keys of Balogh (1972, 1988), Balogh and Balogh (1988, 2002), Balogh et al. (2008), Weigmann (2006) and others. Some new species have already been described (Niedbala and Illig 2007), but a wealth of work remains to be done. However, compared to the number of herbivore species above the ground, the number of oribatid mite species in soil is low as indicated by, e.g., the >1,000 species of geometrid moths found at the studied tropical montane rainforests (Brehm et al. 2003; Fiedler et al. 2008; see Chap. 6). This supports suggestions that compared to above the ground the latitudinal gradient in belowground animals is less pronounced (de Deyn and van der Putten 2005; M. Maraun, unpubl. data).

7.4 Conclusions

In this chapter we showed that in a comparison of forest and the anthropogenic replacement system, AM fungal richness was not greatly changed, even though there were strong changes in the associated plant community. It would be interesting to compare such patterns also for other soil biota groups such as other symbionts or decomposer microorganisms.

In a comparison of altitudinal patterns at three elevations, it became evident that the diversity of belowground animals, such as protists and microarthropods, does not approach that of aboveground animals, such as herbivores. Some groups showed strong altitudinal patterns (oribatids), whereas others (testate amoebae) did not.

Overall, the results highlight that considerably more effort is necessary to understand the relationship between above- and belowground diversities in these tropical forests: what is a hotspot for some organism groups does not necessarily have to be equally "hot" for others, but why this is so is not yet clear. Perhaps it is due to cryptic diversity below the level currently examined, but maybe there are also system-inherent reasons that explain why symbiont or decomposer subsystem diversity does not track the diversity of the aboveground component.

References

Aldrich-Wolfe L (2007) Distinct mycorrhizal communities on new and established hosts in a transitional tropical plant community. Ecology 88:559–566

Allen EB, Rincón E, Allen MF, Pérez-Jimenez A, Huante P (1998) Disturbance and seasonal dynamics of mycorrhizae in a tropical deciduous forest in Mexico. Biotropica 30:261–274

Balogh J (1972) The oribatid genera of the world. Hungarian Natural History Museum

Balogh P (1988) Oribatid mites from Ecuador (Acari) 1. Acta Zool Hung 34:321–338

Balogh J, Balogh P (1988) Oribatid mites of the neotropical region. Elsevier

Balogh J, Balogh P (2002) Identification key to Oribatid mites of the Extra-Holarctic regions. Well-Press

Balogh P, Gergocs V, Farkas E, Farkas P, Kocsis M, Hufnagel L (2008) Oribatid assamblages of tropical high mountains on some points of the Gondwana-Bridge – a case study. Appl Ecol Environ Res 6:127–158

Bamforth SS (2007) Protozoa from aboveground and ground soils of a tropical rain forest in Puerto Rico. Pedobiologia 50:515–525

Bastias BA, Huang ZQ, Blumfield T, Xu Z, Cairney JWG (2006) Influence of repeated prescribed burning on the soil fungal community in an eastern Australian wet sclerophyll forest. Soil Biol Biochem 38:3492–3501

Beck E, Harting K, Roos K (2008a) Forest clearing by slash and burn. In: Beck E, Bendix J, Kottke I, Makeschin F, Mosandl R (eds) Gradients in a tropical mountain ecosystem of Ecuador. Ecological studies, vol 198. Springer, Berlin, pp 371–374

Beck E, Makeschin F, Haubrich F, Richter M, Bendix J, Valerezo C (2008b) The ecosystem (Reserva Biológica San Francisco). In: Beck E, Bendix J, Kottke I, Makeschin F, Mosandl R (eds) Gradients in a tropical mountain ecosystem of Ecuador. Ecological studies, vol 198. Springer, Berlin, pp 1–13

Bever JD, Morton JB, Antonovics J, Schultz PA (1996) Host-dependent sporulation and species diversity of arbuscular mycorrhizal fungi in a mown grassland. J Ecol 84:71–82

Bonnet L (1966) Le peuplement thecamoebien de quelques sols du Chili. Protistologica 2:113–140

Bradford MA, Jones TH, Bardgett RD, Black HIJ, Boag B, Bonkowski M, Cook R, Eggers T, Gange AC, Grayston SJ, Kandeler E, McCaig AE, Newington JE, Prosser JI, Setälä H, Staddon PL, Tordoff GM, Tscherko D, Lawton JH (2002) Impacts of soil faunal community composition on model grassland ecosystems. Science 298:615–618

Brehm G, Homeier J, Fielder K (2003) Beta dieversity of geometrid moths (Lepidoptera: Geometridae) in an Andean montane rain forest. Divers Distrib 9:351–366

Clapp JP, Young JPW, Merryweather JW, Fitter AH (1995) Diversity of fungal symbionts in arbuscular mycorrhizas from a natural community. New Phytol 130:259–265

Coleman DC, Crossley DA Jr, Hendrix PF (2004) Fundamentals of soil ecology, 2nd edn. Elsevier, San Diego

Cuenca G, de Andrade Z, Escalante G (1998) Diversity of glomalean spores from natural, disturbed and revegetated communities growing on nutrient-poor tropical soils. Soil Biol Biochem 30:711–719

De Deyn GB, Van der Putten WH (2005) Linking aboveground and belowground diversity. Trends Ecol Evol 20:625–633

Fiedler, K, Brehm, G, Hilt, N, Süßenbach, D, Häuser, CL (2008) Variation of diversity patterns across moth families along a tropical altitudinal gradient. In: Beck E, Bendix J, Kottke I, Makeschin F, Mosandl R (eds) Gradients in a tropical mountain ecosystem of Ecuador. Ecological studies, vol 198. Springer, Berlin, pp 167–179

Fischer CR, Janos DP, Perry DA, Linderman RG, Sollins P (1994) Mycorrhiza inoculum potentials in tropical secondary succession. Biotropica 26:369–377

Gavito ME, Pérez-Castillo D, González-Monterrubio CF, Vieyra-Hernández T, Martínez-Trujillo M (2008) High compatibility between arbuscular mycorrhizal fungal communities and seedlings of different land use types in a tropical dry ecosystem. Mycorrhiza 19:47–60

Göker M, García-Blázquez G, Voglmayr H, Tellería MT, Martín MP (2009) Molecular taxonomy of phytopathogenic fungi: a case study in Peronospora. PLoS One 29:e6319

Guadarrama P, Castillo-Argüero S, Ramos-Zapata JA, Camargo-Ricalde SL, Álvarez-Sánchez J (2008) Propagules of arbuscular mycorrhizal fungi in a secondary dry forest of Oaxaca. Mexico Rev Biol Trop 56:269–277

Günter S, Gonzalez P, Alvarez G, Aguirre N, Palomeque X, Haubrich F, Weber M (2009) Determinants for successful reforestation of abandoned pastures in the Andes: soil conditions and vegetation cover. For Ecol Manage 258:81–91

Hackl E, Pfeffer M, Donat C, Bachmann G, Zechmeister-Boltenstern S (2005) Composition of the microbial communities in the mineral soil under different types of natural forest. Soil Biol Biochem 37:661–671

Haug I, Wubet T, Weiß M, Aguirre N, Weber M, Günter S, Kottke I (2010) Species-rich but distinct arbuscular mycorrhizal communities in reforestation plots on degraded pastures and in neighboring pristine tropical mountain rain forest. Trop Ecol 51:125–148

Heidemann K, Scheu S, Ruess L, Maraun M (2011) Molecular detection of nematode predation and scavenging in oribatid mites: laboratory and field experiments. Soil Biol Biochem 43:2229–2236

Homeier J, Werner FA (2007) Spermatopyta checklist – Reserva Biologica San Francisco (Prov. Zamora-Chinchipe, S. Ecuador). Ecotrop Monogr 4:15–58

Illig J, Schatz H, Scheu S, Maraun M (2008) Decomposition and colonization by micro-arthropods of two litter types in a tropical montane rain forest in southern Ecuador. J Trop Ecol 24:1–11

Iost S, Makeschin F, Aiby M, Haubrich F (2008) Biotic soil activities. In: Beck E, Bendix J, Kottke I, Makeschin F, Mosandl R (eds) Gradients in a tropical mountain ecosystem of Ecuador. Ecological studies, vol 198. Springer, Berlin, pp 217–227

Janos DP (1980) Mycorrhizae influence tropical succession. Biotropica 12:56–64

Johnson NC, Wedin DA (1997) Soil carbon, nutrients, and mycorrhizae during conversion of dry tropical forest to grassland. Ecol Appl 7:171–182

Kempson D, Lloyd M, Ghelardi R (1963) A new extractor for woodland litter. Pedobiologia 3:1–21

Kottke I, Haug I (2004) The significance of mycorrhizal diversity of trees in the tropical mountain forest of southern Ecuador. Lyonia 7:49–56

Kounda-kiki C, Vaculik A, Ponge JF, Sarthough C (2004) Soil and arthopods in a developmental succession on the Nouragues inselberg (French Guiana). Biol Fertil Soils 40:119–127

Krashevska V, Bonkowski M, Maraun M, Scheu S (2007) Testate amoebae (protista) of an elevational gradient in the tropical mountain rain forest of Ecuador. Pedobiologia 51:319–331

Krashevska V, Maraun M, Ruess L, Scheu S (2010) Carbon and nutrient limitation of soil microorganisms and microbial grazers in a tropical montane rain forest. Oikos 119:1020–1028
Krashevska V, Maraun M, Scheu S (2012a) How does litter quality affect the community of soil protists (testate amoebae) of tropical montane rainforests? FEMS Microbiol Ecol 80:603–607
Krashevska V, Sandmann D, Maraun M, Scheu S (2012b) Consequences of exclusion of precipitation on microorganisms and microbial consumers in montane tropical rain forests. Oecologia 170:1067–1076
Kuptz D, Grams TEE, Günter S (2010) Light acclimation of four native tree species in felling gaps within a tropical mountain rainforest. Trees 24:117–127
Lee J, Lee S, Young JPW (2008) Improved PCR primers for the detection and identification of arbuscular mycorrhizal fungi. FEMS Microbiol Ecol 65:339–349
Maldonado JD, Tainter FH, Skipper HD, Lacher TE (2000) Arbuscular mycorrhiza inoculum potential in natural and managed tropical montane soils in Costa Rica. Trop Agric 77:27–32
Maraun M, Scheu S (2000) The structure of oribatid mite communities (Acari:oribatida) patterns, mechanisms and implications for future research. Ecography 23:374–383
Maraun M, Illig J, Sandmann D, Krashevska V, Norton RA, Scheu S (2008) Soil fauna. In: Gradients in a tropical mountain ecosystem of Ecuador. In Beck E, Bendix J, Kottke I, Makeschin F, Mosandl R (eds) Gradients in a tropical mountain ecosystem of Ecuador. Ecological studies, vol 198. Springer, Berlin, pp 181–192
Muthukumar T, Sha L, Yang X, Cao M, Tang J, Zheng Z (2003) Mycorrhiza of plants in different vegetation types in tropical ecosystems of Xishuangbanna, southwest China. Mycorrhiza 13:289–297
Niedbala W, Illig J (2007) Ptyctimous mites (Acari:Oribatida) from the Ecuadorian rainforest. J Nat Hist 41:771–777
Peters T, Diertl K-H, Gawlik J, Rankl M, Richter M (2010) Vascular plant diversity in natural and anthropogenic ecosystems in the Andes of southern Ecuador. Mt Res Dev 30:344–352
Picone C (2000) Diversity and abundance of arbuscular-mycorrhizal fungus spores in tropical forest and pasture. Biotropica 32:734–750
Plowman KP (1981) Distribution of Cryptostigmata and Mesostigmata (Acari) within the litter and soil layers of two subtropical forests. Austral J Ecol 6:365–374
Scheu S (2003) Effects of earthworms on plant growth: patterns and perspectives. Pedobiologia 47:1–11
Scheu S, Theenhaus A, Jones TH (1999) Links between the detritivore and the herbivore system: effects of earthworms and Collembola on plant growth and aphid development. Oecologia 119:541–551
Scheu S, Ruess L, Bonkowski M (2005) Interactions between microorganisms and soil micro- and mesofauna. In: Buscot F, Varma A (eds) Microorganisms in soils: roles in genesis and functions. Soil biology, vol 3. Springer, Berlin, pp 253–275
Scheu S, Illig J, Eissfeller V, Krashevska V, Sandmann D, Maraun M (2008) The soil fauna of a tropical mountain rainforest in southern Ecuador: structure and functioning. In: Gradstein SR, Homeier J, Gansert D (eds) The tropical mountain forest. Patterns and processes in a biodiversity hotspots. Biodiversity and ecology series, vol 2. Universitätsverlag Göttingen, pp 79–96
Schneider K, Migge S, Norton RA, Scheu S, Langel R, Reineking A, Maraun M (2004) Trophic niche differentiation in soil microarthropods (Oribatida, Acari): evidence from stable isotope ratios ($^{15}N/^{14}N$). Soil Biol Biochem 36:1769–1774
Stürmer SL, Siqueira JO (2011) Species richness and spore abundance of arbuscular mycorrhizal fungi across distinct land uses in Western Brazilian Amazon. Mycorrhiza 21:255–267
Van der Heijden MGA, Boller T, Wiemken A, Sanders IR (1998) Different arbuscular mycorrhizal fungal species are potential determinants of plant community structure. Ecology 79:2082–2091
Wanner M (1989) Zur Morphologie und Ökologie von Thekamöben (Protozoa: Rhizopoda) in Süddeutschen Wäldern. PhD Thesis, Universität Ulm
Wardle DA (2002) Communities and ecosystems: linking the aboveground and belowground components. Princeton University Press, Princeton, 392 pp

Wardle DA, Bonner KI, Barker GM (2002) Linkages between plant litter decomposition, litter quality, and vegetation responses to herbivores. Funct Ecol 16:585–595

Weigmann G (2006) Hornmilben (Oribatida). In: Dahl F (ed) Die Tierwelt Deutschlands und der angrenzenden Meeresteile, vol 76. Goecke & Evers, Keltern, p 520

Wilkinson D, Mitchell EAD (2010) Testate amoebae and nutrient cycling; with particular reference to soil. Geomicrobiol J 27:520–533

Chapter 8
Plant Diversity and Its Relevance for the Provision of Ecosystem Services

Jürgen Homeier, Florian A. Werner, Julia Gawlik, Thorsten Peters, Karl-Heinz J. Diertl, and Michael Richter

8.1 Introduction: Ecosystem Services Related to Plant Species Composition and Diversity

Biodiversity is rapidly declining in tropical ecosystems, and this decline has important implications for the functioning of the ecosystems and the services that they provide (Balvanera et al. 2006; Isbell et al. 2011; Millenium Ecosystem Assessment 2005). Tropical forests can provide a vast array of products, and they regulate many natural processes, e.g., climate regulation, maintenance of air and water quality, nutrient cycling, soil formation, and the sequestration and storage of carbon. They are also a repository for genetic resources, and in addition provide recreational value. Logging and pressures from agriculture or infrastructure development are resulting in the large-scale loss of forested land, and as a result a decline in the provision of ecosystem services (Millenium Ecosystem Assessment 2005).

Ecosystem services are in most cases not related to single plant species, but to the ecosystem as a whole (Lyons et al. 2005). Complex interactions between the components of biodiversity and physical and chemical environmental factors determine the quantity, quality, and stability of ecosystem services (Mace et al. 2012). However, these interactions are commonly poorly understood, especially with regard to diverse tropical ecosystems. The first step towards understanding them is to elucidate how environmental parameters and disturbance regimes drive species

J. Homeier (✉)
Albrecht von Haller Institute of Plant Sciences, University of Göttingen, Untere Karspüle 2, 37073 Göttingen, Germany
e-mail: jhomeie@gwdg.de

F.A. Werner
Institute of Biology and Environmental Sciences, University of Oldenburg, PO Box 2503, 26111 Oldenburg, Germany

J. Gawlik • T. Peters • K.-H.J. Diertl • M. Richter
Institute of Geography, University of Erlangen, Kochstraße 4/4, 91054 Erlangen, Germany

J. Bendix et al. (eds.), *Ecosystem Services, Biodiversity and Environmental Change in a Tropical Mountain Ecosystem of South Ecuador*, Ecological Studies 221,
DOI 10.1007/978-3-642-38137-9_8,

composition and diversity dynamics in tropical ecosystems. In addition, it is important to explore functional properties of species or species groups and their contributions to ecosystem functioning (de Bello et al. 2010; Cadotte et al. 2011).

This chapter summarizes how plant species composition and richness in the San Francisco Reserve and its surroundings are driven by elevation, topography, and different types of natural disturbances (landslides, canopy gaps). In addition, the effects of human land use (active and abandoned pastures) and forest fragmentation are discussed.

8.2 Materials and Methods

Plant species composition was studied according to a variety of different methods depending on the type of vegetation and research question to be addressed. To study the composition and species richness of trees along an altitudinal gradient, 17 permanent plots of 400 m^2 size each were established at altitudes ranging from 1,850 to 2,570 m a.s.l. in the mature natural forest of the Reserva Biologica San Francisco (RBSF, Homeier et al. 2010). In addition, 18 plots (matrix plots) of the same size were installed at 2,000 m a.s.l. to study the effects of topography and related soil parameters on tree species composition and richness. These plots encompassed three different topographic positions (lower slopes, mid-slopes, and upper slopes), with six replicates located on each position (see Chap. 10). All plots were covered by mature, closed-canopy forest representative of the respective topographic situation. Tree regeneration in natural forest canopy gaps was studied in the lower montane forest by comparing the influence of the topographic position (upper slopes vs. lower slopes) on the regeneration (see Homeier and Breckle 2008 for details).

The natural vegetation of the lower montane forests (2,000–2,100 m a.s.l.) and the upper montane forests (2,400–2,500 m a.s.l.) was compared with anthropogenically modified vegetation found at the same altitudes. Terrestrial vascular plant species and their ground coverage (in %) were recorded within eight transects of 100 m^2 size each (50 m $\times$ 2 m) at each site and elevation level (32 transects in total). Eight additional transects, two each, were laid out on active pastures, abandoned pastures, bracken fern areas, and pine afforestations within an altitudinal range of 1,900–2,200 m a.s.l. (see Diertl 2010 for detailed inventory methods).

In addition to terrestrial plants, vascular epiphytes were studied by recording the occurrence of plant individuals on entire remnant trees in pastures and on canopy trees of similar size and the same species in closed forests. Access to the canopy was achieved by single-rope climbing technique (see Werner et al. 2005 for details).

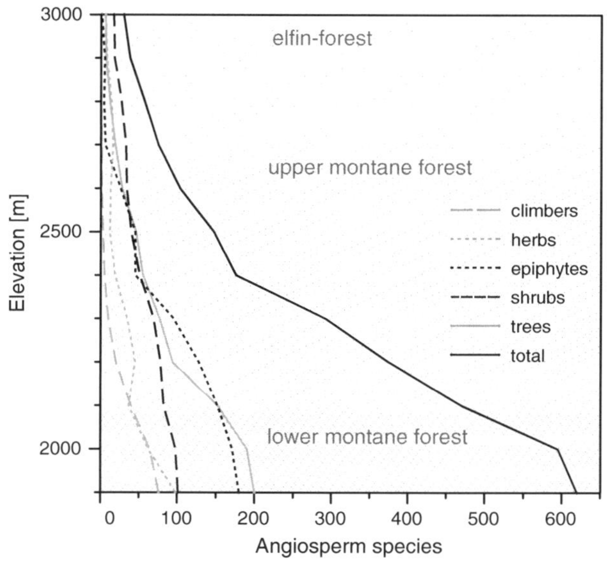

Fig. 8.1 Angiosperm species richness of different plant life forms in natural forest habitats along an altitudinal gradient (drawn with data from Homeier and Werner 2007). Forest types according to Homeier et al. (2008)

8.3 Results and Discussion

8.3.1 Altitudinal and Topographical Gradients as Drivers of Species Richness in the Rio San Francisco Valley

Angiosperm species richness in the San Francisco Valley generally decreases with altitude (Fig. 8.1). Although the existing inventory of approximately 1,200 angiosperm species (Homeier and Werner 2007) is far from being complete, general altitudinal patterns of species composition are not expected to change significantly with additional species records. The patterns of life form composition along the altitudinal gradient are rather similar at each altitude, although an absolute reduction in the number of species is evident. The decrease in species richness of shrubs and herbs that are the dominant life forms of the uppermost elevations is lower than that of epiphytes or trees.

Ravine and lower slope forests are generally more productive (as expressed by tree basal area growth or fine litter production) and richer in tree species than are upper slopes and ridges (Homeier et al. 2010). Soil nutrient concentration appears to be a major factor in determining differences in forest structure and tree species composition within the study area (Homeier et al. 2010; Wolf et al. 2011). The differences in stand structure are principally related to tree species composition. Ravine forest tree species appear to be unable to recruit on the poor, acidic soils covering the upper slopes and ridges, whereas the slow-growing ridge specialists appear to be poorly able to compete for light with the faster-growing ravine species (Homeier 2008). There is a surprisingly small number of tree species shared between lower slopes and upper slopes, indicating that most species are restricted in their occurrence to specific sites.

Topographical gradients with their array of edaphic, climate, and landform parameters thus contribute to the variety of (micro-) habitats and increase vegetation heterogeneity as well as plant species richness of the tropical montane forests being investigated in the present study, as has also been shown for other montane forests (e.g., Webb et al. 1999). The changes in species composition and forest structure detected along topographical gradients often resemble those observed along altitudinal gradients (e.g., Webb et al. 1999; Takyu et al. 2002).

8.3.2 Forest Dynamics in Response to Natural Disturbance

The highly diverse natural forests of the study area harbor a vast number of plant species adapted to specific ecological niches within the spatially heterogeneous and temporally variable site conditions. Natural disturbance events are followed by distinct successional stages, each of which is characterized by a specific plant composition ranging from early pioneers on recent landslides or in canopy gaps to climax species in mature closed-canopy forest.

Two major natural disturbance regimes affect the study area, and each has long-lasting effects on the structure and the species composition of the vegetation. Tree and branch falls result in gaps in the forest canopy that change the light regime on the forest floor and the microclimate in the understory. The tree species composition of different successional stages on the ridges and upper slopes of the lower montane forest is rather uniform, and typical pioneer species are completely lacking (Homeier 2008; Homeier and Breckle 2008). In contrast, the recovering forest on lower slopes and ravine sites is characterized by the presence of fast-growing pioneer taxa such as *Cecropia* spp. (Urticaceae), *Heliocarpus americanus* (Malvaceae), and *Piptocoma discolor* (Asteraceae) which appear to be restricted to more fertile soils. Hence, the absence of pioneer trees on the upper slopes and ridges is likely to be due to unfavorable soil conditions, since the immediate vicinity of the two forest types makes dispersal constraints unlikely. Differences in tree composition and population dynamics in these two neighboring forest types are maintained during the succession from young towards mature forest.

While the stunted upper montane forests exhibit fewer canopy gaps than do lower altitude forests, they are susceptible to the second natural disturbance factor of landslides that have a severe impact on these forests (Restrepo et al. 2009; Muenchow et al. 2012). The duration required for forest recovery and the reestablishment of the original plant species composition after a landslide in the study area is highly dependent on the elevation at which the landslide took place and on the size and aspect of the landslide (Bussmann et al. 2008; Ohl and Bussmann 2004). However, successions setting on landslide sites always start with mosses and lichens covering the bare soil. The first vascular plants establish a foothold after 5–10 years, with Lycopodiaceae and *Baccharis genistelloides* (Asteraceae) being common early pioneers. Later, various fern species (especially Gleicheniaceae), orchids, grasses, and shrubs (Asteraceae, Ericaceae, and Melastomataceae) appear before first tree species become established.

Both canopy gaps and landslides contribute to the spatial heterogeneity of environmental conditions in the study area and thus drive ecosystem dynamics and species richness. It takes no longer than a few decades for open areas to have been reclaimed by closed-canopy forest. The large local species pool and the presence of species well adapted to the natural disturbance events constitute the basis for the resilience of this unique ecosystem and also for its resistance to invading alien plant species.

8.3.3 Land Use Patterns as Determinants of Secondary Vegetation

Ecuador is characterized by one of the highest deforestation rates in South America (FAO 2005; Mosandl et al. 2008). Today approximately 48 % of the montane forests below 2,200 m a.s.l. and at least 6 % of the higher altitude natural vegetation in the Rio San Francisco valley (Göttlicher et al. 2009) have been replaced by man-made ecosystems (Fig. 8.2; see also Chap. 2).

Topographic gradients are of great relevance for understanding the regeneration of vegetation after human disturbance. Accessible terrains on the lower slopes rising from the Loja-Zamora road (see Fig. 1.1c) into the mountains have been transformed into pastures by seeding or planting forage plant species. Agricultural crops (maize) play a subordinate role on deforested areas and are cultivated at most for 1–2 growing seasons subsequent to slash-and-burn practices. Forests on the upper slopes and ridges are burned frequently and repeatedly, but rarely taken under cultivation. The common forage species in the RBSF are *Setaria sphacelata*, *Melinis minutiflora*, *Pennisetum clandestinum*, *Holcus lanatus*, and *Axonopus compressus* (all Poaceae). Except for the native *A. compressus*, all of the forage grasses have been introduced; *S. sphacelata* being the most recently introduced African forage grass in the area. Due to its vigorous growth under a wide range of environmental conditions, this tussock grass has increased its share of ground cover over the past 20 years despite its low nutritional value (e.g., protein content).

Current land use practices have a severe impact on the natural ecosystem. This is clearly shown by a plant composition that is completely different to that characteristic of the natural forests and lacks the vast majority of the forest species. Figure 8.3 gives an overview of the family composition of the natural forest vegetation compared to secondary site vegetation at two different altitudes.

Only 9.9 % of the natural forest species are found on sites currently or previously under human use. Not more than 2.5 % of the forest species are found on pastures being currently grazed, where species richness depends strongly on the density of

Fig. 8.2 Tropical montane forest and cattle pastures in the Rio San Francisco valley

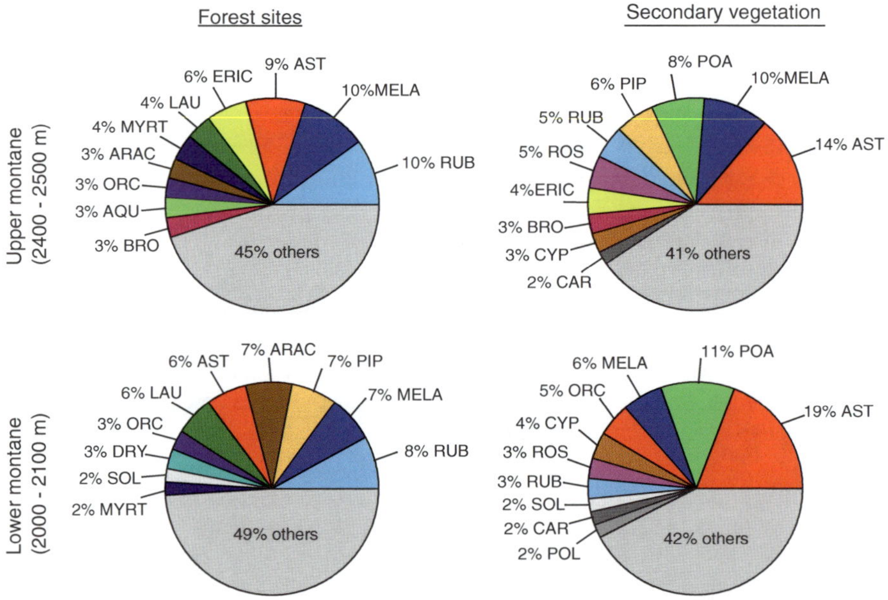

Fig. 8.3 Plant family composition of natural forest vegetation and secondary vegetation at two different elevations (redrawn after Diertl 2010 and Peters et al. 2010). Angiosperm families: AQUifoliaceae, ARACeae, ASTeraceae, BROmeliaceae, CARyophyllaceae, CYPeraceae, ERICaceae, LAUraceae, MELAstomataceae, MYRTaceae, ORChidaceae, PIPeraceae, POAceae, ROSaceae, RUBiaceae, SOLanaceae. Pteridophyte families: DRYopteridaceae, POLypodiaceae

the dominant forage grass species. The dominance of the highly competitive *S. sphacelata* results in a low variety of accompanying herbaceous species (an average of 13 species) compared to the more traditional *Melinis-*, *Pennisetum-*, *Holcus-*, and *Axonopus* pastures (average of 33 species). All together, 245 species

Table 8.1 Mean numbers of vascular plant species found in various land use situations at 1,900–2,200 m a.s.l. in comparison with the natural forest of the study area, together with the corresponding effective Shannon diversity D (Jost 2006) and evenness (natural forest data in the last column are derived from Diertl 2010)

	Active pastures	Abandoned pastures	Bracken fern areas	Pine afforestation	Forest
No of species [100 m^{-2}]	41	22	38	47	125
Shannon-diversity D	8.67	3.13	13.20	14.88	54.60
Evenness	0.58	0.37	0.72	0.69	0.83

of vascular plants from 169 genera and 73 families were recorded for pastures situated between 2,000 and 2,200 m a.s.l. The families exhibiting the most species were the Asteraceae with 27 genera and 45 species, the Poaceae (16 genera, 24 species), and the Melastomataceae (6 genera, 15 species). Forty six percent of the species were herbs, 27 % shrubs, 14 % grasses, 7 % climbers and 6 % trees.

Common woody pioneer species are *Tibouchina laxa*, *T. lepidota* (both Melastomataceae), and *Piptocoma discolor*. Asteraceae shrub species such as *Ageratina dendroides* and *Baccharis genistelloides* occur very frequently across all secondary vegetation types; whereas Melastomataceae such as *Monochaetum lineatum* and *Brachyotum benthamianum*, as well as the Ericaceae shrubs *Gaultheria erecta* and *Bejaria aestuans*, are characteristic for bracken fern-dominated areas on previously burned sites. The highest average species density per 100 m^2 was found in a pine afforestation (47 species), the lowest being detected on abandoned pastures (22) (Table 8.1). Because of the high competitiveness of the forage grass *S. sphacelata*, the numbers of species found on the abandoned pastures are consistently very low.

Floristically, the vast majority of the plant species growing on man-made sites are Andean elements, and only few are aliens. Of the latter, the most important species are cultivated for forage (of African origin: *S. sphacelata*, *M. minutiflora*, *P. clandestinum*, of European origin: *H. lanatus*, *Trifolium repens*) or for forestry (*Pinus patula* from Mexico). While none of these species is invasive in the study area, the cosmopolitan bracken fern *Pteridium arachnoideum* covers large areas that had previously been burned. However, the fern is restricted to open landscapes, as it does not invade natural forest stands.

8.3.4 Secondary Succession Processes in Anthropogenic Vegetation

The course of secondary successions on pastures is strongly influenced by the abundance of fodder grass species. The dense canopy and compact root network of the dominant species *S. sphacelata* prevent rapid succession from taking place upon pasture abandonment.

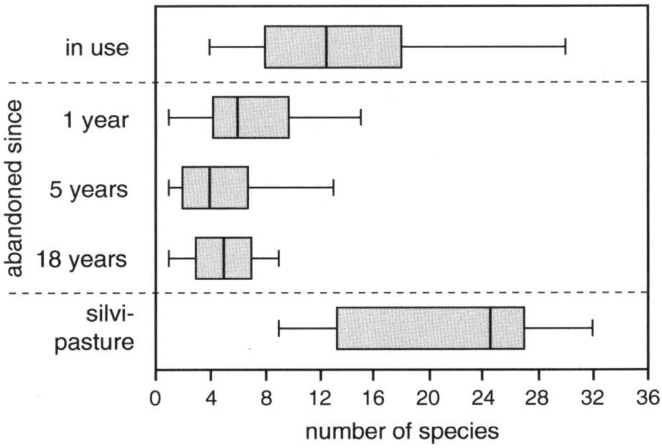

Fig. 8.4 Number of species per 10 m^2 found on *Setaria* pastures under different land use conditions (active use, abandoned, and silvipasture). The boxes represent the medial 50 % (*lower* to *upper quartile*) of each sample, *bold vertical lines* show the median, and the *bars* illustrate the minimum and maximum values of the respective samples

Figure 8.4 shows the number of species found per 10 m^2 on *Setaria* pastures under different land use conditions. The already very small number of herb species present under active pasture conditions decreases continuously during later stages of abandonment, while *S. sphacelata* attains a ground coverage of almost 100 % and reaches a mean height of 1–1.5 m. The vegetation of abandoned *Setaria* pastures is still very poor in species number even after 18 years of fallow. The most frequent shrub species are then *Munnozia senecionides, Austroeupatorium inulifolium, Ageratina dendroides,* and *Lepidaploa canescens* (all Asteraceae). Species of the genus *Rubus* are also common, while bracken fern is not very abundant (mean coverage of about 3 %). In contrast, silvipastures with remnant forest trees exhibit more species than do the other pasture sites (Fig. 8.4). These pastures were not established by slash-and-burn practices, but by lumbering trees without previous burning.

This lumbering leads to a completely different pasture species composition including woody elements such as *Miconia, Hieronyma,* and various Lauraceae that do not occur on previously burned sites. The higher species numbers on silvipastures are not due to herbs, but rather to tree seedlings. These sites accordingly provide better initial conditions for a rapid succession than do the common pastures.

The fires set for slash-and-burn practices do not restrict themselves to their target areas. As a consequence, a bracken fern stage characterized by the high abundance of *P. arachnoideum* establishes itself on burned-out ridges and steep slopes that are not suitable for agricultural use due to their low soil nutrient content. Repeated burning stabilizes the dominance of bracken, since the subterranean rhizomes of this fern tolerate high temperatures. In addition, fire removes the poorly decomposable dead fronds of bracken, and thus removes any restrictions to the propagation of the fern by self-shading. Increasing invasion of bracken fern similarly takes place on active pastures after repeated burning (Roos et al. 2010).

8.3.5 Fragmentation Effects: Epiphytes on Remnant Trees as a Model System

Forest fragmentation is a common phenomenon in most tropical montane regions. During the last decades, the size of intact forest fragments tended to decline rapidly, dividing populations and creating distinct edge habitats. In the face of global climate change, even narrow bands of agricultural matrix may threaten the biodiversity of forests by hindering compensatory range shifts, such as the upslope migration of forest species on the deforested Andean foothills (Bush 2002). Understanding how forest organisms are affected by habitat fragmentation thus requires information as to how these organisms respond to all of the factors that affect landscapes (Gascon et al. 1999).

Isolated trees (ITs) are a common element of anthropogenic land use matrices throughout the tropics. They constitute keystone structures by offering refuge, enhancing connectivity, and providing nuclei for regeneration (Wolf 2005; Manning et al. 2006; Zahawi and Augspurger 2006). ITs can be viewed as the "ultimate," smallest possible forest fragment (Gove et al. 2009), and they thus provide an easily replicable model system for fragmentation studies. ITs represent small habitat islands for forest organisms that are likely to experience constrained dispersal. Being immediately surrounded by open land, they experience pronounced physical edge effects (elevated light levels, wind turbulence, temperatures, and enhanced evapotranspiration rates; Laurance 2004).

With well over 1,200 species of lichens, bryophytes, and vascular plants (Liede and Breckle 2007), epiphytes (including epiphylls) constitute the majority of plant species in the RBSF area. Patterns of IT epiphyte diversity have been studied for all of the three principal groups of epiphytes (see above; Nöske et al. 2008). It is noteworthy that lichen species richness found on ITs was not lower than that found on trees of the closed forest. The species composition on the ITs was also only moderately distinct from that on the forest trees.

In contrast, the epiphytic bryophyte and particularly vascular plant species found on ITs were much smaller in number than those found on forest trees and showed strong compositional differences in the two habitats (Werner et al. 2005; Nöske et al. 2008; unpublished data). A common pattern on isolated trees was a spatial homogenization of epiphyte species. Most epiphyte species in closed-canopy forests occupy narrow vertical niches that offer a suitable trade-off between light and humidity conditions. The microclimate gradient driving this stratification is nonexistent on isolated trees (Nöske 2005; Werner and Gradstein 2008), and typical upper canopy species are widely distributed among the crowns of isolated trees. The distance of ITs to the forest as a source of diaspores was not significantly correlated with species richness per isolated tree for any of the three taxonomic groups of epiphytes (Nöske et al. 2008), suggesting that dispersal limitations were of minor importance.

The processes that lead to such pronounced changes in species assemblage structure and diversity have been studied for vascular epiphytes. Well-established individual plants (late juvenile to adult stages) on the trunks of isolated trees in a fresh clear-cut were tagged and observed over a period of 3 years. More than 50 % of the individuals died within the year following the isolation of their host trees. The mean cumulative mortality of the individuals on the ITs was 72 %, which is significantly higher than that observed in undisturbed forest (11 %). While the mortality of nearly all of the major epiphyte families was significantly higher on the IT than on the forest trees, it was most dramatic with regard to ferns. Plants surviving on ITs generally showed a reduced maximum leaf length (Werner 2011). A parallel experimental study on recruitment indicated that rates of early epiphyte establishment on isolated trunks of *Piptocoma discolor* were 90 % lower than those observed in forest. Even after statistically removing the effect of low plant abundance by rarefaction, the number of epiphyte species that were able to newly establish was significantly lower on IT trunks than it was in the closed forest. Although a broad range of taxa were able to establish in the forest, only few xerotolerant groups (especially tank bromeliads and a desiccation-tolerant polypodioid fern) managed to establish on IT trunks.

These results prove that increased mortality of resident epiphytes as well as impaired recruitment of new epiphytes affect the composition and species richness of vascular epiphyte assemblages on isolated trees. The poor recruitment in terms of both species and individual numbers is hence unlikely to compensate for the strongly enhanced mortality of resident species, and thus results in long-term impoverishment (Fig. 8.5, see also Nöske et al. 2008; Köster et al. 2009). The high rates of mortality and the taxonomically skewed recruitment suggest that increased exposure to light and wind and the resulting increase in desiccation stress are the key drivers of epiphyte diversity on ITs rather than dispersal constraints.

Caution should, however, be taken in generalizing these results. The resilience of epiphyte communities may vary widely with local climate conditions (Werner et al. 2012), and can be substantially higher in perhumid or strongly arid areas (e.g., Werner and Gradstein 2009; Larrea and Werner 2010). On the other hand, most shade- or moisture-demanding taxa of any local community are unlikely to flourish in the multiple edge environment of ITs (Poltz and Zotz 2011). This is commonly indicated by the reduced β-diversity of IT assemblages (e.g., Hietz 2005; Larrea and Werner 2010). Plant responses similar to those documented for ITs in the San Francisco Valley can also be expected to occur along forest edges and in degraded forest. Consequently, the retention of scattered trees and narrow strips or small fragments of forest on clear-cuts are unlikely to be sufficient tools for the conservation of tropical epiphyte diversity.

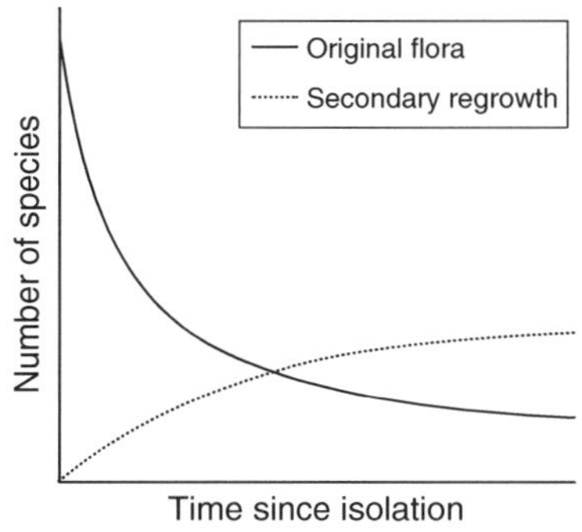

Fig. 8.5 A conceptual model of the extinction and (re-)colonization dynamics of epiphyte species on isolated trees remaining upon the conversion of closed forest into agricultural matrices. Since only a limited number of local species may be adapted well enough to dwell under a high level of exposure, recolonization will often not be able to compensate the loss of the original epiphyte flora

8.3.6 Expected Effects of Reduced Plant Functional Diversity

Changes in biodiversity are accompanied by changes in plant functional diversity (Diaz et al. 2007). Our knowledge about plant functional traits and the role of individual species in the montane forests of the study area is still very limited. We can therefore only assume that the immense loss of species richness associated with land use change results in a drastic deterioration of ecosystem services on the landscape level. Recent studies (e.g., Isbell et al. 2011) suggest that a high level of plant diversity is needed to maintain ecosystem services, and that even the loss of less common species can have significant effects on the ecosystem level (Lyons et al. 2005). Because of the large species pool existing in the study area, one might expect a high complementarity of functions over the range of species, and that insignificant species losses would have no noticeable short-term effects on ecosystem functioning. However, it is beyond question that the drastic species loss resulting from forest fragmentation and current unsustainable land use practices is accompanied by a decreasing resistance to invasive species (e.g., the example of bracken fern in frequently burned areas) and a diminution of ecosystem services.

The average annual total value of ecosystem services provided by tropical forests is estimated to amount to 6,120 $ ha^{-1} (TEEB 2009). According to Göttlicher et al. (2009), almost 48 % of the forests in the study area situated below 2,200 m a.s.l. (lower montane forest, originally amounting to a total of ~7,480 ha) have already been anthropologically transformed. Of this area, 15.4 % is being actively used as pasture, 10.6 % is covered by bracken fern and 21.7 % is overgrown by shrub-dominated succession. Assuming that the average value of the ecosystem services generated by this transformed area is reduced by approximately 50 % (according to Portela and Rademacher 2001), one may estimate an annual loss of 10.9 million $ for that relatively small area. This calculation is only of theoretical value since there is no market for ecosystem services. However, it is hard to imagine that current agricultural incomes (see Chap. 17) will be able to compensate for this loss.

8.4 Conclusion: Implications for Conservation

Managing the montane ecosystems of South Ecuador in a sustainable fashion is a key challenge for the future. Basic knowledge about ecosystem functioning is still scarce for the region, and the study of plant functional traits should receive increasing attention, since these traits are the main attribute by which plants influence ecosystem functioning and thus ecosystem services.

The extent to which the locations that are most valuable for ecosystem services coincide with those that support the highest biodiversity is of critical importance for designing conservation and management strategies. In the study region, high biodiversity commonly coincides with relatively poor soils that are not appropriate for agriculture. This means that the chances for conserving the ecosystems of interest in the study area are good, since the benefits of converting areas of high biodiversity to agricultural use are low. On the other hand, conservation can only be achieved by increasing the value of forest ecosystems to the local communities. It is therefore necessary to define property rights and to establish suitable "Payment for ecosystem services" (PES) compensation schemes.

References

Balvanera P, Pfisterer AB, Buchmann N, He J-S, Nkashizuka T, Raffaelli D, Schmid B (2006) Quantifying the evidence for biodiversity effects on ecosystem functioning and services. Ecol Lett 9:1146–1156

Bush MB (2002) Distributional change and conservation on the Andean flank: a paleoecological perspective. Glob Ecol Biogeogr 11:463–473

Bussmann RW, Wilcke W, Richter M (2008) Landslides as important disturbance regimes – causes and regeneration. In: Beck E, Bendix J, Kottke I, Makeschin F, Mosandl R (eds) Gradients in a tropical mountain ecosystem of Ecuador, vol 198, Ecological studies. Springer, Berlin, pp 319–330

Cadotte MW, Carscadden K, Mirotchnick N (2011) Beyond species: functional diversity and the maintenance of ecological processes and services. J Appl Ecol 48:1079–1087

de Bello F, Lavorel S, Díaz S, Harrington R, Cornelissen JHC, Bardgett RD, Berg MP, Cirpriotti P, Feld CK, Hering D, Martins da Silva P, Potts SG, Sandin L, Sousa JP, Storkey J, Wardle DA, Harrison PA (2010) Towards an assessment of multiple ecosystem processes and services via functional traits. Biodivers Conserv 19:2873–2893

Diaz S, Lavorel S, de Bello F, Quetier F, Grigulis K, Robson TM (2007) Incorporating plant functional diversity effects in ecosystem service assessments. Proc Natl Acad Sci USA 104:20684–20689

Diertl K-HJ (2010) Pflanzendiversität entlang eines Höhengradienten in den Anden Südecuadors. Dissertation thesis, Universität Erlangen-Nuremberg, 277 p. (http://www.opus.ub.uni-erlangen.de/opus/volltexte/2011/2234/pdf/KARL_HEINZDIERTLDISSERTATION.PDF)
FAO (2005) State of the world's forests 2005. Food and Agriculture Organization of the United Nations, Rome
Gascon C, Lovejoy TE, Bierregaard RO, Malcolm JR, Stouffer PC, Vasconcelos HL, Laurance WF, Zimmerman B, Tocher M, Borges S (1999) Matrix habitat and species richness in tropical forest remnants. Biol Conserv 91:223–229
Göttlicher D, Obregon A, Homeier J, Rollenbeck R, Nauss T, Bendix J (2009) Landcover classification in the Andes of southern Ecuador using ETM + and MODIS data as a basis for multi-scale SVAT modeling. Int J Remote Sens 30:1867–1886
Gove AD, Majer JC, Rico-Gray V (2009) Ant assemblages in isolated trees are more sensitive to species loss and replacement than their woodland counterparts. Basic Appl Ecol 10:185–197
Hietz P (2005) Conservation of vascular epiphyte diversity in a Mexican coffee plantation. Conserv Biol 19:391–399
Homeier J (2008) The influence of topography on forest structure and regeneration dynamics in an Ecuadorian montane forest. In: Gradstein SR, Homeier J, Gansert D (eds) The tropical mountain forest – patterns and processes in a biodiversity hotspot, vol 2, Biodiversity and ecology series. Göttingen Centre for Biodiversity and Ecology, Göttingen, pp 97–107
Homeier J, Breckle S-W (2008) Gap-dynamics in a tropical lower montane forest in South Ecuador. In: Beck E, Bendix J, Kottke I, Makeschin F, Mosandl R (eds) Gradients in a tropical mountain ecosystem of Ecuador, vol 198, Ecological studies. Springer, Berlin, pp 311–317
Homeier J, Werner FA (2007) Spermatophyta checklist Reserva Biológica San Francisco (Prov. Zamora-Chinchipe, S-Ecuador). In: Liede-Schumann S, Breckle S-W (eds) Provisional checklists of flora and fauna of the San Francisco valley and its surroundings, vol 4, Ecotropical Monographs. Estación Científica San Francisco, Southern Ecuador, pp 15–58
Homeier J, Werner FA, Gradstein SR, Breckle S-W, Richter M (2008) Potential vegetation and floristic composition of Andean forests in South Ecuador, with a focus on the RBSF. In: Beck E, Bendix J, Kottke I, Makeschin F, Mosandl R (eds) Gradients in a tropical mountain ecosystem of Ecuador, vol 198, Ecological studies . Springer, Berlin, pp 87–100
Homeier J, Breckle S-W, Günter S, Rollenbeck RT, Leuschner C (2010) Tree diversity, forest structure and productivity along altitudinal and topographical gradients in a species-rich Ecuadorian montane rainforest. Biotropica 42:140–148
Isbell F, Calcagno V, Hector A, Connolly J, Harpole WS, Reich PB, Scherer-Lorenzen M, Schmid B, Tilman D, van Ruijven J, Weigelt A, Wilsey BJ, Zavaleta ES, Loreau M (2011) High plant diversity is needed to maintain ecosystem services. Nature 477:199–203
Jost L (2006) Entropy and diversity. Oikos 113:363–375
Köster N, Friedrich K, Nieder N, Barthlott W (2009) Conservation of epiphyte diversity in an Andean landscape transformed by human land use. Conserv Biol 23:911–919
Larrea M, Werner FA (2010) Response of vascular epiphyte diversity to different land-use intensities in a neotropical montane wet forest. For Ecol Manage 260:1950–1955
Laurance WF (2004) Forest-climate interactions in fragmented tropical landscapes. Philos Trans R Soc Lond B Biol Sci 359:345–352
Liede-Schumann S, Breckle S-W (eds) (2007) Provisional Checklists of Flora and Fauna of the San Francisco valley and its surroundings, vol 4, Ecotropical Monographs. Estación Científica San Francisco, Prov. Zamora-Chinchipe, Southern Ecuador
Lyons KG, Brigham CA, Traut BH, Schwartz MW (2005) Rare species and ecosystem functioning. Conserv Biol 19:1019–1024
Mace GM, Norris K, Fitter AH (2012) Biodiversity and ecosystem services: a multilayered relationship. Trends Ecol Evol 27:19–26
Manning AD, Fischer J, Lindenmayer DB (2006) Scattered trees are keystone structures – Implications for conservation. Biol Conserv 132:311–321

Millenium Ecosystem Assessment (2005) Ecosystems and human well-being: biodiversity synthesis. World Resources Institute, Washington, DC
Mosandl R, Günter S, Stimm B, Weber M (2008) Ecuador suffers the highest deforestation rate in South America. In: Beck E, Bendix J, Kottke I, Makeschin F, Mosandl R (eds) Gradients in a tropical mountain ecosystem of Ecuador, vol 198, Ecological studies. Springer, Heidelberg, pp 37–40
Muenchow J, Brenning A, Richter M (2012) Geomorphic process rates of landslides along a humidity gradient in the tropical Andes. Geomorphology 139(140):271–284
Nöske N (2005) Effekte anthropogener Störung auf die Diversität kryptogamischer Epiphyten (Flechten, Moose) in einem Bergregenwald in Südecuador. Dissertation, Universität Göttingen, 137 pp
Nöske NM, Hilt N, Werner FA, Brehm G, Fiedler K, Sipman HJM, Gradstein SR (2008) Disturbance effects on epiphytes and moths in a montane forest in Ecuador. Basic Appl Ecol 9:4–12
Ohl C, Bussmann R (2004) Recolonisation of natural landslides in tropical mountain forests of Southern Ecuador. Feddes Repert 115:248–264
Peters T, Diertl K-H, Gawlik J, Rankl M, Richter M (2010) Vascular plant diversity in natural and anthropogenic ecosystems in the Andes of southern Ecuador. Mt Res Dev 30:344–352
Poltz K, Zotz G (2011) Vascular epiphytes on isolated pasture trees along a rainfall gradient in the lowlands of Panama. Biotropica 43:165–172
Portela R, Rademacher I (2001) A dynamic model of patterns of deforestation and their effect on the ability of the Brazilian Amazonia to provide ecosystem services. Ecol Modell 143:115–146
Restrepo C, Walker LR, Shiels AB, Bussmann R, Claessens L, Fisch S, Lozano P, Negi G, Paolini L, Poveda G, Ramos-Scharrón C, Richter M, Velázquez E (2009) Landsliding and its multiscale influence on mountainscapes. Bioscience 59:685–698
Roos K, Rollenbeck R, Peters T, Bendix J, Beck E (2010) Growth of tropical bracken (Pteridium arachnoideum): response to weather variations and burning. Invasive Plant Sci Manage 3:402–411
Takyu M, Aiba S-I, Kitayama K (2002) Effects of topography on tropical lower montane forests under different geological conditions on Mount Kinabalu, Borneo. Plant Ecol 159:35–49
TEEB (2009) TEEB climate issues update: the economics of ecosystems and biodiversity. September 2009
Webb EL, Stanfield BJ, Jensen ML (1999) Effects of topography on rainforest tree community structure and diversity in American Samoa, and implications for frugivore and nectarivore populations. J Biogeogr 26:887–897
Werner FA (2011) Reduced growth and survival of vascular epiphytes on isolated remnant trees in a recent tropical montane forest clear-cut. Basic Appl Ecol 12:172–181
Werner FA, Gradstein SR (2008) Seedling establishment of vascular epiphytes on isolated and enclosed forest trees in an Andean landscape, Ecuador. Biodivers Conserv 17:3195–3207
Werner FA, Gradstein SR (2009) Diversity of dry forest epiphytes along a gradient of human disturbance in the tropical Andes. J Veg Sci 20:59–68
Werner FA, Homeier J, Gradstein SR (2005) Diversity of vascular epiphytes on isolated remnant trees in the montane forest belt of southern Ecuador. Ecotropica 11:21–40
Werner FA, Köster N, Kessler M, Gradstein SR (2012) Is the resilience of epiphyte assemblages to human disturbance a function of local climate? Ecotropica 17:15–20
Wolf JHD (2005) The response of epiphytes to anthropogenic disturbance of pine-oak forests in the highlands of Chiapas, Mexico. For Ecol Manage 212:376–393
Wolf K, Veldkamp E, Homeier J, Martinson GO (2011) Nitrogen availability links forest productivity, soil nitrous oxide and nitric oxide fluxes of a tropical montane forest in southern Ecuador. Global Biogeochem Cycles 25, GB4009
Zahawi RA, Augspurger CK (2006) Tropical forest restoration: tree islands as recruitment foci in degraded lands of Honduras. Ecol Appl 16:464–478

Chapter 9
Supporting, Regulating, and Provisioning Hydrological Services

Lutz Breuer, David Windhorst, Andreas Fries, and Wolfgang Wilcke

9.1 Water-Related Ecosystem Services

Water is involved in all types of ecosystem services including supporting, provisioning, regulating, and cultural ones (Millennium Ecosystem Assessment 2005). Many of these ecosystem services are intrinsically coupled and water is a major coupler, either directly by the interaction with water or indirectly through transporting processes (Brauman et al. 2007). There are a number of studies about land-use and management changes and their effect on hydrological processes and fluxes. However, there is no comprehensive overview of their impact on water-related ecosystem services for tropical montane ecosystems.

The maintenance of the water cycle belongs to the supporting services of the forest ecosystem. The water cycle provides water for primary production in all ecosystems and further effectively triggers soil formation and nutrient cycling. Water also serves as a habitat for aquatic organisms (Brauman et al. 2007). Furthermore, water plays an indirect, proximate role as a driver for services provided by other parts of the ecosystem. Water erosion not only reshapes landscapes and leads to substantial soil losses, but also influences the carbon budget (Lal 2003).

In a recent study on the ecological and economic costs of carbon sequestration by tree plantations, it was shown that a multitude of trade-offs between groundwater recharge, streamflow, salinity, and acidity of soils may occur (Jackson et al. 2005).

L. Breuer (✉) • D. Windhorst
Institute for Landscape Ecology and Resources Management, University of Gießen, Gießen, Germany
e-mail: lutz.breuer@umwelt.uni-giessen.de

A. Fries
Faculty of Geography, University of Marburg, Marburg, Germany

W. Wilcke
Geographic Institute of the University of Bern (GIUB), University of Bern, Bern, Switzerland

J. Bendix et al. (eds.), *Ecosystem Services, Biodiversity and Environmental Change in a Tropical Mountain Ecosystem of South Ecuador*, Ecological Studies 221,
DOI 10.1007/978-3-642-38137-9_9, © Springer-Verlag Berlin Heidelberg 2013

A further example is the soil moisture dependence of the carbon and nitrogen cycle within soils which in turn determine ecosystem services such as the regulation of trace gas emissions or carbon sequestration. However, N_2O and NO emissions measured at different elevations in the San Francisco valley did not reveal a correlation with the measured soil moisture content (Wolf et al. 2011). Water also acts as a resource that is obtained from the landscapes for human consumption, industrial processes, or agricultural utilization and therefore is a provisioning service. In tropical montane regions in Ecuador, this service most often is related to hydropower generation, mainly in the Páramo (Buytaert et al. 2006). Further to the already existing hydropower stations, there is a large development potential, especially for medium and small hydropower plants in the Andes. A second important provisioning service in the Andean region of Ecuador is related to freshwater fish production. Various tilapia and carp species as well as rainbow trout are commonly grown in Ecuador (Schwarz and Schwarz 2007).

The timing of runoff, flooding, and groundwater recharging depends on the water storage potential and dynamic of the system which is therefore a regulating service. Regulating services of the forest are linked to land-use change and type and intensity of land use. A conversion from forests to cropland or pastures can substantially alter the water regulation function of the ecosystem (see Chap. 20).

Cultural water-related ecosystem services include recreational, esthetic, spiritual, religious, educational, and scientific factors. The San Francisco catchment may offer a high development potential for tourism given the wide-stretching coverage of the Podocarpus National Park in the basin (see Fig. 1.1). But currently there is hardly any infrastructure developed in the region to support this. Water also does not play a dominant role in local societies with regard to spiritual or religious aspects. However, indigenous people (Shuar) living close to the Rio San Francisco Valley worship cascades as sacred places and take spiritual benefits from those (Harner 1984). From an educational and scientific point of view, recent and ongoing research at the Estación Científica San Francisco (ECSF) as well as the commitment of the Foundation Nature and Culture International (NCI) is a significant component of cultural ecosystem services in the region.

This chapter focuses on supporting and regulating services of tropical montane rain forest ecosystems of the San Francisco catchment in the Andes of southern Ecuador. Furthermore, we shortly address the provisioning services with respect to hydropower generation.

9.2 Supporting Service

Water balances in tropical mountainous areas are first of all characterized by the overall high precipitation inputs. As a result of spatially heterogeneous precipitation, point meteorological measurements in the field using classic tipping buckets or rainfall totalizers are hardly representative for whole catchments of the size of that of the Rio San Francisco. Merging data from weather radar, scatterometers, field

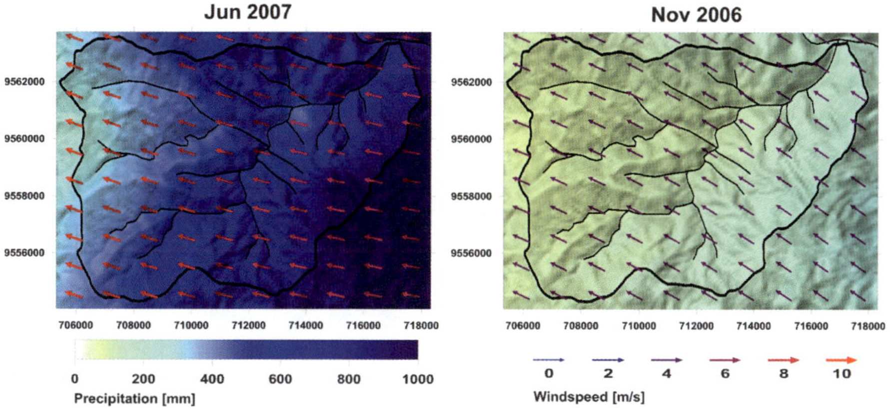

Fig. 9.1 Monthly rainfall sums for the rainy (June 2007) and dry (November 2006) season and predominant wind field for the Rio San Francisco catchment area. The geographic extend of the catchment area is indicated by the latitude and longitude coordinates along the *x*/*y* axes

observations from a set of automatic stations, totaling rain gauges, and state-operated meteorological stations resulted in new insights into atmospheric water inputs to the San Francsico catchment (Rollenbeck and Bendix 2011). For the period 2002–2008, Rollenbeck and Bendix (2011) showed that precipitation is mainly resulting from humid air advectively transported from the Amazon. This type of atmospheric condition mainly leads to long lasting rainfall with little intensity. Occasionally, however, this general pattern is heavily modified by other small- and large-scale processes such as convective clouds or anabatic and katabatic wind systems.

For two years, high resolution rain radar data (100 m resolution), blended with rain gauges measurements, could be used to provide spatial rainfall maps of the catchment area of the Rio San Francisco (for methodology refer to Rollenbeck & Bendix 2011). Figure 9.1 shows the result for a rainy (June 2007) and dry season month (Nov 2006). It is clearly shown that the shelter of topography against the predominant circulation from E-SE leads to a spatial differentiation of rainfall in the catchment of the Rio San Francisco in both seasons. The easternmost part, particularly at higher altitudes, gains most rainfall while the north facing slopes in the catchment and the westernmost part in general receive less rainfall. This gradient is very steep during the rainy season but less pronounced during the dry season.

In the lower regions of the San Francisco valley, precipitation is to a large extent (95 %) composed of rainfall, while fog (and horizontal rain) precipitation becomes more and more important at higher elevations. Under extreme condition the contribution of fog precipitation can exceed 40 % of the total water input (Bendix et al. 2008). Because of a strong altitudinal gradient annual precipitation is highly variable in space, amounting to annual totals around 2,300 mm at the ECSF (1,800 m a.s.l.) and up to 6,700 mm at the Cerro del Consuelo (3,800 m a.s.l.).

However, the postulated altitudinal gradient of precipitation cannot be found throughout the entire San Francisco catchment. For example, along the east to west direction precipitation does not necessarily increase with altitude (Fig. 9.1). More evident gradients that increase with altitude and that impact the water cycle are found for temperature, wind speed, and irradiation (Beck et al. 2008). These, together with at the same time decreasing canopy height and leaf area index (LAI), lead to higher potential evapotranspiration at lower elevations. However, as relative humidity is often high at all elevations and especially under cloudy conditions in the upper parts of the San Francisco catchment, the water vapor deficit is generally low. Hence, the water vapor deficit is leveling the potential altitudinal gradient of evapotranspiration. In addition to climatic boundary conditions relevant for potential evapotranspiration rates, soil water status is responsible for the actual evapotranspiration rates.

Reported values of actual evapotranspiration losses of rainforests, and tropical mountain rainforests in particular, are limited. In addition to the heterogeneous stand characteristics that drive evapotranspiration, differences in the methods to quantify actual evapotranspiration rates are substantial. In a recent review provided by Bruijnzeel et al. (2011), the San Francisco catchment belongs to the forest type of "lower montane rain forest little affected by fog." Within this group evapotranspiration rates reported for the study area are placed at the top end of all reported evapotranspiration losses, mainly attributable to the very high interception losses (Fig. 9.2, Fleischbein et al. 2006). These values are 67–188 % higher than the average of other reported interception losses for this group (Bruijnzeel et al. 2011). We assessed evapotranspiration by a catchment water budget approach, where precipitation P equals the sum of evapotranspiration and runoff Q. This approach is particularly sensitive to hydrometric errors, especially attributable to flawed discharge measurements and discrepancies between surface and subsurface catchment area (Wilson et al. 2001). The latter, however, only affects our estimate of the contribution of transpiration to evapotranspiration while the interception loss was directly measured with a high number of throughfall collectors (Wullaert et al. 2009). In addition, only long-term measurements of decades and longer should be used due to the influence of soil and groundwater storage changes between years. Our 12-year record of the elements of the water budget is the longest we are aware of in tropical montane forests and illustrates that there is relatively little interannual variation in the individual fluxes as reflected by low standard deviations of <15 % of the means of most fluxes except for the relatively minor fluxes of stemflow (24 %) and transpiration (49 %) (Fig. 9.2). The catchment water balance approach further assumes that groundwater losses are known (or can be excluded) to close the water balance (Wilson et al. 2001). In an ^{18}O tracer study, we did not detect hints towards the presence of a deep aquifer which contributes to the aboveground water cycling (Goller et al. 2005).

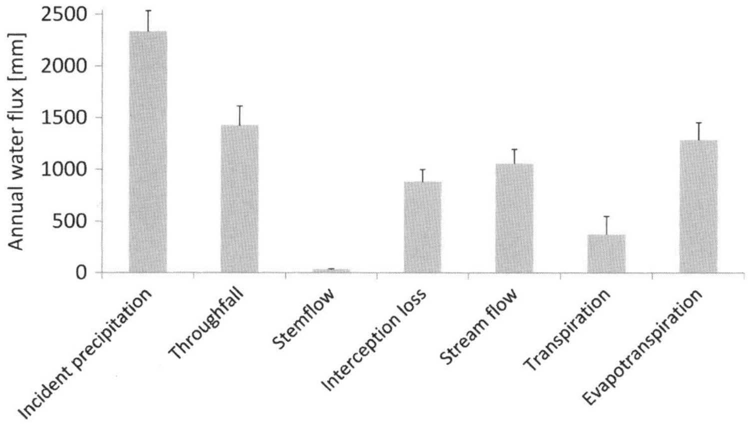

Fig. 9.2 Mean annual water fluxes of an approx. 9 ha large micro-catchment in the Reserva Biologica San Francisco (RBSF) area between 1999 and 2010. Direct evaporation from the forest floor was neglected. Error bars indicate standard deviations

9.3 Regulating Service

Discharge in the Rio San Francisco is highly variable in time and space (Crespo et al. 2011; 2012; Plesca et al. 2012). In general, the timing of runoff is highly correlated with precipitation, i.e., responses in rainfall runoff are very quick (Fig. 9.3). Spatial variability is due to the heterogeneous precipitation patterns (Rollenbeck and Bendix 2011) that can lead to a significant rainfall-runoff reaction in only parts of the catchment. As an example we compare discharge at two sub-catchment gauging stations in the San Francisco catchment with rainfall measurements within one of the sub-catchments. Precipitation shown in Fig. 9.3 is derived from a gauging station close to the Quebrada Ramón. As expected Quebrada Ramón responds to nearly all precipitation events immediately, in contrast to the discharge peaks of Quebrada Zurita, which are not always associated with precipitation events recorded close to Quebrada Ramón (e.g., in the end of July 2007). The other way around, moderate precipitation sums in early August 2007 are in no relation with large discharges in Quebrada Zurita recorded at the same time. A second example is related to the largest discharge event in the past years, occurring on October 11, 2008, which was most likely associated with local, but heavy rainfall around El Tiro. This local rainfall event lead to severe discharge volumes estimated to exceed 500–600 $m^3\ s^{-1}$ close to the ESCF (Honer 2010). However, even such large floods do not evoke losses in lives or economic values as the catchment is poorly inhabited.

Despite the tight connection to rainfall events, tracer and stable water isotope analyses confirmed a significant contribution of old water to discharge generation, most likely from deeper aquifers. The latter is confirmed by analyses of geogenic tracers measured in a variety of sub-catchments in the San Francisco (Bücker et al. 2010;

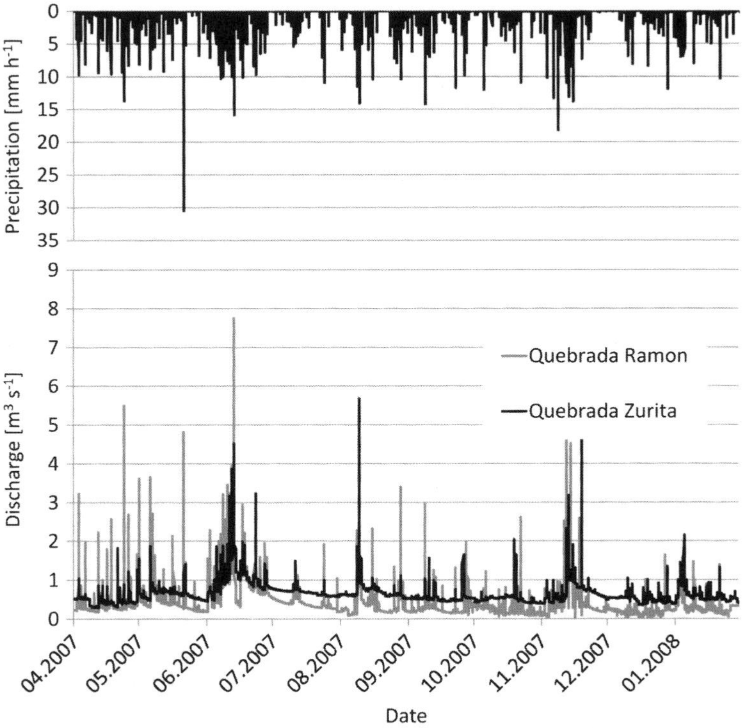

Fig. 9.3 Hourly discharge for Quebrada Ramon (forested) and Quebrada Zurita (land-use mix of forest, shrubs, pasture) and precipitation (gauging station TS1, located close to the headwater of Quebrada Ramon)

Crespo et al. 2012). They found that concentrations of water solutes at several catchment outlets were significantly different from concentrations of water sources within these catchments such as precipitation, soil water, and water from seeps in bedrock cracks, indicating an additional source of streamwater, most likely deeper aquifers. Further, mean transit time estimations in 10 sub-catchments of the San Francisco using stable water isotopes revealed an unexpected long transit time of water between 230 and 310 days (Crespo et al. 2012), despite the very reactive, flashy hydrographs observed in the San Francisco. This observation might seem contradictory at first sight to findings for the micro-catchments (Goller et al. 2005) where near-surface flowing event water of short transit time has shown to substantially contribute to discharge generation. Some studies have shown the increase of transit time with increasing catchment size; however, this general dependency is currently critically discussed (Hrachowitz et al. 2009).

The only explanation of the very fast response of discharge to rainfall with at the same time long mean transit time of waters is by a rapid mobilization of old (pre-event) water. Many explanations of such observations have been given (Kirchner 2003), but so far no universal mechanism has been identified and there

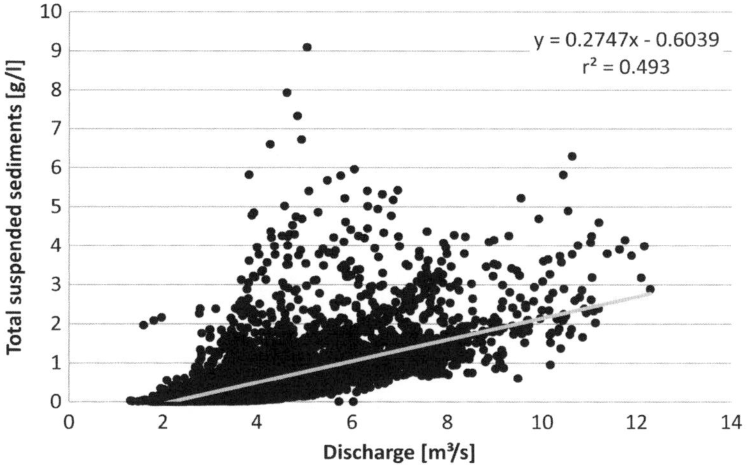

Fig. 9.4 Correlation and regression between discharge and total suspended sediment (TSS) concentration at the outlet of the San Francisco (75 km^2) upstream of the village of Sabanilla

is still a need to test the different explanations of this so-called old-water paradox for the San Francisco.

Despite the aforementioned indications that deeper aquifers contribute to streamflow, no information is available about the amount of aquifer recharge nor are there any hydro-geological information available for the region or the San Francisco valley in particular.

The high precipitation amounts also contribute to catchment-wide occurring landslides at all land-use types in the San Francisco catchment. It has been shown that these landslides severely impact the development of soils (Wilcke et al., 2003; Huwe et al. 2008; Liess et al. 2010) as well as biotic or geomorphic processes (Restrepo et al. 2009). Despite the importance of landslides for the landscape, measurements of suspended sediments have only been started recently. In general, discharge is correlated with sediment concentrations (Fig. 9.4), indicating the importance of discharge volume for sediment loads. However, we also observed suspended sediment peaks that are not directly linked to large peak flow. One likely reason is that local natural or anthropogenically introduced landslides are responsible for the export of large sediment loads while at the same time no distinct precipitation peaks occur.

9.4 Provisioning Service

Provisioning services of water resources in the San Francisco are limited in view of the low number of inhabitants: hardly any water is needed for extractive water use in the agricultural, industrial, commercial or municipal sector. The most striking

Table 9.1 Expected future trends in human use and benefits of water-related ecosystem services in the San Francisco (definitions in agreement with MA 2005)

	Human use	Benefits	Note
Provisioning service: fresh water	↑	↑	Hydropower generation will increase, but water resources will not be heavily impacted; aquaculture is of no economic importance in RBSF
Regulating service: water purification and waste treatment	→	→	Water quality will remain stable and not be impacted by local population
Regulating service: water regulation	→	→	Assumed minor land-use change is not expected to alter timing and magnitude of runoff significantly

For provisioning services, human use increases if human consumption of service increases; for regulating services, human use increases if number of people affected by service increases. For provisioning services, benefits are judged to be degraded if current use exceeds sustainable levels. For regulating services, degradation means a reduction in benefits obtained from the service, either through a change in or through human pressures on the service exceeding its limits. Supporting services are not directly used, and changes influence the supply of all other services. If indirect effects are listed, supporting services would be double counted. ↑ increasing/enhancing, → no change

man-made infrastructure in the San Francisco valley is—apart from the main road connecting Loja and Zamora—a hydropower plant. Located upstream, close to the small village of Sabanilla, this small-sized hydropower plant produces electricity for the Loja and Zamora provinces in the south of Ecuador. To increase power production, water from the San Francisco headwater and the Quebrada San Ramón are diverted to the Qubrada Milagro via channels. The plant is operating since 1957 and generates 3 MW at the moment. A recent study analyzed the possibility to increase power generation in future years (EERSSA 2010).

9.5 Future Trends

Based on the general directions of changes in ecosystem services as identified by Millennium Ecosystem Assessment (2005), we summarized likely trends in human use and degradation of water-related ecosystem services in the San Francisco in Table 9.1. As can be seen, we assumed no decreasing trend in benefits. Also local human use of water is not being expected to increase. The only change we envisage is the increase of hydropower generation for utilization outside of the San Francisco catchment.

9.6 Conclusion

Water-related ecosystem services in the San Francisco are mainly related to ecosystem functioning, supporting nutrient and matter cycling and regulating the water cycle. Primarily due to the low population density, provisioning services are less important in the research area. However, given the dramatic increase of settlements in many parts of Ecuador as well as the pressure to produce food, feed, and timber, one has to be aware of the future impact of these drivers on ecosystem functioning and associated services. Last but not least, this pressure might be increased by superimposed climate change effects.

We conclude from the current environmental situation that water-related ecosystem services are not severely threatened by future global change impact in the area. However, a further assessment of the potential impact of land use and climate change should include model-based scenario evaluation. The reader is referred to Chaps. 19 and 20 of this book which include such scenario evaluations.

Acknowledgments We thank Amelie Bücker, Patricio Crespo and Edison Timbe as well as many undergraduate and graduate students for their extended support during field work.

References

Beck E, Mosandl R, Richter M, Kottke I (2008) The investigated gradients. In: Beck E, Bendix J, Kottke I, Makeschin F, Mosandl R (eds) Gradients in a tropical mountain ecosystem of Ecuador. Springer, Berlin, pp 55–62

Bendix J, Rollenbeck R, Fabian P, Emck P, Richter M, Beck E (2008) Climate. In: Beck E, Bendix J, Kottke I, Makeschin F, Mosandl R (eds) Gradients in a tropical mountain ecosystem of Ecuador. Springer, Berlin, pp 281–290

Brauman KA, Daily GC, Duarte TK, Mooney HA (2007) The nature and value of ecosystem services: an overview highlighting hydrologic services. Annu Rev Environ Resour 32:67–98

Bruijnzeel L, Mulligan M, Scatena FN (2011) Hydrometeorology of tropical montane cloud forests: emerging patterns. Hydrol Proc 25:465–498. doi:10.1002/hyp.7974

Bücker A, Crespo P, Frede HG, Vaché K, Cisneros F, Breuer L (2010) Identifying controls on water chemistry of tropical cloud forest catchments: combining descriptive approaches and multivariate analysis. Aquat Geochem 16(1):127–149. doi:10.1007/s10498-009-9073-4

Buytaert W, Célleri R, De Bièvre B, Cisneros F, Wyseure G, Deckers J, Hofstede R (2006) Human impact on the hydrology of the Andean páramos. Earth Sci Rev 79:53–72

Crespo P, Feyen J, Buytaert W, Bücker A, Breuer L, Frede H-G, Ramírez M (2011) Identifying controls of the rainfall-runoff response of small catchments in the tropical Andes (Ecuador). J Hydrol 407(1–4):164–174. doi:10.1016/j.jhydrol.2011.07.021

Crespo P, Bücker A, Feyen J, Vaché KB, Frede H-G, Breuer L (2012) Preliminary evaluation of the runoff processes in a remote montane cloud forest basin using Mixing Model Analysis and Mean Transit Time. Hydrol Proc 26:3896–3910. doi:10.1002/hyp.8382

EERSSA (2010) Investment plan 2011–2020. Empresa electrica regional del sur SA. Technical Report (in Spanish)

Fleischbein K, Wilcke W, Valarezo C, Zech W, Knoblich K (2006) Water budgets of three small catchments under montane forest in Ecuador: experimental and modelling approach. Hydrol Proc 20:2491–2507

Goller R, Wilcke W, Leng MJ, Tobschall HJ, Wagner K, Valarezo C, Zech W (2005) Tracing water paths through small catchments under a tropical montane rain forest in south Ecuador by an oxygen isotope approach. J Hydrol 308:67–80

Harner MJ (1984) The Jívaro: people of the sacred waterfalls. University of California Press, Berkeley

Honer H (2010) Hydrological and geomorphological processes of an extreme flood in the Rio San Francisco Valley, South Ecuador. Diploma Thesis, Albert-Ludwigs-Universität Freiburg, 256 pp

Hrachowitz M, Soulsby C, Tetzlaff D, Dawson JJC, Malcolm IA (2009) Regionalization of transit time estimates in montane catchments by integrating landscape controls. Water Resour Res 45, W05421. doi:10.1029/2008WR007496

Huwe B, Zimmermann B, Zeilinger J, Quizhpe M, Elsenbeer H (2008) Gradients and patterns of soil physical parameters at local, field and catchment scales. In: Beck E, Bendix J, Kottke I, Makeschin F, Mosandl R (eds) Gradients in a tropical mountain ecosystem of Ecuador. Springer, Berlin, pp 391–402

Jackson RB, Jobbagy EG, Avissar R, Roy SB, Barrett DJ, Cook CW, Farley KA, le Maitre DC, McCarl BA, Murray BC (2005) Trading water for carbon with biological carbon sequestration. Science 310:1944–1947

Kirchner JW (2003) A double paradox in catchment hydrology and geochemistry. Hydrol Proc 17:871–874

Lal R (2003) Soil erosion and the global carbon budget. Environ Int 29:437–450

Liess M, Glaser B, Huwe B (2010) Digital soil mapping in southern Ecuador. Erdkunde 63 (4):309–319

Millennium Ecosystem Assessment (2005) Ecosystems and human well-being: synthesis. Island Press, Washington, DC

Plesca I, Timbe E, Exbrayat JF, Windhorst D, Kraft P, Crespo P, Vaché KB, Frede H-G, Breuer L (2012) Model intercomparison to explore catchment functioning: results from a remote montane tropical rainforest. Ecol Model. doi:10.1016/j.ecolmodel.2011.05.005

Restrepo C, Walker LR, Shiels AB, Bussmann R, Claessens L, Fisch S, Lozano P, Negi G, Paolini L, Poveda G, Ramos-Scharrón C, Richter M, Velázquez E (2009) Landsliding and its multiscale influence on mountainscapes. Bioscience 59:685–698

Rollenbeck R, Bendix J (2011) Rainfall distribution in the Andes of southern Ecuador derived from blending weather radar data and meteorological field observations. Atmos Res 99:277–289

Schwarz LG, Schwarz L (2007) Freshwater fish seed resources in Ecuador. Assessment of freshwater fish seed resources for sustainable aquaculture. FAO Fisheries Technical Paper No. 501, pp 233–240

Wilcke W, Valladarez H, Stoyan R, Yasin S, Valarezo C, Zech W (2003) Soil properties on a chronosequence of landslides in montane rain forest, Ecuador. Catena 53:79–95

Wilson KB, Hanson PJ, Mulholland PJ, Baldocchi DD, Wullschleger SD (2001) A comparison of methods for determining forest evapotranspiration and its components: sap-flow, soil water budget, eddy covariance and catchment water balance. Agric For Meteorol 106:153–168

Wolf K, Veldkamp E, Homeier J, Martinson GO (2011) Nitrogen availability links forest productivity, soil nitrous oxide and nitric oxide fluxes of a tropical montane forest in southern Ecuador. Glob Biogeochem Cycle 25, GB4009. doi:10.1029/2010GB003876

Wullaert H, Pohlert T, Boy J, Valarezo C, Wilcke W (2009) Spatial throughfall heterogeneity in a montane rain forest in Ecuador: extent, temporal stability and drivers. J Hydrol 377:71–79

Chapter 10
The Carbon Balance of Tropical Mountain Forests Along an Altitudinal Transect

Christoph Leuschner, Alexandra Zach, Gerald Moser, Jürgen Homeier, Sophie Graefe, Dietrich Hertel, Bärbel Wittich, Nathalie Soethe, Susanne Iost, Marina Röderstein, Viviana Horna, and Katrin Wolf

10.1 Introduction

The storage and sequestration of carbon is one of the most important ecosystem services provided by forests. Although tropical forests cover only about 12 % of the land surface of the earth, it is estimated that they account for about 25 % of the world's biomass carbon and contribute c. 40 % of terrestrial net primary production (Cleveland et al. 2011; Townsend et al. 2011). Consequently, tropical forests play

C. Leuschner (✉) • A. Zach • J. Homeier • D. Hertel • B. Wittich • M. Röderstein • V. Horna
Plant Ecology, Albrecht von Haller Institute of Plant Sciences, University of Göttingen, Untere Karspüle 2, 37073 Göttingen, Germany
e-mail: cleusch@gwdg.de

G. Moser
Plant Ecology, Albrecht von Haller Institute of Plant Sciences, University of Göttingen, Untere Karspüle 2, 37073 Göttingen, Germany

Plant Ecology, University of Gießen, Heinrich-Buff-Ring 26, 35392 Gießen, Germany

S. Graefe
Plant Ecology, Albrecht von Haller Institute of Plant Sciences, University of Göttingen, Untere Karspüle 2, 37073 Göttingen, Germany

Tropical Silviculture, Burckhardt Institute, University of Göttingen, Büsgenweg 1, 37077 Göttingen, Germany

N. Soethe
Botany and Landscape Ecology, University of Greifswald, Grimmer Str. 88, 17487 Greifswald, Germany

S. Iost
Institute for World Forestry, Johann Heinrich von Thünen-Institut (vTI), Federal Research Institute for Rural Areas, Forestry and Fisheries, Hamburg, Leuschnerstr. 91, 21031 Hamburg, Germany

K. Wolf
Soil Science and Forest Nutrition, Büsgen Institute, University of Göttingen, Büsgenweg 2, 37077 Göttingen, Germany

J. Bendix et al. (eds.), *Ecosystem Services, Biodiversity and Environmental Change in a Tropical Mountain Ecosystem of South Ecuador*, Ecological Studies 221,
DOI 10.1007/978-3-642-38137-9_10,

an outstanding role in the global C cycle and a solid understanding of the C pools and fluxes with carbon assimilation, net primary production and decomposition in this biome is of paramount importance for predicting changes in the earth's climate and to adopt suitable mitigation strategies for reducing global warming (Grace and Meir 2009).

In the last two decades, our knowledge about the tropical forest carbon cycle has considerably increased due to several large-scale biomass inventories and growth analyses in neotropical and paleotropical forest plot networks (e.g. Malhi et al. 2004; Slik et al. 2010) and the application of the eddy covariance technique on towers for measuring ecosystem-level CO_2 fluxes in tropical old-growth forests (Grace et al. 1995). One prominent result is that, in the past two decades, intact tropical lowland forests apparently represented CO_2 sinks in moist years (overview in Malhi 2010), while they often were CO_2 sources in dry years (Davidson et al. 2012). Another surprising outcome of global carbon flux data bases is that the net primary production (NPP) of forest ecosystems appears to be independent from mean annual temperature (MAT) in regions with MAT $> 10\,°C$, i.e. in tropical and subtropical climates (Luyssaert et al. 2007). This view is challenged by an earlier compilation of data from tropical moist forests along elevation transects by Raich et al. (2006) who reported a mean increase in aboveground NPP (ANPP) by 0.66 Mg C ha^{-1} $year^{-1}$ per K temperature increase. However, these authors provided no data on belowground NPP (root production) but gave only an indirect estimate of belowground C allocation from data on litter fall and soil respiration. The C transfer to the roots increased with increasing MAT as well, indicating an increase in total NPP with higher temperatures. However, without more reliable data on root production, this conclusion must remain vague. Another surprising outcome of the data compilation by Luyssaert et al. (2007) is the lacking precipitation dependence of NPP above a threshold of about 1,500 mm $year^{-1}$ indicating that neither temperature nor precipitation should be important controls of NPP in tropical moist forests. However, the scatter in the data of this analysis was large and total NPP (above- and belowground) consists of a number of components, which are difficult to measure, often are estimated and thus are likely to introduce a considerable bias in the calculation of gross primary production (GPP), NPP and net ecosystem production (NEP) in tropical forests. So far, only very few studies have attempted to measure all relevant C stores and C fluxes in tropical forests completely or near-completely which leaves a considerable uncertainty in current data analyses and syntheses with respect to the tropical forest carbon cycle (Clark et al. 2001).

The situation is even less satisfying for tropical mountain forests, which once covered the mountains of the tropics from about 1,000 to more than 4,000 m elevation and that have been found to store considerable C amounts in biomass and soil (e.g. Kitayama and Aiba 2002; Benner et al. 2010). Due to their poor accessibility and the often steep slopes in rugged terrain, carbon inventories and the measurement of C fluxes is often more difficult than in tropical lowland forests. For example, using eddy covariance technique is rarely an option in tropical mountain forests. However, these forests are not only important in the cycles of C and water (McJannet et al. 2010) and are havens of a unique biodiversity (Gentry 2001;

Bruijnzeel et al. 2010), they also offer the rare opportunity to examine the C cycle of tropical forests along well-defined temperature gradients (Girardin et al. 2010), which may help to understand the temperature dependence of forest NPP (Malhi 2012; Malhi et al. 2010) and to test the more recent conclusions on the apparent temperature insensitivity of tropical forest productivity drawn from global data surveys (Luyssaert et al. 2007).

This contribution summarises extensive field work in pristine tropical pre-montane to upper montane forests conducted over the past 10 years in an altitudinal transect from 1,000 to 3,000 m a.s.l. on the eastern slope of the southern Ecuadorian Andes. For the first time, it was attempted to measure all major C pools and C fluxes in a set of forest plots at different elevations, including the C pools in above- and belowground biomass, and the fluxes associated with photosynthesis, stem wood growth, leaf and fine litter production, fine and coarse root production, and autotrophic (tree) and soil respiration by applying up-to-date ecophysiological techniques in a considerable number of plots and tree species under field conditions.

The main aims of the study were (1) to analyse altitudinal changes in forest above- and belowground biomass and forest structure using a large number of plots (59), (2) to close gaps in our knowledge on rarely studied components of the C cycle in tropical forests (in particular root production, autotrophic respiration and photosynthesis) and (3) to search for altitudinal trends in GPP, NPP, NEP and respiration components, which could give hints on their temperature dependence.

This chapter compiles a wealth of data on C cycle components that have been published or are being prepared for publication (see references list), in order to reach at a comprehensive synthesis on the C balance of the forests between 1,000 and 3,000 m elevation.

10.2 Materials and Methods

10.2.1 Climate, Geology and Vegetation of the Study Region

The study region close to the equator on the eastern slope of the Andes has a humid tropical montane climate with MAT ranging between 19 °C at 1,000 m and 9 °C at 3,000 m a.s.l. (Bendix et al. 2008; Emck 2007, see Table 10.1). The soils along the slope are generally acidic with increasing amounts of humus material on top of the soil towards higher elevations (Moser et al. 2011, see also Wolf et al. 2011). The availability of N in the densely rooted uppermost organic layer markedly decreases along the transect as is shown by large reductions in gross N mineralisation rate and in the amount of KCl-extractable inorganic N in the organic layers from 1,050 to 3,060 m indicating a slowing down of decomposition with increasing elevation. The species-rich premontane to upper montane forest communities are described in detail in Homeier et al. (2008) and Chap. 8. All stands are located in protected forest sections with a representative stand structure and no marked signs of human impact or recent natural disturbance.

Table 10.1 Location and climatic, edaphic and stand structural characteristics of the five intensive study plots (adopted from Moser et al. 2011)

Plot no.	1	2	3	4	5
Coordinates	S 04°06′54″ W 78°58′02″	S 04°06′42″ W 78°58′20″	S 03°58′35″ W 79°04′65″	S 03°59′19″ W 79°04′55″	S 04°06′71″ W 79°10′58″
Elevation [m a.s.l.]	1,050	1,540	1,890	2,380	3,060
Inclination [°]	26	10	31	28	27
Air temperature [°C] (min–max)	19.4 (11.5–30.2)	17.5 (11.2–26.7)	15.7 (7.9–29.4)	13.2 (7.0–25.1)	9.4 (3.1–18.8)
Rainfall [mm year^{-1}]	c. 2,230	c. 2,300	c. 1,950	c. 5,000	c. 4,500
Soil types	Alumic Acrisol	Alumic Acrisol	Gleyic Cambisol	Gleyic Cambisol	Podzol
Organic layer thickness [mm]	48	243	305	214	435
Soil moisture [vol.%]	29.7 (15.3–38.5)	30.3 (20.4–43.5)	35.4 (27.4–44.7)	44.7 (35.7–48.7)	49.1 (39.5–59.5)
Soil pH [$CaCl_2$]	3.9	3.9	3.5	3.3	2.9
Soil C/N [Oi horizon, g g^{-1}]	22	29	28	46	63
KCl-extract. N_{inorg}: Oi layer [μg g^{-1}]	1,180 (329–2,238)	219 (76–953)	234 (34–574)	17 (12–209)	6 (4–19)
KCl-extract. N_{inorg}: 0–10 cm min. soil [μg g^{-1}]	45 (43–69)	19 (14–27)	73 (35–78)	26 (11–44)	38 (19–46)
Pool of KCl-extract. N_{inorg} in org. layers[g N m^{-2}]	2.9	2.5	3.3	3.1	1.1
Gross N mineralisation: Oi layer [μg g^{-1} day^{-1}]	159 (80–213)	n.d.	115 (107–120)	n.d.	23 (0–32)
Gross N mineralisation: 0–10 cm min. soil [μg g^{-1} day^{-1}]	0 (<0–48)	n.d.	8 (3–13)	n.d.	13 (9–17)
Stem density (>5 cm DBH) [ha^{-1}]	968	2,167	2,245	2,512	5,613
Canopy height [m]	31.8	21.7	18.9	12.0	9.0

Mean annual air temperature and relative air humidity measured at 1.5 m height inside the stands, soil moisture in 10 cm depth of the mineral soil; given are annual means, minimum and maximum (in brackets) for the period April 2003–March 2004; bedrock types after Litherland et al. (1994); soil classification (FAO system), pH($CaCl_2$) of the mineral topsoil (0–30 cm), C/N ratio of the organic layer (L/Of1) after Iost (2007); rainfall data after P. Emck & M. Richter and own measurements (3-year means, unpublished); KCl-extractable inorganic N measured in April 2004 by Iost (2007), (median and range), gross N mineralisation according to the ^{15}N isotopic pool dilution approach (three plots only), the pool of KCl-extract. N in the organic layers is a rough estimate derived from N_{inorg} concentrations and humus mass; data on forest structure were measured for 80 trees per plot (after Moser et al. 2008)

10.2.2 Study Plots

In the transect of about 30-km length and 2,000-m elevation distance, a set of 59 study plots of 20 m × 20 m size was identified in a stratified random selection procedure. Three plots served for the very intensive carbon cycle measurement programme (all C fluxes including photosynthesis and respiration measurements; plots # 1, 3 and 5), five for the spatially more extended intensive measurement program (net primary production including root production; plots # 1, 2, 3, 4 and 5) and 54 additional plots for the analysis of the spatial variability of selected biomass and productivity parameters in the rugged landscape ("matrix plots"). Each 18 of the 54 matrix plots were located at 1,000, 2,000 and 3,000 m elevation with each six plots being assigned to upper slope, mid-slope and lower slope positions to cover the variable topography of the rugged terrain. The intensive study plots # 1–5 ("core plots") were located on mid-slope positions. The biomass and soil C pools were inventoried in all plots. All stands were selected randomly in closed stands without larger gaps (>2 m) that met the above-mentioned criteria of elevation and slope position.

It is important to note that all C pool and C flux data presented here refer to the projected horizontal area, i.e. the original data were corrected for slope angle.

10.2.3 Methods for Determining Biomass and Soil C Pools and Components of Productivity

Table 10.2 lists the principal methods used to measure the above- and belowground biomass stocks and their C pools and the productivity components. For further details, see the publications listed in this table. A few details are given below for the photosynthesis measurements and the calculation of carbon gain.

A stand-level estimate of gross photosynthesis was obtained from measured A_{max}, leaf dark respiration and the Leaf Area Index (LAI) and incident global radiation data of the stands (Table 10.2). Because most of the canopy leaves are not exposed to full sunlight in dense tropical forests, we followed Mercado et al. (2006) who calculated the fraction of shaded leaves to account for 70–85 % of total leaf area in a mature tropical lowland forest in Brazil with an LAI of 5.7. For the dense stands at 1,050 and 1,890 m, we assumed that only 20 % of the leaf area is exposed to more or less full sunlight, while 80 % were assumed to be shade leaves. In the elfin forest at 3,060 m with a much smaller LAI, in contrast, we estimated that half of the leaves were sun leaves. We reduced the measured sun-canopy A_{max} figures by 35 % following Strauss-Debenedetti and Bazzaz (1996) and used these values to extrapolate to the assimilation of the shade canopy. To account for the effect of light limitation, we assumed that sun and shade leaves operated with light saturation during the sunshine hours but reached only $0.3 \times A_{max}$ in the remaining overcast or rainy hours (model 1). In a second approach (model 2), we assumed that

Table 10.2 Approaches and measuring conditions for investigating the biomass and productivity parameters in the 5 (3) intensive study plots (INT) and the 54 matrix plots (MAT)

Parameter measured	Plot type	Methods and measuring conditions	Replicates per plot	References
Live biomass				
Trunk and branch biomass	INT	Allometric equation (DHB, height, wood density) of Chave et al. (2005)	80 trees >10 cm DBH	Moser et al. (2008, 2011)
Trunk and branch biomass	MAT	Allometric equation (DHB, height, wood density) of Chave et al. (2005)	All trees ≥5 cm DBH in 0.04 ha-plots	
Liana biomass	MAT	Allometric equations (DHB), average of Lü et al. (2009), Schnitzer et al. (2006) and Sierra et al. (2007)	All liana stems ≥1 cm DBH in 0.04 ha-plots	
Biomass of epiphytes and ground vegetation	None	Literature data from tropical forests with similar structure (see Table 10.7)	None	Werner et al. (2012), Sierra et al. (2007), Gibbon et al. (2010), Vieira et al. (2011)
Standing leaf biomass	INT	Annual leaf litter production × mean leaf lifespan	12 litter traps, 254–666 marked leaves	Moser et al. (2007)
Leaf area index (1)	INT	LAI-2000 Plant Canopy Analyzer (Li-Cor)	Taken at 10 random locations	Moser et al. (2007)
Leaf area index (2)	INT	Leaf litter production + mean leaf lifespan + mean SLA	Data from 12 litter traps	Moser et al. (2007)
Fine root biomass (<2 mm in diameter)	INT	Soil coring to 30 cm depth (diameter: 3.5 cm); life/dead separation under microscope	20 locations per plot	Moser et al. (2010), Hertel and Leuschner (2002)
Coarse and large root biomass (>2 mm–30 cm in diameter)	INT	Excavation to 50 cm, live/dead separation; root stumps not covered	12–16 soil pits	Soethe et al. (2007)
Productivity				
Stem wood increment	INT	Dendrometer tapes (D1, UMS, Munich) read monthly	80 trees >10 cm DBH	Moser et al. (2011)
Stem wood increment	MAT	Repeated annual DBH measurements	All trees >10 cm in 0.04 ha-plot	

(continued)

Table 10.2 (continued)

Parameter measured	Plot type	Methods and measuring conditions	Replicates per plot	References
Leaf litter fall	INT	Litter traps (0.25 m^2) sampled every 3–6 weeks	12, randomly placed	Moser et al. (2007), Moser et al. (2011)
Leaf litter fall	MAT	Litter traps (0.36 m^2) sampled every 2–4 weeks	6, randomly placed	
Leaf lifespan	INT	Leaf survivorship curves of 10–15 understorey trees	254–666 leaves per plot	Moser et al. (2007)
Fine root production	INT	Minirhizotron observation to 40 cm depth, CI-600 root growth scanner (Washington, USA)	10 tubes per plot	Graefe et al. (2008a, b), Moser et al. (2010)
Coarse and large root growth	INT	Dendrometer tapes (D1, UMS) on roots >3 cm and <32 cm diameter, read every 3 months	20 root segments	Soethe et al. (2007)
Gross photosynthesis (trees)	INT	LI-6400 (Li-Cor), A_{max} at ambient T and [CO_2] of light-exposed leaves, 4–20 m tall trees (lower sun canopy), light response curves (see Table 10.3)	10–15 species per plot (62 species in total), each 3 leaves on 1 tree per species	Wittich et al. (2012), Zach (2008)
Leaf dark respiration (trees)	INT	LI-6400, sun-exposed leaves, 2–5 min acclimation to darkness, shade leaf respiration: Veneklaas and Poorter (1998)	10–15 species per plot (40 species in total), 1 tree per species	Wittich et al. (2012)
Stem and coarse root respiration	INT	6-chamber respiration system ANARESY 2 (Walz, Germany), LI-7000 analyser, branch respiration estimated	13–16 species (stem respiration) and 4–8 coarse roots per plot	Zach et al. (2008, 2010)
Soil respiration	INT	Closed chamber method (EGM-4 IRGA, PP systems, UK), correction for root decomposition	16 per plot, root trenching for estimating root respiration	Iost (2007)
Soil organic carbon	MAT	Organic layer + mineral soil to 50 cm depth, corrected for bulk density	Soil pits in close vicinity of the plots (1,000 m: 14, 2,000 m: 16 and 3,000 m: 12)	

For further details see the publications listed
DHB diameter at breast height, *SLA* specific leaf area

Table 10.3 Parameters characterising the radiation climate and photosynthetic capacity in the canopies of the plots # 1, 3 and 5 and estimated annual gross photosynthesis according to two different models (photosynthesis data after Wittich et al. unpubl. radiation data from Emck 2007)

Plot no.	1	3	5
Elevation [m]	1,050	1,890	3,060
LAI (litter production + leaf lifespan + SLA)	6.0	5.7	2.2
LAI (LAI-2000 measurement)	5.1	3.9	2.9
Proportion of LAI in sun canopy (estim.) [%]	20	20	50
Proportion of LAI in shade canopy (estim.) [%]	80	80	50
Mean global radiation[a] [W m^{-2}]	285[b]	348	360
Mean PPFD[a,c] [μmol photons m^{-2} s^{-1}]	610[b]	740	765
Sunshine duration [% of daytime period]	18	27	30
A_{max} of sun leaves [μmol CO_2 m^{-2} s^{-1}]	6.9 (±0.6)	8.6 (±0.5)	5.3 (±0.7)
Light compensation point of sun leaves [μmol photons m^{-2} s^{-1}]	601	696	620
A_{max} of shade leaves (estimate)[d] [μmol CO_2 m^{-2} s^{-1}]	c. 4.5	c. 5.9	c. 3.7
Leaf dark respiration in daylight hours[e] [Mg C ha^{-1} $year^{-1}$]	5.6	4.0	3.5
Gross photosynthesis (annual total)—Model 1[f] [Mg C ha^{-1} $year^{-1}$]	26.0	27.2	16.0
Gross photosynthesis (annual total)—Model 2[g] [Mg C ha^{-1} $year^{-1}$]	26.0	23.7	12.5

[a]Mean of daytime hours

[b]Meteorological station El Libano at 1,970 m above plot # 1

[c]Estimated by assuming that 1 mol photons is equivalent to 0.235 J and 50 % of incident global radiation is in the PhAR range

[d]Assuming an A_{max} reduction by 35 % relative to the sun leaves according to empirical data of Strauss-Debenedetii and Bazzaz (1996)

[e]Based on R_D means

[f]Assuming photosynthesis at light saturation during sunshine hours and a mean rate of 0.3 × A_{max} during overcast periods in all three stands

[g]Assuming photosynthesis at light saturation during sunshine hours and at mean rates of 0.3, 0.2 or 0.1 × A_{max} during overcast periods in the 1,050, 1,890 and 3,060 m stands, respectively

photosynthesis is, in addition to light limitation, temporarily reduced by unfavourable temperatures at higher elevations and thus set the reduction factor to 0.3, 0.2 or 0.1 × A_{max} in the stands at 1,050, 1,890 and 3,060 m, respectively. We added the measured leaf dark respiration to the sum of CO_2 net assimilation (12 h day^{-1}, 365 days) for obtaining annual gross photosynthesis (see Table 10.3). Clearly, the calculated daily and annual sums of carbon influx into the forest stands are only rough estimates, but they may be useful for characterising the altitudinal trend because two influential variables used for upscaling (A_{max} and LAI) were measured with quite a large effort.

To account for differences in dark respiration (R_D) of sun and shade leaves, we used the proportional difference in R_D measured by Veneklaas and Poorter (1998) in tropical tree seedlings grown either under high (1,000 μmol photons m^{-2} s^{-1}) or low light (100 μmol photons m^{-2} s^{-1}, 53 % lower). In an attempt to calculate annual totals of foliage dark respiration, we assumed that nighttime leaf respiration (R_N) is roughly equal to the measured daytime respiration rates (R_D). Such a rough approximation may be justified because, in these forests, temperatures at night are typically 5–7 K lower than the daytime temperatures when the R_D measurements were conducted. On the other hand, R_D may be only 50 % or less of R_N at equal temperatures due to refixation of CO_2 (Evans et al. 2004).

We did not measure branch and twig wood respiration (R_B) but added a term of the same size of wood respiration, thereby accounting for the observation that stem and branch wood respiration in forests may be equally large (e.g., Ryan et al. 1995). Cavaleri et al. (2006) found wood elements <10 cm in diameter to account for 70 % of the total wood CO_2 efflux in various plant functional groups (trees, lianas and palms) in a tropical lowland forest indicating that our assumed figure may even underestimate reality.

10.3 Results and Discussion

10.3.1 Carbon in Biomass and Soil

Data on C pools in biomass (above- and belowground) and soil organic matter are compiled for the five intensive study plots in Tables 10.4 and 10.5. In addition, Table 10.6 presents means, standard errors and ranges of aboveground biomass (and wood and litter production) for the 54 matrix plots at 1,000, 2,000 and 3,000 m elevation. The matrix plot data may help to place the results from the intensive plots in a broader context and to extrapolate from the plot to the landscape level. In the synthetic Table 10.7, data from the intensive plots and the matrix plots are combined to generate an as accurate as possible estimate of the carbon stored in all relevant biomass and soil fractions of South Ecuadorian pre-montane (~1,000 m a.s.l.), montane (~2,000 m) and upper montane forests (~3,000 m).

Accordingly, total aboveground C (AGC) markedly decreases from 1,000 to 3,000 m (from 128 to 70 Mg C ha^{-1}), while belowground C (BGC) seems to remain invariant with elevation or shows a slight increase (from 242 to 270 Mg C ha^{-1}; tree stumps are not included in these figures). With 190–240 Mg C ha^{-1}, the soil organic carbon (SOC) pool (determined to a mineral soil depth of 50 cm plus organic layer; Iost 2007) was much larger at all three elevations than SOC pools reported from lowland forests: For the 0–100 cm profile, roughly 100 Mg C ha^{-1} are a typical estimate for tropical soils (mainly under forest) in Central Africa (82–84 Mg C ha^{-1}; Batjes 2008), Amazonia (98 Mg C ha^{-1}, Batjes and Dijkshoorn 1999; 103 Mg C ha^{-1}, De Moraes et al. 1995) and Southeast Asia (102–124 Mg C ha^{-1}, Chen et al. 2005; 55 Mg C ha^{-1}, Yonekura et al. 2010). Given that our figures cover only the 0–50 cm profile, the montane forests in South Ecuador stored more than twice the SOC amount than lowland forest soils.

Table 10.4 Above- (AGB) and belowground biomass (BGB) and above- (ANPP) and belowground production fractions (BNPP) of the core study plots # 1–5 (after Moser et al. 2011)

	Dry mass					Carbon				
Plot no.	1	2	3	4	5	1	2	3	4	5
Elevation [m a.s.l.]	1,050	1,540	1,890	2,380	3,060	1,050	1,540	1,890	2,380	3,060
Biomass [Mg DM ha^{-1}]						Biomass [Mg C ha^{-1}]				
Tree leaves	5.9	6.8	6.6	6.9	4.8	2.9	3.5	3.4	3.5	2.4
Stem and branch wood	278.3	159.5	163.2	94.6	108.6	134.7	77.4	79.2	45.1	51.7
Total AGB	**284.1**	**166.3**	**169.8**	**101.5**	**113.4**	**137.6**	**80.8**	**82.5**	**48.6**	**54.1**
Coarse roots	29.4	30.7	19.9	32.9	51.9	14.2	14.9	9.7	15.7	24.7
Fine roots[a]	2.7	5.6	6.2	6.3	10.8	1.2	2.7	2.8	3.0	5.4
Total BGB	**32.1**	**36.3**	**26.1**	**39.2**	**62.8**	**15.5**	**17.6**	**12.5**	**18.7**	**33.1**
Total tree biomass	**316.2**	**202.7**	**195.9**	**140.7**	**176.1**	**153.1**	**98.4**	**95.0**	**67.3**	**87.3**
Production [Mg DM ha^{-1} year^{-1}]						Production [Mg C ha^{-1} year^{-1}]				
Tree leaves	5.05	5.06	4.96	2.64	1.79	2.54	2.57	2.51	1.34	0.92
Reproductive organs	0.91	0.43	0.36	0.11	0.07	0.45	0.21	0.18	0.05	0.03
Twigs	1.13	0.93	0.88	0.39	0.39	0.55	0.45	0.43	0.19	0.19
Epiphyte biomass	0.28	0.58	0.25	0.25	0.21	0.14	0.29	0.13	0.13	0.11
Bamboo biomass	0.00	0.00	1.55	0.12	0.20	0.00	0.00	0.78	0.06	0.10
Other fine litter compon.	0.41	0.34	0.34	0.14	0.04	0.21	0.17	0.17	0.07	0.02
Total fine litter	7.78	7.34	8.34	3.65	2.70	3.89	3.69	4.20	1.84	1.37
Stem and branch wood	2.82	1.58	0.87	0.22	0.14	1.37	0.77	0.42	0.11	0.07
Total ANPP	**10.60**	**8.92**	**9.21**	**3.87**	**2.84**	**5.26**	**4.46**	**4.62**	**1.95**	**1.44**
Coarse roots	0.17	0.79	0.23	0.23	0.89	0.08	0.38	0.11	0.11	0.42
Fine roots[b]	2.28	3.16	2.97	3.72	9.40 (4.42)	1.06	1.50	1.35	1.79	4.70 (2.21)
Total BNPP[b]	**2.45**	**3.95**	**3.20**	**3.95**	**10.29 (5.31)**	**1.14**	**1.88**	**1.46**	**1.90**	**5.12 (2.66)**
Total NPP[b]	**13.05**	**12.84**	**12.41**	**7.85**	**12.96 (8.15)**	**6.40**	**6.34**	**6.08**	**3.85**	**6.56 (4.10)**

The plots # 1, 3 and 5 are the intensive C flux study plots

[a]Data for the plots # 3, 4 and 5 after Leuschner et al. (2007)

[b]For plot # 5, extrapolated fine root production estimates are given in brackets

Table 10.5 Estimates of stand leaf biomass and leaf area index (LAI) for the five core study plots as obtained with each two different approaches (see text; from Moser et al. 2007)

	Plot no.	1	2	3	4	5
Approach	Elevation [m]	1,050	1,540	1,890	2,380	3,060
Stand leaf biomass [Mg C ha^{-1}]	Leaf litter production + leaf lifespan	3.4 (±0.2)^{a}	4.1 (±0.02)^{b}	5.0 (±0.4)^{b}*	2.6 (±0.2)^{c}*	1.8 (±0.2)^{d}
	LAI2000 + SLA data	2.9 (±0.06)^{a}	3.5 (±0.03)^{b}	3.3 (±0.1)^{b}	3.5 (±0.09)^{b}	2.4 (±0.2)^{a}
Leaf area index	Leaf litter production + leaf lifespan + SLA	6.0 (±0.4)^{a}	5.4 (±0.4)^{a}	5.7 (±0.5)^{a}*	2.8 (±0.2)^{b}*	2.2 (±0.2)^{c}
	LAI-2000	5.1 (±0.1)^{a}	4.6 (±0.1)^{b}	3.9 (±0.2)^{c}	3.6 (±0.1)^{c}	2.9 (±0.3)^{d}

Different letters indicate significant differences between the plots, *significant differences between the two approaches (one-way ANOVA, Tukey HSD test at $p < 0.05$)

Table 10.6 Some parameters characterising stand structure and productivity of the 54 matrix plots at the three elevations (18 plots per elevation level each)

Elevation [m]	1,000 (1,020–1270)	2,000 (1,910–2,090)	3,000 (2,800–2,900)
Stem density (DBH ≥ 5 cm) [ha^{-1}]	1,707 ± 111[a] (1,327–1,884)	2,486 ± 196[b] (1,915–2,973)	2,804 ± 320[b] (1,949–3,663)
Tree basal area (DBH ≥ 5 cm) [$m^2\ ha^{-1}$]	46 ± 6 (31–56)	48 ± 4 (43–56)	41 ± 2 (34–47)
Stem and branch wood biomass (DBH ≥ 5 cm) [Mg C ha^{-1}]	109 ± 9[a] (84–121)	104 ± 13[a] (70–113)	60 ± 4[b] (49–66)
Stem and branch wood biomass increment (DBH ≥ 10 cm) [Mg C ha^{-1} $year^{-1}$]	1.6 ± 0.1[a] (1.1–2.1)	1.3 ± 0.2[a] (0.7–1.6)	0.7 ± 0.1[b] (0.5–0.9)
Leaf litter production [Mg C ha^{-1} $year^{-1}$]	2.9 ± 0.2[a] (2.6–3.2)	3.4 ± 0.2[b] (2.8–4.0)	1.9 ± 0.1[c] (1.7–2.0)

The labels 1,000, 2,000 and 3,000 m stand for the elevation ranges indicated below. Given are means ± standard errors and the range of second to third quartile. C concentrations for wood biomass and leaf litter were taken from Table 10.4. Different letters indicate significant differences among elevations (one-way ANOVA, Tukey HSD test at $p \geq 0.05$)
DHB diameter at breast height

Assuming Amazonian lowland forests to store on average about 160 Mg C ha^{-1} in aboveground biomass (Malhi et al. 2006) and adding a root biomass estimate by assuming a root:shoot ratio of 0.235 for tropical moist forests (Mokany et al. 2006) plus 100 Mg C ha^{-1} for the SOC pool, an ecosystem C pool of about 300 Mg C ha^{-1} is obtained for lowland forests, which is less than the estimate of 319–369 Mg C ha^{-1} for the pre-montane to upper montane forests in Ecuador. It may well be that certain other tropical mountain forests contain even larger ecosystem C pools than those reported here since Raich et al. (2006) found SOC pools up to >400 Mg C ha^{-1} (0–100 cm, excluding surface litter), which is more than that found in the Ecuadorian mountain forests. However, lower SOC figures have also been reported (e.g. 118 Mg C ha^{-1} in Peruvian treeline forests, Zimmermann et al. 2010). One likely reason for high belowground/aboveground ratios in C storage in tropical mountain forests is low N (and perhaps P) availability in many montane and upper montane forests (Moser et al. 2011, Chap. 23). Thus, we conclude that the tropical mountain forests of this study represent similarly important, or even more important, carbon stores than tropical lowland forests despite the markedly smaller wood biomass at higher elevations.

10.3.2 Carbon Fluxes

Our estimates of gross primary production (GPP, i.e. NPP + R_{aut}, Table 10.8: lines 38 and 39) ranged between 25.5 and 14.1 Mg C ha^{-1} $year^{-1}$ for the stands at 1,050 and 1,890 m, respectively, which is markedly lower than the GPP mean given by

Table 10.7 Carbon pools [Mg C ha^{-1}] in biomass and soil in forests at 1,000, 2,000 and 3,000 m elevation. Values in parentheses (lines 3, 5 and 6) are estimates

Elevation [m]	1,000	2,000	3,000
Leaves[a]	2.9	3.4	2.4
Stem and branch wood[b]	109	104	61
Epiphytes[c]	(2.5)	2.1	(1.5)
Lianas	2.5	1.7	0.4
Ground vegetation[d,e,f]	(0.3)[d]	1.5[f]	(2.7)[e]
Coarse woody debris[d,e,g,h]	(2.7–19.1)[d,g]	4.3[h]	(1.8)[e]
Coarse roots[a]	14.2	9.7	24.7
Fine roots[a]	1.8	2.8	5.4
SOC[b,i]	226	191	241
Total	370.1	319.5	339.9
Total aboveground (AGC)	128.1 (35 %)	117.0 (37 %)	69.8 (21 %)
Total belowground (BGC)	242 (65 %)	203.5 (63 %)	270.1 (79 %)
C in biomass	133.2 (36 %)	125.2 (39 %)	98.1 (29 %)

[a]Intensive study plots # 1, 3 and 5
[b]Matrix plots; DBH $\geq$ 5 cm, means of $n = 18$ plots each
[c]Werner et al. (2012): data from Puerto Rico (pre-montane forest, 930–1,015 m a.s.l.), from the San Francisco (S Ecuador) study site (lower montane forest, 2,050–2,150 m a.s.l.) and from Costa Rica (upper montane forest, 2,900 m a.s.l.), carbon fraction in dry epiphyte biomass was estimated to be 0.45
[d]Sierra et al. (2007), data from Colombia (pre-montane forest, ~1,000 m a.s.l.): CWD > 2 cm diameter, ground vegetation included herbs and all woody plants with DBH < 1 cm
[e]Gibbon et al. (2010), data from Peru (upper montane forest, >3,000 m a.s.l.): CWD > 10 cm diameter, ground vegetations included shrubs and bamboo
[f]No data available, figures estimated
[g]Vieira et al. (2011): data from Brazil (montane Atlantic forest, 1,027–1,070 m a.s.l.): CWD > 2 cm diameter
[h]Wilcke et al. (2005): data from the San Francisco study site (S Ecuador, lower montane forest, 1,900–2,180 m a.s.l.): CWD > 10 cm diameter
[i]Mineral soil 0–50 cm depth plus organic layer

Luyssaert et al. (2007) for tropical humid evergreen forests at low elevations (35.51 ± 1.60 Mg C ha^{-1} $year^{-1}$, $n = 6$); our GPP figures are closer to the means given by these authors for temperate evergreen and deciduous forests (17.62 and 13.75 Mg C ha^{-1} $year^{-1}$). The annual total of gross photosynthesis (30.4–32.0 Mg C ha^{-1} $year^{-1}$) calculated by Malhi et al. (1999, 2012) for an Amazonian lowland forest is also considerably larger than the GPP value of our pre-montane and montane forests. One explanation for the comparatively low calculated gross primary production in the Ecuadorian mountain forests is that our NPP figure is an underestimate because it does not include all relevant components such as herbivory and root exudation. However, a probably equally important cause is the lower temperature in pre-montane and montane elevation (19 and 16 °C compared to 23–24 °C in the lowlands), which must result in a smaller GPP than in lowland forests due to lower autotrophic respiration rates. According to the data compilation of Luyssaert et al. (2007), tropical moist forests at low elevations have a mean autotrophic respiration rate of

Table 10.8 Summary table of C fluxes [Mg C ha^{-1} $year^{-1}$] in the three intensive study plots # 1, 3 and 5

Plot no.	1	3	5
Elevation [m]	1,050	1,890	3,060
Aboveground (AG)			
(1) Leaf area index [m^2 m^{-2}] (LAI)	5.1	3.9	2.9
(2) Sun leaf A_{max} [µmol CO_2 m^{-2} s^{-1}] (A_{max})	6.9	8.6	5.3
(3) Estimate of annual gross photosynthesis[a,b]	26.0	23.7	12.5
(4) Production of stem and branch wood (P_W)	1.37	0.42	0.07
(5) Production of tree leaves (P_L)	2.54	2.51	0.92
(6) Production of reproductive organs (P_{Rep})	0.45	0.18	0.03
(7) Production of twigs (P_{Tw})	0.55	0.43	0.19
(8) Production of epiphyte biomass (P_E)	0.14	0.13	0.11
(9) Production of bamboo biomass (P_{Ba})	0	0.78	0.10
(10) Other fine litter components (P_{oth})	0.21	0.17	0.02
(11) Total fine litter production (P_{FL})	3.89	4.20	1.37
(12) Dark respiration of tree foliage (R_D)	5.60	3.96	3.38
(13) Respiration of reproductive organs (R_{Rep})	n.d.	n.d.	n.d.
(14) Respiration of branches and twigs[c] (R_B)	1.86	1.20	0.43
(15) Respiration of stems (R_W)	1.86	1.20	0.43
(16) Tree AG NPP (4) + (5) + (6) + (7) ($ANPP_T$)	4.91	3.54	1.21
(17) Total AG NPP (16) + (8) + (9) + (10) ($ANPP_{tot}$)	5.26	4.62	1.44
(18) Total AG C efflux[d] (12) + (13) + (14) + (15)	9.32	6.36	4.24
(19) Total AG C consumption (17) + (18)	14.58	10.98	5.68
(20) C transfer to roots (39) − (17) − (18)	>10.90	>6.90	>3.52
Belowground (BG)			
(21) Production of fine roots (P_{FR})	1.06	1.35	2.21[e]
(22) Production of coarse and large roots (P_{CR})	0.08	0.11	0.42
(23) Respiration of fine roots[f] (R_{FR})	5.36	1.44	0.30
(24) Respiration of coarse and large roots (R_{CR})	0.46	0.22	0.42
(25) Tree BG NPP[g] (21) + (22) ($BNPP_T$)	1.14	1.46	2.63
(26) Root respiration (23) + (24) (R_R)	5.82	1.66	0.72
Soil			
(27) Total soil respiration (R_{soil})	13.03	9.32	3.83
(28) Heterotrophic respiration (27)−(26) (R_{het})	7.21	7.66	3.11
(29) Fine root litter production[h] (L_{FR})	1.06	1.35	2.21
(30) Coarse root litter production (L_{CR})	n.d.	n.d.	n.d.
(31) AG tree litter prod. (5) + (6) + (7) (L_T)	3.54	3.12	1.14
(32) Non-tree fine litter (8) + (9) + (10) (L_{NT})	0.35	1.08	0.13
(33) SOM change (29) + (30) + (31) + (32) − (28) (ΔSOM)	−2.26	−2.11	+0.37
(34) Total autotrophic respiration (18) + (26) (R_{aut})	15.14	8.02	4.96
(35) Ecosystem respiration (28) + (34) (R_{eco})	22.35	15.68	8.07
(36) Tree NPP (16) + (25) (NPP_T)	6.05	5.00	3.84
(37) NPP (17) + (25) (NPP)	6.40	6.08	4.07
(38) GPP (37) + (34) (GPP)	>21.54	>14.10	>9.03
(39) GPP ~ 1.14 × R_{eco}[i] (GPP_{est})	25.48	17.88	9.20
(40) NEP[j] (38) − (35) (NEP)	−0.81	−1.58	0.96

(continued)

Table 10.8 (continued)

Plot no.	1	3	5
Elevation [m]	1,050	1,890	3,060
(41) Percent NPP of GPP_{est} (37)/(39)[i]	25	34	44
(42) Percent P_W of NPP_T (4)/(36)	23	8	2
(43) Percent root transfer of GPP_{est} (20)/(39)	>43	>39	>38

n.d. not determined
[a]Only trees, does not include epiphytes, lianas and understorey
[b]Output of model 2 assuming photosynthesis at light saturation during sunshine hours and at mean rates of 0.3, 0.2 or 0.1 × A_{max} during overcast periods in the 1,050, 1,890 and 3,060 m stands
[c]Assumed to be equal to stem respiration
[d]Aboveground autotrophic respiration (only trees)
[e]Extrapolated from the trend line over the plots # 1–4 (see Table 10.4)
[f]Soil respiration in trenched plots corrected for root decomposition
[g]C transfer to mycorrhiza and root exudation not considered
[h]Assumed to equal fine root production
[i]Assuming a R_{eco}/GPP ratio of 0.88 for tropical moist forests (Luyssaert et al. 2007)
[j]Minimum values due to the underestimation of NPP and GPP; NEP figures based on GPP_{est} are larger and in all plots positive

c. 23.2 Mg C ha^{-1} $year^{-1}$, which is about 50 % larger than our R_{aut} estimate for the stand at 1,050 m with a 4–5 °C lower MAT. The significant temperature effect on gross primary production (GPP) is also visible in the much lower GPP value of the uppermost stand at 3,060 m, which apparently assimilated CO_2 at less than half the rate (c. 9.2 Mg C ha^{-1} $year^{-1}$) than did the stand at 1,050 m.

We estimated the gross photosynthesis of the tree canopies in its annual total using the A_{max}, R_D and LAI data of the stands. Extrapolating photosynthesis and leaf respiration to the stand level is sensitive to variation in LAI and thus to the method used for leaf area measurement (Cavaleri et al. 2006). We preferred the optical LAI figures (LAI-2000 system) for calculation, because the values derived for the 1,890 and 3,060 m stands appeared to be more realistic than the biomass-related LAI data that require an estimate of mean leaf longevity. The LAI estimate of 5.1 for the 1,050-m stand agrees well with leaf area indices of 5–6 reported from tropical lowland forests (Malhi et al. 1999; Grace and Meir 2009), given that we found a LAI decrease by roughly 1 unit per km elevation increase along the transect (see also Moser et al. 2007 and Unger et al. 2013).

While the absolute amount of CO_2 assimilated by the tree canopies must remain relatively uncertain (estimated at 12.5–26.0 Mg C ha^{-1} $year^{-1}$ in the three stands, Table 10.8: line 3), it is safe to conclude that gross photosynthesis is substantially reduced from 2,000 to 3,000 m elevation in this transect due to large decreases in LAI and photosynthetic capacity (Table 10.3).

We attempted to analyse the carbon balance of the three forests with a bottom-up approach, because it was not possible to conduct stand-level measurements of total CO_2 influx and efflux using eddy covariance or microclimatological gradient techniques in the mountainous terrain. This has the consequence that the upscaled growth and respiration figures cannot be checked against independent stand-level data and certain fluxes calculated in this study must therefore be treated as rough estimates. For example, the GPP calculation is likely biased by errors in the upscaling process from organ-level respiration measurements to the stand level (Zach et al. 2010). Further, no empirical data exist for the respiration of branches and twigs, which may have higher CO_2 efflux rates than stems of the same diameter (Cavaleri et al. 2006; Robertson et al. 2010). Similarly, our "bottom-up" approach of calculating NPP faces a number of shortcomings; a major uncertainty is the fine root production figure especially of the uppermost plot (# 5), which may be an overestimate given the low root respiration rates measured in this stand (compare Moser et al. 2010). The direct observation of fine root growth and death by the minirhizotron technique in the 3,060 m stand may have included periods in which a steady state of root production and mortality had not yet been reached, thus resulting in the overestimation of root production. Furthermore, extrapolating root growth from the topsoil to lower horizons might introduce an additional error. However, most other studies on root production in tropical forests used even less reliable approaches such as the ingrowth core method or indirect estimates of fine root production (Raich et al. 2006), which similarly questions the accuracy of NPP and GPP figures that have been used for calculating biome means (e.g. the database in Luyssaert et al. 2007). Similar to most other related investigations, we ignored a number of C consuming processes that are notoriously difficult to measure such as root exudation, C transfer to mycorrhizal hyphae, herbivory and the release of volatile organic compounds, which would increase our NPP figures.

A strength of our study is that the NPP, respiration and photosynthesis figures base on intensive measuring campaigns covering a large number of tree individuals and species, thus generating a more comprehensive data base than is typically available in carbon cycle studies in tropical forests. Moreover, the focus of this study was primarily on altitudinal trends in carbon cycle components and a relative comparison of the C balance of the stands. The data from the 54 additional matrix plots at 1,000, 2,000 and 3,000 m elevation are useful for validating altitudinal trends for a number of key carbon pool and flux parameters including aboveground biomass and wood production.

The data from the 54 matrix plots at variable slope positions show that the results from the three intensive study plots # 1, 3 and 5 give correct altitudinal trends for biomass and productivity along the transect (Table 10.6). However, the three plots are partly deviating from the landscape means of biomass and productivity as they appear from the matrix plots at 1,000, 2,000 and 3,000 m elevation. The stand # 1 possessed a higher than average wood biomass due to a relatively low number of stems, which had particularly large diameters. In contrast, the stands # 3 and 5 at 2,000 and 3,000 m had smaller than average wood biomasses and lower wood and leaf production figures, which is a consequence of a relatively low basal area in plot # 3 and a high density of small-diameter stems in plot # 5.

From the difference between GPP and the C consumption by aboveground tree organs (growth and respiration), we estimated a C transfer to the roots of c. 10.9–3.5 Mg C ha^{-1} $year^{-1}$ for the stands # 1–5, which equals 43–38 % of GPP (lines 20 and 43 in Table 10.8). The order of magnitude agrees well with data of Raich et al. (2006) on the estimated belowground C allocation in tropical evergreen forests growing at mean annual temperatures of ~10–20 °C (c. 1.8–10.0 Mg C ha^{-1} $year^{-1}$). Given that the estimated belowground transfers in Ecuador are minimum fluxes, the figures also are in accordance with the value (45 %) given by Malhi et al. (1999) for a tropical lowland forest. The estimated belowground C transfers in our three stands exceed the sum of measured root production and root respiration by 0.17 to >3.5 Mg C ha^{-1} $year^{-1}$. Possible causes for this discrepancy are errors in the calculation of autotrophic respiration at the stand level and the existence of quantitatively important carbohydrate fluxes from the roots to the rhizosphere, which are not included in our calculation.

10.3.3 Altitudinal Trends

We found several changes in C cycle components with increasing elevation that may be direct effects of the temperature decrease or are caused by other factors that also change with altitude. GPP, NPP and NEP showed large and continuous decreases from 1,050 m (19 °C) to 3,060 m (9 °C); we calculated a GPP decrease by about 1.25 Mg C ha^{-1} $year^{-1}$ and an NPP decrease by about 0.23 Mg C ha^{-1} $year^{-1}$ per K temperature decrease. The NPP reduction was particularly large between 2,000 and 3,000 m, i.e., with a decrease in mean temperature from 16 °C to 9 °C. In an altitudinal gradient in Peru, Girardin et al. (2010) found a relatively small NPP reduction between 1,000 and 3,000 m, but a drop to the half from 210 to 1,000 m elevation.

The results from tropical altitudinal transects indicate in general a positive relation between temperature and the NPP of tropical moist forests, thereby contradicting Luyssaert et al. (2007) who concluded that forest NPP does not increase at temperatures above 10 °C. However, their data from tropical forests showed a considerable scatter and included a few tropical seasonal forests with less than 1,500 mm rain per year. There is the possibility that a temperature effect on tropical forest NPP (above- and belowground) is only detectable in samples restricted to stands with defined moisture status as is the case in the moist mountain forests of this study. Moser et al. (2011) and Wittich et al. (2012) suggested that the main reason for the NPP decrease is N shortage at high elevations, which limits stand leaf area and A_{max} while reduced temperature should affect productivity mainly indirectly through a negative effect on N supply.

Marked decreases with elevation were also detected for autotrophic and heterotrophic respiration and all organ-specific respiration fluxes. While stem and root respiration on a tissue surface area basis decreased with elevation (see also Robertson et al. 2010), this was not the case with leaf dark respiration per leaf area, which remained invariant between 1,000 and 3,000 m; this is primarily a consequence of the large SLA decrease along the transect (Leuschner et al. unpubl.). The efflux of CO_2 from the soil (total soil respiration as the sum of root respiration and soil heterotrophic respiration) decreased to less than a third from

1,050 to 3,060 m in the intensive study plots and to the half in the 54 matrix plots. The upslope reduction in fine root respiration appeared to be even larger than the reduction in total soil respiration. Given the more than fourfold increase in standing fine root biomass between pre-montane and upper montane elevation, the decrease in fine root respiration implies that fine root relative growth rate and nutrient and water uptake must be much lower in the 3,060 m stand than under the higher temperatures at 1,050 m. It appears that the trees in the high-elevation elfin forest are maintaining a very large fine root system for compensating a low resource uptake and growth activity of their roots.

Leuschner et al. (2007), Hertel and Leuschner (2010) and Moser et al. (2011) reported a large shift in C allocation patterns from stem growth to root (mostly fine root) production in this transect between 1,050 and 3,060 m. Indeed, the proportion of stem growth in NPP dropped to a tenth of its value between 1,050 and 3,060 m, while the proportion of C invested in root production increased from 19 to 68 %. In the Peruvian elevation transect, the ratio of above- to belowground NPP stayed more or less constant between 210 and 3,000 m elevation (Girardin et al. 2010), which may point at different soil fertilities in the two Andean transects. A closer look with additional consideration of data on root respiration and relative belowground C transfer (lines 23, 24 and 20 of Table 10.8) shows no altitudinal trend in relative aboveground/belowground C allocation patterns in the Ecuadorian transect. Rather, a marked shift from root respiration to root biomass production occurred, i.e., from root metabolic activity to the production of root structures.

A large altitudinal decrease was also detected for the total amount of aboveground litter production, which reached at 3,060 m only a third of its low-elevation value (see Chap. 23) and was accompanied by a doubling of the amount of organic carbon stored in the organic layer and the mineral topsoil (0–30 cm). These opposing trends indicate a dramatic decrease in decomposition rate between 1,050 and 3,060 m elevation and consequently a rising mean carbon residence time in the soil with increasing altitude.

Our calculation of soil heterotrophic respiration shows that the C cycle in the soil appears not to be balanced; rather, it indicates a deficit of c. 2 Mg C ha^{-1} $year^{-1}$ in the input/output balance of soil organic matter (SOM) for the 1,050 m and 1,890 m stands. If this statement is valid, substantial C losses from the belowground compartment are occurring in the pre-montane and lower montane stands and NEP should take a negative value as is indicated by our NEP estimates derived from the difference between GPP and R_{eco}. Whether this apparent imbalance is indeed caused by SOM losses over longer time spans and the stands at 1,050 and 1,890 m are in fact functioning as CO_2 sources in recent time, or whether the deficit is the result of erroneous estimates of heterotrophic respiration (possible overestimation) or of fine and coarse root mortality (possible underestimation), can only be answered by repeated sampling of the SOM pools over one or two decades.

We conclude that temperature is influencing mountain forest GPP and NPP both directly and indirectly. The most influential temperature effects are probably the altitudinal decline in stem and root respiration rates (but less in leaf respiration), which may be linked to reduced growth rates, and the reduction in heterotrophic respiration in the soil, slowing down decomposition and mineralisation rates. Impaired N supply in the cold high-elevation forests seems to reduce annual carbon gain mainly through a restriction of leaf area expansion, which limits carbon gain.

10.4 Conclusions

Several results of this study are unexpected. First, the ecosystem C pool of the mountain forests was equally large, or even larger, than that of neotropical lowland forests, mainly due to the large C store in the soil. This finding highlights the outstanding role, tropical Andean mountain forests are playing with respect to C storage, an ecosystem service with increasing importance in the light of rising atmospheric carbon dioxide concentrations and temperatures (see Chap. 24). Second, total (above- and belowground) NPP decreased by about 0.23 Mg C ha^{-1} $year^{-1}$ per °C temperature decrease, which contradicts earlier assumptions that forest NPP is insensitive to temperature in tropical and subtropical regions. Marked decreases with elevation were also detected for the stand-level estimates of annual gross photosynthesis and autotrophic and heterotrophic respiration. While fine root production seems to increase with elevation, root respiration decreased, which indicates a marked shift from C investment in root metabolic activity to the production of root structures towards high elevations; we found no altitudinal trend in the relative aboveground/belowground C allocation patterns of the trees. We conclude that altitudinal gradient studies may represent a valuable tool for studying temperature effects on the functioning of tropical forests.

References

Batjes NH (2008) Mapping soil carbon stocks of Central Africa using SOTER. Geoderma 146:58–65

Batjes NH, Dijkshoorn JA (1999) Carbon and nitrogen stocks in the soil of the Amazon region. Geoderma 89:273–286

Bendix J, Rollenbeck R, Fabian P, Emck P (2008) Climate. In: Beck E, Bendix J, Kottke I, Makeschin F, Mosandl R (eds) Gradients in a tropical mountain ecosystem of Ecuador. Springer Verlag, Berlin, pp 63–74

Benner J, Vitousek PM, Ostertag R (2010) Nutrient cycling and nutrient limitation in tropical montane cloud forests. In: Bruijnzeel LA, Scatena FN, Hamilton LS (eds) Tropical montane cloud forests: science for conservation and management. Cambridge University Press, Cambridge, pp 90–100

Bruijnzeel LA, Kappelle M, Mulligan M, Scatena FN (2010) Tropical montane cloud forests: state of knowledge and sustainability perspectives in a changing world. In: Bruijnzeel LA, Scatena FN, Hamilton LS (eds) Tropical montane cloud forests: science for conservation and management. Cambridge University Press, Cambridge, pp 691–740

Cavaleri MA, Oberbauer SF, Ryan MG (2006) Wood CO_2 efflux in a primary tropical rain forest. Glob Change Biol 12:2442–2458

Chave J, Andalo C, Brown S, Cairns MA, Chambers JQ, Eamus D, Folster H, Fromard F, Higuchi N, Kira T, Lescure JP, Nelson BW, Ogawa H, Puig H, Riera B, Yamakura T (2005) Tree allometry and improved estimation of carbon stocks and balance in tropical forests. Oecologia 145:87–99

Chen G-S, Yang Y-S, Xie J-S, Guo J-F, Gao R, Qian W (2005) Conversion of a natural broad leafed evergreen forest into pure plantation forests in a subtropical area: effects on carbon storage. Ann For Sci 62:659–668

Clark DA, Brown S, Kicklighter DW, Chambers JQ, Thomlinson JR, Ni J (2001) Measuring net primary production in forests: concepts and field methods. Ecol Appl 11:356–370

Cleveland CC, Townsend AR, Taylor P, Alvarez-Clare S, Bustamante MMC, Chuyong G, Dobrowski SZ, Grierson P, Harms KE, Houlton BZ, Marklein A, Parton W, Porder S, Reed SC, Sierra CA, Silver WL, Tanner EVJ, Wieder WR (2011) Relationships among net primary productivity, nutrients and climate in tropical rain forest: a pan-tropical analysis. Ecol Lett 14:939–947

Davidson EA, de Araujo AC, Artaxo P, Balch JK, Brown IFC, Bustamante MM, Coe MT, DeFries RS, Keller M, Longo M, Munger JW, Schroeder W, Soares-Filho BS, Souza CM, Wofsy SC (2012) The Amazon basin in transition. Nature 481:321–328

De Moraes JF, Cerri CC, Melillo JM, Kicklighter D, Neill C, Skole DL, Steudler PA (1995) Soil carbon stocks of the Brazilian Amazon Basin. Soil Sci Soc Am J 59:244–247

Emck P (2007) A climatology of South Ecuador with special focus on the major Andean ridge as Atlantic-Pacific climate divide. PhD Thesis, University Erlangen, Germany

Evans JR, Terashima I, Hanba Y, Loreto F (2004) Chloroplast to leaf. In: Smith WK, Vogelmann TC, Critchley C (eds) Photosynthetic adaptation. Chloroplast to landscape. Ecological studies 178. Springer, New York, pp 107–132

Gentry AH (2001) Patrones de diversidad y composicion floristica en los bosques de las montanas neotropicales. In: Kappelle M, Brown AD (eds) Bosques nublados del neotropico. Editorial INBio, Santo Domingo de Heredia, Costa Rica, pp 85–123

Gibbon A, Silman M, Malhi Y, Fisher J, Meir P, Zimmermann M, Dargie G, Farfan W, Garcia K (2010) Ecosystem carbon storage across the grassland–forest transition in the high Andes of Manu National Park, Peru. Ecosystems 13:1097–1111

Girardin CAJ, Malhi Y, Aragao LEOC, Mamani M, Huasco WH, Durand L, Feeley KJ, Rapp J, Silva-Espejo JE, Silman M, Salinas N, Whittaker RJ (2010) Net primary productivity allocation and cycling of carbon along a tropical forest elevational transect in the Peruvian Andes. Glob Change Biol 16:3176–3192

Grace J, Meir P (2009) Tropical rain forests as old-growth forests. In: Wirth C, Gleixner G, Heimann M (eds) Old-growth forests. Function, fate and value. Ecological studies, vol 207. Springer, Berlin, pp 391–408

Grace J, Lloyd J, McIntyre J, Miranda A, Meir P, Miranda H, Moncrieff J, Massheder J, Wright I, Gash J (1995) Fluxes of carbon dioxide and water vapour over an undisturbed tropical forest in south-west Amazonia. Glob Change Biol 1:1–12

Graefe S, Hertel D, Leuschner C (2008a) Estimating fine root turnover in tropical forests along an elevational transect using minirhizotrons. Biotropica 40:536–542

Graefe S, Hertel D, Leuschner C (2008b) Fine root dynamics along a 2000-m elevation transect in South Ecuadorian mountain forests. Plant Soil 313:155–166

Hertel D, Leuschner C (2002) A comparison of four different fine root production estimates with ecosystem carbon balance data in a Fagus-Quercus mixed forest. Plant Soil 239:237–251

Hertel D, Leuschner C (2010) Fine root mass and fine root production in tropical moist forests as dependent on soil, climate, and elevation. In: Bruijnzeel LA, Scatena FN, Hamilton LS (eds) Tropical montane cloud forests: science for conservation and management. Cambridge University Press, Cambridge, pp 428–443

Homeier J, Werner FA, Gradstein SR, Breckle S-W, Richter M (2008) Potential vegetation and floristic composition of Andean forests in South Ecuador, with a focus on the RBSF. In: Beck E, Bendix J, Kottke I, Makeschin F, Mosandl R (eds) Gradients in a tropical mountain ecosystem of Ecuador. Ecological studies, vol 198. Springer, Berlin, pp 87–100

Iost S (2007) Soil respiration, microbial respiration and mineralisation in soils of montane rainforests of southern Ecuador: influence of altitude. PhD Thesis, Technical University of Dresden, Germany. 171 pp. Available at http://nbn-resolving.de/urn:nbn:de:bsz:14-ds-1201126765623-42870

Kitayama K, Aiba S-I (2002) Ecosystem structure and productivity of tropical rain forests along altitudinal gradients with contrasting soil phosphorus pools on Mount Kinabalu, Borneo. J Ecol 90:37–51

Leuschner C, Moser G, Bertsch C, Röderstein M, Hertel D (2007) Large altitudinal increase in tree root/shoot ratio in tropical mountain forests of Ecuador. Basic Appl Ecol 8:219–230

Litherland M, Aspden J, Jemielita R (1994) The metamorphic belts of Ecuador. No. 11 in Overseas Memoir of the British Geological Survey. British Geology Survey, Keyworth

Lü X-T, Tang J-W, Feng Z-L, Li M-H (2009) Diversity and aboveground biomass of lianas in the tropical seasonal rain forests of Xishuangbanna, SW China. Rev Biol Trop 57(1–2):211–222

Luyssaert S, Inglima I, Jung M, Richardson AD, Reichstein M, Papale D, Piao SL, Schulze ED, Wingate L, Matteucci G, Aragao L, Aubinet M, Beers C, Bernhoffer C, Black KG, Bonal D, Bonnefond JM, Chambers J, Ciais P, Cook B, Davis KJ, Dolman AJ, Gielen B, Goulden M, Grace J, Granier A, Grelle A, Griffis T, Grunwald T, Guidolotti G, Hanson PJ, Harding R, Hollinger DY, Hutyra LR, Kolar P, Kruijt B, Kutsch W, Lagergren F, Laurila T, Law BE, Le Maire G, Lindroth A, Loustau D, Malhi Y, Mateus J, Migliavacca M, Misson L, Montagnani L, Moncrieff J, Moors E, Munger JW, Nikinmaa E, Ollinger SV, Pita G, Rebmann C, Roupsard O, Saigusa N, Sanz MJ, Seufert G, Sierra C, Smith ML, Tang J, Valentini R, Vesala T, Janssens IA (2007) CO_2 balance of boreal, temperate, and tropical forests derived from a global database. Glob Change Biol 13:2509–2537

Malhi Y (2010) The carbon balance of tropical forest regions, 1990–2005. Curr Opin Environ Sustain 2:237–244

Malhi Y (2012) The productivity, metabolism and carbon cycle of tropical forest vegetation. J Ecol 100:65–75

Malhi Y, Baldocchi DD, Jarvis PG (1999) The carbon balance of tropical, temperate and boreal forests. Plant Cell Environ 22:715–740

Malhi Y, Baker TR, Phillips OL, Almeida S, Alvarez E, Arroyo L, Chave J, Czimczik CI, Di Fiore A, Higuchi N, Killeen TJ, Laurance SG, Laurance WF, Lewis SL, Montoya LMM, Monteagudo A, Neill DA, Nunez VP, Patino S, Pitman NCA, Quesada CA, Salomao R, Silva JNM, Torres LA, Vasquez MR, Terborgh J, Vinceti B, Lloyd J (2004) The above-ground coarse wood productivity of 104 Neotropical forest plots. Glob Change Biol 10:563–591

Malhi Y, Wood D, Baker TR, Wright J, Phillips OL, Cochrane T, Meir P, Chave J, Almeida S, Arroyo L, Higuchi N, Killeen TJ, Laurance SG, Laurance WF, Lewis SL, Monteagudo A, Neill DA, Vargas PN, Pitman NCA, Quesada CA, Salomao R, Silva JNM, Lezama AT, Terborgh J, Martinez RV, Vinceti B (2006) The regional variation of aboveground live biomass in old-growth Amazonian forests. Glob Change Biol 12:1107–1138

Malhi Y, Silman M, Salinas N, Bush M, Meir P, Saatchi S (2010) Introduction: elevation gradients in the tropics: laboratories for ecosystem ecology and global change research. Glob Change Biol 16:3171–3175

McJannet DL, Wallace JS, Reddell P (2010) Comparative water budgets of a lower and an upper montane cloud forest in the Wet Tropics of northern Australia. In: Bruijnzeel LA, Scatena FN, Hamilton LS (eds) Tropical montane cloud forests: science for conservation and management. Cambridge University Press, Cambridge, pp 479–490

Mercado L, Lloyd J, Carswell F, Malhi Y, Meir P, Nobre AD (2006) Modelling Amazonian forest eddy covariance data: a comparison of big leaf versus sun/shade models for the C-14 tower at Manaus. I. Canopy photosynthesis. Acta Amazon 36:69–82

Mokany K, Raison RJ, Prokushkin AS (2006) Critical analysis of root:shoot ratios in terrestrial biomes. Glob Change Biol 12:84–96

Moser G, Hertel D, Leuschner C (2007) Altitudinal change of leaf area and leaf mass in tropical mountain forests – a transect study in Ecuador and a pan-tropical meta-analysis. Ecosystems 10:24–935

Moser G, Röderstein M, Soethe N, Hertel D, Leuschner C (2008) Altitudinal changes in stand structure and biomass allocation of tropical mountain forests in relation to microclimate and soil chemistry. In: Beck E, Bendix J, Kottke I, Makeschin F, Mosandl R (eds) Gradients in a tropical mountain ecosystem of Ecuador. Ecological studies, vol 198. Springer, Berlin, pp 229–242

Moser G, Leuschner C, Röderstein M, Graefe S, Soethe N, Hertel D (2010) Biomass and productivity of fine and coarse roots in five tropical mountain forests stands along an altitudinal transect in southern Ecuador. Plant Ecol Divers 3:151–164

Moser G, Leuschner C, Hertel D, Graefe S, Soethe N, Iost S (2011) Elevation effects on the carbon budget of tropical mountain forests (S Ecuador): the role of the belowground compartment. Glob Change Biol 17:2211–2226

Raich JW, Russell AE, Kitayama K, Parton WJ, Vitousek PM (2006) Temperature influences carbon accumulation in moist tropical forests. Ecology 87:76–87

Robertson AL, Malhi Y, Farfan-Amezquita F, Aragao LEOC, Espejo JES, Robertson MA (2010) Stem respiration in tropical forests along an elevation gradient in the Amazon and Andes. Glob Change Biol 16:3193–3204

Ryan MG, Gower ST, Hubbard RM, Waring RH, Gholz HL, Cropper WP, Running SW (1995) Woody tissue maintenance respiration of four conifers in contrasting climates. Oecologia 101:133–140

Schnitzer SA, DeWalt SJ, Chave J (2006) Censusing and measuring lianas: a quantitative comparison of the common methods. Biotropica 38(5):581–591

Sierra CA, del Valle JI, Orrego SA, Moreno FH, Harmon ME, Zapata M, Colorado GJ, Herrera MA, Lara W, Restrepo DE, Berrouet LM, Loaiza LM, Benjumea JF (2007) Total carbon stocks in a tropical forest landscape of the Porce region, Colombia. For Ecol Manage 243:299–306

Slik JWF, Aiba S-I, Brearley FQ, Cannon CH, Forshed O, Kitayama K, Nagamasu H, Nilus R, Payne J, Paoli G, Poulsen AD, Raes N, Sheil D, Sidiyasa K, Suzuki E, van Valkenburg JLCH (2010) Environmental correlates of tree biomass, basal area, wood specific gravity and stem density gradients in Borneo's tropical forests. Glob Ecol Biogeogr 19:50–60

Soethe N, Lehmann J, Engels C (2007) Carbon and nutrient stocks in roots of forests at different altitudes in the Ecuadorian Andes. J Trop Ecol 23:319–328

Strauss-Debenedetti S, Bazzaz F (1996) Photosynthetic characteristics of tropical trees along successional gradients. In: Mulkey SS, Chazdon RL, Smith AP (eds) Tropical forest plant ecophysiology. Chapmann & Hall, New York, pp 162–186

Townsend AR, Cleveland CC, Cory C et al (2011) Multi-element regulation of the tropical forest carbon cycle. Front Ecol Environ 9:9–17

Unger M, Homeier J, Leuschner C (2013) Relationships among leaf area index, below-canopy light availability and tree diversity along a transect from tropical lowland to montane forests in NE Ecuador. Trop Ecol 54(1):33–45

Veneklaas EJ, Poorter L (1998) Growth and carbon partitioning of tropical tree seedlings in contrasting light environments. In: Lambers H, Poorter H, van Vuuren MMI (eds) Inherent variation in plant growth. Physiological mechanisms and ecological consequences. Backhuys, Leiden, pp 337–361

Vieira SA, Alves LF, Duarte-Neto PJ, Martins SC, Veiga LG, Scaranello MA, Picollo M, Camargo PB, do Carmo JB, Sousa Neto E, Santos FAM, Joly CA, Martinelli LA (2011) Stocks of carbon and nitrogen and partitioning between above- and belowground pools in the Brazilian coastal Atlantic Forest elevation range. Ecol Evol 1:421–434

Werner FA, Homeier J, Oesker M, Boy J (2012) Epiphytic biomass of a tropical Andean forest varies with topography. J Trop Ecol 28:23–31

Wilcke W, Hess T, Bengel C, Homeier J, Valarezo C, Zech W (2005) Coarse woody debris in a montane forest in Ecuador: mass, C and nutrient stock, and turnover. For Ecol Manage 205:139–147

Wittich B, Horna V, Homeier J, Leuschner C (2012) Altitudinal change in the photosynthetic capacity of tropical trees: a case study from Ecuador and a pantropical literature analysis. Ecosystems 15:958–973

Wolf K, Veldkamp E, Homeier J, Martinson GO (2011) Nitrogen availability links forest productivity, soil nitrous oxide and nitric oxide fluxes of a tropical montane forest in southern Ecuador. Global Biogeochem Cycles 25, GB4009

Yonekura Y, Ohta S, Kiyono Y, Ahsa D, Morisada K, Tanaka N, Kanzaki M (2010) Changes in soil carbon stocks after deforestation and subsequent establishment of "Imperata" grassland in the Asian humid tropics. Plant Soil 329:495–507

Zach A (2008) Carbon release from woody parts of trees along an elevation gradient in a tropical montane moist forest of southern Ecuador. PhD Thesis, University of Göttingen, Germany. 135 pp. Available at http://webdoc.sub.gwdg.de/diss/2008/zach/

Zach A, Horna V, Leuschner C (2008) Elevational change in woody tissue CO_2 efflux in a tropical mountain rain forest in southern Ecuador. Tree Physiol 28:67–74

Zach A, Horna V, Leuschner C (2010) Patterns of wood carbon dioxide efflux across a 2000-m elevation transect in an Andean moist forest. Oecologia 162:127–137

Zimmermann M, Meir P, Silman MR, Fedders A, Gibbon A, Malhi Y, Urrego DH, Bush MB, Feeley KJ, Garcia KC, Dargie GC, Farfan WR, Goetz BP, Johnson WT, Kline KM, Modi AT, Rurau NMQ, Staudt BT, Zamora F (2010) No differences in soil carbon stocks across the tree line in the Peruvian Andes. Ecosystems 13:62–74

Chapter 11
Current Regulating and Supporting Services: Nutrient Cycles

Wolfgang Wilcke, Jens Boy, Ute Hamer, Karin Potthast, Rütger Rollenbeck, and Carlos Valarezo

11.1 Introduction

Essential mineral nutrients including N, P, K, Ca, and Mg are the basis for any plant growth (Marschner 1995). Usually, there are two major nutrient sources for native old-growth forests: the weathering of soil and its parent material and the atmosphere from which N enters the forest by microbial fixation (frequently in symbiosis with plants) and all nutrients via wet or dry deposition (Likens and Bormann 1995). Depending on the needs of the vegetation and microorganisms, the released and deposited nutrients are taken up and incorporated into the internal nutrient cycle via fine and coarse litterfall, throughfall, and stemflow or exported via the stream, by erosion, or in gaseous form. The deposition of some nutrients may be associated with acids (e.g., HNO_3 released from forest fires, Boy et al. 2008a). Furthermore, the deposition of NH_4^+ can indirectly acidify the soil if NH_4^+ is nitrified and the resulting NO_3^- leached (Matson et al. 1999). Soil acidification usually depletes the buffer capacity and the base metal pool and thus has a direct effect on nutrient

W. Wilcke (✉)
Geographic Institute of the University of Bern (GIUB), University of Bern, Hallerstrasse 12, 3012 Bern, Switzerland
e-mail: wolfgang.wilcke@giub.unibe.ch

J. Boy
Institute of Soil Science, Leibniz University Hannover, Herrenhäuser Str. 2, 30419 Hannover, Germany

U. Hamer • K. Potthast
Institute of Soil Science and Site Ecology, Dresden University of Technology, Pienner Str. 19, 01737 Tharandt, Germany

R. Rollenbeck
Phillips University Marburg, Deutschhausstr. 10, 35032 Marburg, Germany

C. Valarezo
Dirección General de Investigaciones, Universidad Nacional de Loja, Loja, Ecuador

J. Bendix et al. (eds.), *Ecosystem Services, Biodiversity and Environmental Change in a Tropical Mountain Ecosystem of South Ecuador*, Ecological Studies 221,
DOI 10.1007/978-3-642-38137-9_11, © Springer-Verlag Berlin Heidelberg 2013

cycling (Likens and Bormann 1995). The regulation of the nutrient cycles is seen as a "supporting ecosystem service" of the study forest by the Millennium Ecosystem Assessment (2005) which will respond to expected environmental changes in the study area including climate change, increasing nutrient deposition, and land-use change (cf. Chap. 4 and Fig. 4.1).

While tropical lowland forests may grow on soils which are extremely depleted or devoid of weatherable minerals and therefore cannot obtain nutrients from weathering (Jordan 1985) tropical mountain forests usually grow on geologically younger parent materials and soils still supply plant nutrients via weathering (Roman et al. 2010). Furthermore, in contrast to most tropical lowland forests, the mountain forest soils are covered by thick organic layers harboring large stocks of all mineral nutrients (Tanner et al. 1998; Wilcke et al. 2002; Roman et al. 2010) which are released by mineralization of the organic matter. In spite of low mineralization rates as indicated by the accumulation of thick organic layers, at our study site, there is still a high release of N (ca. 600–800 kg ha^{-1} a^{-1}) and P (ca. 30–70 kg ha^{-1} a^{-1}) which is roughly 3–6 times the N and P fluxes with litterfall suggesting that a general N and P limitation is not likely (Wilcke et al. 2002). Nevertheless, N and P additions in the Nutrient Management Experiment (NUMEX) resulted in various responses of the vegetation (cf. Chap. 23 and Fig. 23.2) and N and P were almost completely retained in the system (Wullaert et al. 2010).

Soil fertility decreases with increasing altitude mainly because of increasingly anaerobic conditions and decreasing temperatures resulting in even thicker, peat-like organic matter accumulations on top of the mineral soil (Marrs et al. 1988; Grieve et al. 1990; Wilcke et al. 2008a, b). Therefore, soil moisture regime is a major control of soil fertility in tropical mountain forests (Benner et al. 2010). Furthermore, nutrient export via stream water is related with depth of water percolation in soil determined by the soil water content at the time of a rainfall event and size of the event (Boy et al. 2008b).

An important natural component of the ecosystem dynamics in the study area are usually shallow landslides removing all biomass and the organic layers including all nutrients contained therein which are deposited in the foot area of the landslide and released by mineralization (Wilcke et al. 2003; Richter et al. 2008, cf. Chap. 12). Thus, nutrients are not lost from the system as a whole but spatially redistributed unless the slides end directly in the draining river of the study area.

By the replacement of the native forest with pastures—which is common in the study region (see Fig. 2.2)—particularly nutrients in the standing biomass and the whole organic layer are removed. The conversion is frequently realized by slash and burn practices resulting in huge nutrient losses to the atmosphere and by subsequent erosion. The alkaline ashes remaining after forest burning increase the pH of the upper mineral soil and the availability of P, Ca, Mg, and N in the short term (Makeschin et al. 2008). This increased nutrient availability favors the growth of the pasture grass which is planted directly after burning. As a consequence of the dense grass roots, the mineral topsoil is enriched in organic matter (Potthast et al. 2012). The pastures additionally receive direct nutrient inputs via the frequent intentional vegetation burning and the feces of the animals, while fertilizer application is not common in the study region. Furthermore, hydrological flow conditions are changed resulting in more superficial water flow and erosion.

The plant and microbial communities of native old-growth forests are usually adapted to the local nutrient supply and any changes of the nutrient status of the ecosystems can translate into changed community compositions (Wassen et al. 2005; Potthast et al. 2012). In particular, nutrient inputs threaten the biodiversity (Sala et al. 2000; Wassen et al. 2005). If deposited nutrients are not retained in the ecosystem, they will reach ground and surface waters which might deteriorate the water quality and thus an important provisioning ecosystem service (Millennium Ecosystem Assessment 2005, Chap. 4). The latter could also occur if the precipitation regime changed towards a higher frequency of strong rain events resulting in increased fast nutrient export via near-surface flow (Goller et al. 2005; Boy et al. 2008b).

Thus, the regulatory function of the study forest with respect to nutrient cycles as supporting ecosystem service consists of the degree of nutrient retention in the internal cycle between organic layer and forest canopy. Nutrient retention is driven by the biological demand for nutrients which increases with increasing above- and belowground biomass production. Biomass production, another provisioning ecosystem service (Millennium Ecosystem Assessment 2005, cf. Chap. 4 and Fig. 4.1), is in turn driven by environmental conditions of which nutrient availability and climate (which are not entirely independent from each other) are major components. Nutrient availability depends on size of nutrient deposition, extent of acidification governing microbial activity, turnover times of the organic layers, and water flow regime in soil. Interferences in the old-growth forest or complete conversion to pasture likely change the nutrient availability. In this chapter, we focus on (1) deposition from the atmosphere, (2) soil moisture regime, and (3) conversion of forest into pasture because these drivers of the nutrient cycle are particularly prone to environmental change in the near future.

11.2 Methods

Our main long-term observation site is an approx. 9 ha-large micro-catchment between 1,900 and 2,200 m a.s.l. on 30–50° steep slopes on the north-facing flank of the San Francisco valley (MC2, see Fig. 1.2). The underlying bedrock is interbedding of paleozoic phyllites, quartzites, and metasandstones, from which mainly Cambisols and at higher positions Histosols developed (IUSS Working Group WRB 2006). To measure incident precipitation (on clearings near MC2), throughfall, stemflow, and stream flow, we used standard devices described in detail in Fleischbein et al. (2005, 2006). Briefly, incident precipitation and throughfall were measured with Hellmann-type collectors (five samplers at 2–4 incident precipitation gaging sites and 60 throughfall samplers at three throughfall gaging sites in MC2), stemflow with a polyurethane foam collar around five representative large trees, and surface flow with a manually calibrated V-shaped weir at the outlet of MC2. Measurements were made in at least weekly intervals since April 1998. Annual surface water flow was modeled between 1998 and 2003 (Fleischbein et al. 2006; Boy et al. 2008b) and between 2003 and 2008 estimated by assuming that the water level measured manually once or twice weekly is representative for the whole week preceding the measurement. This generated a small error with respect to the water budget because of only a few strong rainfall events (Boy et al. 2008b). Soil solution was collected below the organic layer and at the 0.15 and 0.30 m depths in the mineral soil with three lysimeters at each depth of each of the three measurement sites where also throughfall was collected. All solutions were bulked to a composite sample per measurement site prior to chemical characterization. Soil matric potentials were measured with three replicate tensiometers at each of the 0.15 and 0.30 m soil depths which were manually read 1–2 times per week. The deposition measurements in MC2 were complemented by deposition measurements at 2,270, 2,660, and 3,180 m a.s.l. with similar rain totalisators like in MC2 and additional mesh–grid fog collectors (Schemenauer and Cereceda 1994).

To determine deposition rates and concentrations and fluxes of elements in rainfall, throughfall, and stemflow, we multiplied water fluxes with volume-weighted mean concentrations determined by Continuous Flow Analysis (NH_4^+, NO_3^-, total N, PO_4^{3-}, total P, and Cl^-) and Atomic Absorption Spectrometry (K, Ca, Mg). The dry deposition was estimated with the canopy budget model of Ulrich (1983) using Cl^- as assumed inert tracer.

The pasture study sites are located on the opposite valley slope of MC2 and about 4 km east of MC2 on the same main ridge (see Fig. 1.2). At these sites, pH, concentrations of exchangeable base metals (Ca, Mg, K, Na) and NH_4F-extractable P, and stocks of C, N, P, and S in the upper 0–0.2 m of the mineral soil were measured with standard methods. Furthermore, microbial biomass, community composition of the microorganisms, and soil respiration rates were determined (Iost 2007; Potthast et al. 2012).

11.3 Results and Discussion

11.3.1 Deposition from the Atmosphere

In the study area, there is a seasonal cycle of acid and N inputs which can be attributed to the forest fires in the Amazon basin (Boy et al. 2008a, Fig. 11.1a). The acid deposition depleted base metals from all functional strata. Together with the nitrification of deposited NH_4^+, this acidified the main rooting zone (i.e., the soil organic layer). Base metals were partly replenished by an overseasonal cycle of elevated Ca and Mg deposition which was attributed to Sahara dust transport across the entire Amazon basin with the dominating northeasterly trade winds during strong La Niña conditions (Boy and Wilcke 2008). La Niña enforces the

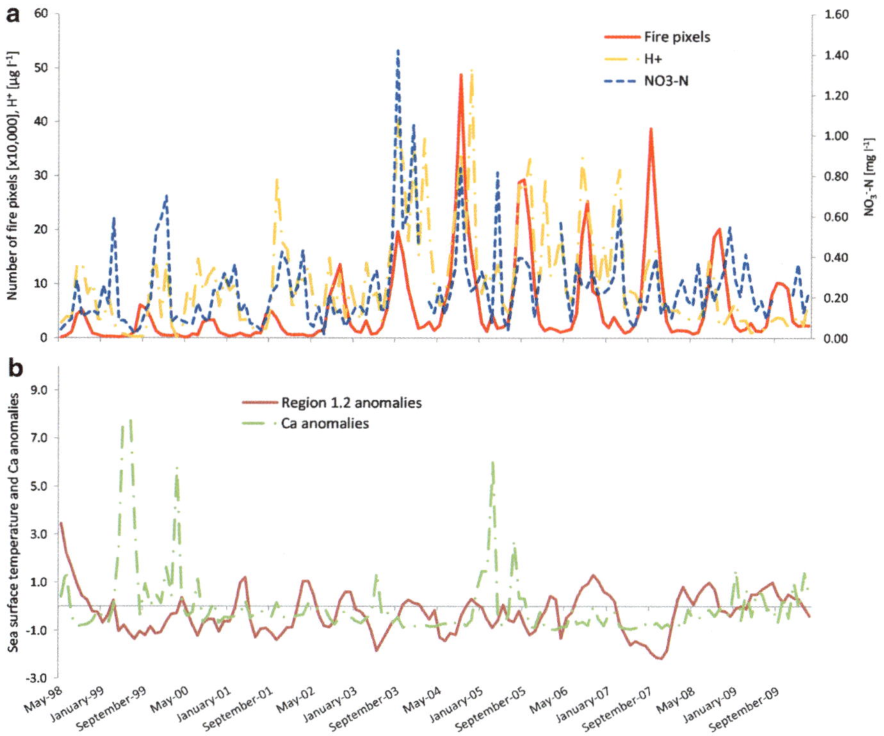

Fig. 11.1 Course of (**a**) monthly fire pixel counts of NOAA 12 (National Oceanic and Atmospheric Administration) in Amazonia and northern South America along the daily Hysplit trajectories (radius 25 km around the actual parcel location) and of the monthly concentrations of H^+ and NO_3^-–N in bulk incident precipitation from 1998 to 2010 and (**b**) the monthly sea surface temperate (SST) anomalies in the Pacific in front of the Ecuadorian coast and anomalies of Ca deposition (i.e., monthly deposition divided by overall monthly mean). In Boy and Wilcke (2008), a relationship between the SST anomalies in region 3.4 and Ca concentrations in rainfall was shown. However, particularly for the last 3 years the coincidence between SST anomalies of region 1.2 and anomalies of Ca deposition was more evident

trade-wind system and shifts the intertropical convergence zone southwards, channeling a larger portion of Saharan aerosols to the Amazon basin which otherwise would be deposited in the Caribbean (Kaufman et al. 2005). Although base metal-rich Saharan aerosols were found to reach the Amazon basin before (Swap et al. 1992) a transport to far windward sections of the wet Amazon basin seemed unlikely because of high scavenging probability along the trajectories. Despite this, we found La Niña increasing the frequency of storm events over Amazonia without yielding considerably higher monthly rainfall volumes. The resulting dry spells allowed for aerosol transport to the outer rim of the Amazon basin, suggesting a far higher impact to forests further upwind the dust passage in spite of cumulatively wetter conditions than previously assumed (Boy and Wilcke 2008, Fig. 11.1b). Thus, there were changing acid and base depositions as reflected

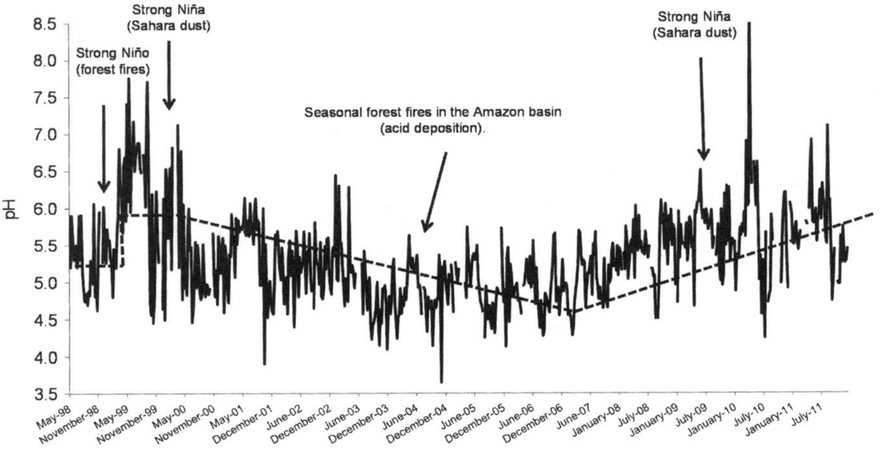

Fig. 11.2 Course of weekly mean pH (calculated from H^+ activities) in rainfall at the 2–4 gaging stations in the study area (see Fig. 1.2) from 1998 to 2011 and suggested interpretation of the temporal development of the pH

by the pH of rainfall (Fig. 11.2). As both forest fires and the El Niño Southern Oscillation (ENSO) cycle are prone to changes because of climate change, it can be expected that there will be a changed relationship between acid and base deposition.

Deposition showed an altitudinal gradient in the study area following increasing precipitation with altitude (Bendix et al. 2008). Therefore N, P, and K depositions increased strongly with altitude (Rollenbeck et al. 2008). However, independent of the higher precipitation at higher altitudes, typical byproducts of biomass burning like SO_4^{2+}, NH_4^+, and NO_3^- also showed elevated concentrations in water of light rainfall and fog at higher altitudes. This further supported the hypothesis of long-range transport of forest fire emissions by clouds from the Amazon basin. Consequently, a potentially increasing fire frequency in Amazonia would affect nutrient cycling of the Andean mountain forest (cf. Sect. 21.3.2 for other expected changes of deposition).

In spite of the high N availability in soil, results of a fertilization experiment in the study area (NUMEX, cf. Chap. 23 and see Fig. 1.2) revealed that up to 15 % of the applied 75 kg ha^{-1} N during the first 1.5 years of the experiment was cumulatively recycled with litterfall (Wullaert et al. 2010). Therefore, N deposition represents a remote fertilization of the study forest although it is still unclear if N causes a direct growth effect or if the growth effect results from the deposition of another nutrient and simultaneously increased N uptake to balance nutrient requirements (Boy and Wilcke 2008). The deposited reactive N can readily be used by the plants. Nevertheless, export of NO_3^- with litter leachate increased significantly during the first 10 years of our observation period suggesting that not the entire pool of available N is used by the organism community in the studied ecosystem (Fig. 11.3).

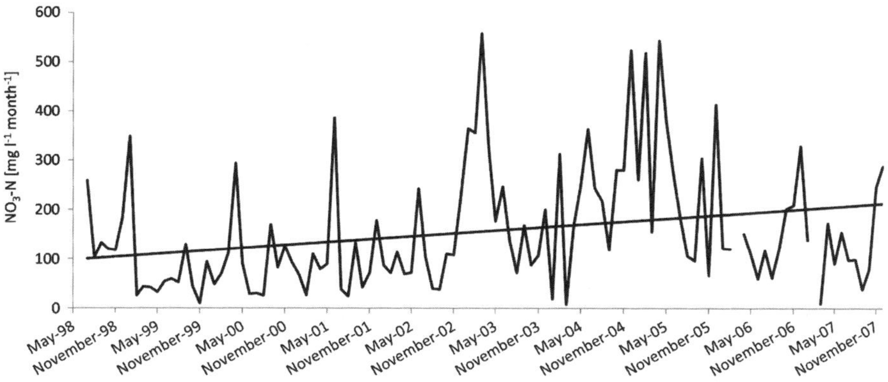

Fig. 11.3 Course of monthly NO_3^-–N export with litter leachate from 1998 to 2007. The regression line is shown for illustration purposes. The trend is significant according to the Seasonal Mann-Kendall test ($\tau = 0.16$, $p = 0.031$; Hirsch et al. 1982)

11.3.2 Soil Moisture Regime

Between 1998 and 2010, we observed increasingly dry conditions in the soils of the study site as illustrated by decreasing frequencies of observed soil water saturation by tensiometry (Fig. 11.4). The decrease was more pronounced in the topsoil than in the subsoil. Increasingly dry conditions will favor the aeration of the soil likely resulting in accelerated organic matter turnover times and thus increased nutrient release. It is frequently hypothesized that slow nutrient release (because of reduced microbial activity as a consequence of persisting water logging) is a main reason for limited nutrient supply in tropical mountain forests (Benner et al. 2010). This is reflected by the increasing thickness of the organic layer with increasing altitude and associated precipitation while temperature decreases (Wilcke et al. 2008a, b). As a consequence, more N might be released by mineralization which likely contributes to explain the increasing NO_3^- export from the organic layer (Fig. 11.3), where most roots are located, to the mineral soil. On the other hand, gaseous N losses by denitrification might be reduced if soil conditions become more oxidizing.

The work of Boy et al. (2008b) has shown that the nutrient concentrations in stream water depend on depth of water percolation through the soil. The concentrations of N, K, Ca, and Mg were highest during peak flow associated with fast near-surface lateral flow because of the accumulation of these nutrients near the surface of the soil, mainly in the organic layer. In contrast, highest P concentrations occurred during baseflow conditions because most P was stored in the deeper mineral soil. Near-surface lateral flow in soil was the most important export path of N and also accounted for about 1/3 of the export of base metals (K, Ca, Mg). Thus, any climate change resulting in higher frequencies of storm flow events causing near-surface flow will likely enhance the loss of N and base metals.

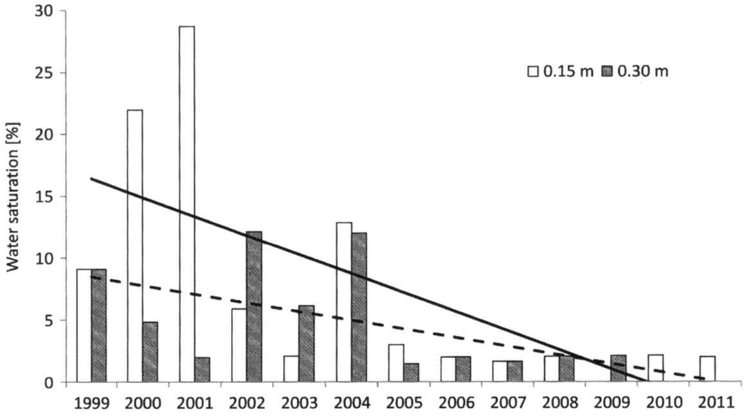

Fig. 11.4 Frequency of matric potentials of 0 indicating water-saturated soil [in % of total number of measurements], 11–102 measurements per soil depth and year at 3–5 stations at the 0.15 and 0.30 m mineral soil depths from 1999 to 2011. The regression lines are shown for illustration purposes

11.3.3 Conversion of Forest into Pasture

Land-use changes have significant impacts on the soil system, especially on rates of soil nutrient cycling (McGrath et al. 2001). In our study area, conversion of tropical mountain forest to active pasture resulted in the loss of the soil organic layer but induced an increase in the stocks of organic C and total N of the upper mineral soil of 50-year-old pastures (Makeschin et al. 2008). Stocks of total P and S were higher in younger pastures (17-year-old, own unpublished results). On the established pastures, the upper mineral soil (0–20 cm depth) plays the most important role in terms of nutrient cycling and is characterized by increased nutrient cycling rates compared with forest (Potthast et al. 2012).

Soil chemical properties and soil microorganisms are sensitive indicators of nutrient cycling rates (Swift et al. 1979; Six et al. 2006). The input of alkaline ashes from forest burning induced a significant increase in the pH value and in the exchangeable calcium of the mineral topsoil (Makeschin et al. 2008; Potthast et al. 2012). Both have been identified as important drivers of the observed changes in soil microbial community composition (Potthast et al. 2012). In addition, a huge stock of fine roots of the pasture grass *Setaria sphacelata* (Schumach.) Stapf & C. E. Hubb. ex M. B. Moss together with a high-quality, N-rich litter represented beneficial conditions for soil microbes (Potthast et al. 2012). The enhanced substrate and nutrient availability for soil microbes favored microbial growth and microbial activity in pasture compared to forest soils leading to faster cycling rates of organic matter. In comparison to forest litter (Iost 2007; Illig et al. 2008) grass litter showed a 2.5 times faster decomposition rate (Potthast et al. 2012). In situ measurements of soil respiration rates revealed that the pasture soil emitted approximately 2 Mg ha^{-1} a^{-1} more CO_2–C (Potthast et al. 2012) than the forest soil (Iost et al. 2008). The amounts of soil N supplied to plants and microbes increased following pasture establishment as indicated by increased rates of gross N mineralization and gross NH_4^+ consumption in the mineral soil (Potthast et al. 2012). Although the NH_4F-extractable organic P pool increased, too, in active pastures compared to forest (own unpublished results), the pasture grass *S. sphacelata* is still N- and P-limited (cf. Chap. 23). This is one reason for the degradation and subsequent abandonment of pastures (cf. Chap. 15).

11.4 Conclusions

Major external drivers of the supporting ecosystem service "nutrient cycling" in the south Ecuadorian tropical mountain forest are nutrient and acid deposition, the soil moisture regime, and land-use change. Acid and nutrient deposition is driven by two different climate-related sources: vegetation fires in the Amazon and Sahara dust deposition coupled to the ENSO cycle. Furthermore, it is expected that nutrient deposition, particularly of N generally increases in the tropics because of economic growth including agricultural intensification with higher mineral fertilizer use (Galloway et al. 2004, 2008). The soil moisture regime which is directly related to amount and distribution of precipitation controls oxidation processes such as organic matter mineralization. Thus, increasingly dry conditions—as observed in the last decade—will enhance organic matter turnover releasing CO_2 which feeds back on the climate system while improving the nutrient supply to plants and stimulating nutrient export to below the main rooting zone. The soil moisture regime also drives nutrient cycles via its control of the depth of water flow paths through soil which depends on soil water content and intensity of rainfall. While the acid and nutrient deposition will affect the whole altitudinal range because the deposition reaches our study area via long-range transport in the atmosphere, the increasing dryness will—at least for some time—be likely restricted to the lower altitudes. Land-use change from forest to pasture drives nutrient cycling mainly by the introduction of alkaline ashes and the input of grass-derived organic matter characterized by its higher susceptibility to microbial decomposition. The resulting increases in soil pH, nutrient, and substrate supply in the upper mineral soil not only changed the microbial community composition but also enhanced microbial growth and activity leading to increased nutrient cycling rates. To maintain fertility of these pasture soils and to prevent severe weed-infestation (bracken fern, cf. Chap. 15), a sustainable pasture management is necessary (cf. Chap. 26).

References

Bendix J, Rollenbeck R, Richter M, Fabian P, Emck P (2008) Climate. In: Beck E, Bendix J, Kottke I, Makeschin F, Mosandl R (eds) Gradients in a tropical mountain ecosystem of Ecuador. Ecological studies, vol 198. Springer, Berlin, pp 63–74

Benner J, Vitousek PM, Ostertag R (2010) Nutrient cycling and nutrient limitation in tropical montane cloud forests. In: Bruijnzeel LA, Scatena FN, Hamilton LS (eds) Tropical montane cloud forests, International hydrology series. Cambridge Academic Press, Cambridge, pp 90–100

Boy J, Wilcke W (2008) Tropical Andean forest derives calcium and magnesium from Saharan dust. Glob Biogeochem Cycle 22, GB1027

Boy J, Rollenbeck R, Valarezo C, Wilcke W (2008a) Amazonian biomass burning-derived acid and nutrient deposition in the north Andean montane forest of Ecuador. Glob Biogeochem Cycle 22, GB4011

Boy J, Valarezo C, Wilcke W (2008b) Water flow paths in soil control element exports in an Andean tropical montane forest. Eur J Soil Sci 59:1209–1227

Fleischbein K, Wilcke W, Goller R, Valarezo C, Zech W, Knoblich K (2005) Rainfall interception in a lower montane forest in Ecuador: effects of canopy properties. Hydrol Proc 19:1355–1371

Fleischbein K, Wilcke W, Valarezo C, Zech W, Knoblich K (2006) Water budgets of three small catchments under montane forest in Ecuador: experimental and modeling approach. Hydrol Proc 20:2491–2507

Galloway JN, Dentener FJ, Capone DG, Boyer EW, Howarth RW, Seitzinger SP, Asner GP, Cleveland CC, Green PA, Holland EA, Karl DM, Michaels AF, Porter JH, Townsend AR, Vörösmarty CJ (2004) Nitrogen cycles: past, present, and future. Biogeochemistry 70:153–226

Galloway JN, Townsend AR, Erisman JW, Bekunda M, Cai Z, Freney JR, Martinelli LA, Seitzinger SP, Sutton MA (2008) Transformation of the nitrogen cycle: recent trends, questions, and potential solutions. Science 320:889–892

Goller R, Wilcke W, Leng M, Tobschall HJ, Wagner K, Valarezo C, Zech W (2005) Tracing water paths through small catchments under a tropical montane rain forest in south Ecuador by an oxygen isotope approach. J Hydrol 308:67–80

Grieve IC, Proctor J, Cousins SA (1990) Soil variation with altitude on volcan Barva, Costa Rica. Catena 17:525–534

Hirsch RM, Slack JR, Smitz RA (1982) Techniques of trend analysis for monthly water quality data. Water Resour Res 18:107–121

Illig J, Schatz H, Scheu S, Maraun M (2008) Decomposition and colonization by micro-arthropods of two litter types in a tropical montane rain forest in southern Ecuador. J Trop Ecol 24:157–167

Iost S (2007) Soil respiration, microbial respiration and mineralisation in soils of montane rainforests of Southern Ecuador: influence of altitude. Faculty of Forest, Geo & Hydro Sciences, Dresden University of Technology, Dresden, Germany, 186 p. http://nbn-resolving.de/urn:nbn:de:bsz:14-ds-1201126765623-42870

Iost S, Makeschin F, Abiy M, Haubrich F (2008) Biotic soil activities. In: Beck E, Bendix J, Kottke I, Makeschin F, Mosandl R (eds) Gradients in a tropical mountain ecosystem of Ecuador. Ecological studies, vol 198. Springer, Berlin, pp 217–228

IUSS Working Group WRB (2006) World Soil Resource Reports No. 103. FAO, Rome

Jordan CF (1985) Nutrient cycling in tropical forest ecosystems: principles and their application in management and conservation. Wiley, Chichester

Kaufman YJ, Koren I, Remer LA, Tanre D, Ginoux P, Fan S (2005) Dust transport and deposition observed from the Terra-Moderate Resolution Imaging Spectroradiometer (MODIS) spacecraft over the Atlantic ocean. J Geophys Res 110:D10S12

Likens GE, Bormann FH (1995) Biogeochemistry of a forested ecosystem, 2nd edn. Springer, Berlin, 159 p

Makeschin F, Haubrich F, Abiy M, Burneo JI, Klinger T (2008). Pasture management and natural soil regeneration. In: Beck E, Bendix J, Kottke I, Makeschin F, Mosandl R (eds) Gradients in a tropical mountain ecosystem of Ecuador. Ecological studies, vol 198. Springer, Berlin, pp 397–408

Marrs RH, Proctor J, Heaney A, Mountford MD (1988) Changes in soil nitrogen-mineralization and nitrification along an altitudinal transect in tropical rain forest in Costa Rica. J Ecol 76:466–482

Marschner H (1995) Mineral nutrition of higher plants, 2nd edn. Academic, London

Matson PA, McDowell WH, Townsend AR, Vitousek PM (1999) The globalization of N deposition: ecosystem consequences in tropical environments. Biogeochemistry 46:67–83

McGrath DA, Smith CK, Gholz HL, de Assis Oliveira F (2001) Effects of land-use change on soil nutrient dynamics in Amazonia. Ecosystems 4:625–645

Millennium Ecosystem Assessment (2005) Ecosystems and human well-being: synthesis. Island Press, Washington, DC

Potthast K, Hamer U, Makeschin F (2012) Land-use change in a tropical mountain rainforest region of southern Ecuador affects soil microorganisms and nutrient cycling. Biogeochemistry. 111:151–167

Richter M, Wilcke W, Bussmann R (2008) Landslides as important disturbance regimes – causes and regeneration. In: Beck E, Bendix J, Kottke I, Makeschin F, Mosandl R (eds) Gradients in a tropical mountain ecosystem of Ecuador. Ecological studies, vol 198. Springer, Berlin, pp 319–330

Rollenbeck R, Fabian P, Bendix J (2008) Matter deposition from remote areas. In: Beck E, Bendix J, Kottke I, Makeschin F, Mosandl R (eds) Gradients in a tropical mountain ecosystem of Ecuador. Ecological studies, vol 198. Springer, Berlin, pp 303–309

Roman L, Scatena FN, Bruijnzeel LA (2010) Global and local variations in tropical montane cloud forest soils. In: Bruijnzeel LA, Scatena FN, Hamilton LS (eds) Tropical montane cloud forests, International hydrology series. Cambridge Academic Press, Cambridge, pp 77–89

Sala OE, Chapin FS III, Armesto JJ, Berlow E, Bloomfeld J, Dirzo R, Huber-Sanwald E, Huenneke LF, Jackson RB, Kinzig A, Leemans R, Lodge DM, Mooney HA, Oesterheld M, Poff NL, Sykes MT, Walker BH, Walker M, Wall DH (2000) Global biodiversity scenarios for the year 2100. Science 287:1770–1774

Schemenauer RS, Cereceda P (1994) A proposed standard fog collector for use in high elevation regions. J Appl Meteorol 33:1313–1322

Six J, Frey SD, Thiet RK, Batten KM (2006) Bacterial and fungal contributions to carbon sequestration in agroecosystems. Soil Sci Soc Am J 70:555–569

Swap R, Garstang S, Greco S, Talbot R, Kallberg P (1992) Saharan dust in the Amazon basin. Tellus 44:133–149

Swift MJ, Heal OW, Anderson JM (1979) Decomposition in terrestrial ecosystems. Studies in ecology, vol. 5. University of California Press, Berkeley and Los Angeles

Tanner EVJ, Vitousek PM, Cuevas E (1998) Experimental investigations of nutrient limitations of forest growth on wet tropical mountains. Ecology 79:10–22

Ulrich B (1983) Interaction of forest canopies with atmospheric constituents: SO_2, alkali and earth alkali cations and chloride. In: Ulrich B, Pankrath J (eds) Effects of accumulation of air pollutants in forest ecosystems. D Reidel, Dordrecht, pp 33–45

Wassen MJ, Venterink HO, Lapshina ED, Tanneberger F (2005) Endangered plants persist under phosphorus limitation. Nature 437:547–550

Wilcke W, Yasin S, Abramowski U, Valarezo C, Zech W (2002) Nutrient storage and turnover in organic layers under tropical montane rain forest in Ecuador. Eur J Soil Sci 53:15–27

Wilcke W, Valladarez H, Stoyan R, Yasin S, Valarezo C, Zech W (2003) Soil properties on a chronosequence of landslides in montane rain forest, Ecuador. Catena 53:79–95

Wilcke W, Yasin S, Schmitt A, Valarezo C, Zech W (2008a) Nutrient status and fluxes at the field and catchment scale. In: Beck E, Bendix J, Kottke I, Makeschin F, Mosandl R (eds) Gradients in a tropical mountain ecosystem of Ecuador. Ecological studies, vol 198. Springer-Verlag, Heidelberg, pp 203–215

Wilcke W, Oelmann Y, Schmitt A, Valarezo C, Zech W, Homeier J (2008b) Soil properties and tree growth along an altitudinal transect in Ecuadorian tropical montane forest. J Plant Nutr Soil Sci 171:220–230

Wullaert H, Homeier J, Valarezo C, Wilcke W (2010) Response of the N and P cycles of an old-growth montane forest in Ecuador to experimental low-level N and P amendments. For Ecol Manage 260:1434–1445

Chapter 12
Natural Landslides Which Impact Current Regulating Services: Environmental Preconditions and Modeling

Jörg Bendix, Claudia Dislich, Andreas Huth, Bernd Huwe, Mareike Ließ, Boris Schröder, Boris Thies, Peter Vorpahl, Julia Wagemann, and Wolfgang Wilcke

12.1 Introduction

Manifold interactions between the abiotic and the biotic environment doubtlessly exist in the complex biodiversity hotspot of the Rio San Francisco valley. Hitherto, it is not unveiled how the natural forest and its biodiversity which regulates (regulating services) the local abiotic conditions (climate, water, soil) is subjected to feedbacks regarding the preservation of species richness. Different hypotheses how interactions and feedbacks between abiotic factors and biota contribute to determine biodiversity are under discussion since decades. Widely accepted in the

J. Bendix (✉) • B. Thies • J. Wagemann
Faculty of Geography, Laboratory for Climatology and Remote Sensing, University of Marburg, Deutschhausstraße 10, 35032 Marburg, Germany
e-mail: bendix@staff.uni-marburg.de

C. Dislich • A. Huth
Department of Ecological Modelling, Helmholtz Centre for Environmental Research – UFZ, P.O. Box 500 136, 04301 Leipzig, Germany

B. Huwe • M. Ließ
Department of Soil Physics, University of Bayreuth, Universitätsstraße 30, 95447 Bayreuth, Germany

B. Schröder
Department of Ecology and Ecosystem Management, Technical University of Munich, Emil-Ramann-Str 6, 85354 Freising-Weihenstephan, Germany

P. Vorpahl
Department of Ecology and Ecosystem Management, Technical University of Munich, Emil-Ramann-Str 6, 85354 Freising-Weihenstephan, Germany

Institute of Earth and Environmental Science, University of Potsdam, Karl-Liebknecht-Str. 24-25, 14476 Potsdam-Golm, Germany

W. Wilcke
Geographic Institute of the University of Bern (GIUB), University of Bern, Hallerstrasse 12, 3012 Bern, Switzerland

J. Bendix et al. (eds.), *Ecosystem Services, Biodiversity and Environmental Change in a Tropical Mountain Ecosystem of South Ecuador*, Ecological Studies 221,
DOI 10.1007/978-3-642-38137-9_12, © Springer-Verlag Berlin Heidelberg 2013

group of hypotheses regarding internal feedbacks controlling biodiversity is the intermediate disturbance hypothesis (IDH) (Connell 1978; Molino and Sabatier 2001) which means that moderate disturbances are fostering the highest degree of species richness. Roxburgh et al. (2004) stressed that the original specification of the IDH requires patchy disturbances. Sheil and Burslem (2003) emphasized that landslides are proven to be one important patchy disturbance type promoting biodiversity above and below ground. It is meanwhile undisputed that landslides are a major factor of natural disturbance in the mountain forest of the study area (Wilcke et al. 2003, Chap. 1). While reducing the overall aboveground biomass, landslides increase the spatial heterogeneity of biomass distribution and thus create distinct habitat types (Dislich and Huth 2012). Particularly plant succession after a landslide and the related above ground species pool of mosses, lichens, vascular plants like orchids and pioneer tree species contribute to the high biodiversity of the mountain rain forest and its resilience against natural disturbances (refer to Chap. 8). Below ground diversity and abundance (e.g., AM fungi) might be affected by landslides, too (refer to Chap. 7).

Profound knowledge on physical interactions between abiotic factors and the forest that are assumed to trigger landslides is mandatory for predicting landslide occurrence probabilities and potential future changes.

The basic factors controlling landslide occurrence in the study area are geology in terms of bedrock material, climate, and topography (Fig. 12.1).

While the geological substrate in the study area is nearly homogenous, the topographic situation is highly variable. In this context, elevation, slope position, steepness, and terrain curvature are the most important factors (Sect. 12.2.1, Fig. 12.1a). Regarding climatic parameters, particularly the abundant rainfall enhances the weight of vegetation and soil and reduces soil strength. Because rainfall generally increases with terrain altitude (Chaps. 1, 19, and 24) elevation is a good proxy for rainfall. High wind speed and resulting dynamic pressure particularly at windward sides at higher altitudes transfer the dynamic stress of trees into the tree root layer and thus are also expected to be important predictors to assess landslide risks (Sect. 12.3.3, Fig. 12.1c). Soil conditions (thickness of the organic and mineral soil layers, soil water logging conditions as indicated by stagnic horizon occurrence probability) are suspected to play a major role and should be considered for landslide prediction (Fig. 12.1b), too. Beyond physical interactions, also chemical interactions might influence the risk for landslides. The role of a specific abiotic–biotic interaction — the relation between soil nutrient availability and fine litter production as a proxy for biomass production and thus vegetation and organic layer weight (Fig. 12.1d) — is discussed in this chapter. Nutrient availability in the soil as an important control of biomass production (influencing the weight of the vegetation) and organic matter degradation (influencing the weight of organic layers) is thus assumed to be an important predictor for landslide probability.

To disentangle the processes responsible for landslide activity, spatial explicit models as presented in this chapter are necessary, which are currently based solely on topographical predictor variables (Sect. 12.2.1, Fig. 12.1a). For a future improvement of the presented model, further spatial input data of relevant climatic

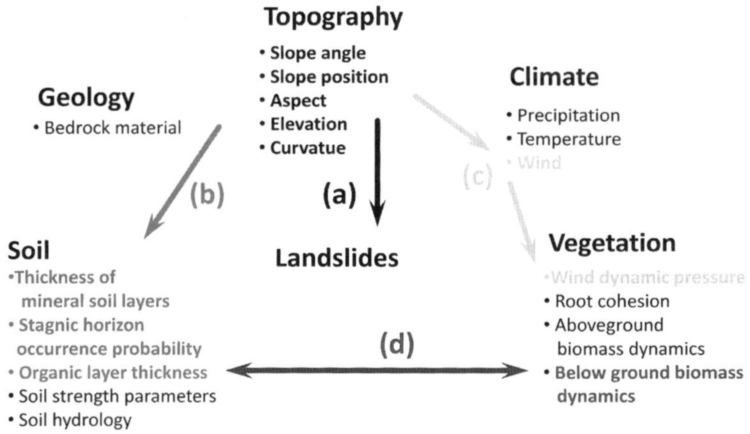

Fig. 12.1 Overview on factors controlling landslide susceptibility in the study area. *Arrows* indicate aspects covered by this chapter: The topographic control on landsliding (**a**), on soil formation (**b**), on the distribution of local wind fields (**c**), and the dependence of organic matter decomposition, organic layer mass, and biomass dynamics (**d**). Future model parameters written in *gray* are developed and discussed in this chapter

and soil predictors as described above are required (Sects. 12.2.2–12.2.4). Most of these data were not available when developing the model described in Sect. 12.2.1. Consequently, this chapter is also devoted to exemplarily present methods to regionalize point-based soil and climate data.

It should be stressed that landslides in a protected, unused pristine mountain forest are not a direct ecosystem service (refer to Chap. 4). However, natural landslide dynamics cause feedbacks to other abiotic and biotic ecosystem components which give reason to expect impacts on several service levels as, e.g., regulation services. On the landscape scale, naturally and anthropogenically induced landslides seem to play a major role in sediment regulation of the catchment, being claimed to be responsible for a quasi-continuous export of sediment loads independent on precipitation peaks (refer to Chap. 9). On the smaller scale, nutrient regulation is clearly affected by landslides. Nutrients are removed with the biomass and the organic layers from the slide area but deposited and concentrated in its foot area (refer to Chap. 11). Regarding carbon regulation, landslides are characterized by reduced tree growth on the slides due to the poor nutrient conditions, thus diminishing aboveground carbon stocks considerably (refer to Chap. 24).

12.2 Methods

12.2.1 The Statistical Landslide Model

Conditions leading to slope failure in the past are likely to cause landslides in the future as well. Thus, inventories of past landslides combined with topographic information and thematic maps of controlling factors are used to train statistical landslide models with multiple predictors. Univariate response curves of these models can provide insights into driving factors of landslides if the following preconditions are met: (1) The model quality (in terms of performance and calibration) is sufficient. (2) Consistency between mechanistic assumptions and training data is maintained. (3) The chosen predictors are interpretable.

(1) Vorpahl et al. (2012) provided a unified framework to train, test and compare different statistical methods. Applying this framework to eight different methods from statistics and machine learning (i.e., generalized linear and additive models, multivariate adaptive regression splines, artificial neural networks, classification tree analysis, random forests, boosted regression trees, and the maximum entropy method), they generated weighted model ensembles.
(2) Vorpahl et al. (2012) maintained consistency between training data and mechanistic assumptions by using a subset of five historical landslide inventories of the RBSF provided by Stoyan (2000) and confined their analysis to landslides that occurred in an area free of anthropogenic interference (Fig. 12.2). Furthermore, they distinguished different functional units of landslides: i.e., initiation, transport, and deposition zones. This distinction is of key importance for an interpretation of univariate model response curves, since linkages between model predictors and actual mechanisms in the distinct functional units differ.
(3) In a case study, Vorpahl et al. (2012) exclusively used terrain attributes derived from a digital elevation model (DEM) as predictors: elevation above sea level (ALT), slope angle, topographic wetness index (TWI), stream power index (SPI), convergence index (CI), topographic position index (TPI) with two different radii (100 m and 500 m), and the aspect. To model landslide initiation as a phenomenon of abiotic–biotic interactions by assessing the importance of abiotic and biotic predictor values in later applications of the method, spatial parameter values as presented in the succeeding sections might be helpful.

12.2.2 Potential Model Parameter: Regionalization of Soil Data

The spatially explicit prediction of histic and stagnic soil horizons is necessary as a major precondition to understand the landslide dynamics in the study area.

Soil regionalization is based on the general concept (e.g., Jenny 1941) that soil genesis and, hence, the soils' distribution throughout the landscape mainly depend on topography, among other parameters. Therefore, topographic parameters can be used as predictors to develop digital maps of various soil attributes.

Soil horizons were assessed by 56 soil profiles and 315 auger sampling points. Key topographical parameters were calculated based on the DEM and implemented area wide as predictors in the software SAGA GIS. To collect a representative dataset, sampling sites were selected according to a 24 terrain classes comprising

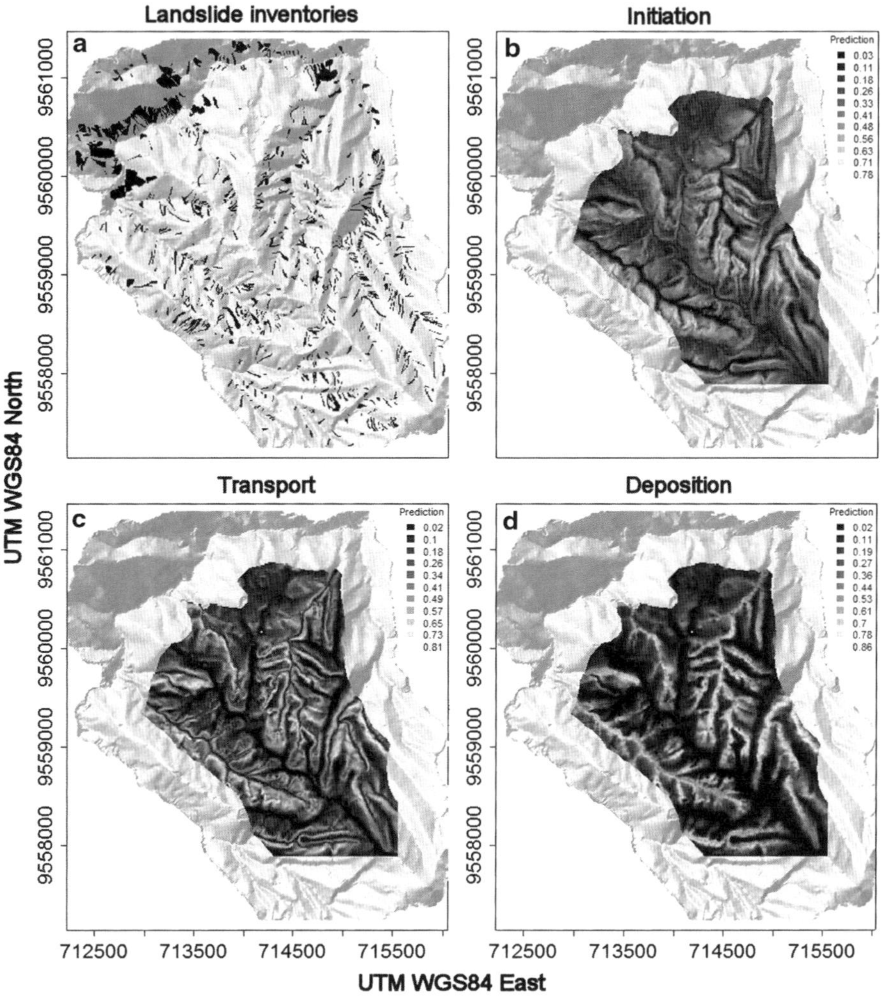

Fig. 12.2 (**a**) Landslide inventories created by evaluation of aerial photographs of five different years (i.e., 1692, 1969, 1976, 1989, and 1998) by Brenning (2005) and landslide susceptibility maps as produced by weighted model ensembles for (**b**) landslide initiation, (**c**) transport, and (**d**) deposition zones (cf. Vorpahl et al. 2012)

sampling design along transects extending along side valley slopes (Liess et al. 2009).

The regionalization as presented here is based on earlier attempts to predict these horizons (Ließ 2011). In comparison to Ließ (2011), an improvement could be achieved by focusing on (1) additional terrain parameter selection and by (2) investigating the dependence on scale as well as (3) the performance of another recursive partitioning method, Random Forest (RF) (Breiman 2001).

Table 12.1 SAGA modules to calculate terrain parameters

Terrain parameter	Module library	Module
Altitude	Terrain analysis—preprocessing	Fill sinks (Planchon/Darboux, 2001)
Slope	Terrain analysis—morphometry	Slope, aspect, curvature
Aspect		
Profile curvature		
Plan curvature		
Convergence index	Terrain analysis—morphometry	Convergence Index (search radius)
Normalized height	Terrain analysis—morphometry	Relative heights and slope positions
Valley depth		
TRI	Terrain analysis—morphometry	Terrain Ruggedness Index
Wind effect	Terrain analysis—morphometry	Wind effect
KRAarea	Terrain analysis—hydrology	Catchment area (flow tracing)
KRAslope		
SWI	Terrain analysis—hydrology	Saga Wetness Index
PISR	Terrain analysis—lighting, visibility	Potential incoming solar radiation—direct insolation

(1) Based on the investigations by Ließ (2011), predictors representing climate (altitude, PISR), water accumulation (curvature, convergence, KRAarea), water discharge (slope, KRAslope), the insulating effect of the heterogeneous geomorphology with the ridge—side valley structure in particular (TRI, normalized height, valley depth) and the wind effect (wind effect, aspect) were selected to model the soil pattern of this area, all calculated by using SAGA GIS (Table 12.1, Böhner et al. 2006).

(2) According to the assumptions of Ließ (2011) that the influence of certain predictors on soil property development is scale dependent, Brown et al. (2004) had reported this for the influence of curvature on soil texture, terrain parameters were calculated for three different GIS raster grid cell sizes (10, 20, 30 m).

(3) Because RF shows a strong dependence on the used dataset used for model development (Ließ et al. 2012), i.e., the terrain parameters used as predictors with the soil parameter as response variable, 100-fold RF calculations of the spatial water stagnation pattern as well as organic layer and stagnic horizon thickness were carried out. For each of the 100 model runs, the used dataset was varied by using 9/10 random Jackknife partitions data subsets of the complete dataset. The 100 models' prediction results were then averaged and displayed as two maps: the mean prediction value of the particular soil parameter and its prediction uncertainty which is represented by the coefficient of variation. Cross validation is applied to the remaining 1/10 of the dataset, which was not used to develop the RF models, for model evaluation.

12.2.3 Potential Model Parameter: Regionalization of Wind Data

Regionalization of meteorological point observations facilitates the analysis of interactions between the abiotic environment and biosphere (e.g., Fries et al. 2009, 2012). Strong wind pressure to forest trees might be one reason fostering landslides and shaping the tree line. Therefore, digital wind speed and dynamic pressure maps are determined using the following procedure: (1) Statistical analysis of wind speed observations using the Weibull density function. (2) Calculation of digital wind speed maps by applying a sheltering factor—algorithm to a DEM. (3) Validation of calculated wind speed using model-independent meteorological stations. (4) Calculation of dynamic pressure maps based on the tropical standard atmosphere and the generated wind speed maps.

(1) Point measurements of hourly wind speed data and the wind direction at 2 m above surface level for a period of 8 years (1999–2006) for five meteorological stations (Cerro, ECSF, El Trio, Paramo, TS1, Fig. 12.3) were analyzed regarding mean and maximum wind speed. According to meteorological conventions (e.g., Weisser 2003), mean and maximum wind speed per 45°-wind direction class are derived from the Weibull density function (50 % and 95 % percentile), where the parameters of the distribution are estimated by the maximum likelihood method.
(2) Wind speed maps are calculated in three steps: First, data of the station Zamora and the highest meteorological station (Paramo) are used to calculate a linear decrease of average and maximum background wind speed with decreasing terrain altitude. Second, the approach of Winstral and Marks (2002) is used to derive the maximum upwind slope parameter which is a measure of topographic shelter or exposure relative to a particular wind direction. The finally determined shelter factor is multiplied with the background wind speed for every pixel, providing the digital wind speed maps for every wind direction class.
(3) Wind speed is extracted for the grid points of the meteorological stations not used for the regionalization and compared to the modeled data. For the most stations (e.g., ECSF, Cerro), the correlation is significant and well-suited, except for the station El Tiro which is known to be strongly influenced by topographic venturi effects not considered by the regionalization method.
(4) Finally maps of average and maximum dynamic pressure are calculated from the wind speed maps and average air density where the latter is derived by blending the tropical standard atmosphere with the DEM.

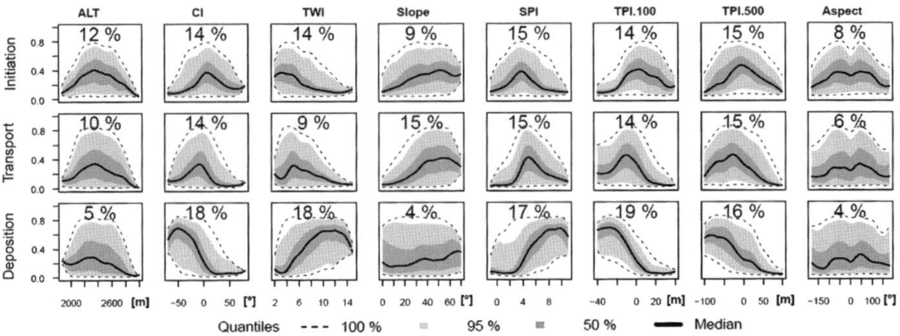

Fig. 12.3 Univariate response curves (*black lines*) and predictor importance scores of weighted ensembles of statistical models. Response quartile ranges are shaded in *gray*. The *curves* in each column show the probability of observing a landslide initiation, transport or deposition zone as a function of a single predictor variable, i.e., elevation above sea level (ALT), the convergence index (CI), indicating small scale concavities ($CI < 0$) or convexities ($CI > 0$), the topographic wetness index (TWI), the slope angle (Slope), the stream power index (SPI), the topographic position index (TPI), describing the difference between local elevation and the mean elevation within two different radii of 100 m (TPI.100) and 500 m (TPI.500), respectively, and the direction of the steepest slope angle (Aspect) (Vorpahl et al. 2012)

12.2.4 Soil Properties and Litterfall

Between 1998 and 2010 we collected data from 12 sites in the study area (one in each of the microcatchments (MC) 1 and 3, three in each of MCs 3 and 5 and the four control sites of Nutrient Manipulation Experiment (NUMEX, for locations see Chap. 1, NUMEX is explained in Chap. 23). Monitoring in MC2 lasted for 12 years, in MCs 1, 3, and 5 for 5 years and in NUMEX for 1 year. At each site, mass of organic layer was determined once by measuring depth and densities of the organic horizons (Oi, Oe, and Oa) and mass of fine litterfall was determined with three- to sixfold replicated $0.3 \times 0.3\ m^2$ to $0.6 \times 0.6\ m^2$ large litter traps in at least monthly resolution. Furthermore, free-draining litter lysimeters just below the organic layer were used to collect litter leachate in weekly to fortnightly resolution in which mineral N (NH_4^+–N + NO_3^-–N) concentrations were determined with a Continuous Flow Analyzer and K, Na, Ca, and Mg concentrations with flame Atomic Absorption Spectrometry.

12.3 Results and Discussion

12.3.1 Statistical Landslide Modeling

With the exception of classification tree analysis all techniques performed comparatively well while being outperformed by weighted model ensembles (refer to Vorpahl et al. 2012 for details). As expected, models trained on different functional units of landslides led to different model outcomes (Fig. 12.2).

Univariate model response curves to changes in predictor values—also called partial dependency plots (Fig. 12.3)—show that landslide deposition zones tend to be located at valley bottoms, indicated by high values of SPI and TWI as well as by negative values of CI and TPI.

Landslides follow the local topography by sliding along shallow ducts in the slope as indicated by a maximum susceptibility for transport zones at slightly negative values of CI and TPI. The response curve for initiation zones to changes in slope indicated an increasing contribution up to ~52°. Even steeper slopes lead to a decrease of landslide susceptibility. This can be attributed to the fact that on extremely steep slopes the soil layer is usually thinner and hence insufficient for landslide initiation.

Model response to elevation above sea level exposed an increasing landslide initiation probability with elevation up to 2,400 m a.s.l. At higher elevations, landslide initiation probability decreases. Rollenbeck (2006) reported an altitudinal increase of average precipitation (from about 2,050 mm a^{-1} at 1,960 m a.s.l. up to 4,400 mm a^{-1} at 3,200 m a.s.l.) in the research area. If rainfall is an important factor, this should hint towards a positive correlation between precipitation and landslide susceptibility which contradicts the above finding. As additional factors at the intersection from dense forest into the Páramo, lower standing biomass and lower inclination may strongly reduce landslide formation. Furthermore, Bussmann et al. (2008) gave a possible explanation for the decreasing landslide susceptibility at higher elevations by a change in soil substrate from slightly metamorphosed clayey/sandy sediments, originating from phyllites, at the lower and intermediate elevations to a more quartzite rich substrate at higher elevations.

Other altitudinal gradients reported for the research area are related to vegetation. The decrease of average tree heights with higher elevations (Bräuning et al. 2008), for example, may cause a reduced contribution of plant biomass to slope instability. Smaller trees are less capable of transferring wind forces into the ground via a turning moment. Soethe et al. (2006a, b) as well as Leuschner et al. (2007) reported an altitudinal change in tree root structure and in the ratio of aboveground to belowground biomass. Thus an increase of root contribution to slope stability at higher elevations can be additionally suspected.

12.3.2 Digital Soil Maps

To predict organic layer thickness, the models based on 20 or 30 m DEM resolution performed better than those using 10 m. Regarding the prediction of the occurrence of a stagnic color pattern, all models using 10 m resolution performed better or equally well than those of lower resolution. Chaplot et al. (2000) found prediction accuracy to be highly dependent on DEM resolution: Regarding the prediction of hydromorphic features 10 m DEM resolution outperformed lower resolutions. Compared to the median r_{xy} resulting from CART methodology and a smaller number of prediction parameters (Ließ 2011), model performance was now

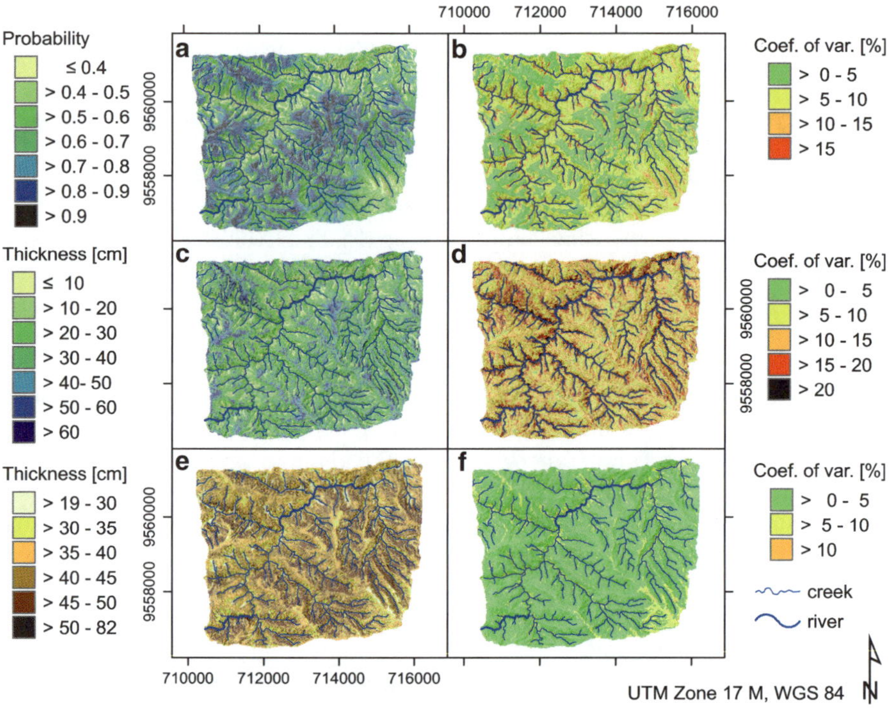

Fig. 12.4 Mean stagnic horizon occurrence probability (**a**) and thickness (**c**) with coefficient of variation (**b**, **d**) and mean organic layer thickness (**e**) with coefficient of variation (**f**) (Overlaid hill shading with light source from north)

improved for all three predicted soil parameters: regarding stagnic horizon occurrence probability it was improved by 0.1 (0.6), regarding horizon thickness it was doubled (0.36), and regarding organic layer thickness it was more than doubled (0.47).

The digital soil map of the stagnic horizon occurrence probability is shown in Fig. 12.4a, b. A low coefficient of variation ($\leq$10 % for >80 % of the area, see Fig. 12.4b) shows that the dataset is well suited to model the stagnic properties pattern within this area. The influence of the relative slope position on the occurrence probability is clearly visible: The exposed mountain ridges between 2,100 and 2,650 m a.s.l. display a very high probability of stagnic soil properties, >0.8, which is decreasing down the side valley slopes to a probability of $\leq$0.4 (minimum = 0.2). The flat platform-like areas on top of the ridges, display a particularly high probability of >0.9. The areas below 2,100 and above 2,650 m a.s.l. are predicted with an overall lower probability. Below 2,100 m a.s.l. the lower bulk density (Ließ et al. 2011) and above 2,650 m a.s.l. the coarser soil texture (Ließ et al. 2012) leads to a higher saturated hydraulic conductivity and therefore less chance for the development of stagnic soil properties. For the development of the model to predict stagnic horizon occurrence probability, all terrain parameters were included.

This confirms the assumption that it is the complex pattern of climate (altitude, PISR), water accumulation (curvature, convergence, KRAarea), water discharge (slope, KRAslope), the insulating effect of the heterogeneous geomorphology with the ridge—side valley structure in particular (TRI, normalized height, valley depth) as well as the wind effect (wind effect, aspect) which lead to the distribution pattern of stagnic soil properties within the investigation area.

The model to regionalize stagnic horizon thickness is less stable than the model to predict the horizon's occurrence probability. This is indicated by the higher values of the coefficient of variation in Fig. 12.4d. Ließ (2011) describes similar results. According to Park and Vlek (2002), soil attributes of which the vertical distribution is strongly determined by pedogenesis or unknown factors are poorly modeled by environmental variables. Accordingly, the frequent change of parent material within one soil profile (Ließ et al. 2012) might be the reason why stagnic horizon thickness cannot be explained by geomorphology alone. The thickest stagnic layers >40 or even >60 cm are found along the mountain ridges, with decreasing thickness while proceeding down side valley slopes.

The low uncertainty of the digital soil map of the organic layer (Fig. 12.4e, f) indicates a stable model. The thickest organic layers are found on mid-slope positions, decreasing towards the creeks and towards the crests. Furthermore, altitude is not among the five most influential predictors of organic layer thickness and there is no correlation between the occurrence of stagnic horizons and organic layer thickness. This is unexpected because in previous work it was shown that the crests had usually thicker organic layers than the valley bottom positions in the study area (Wilcke et al. 2010) in line with reports from a similar forest in Puerto Rico (Silver 1994). Furthermore, studies in Costa Rica (Marrs et al. 1988; Grieve et al. 1990) and at our study site in Ecuador (Schrumpf et al. 2001; Wilcke et al. 2008a, b) have shown that organic layer thickness usually increases with increasing altitude because of decreasing microbial turnover of organic matter with increasing altitude (Benner et al. 2010). Table 10.1 shows a general trend towards increasing organic layer thickness with altitude. However, the transect that was investigated covers a much larger distance (30 km compared to c. 4 km), and spatial data coverage is therefore limited. Taking a closer look, the results of Chap. 10 also do not describe any positive correlation between organic layer thickness and altitude for the altitudinal range between 1,890 and 3,060 m a.s.l. studied here. Finally, it is assumed that soil waterlogging limits organic matter turnover (Schuur and Matson 2001; Roman et al. 2010) which results in the expectation of a positive correlation between the occurrence of waterlogging (as indicated by stagnic horizons) and organic layer thickness. However, there is a considerable variation in organic layer thickness at small scale (Wilcke et al. 2002, 2008b) illustrating that none of altitude, topographic position, and waterlogging alone can explain the entire variability in organic layer thickness.

A possible explanation for the seeming contradictions might be that our dataset is representative for the whole study area and therefore also includes landslide sites with incomplete organic layers which form an important part of the studied forest area (Bussmann et al. 2008). Wilcke et al. (2003) have shown that the full regeneration of the organic layer only occurs at the time scale of a few decades. It also seems likely that waterlogging favors the initiation of landslides because of the associated high soil weight (Ließ et al. 2011). The results in the literature, in contrast, usually refer to undisturbed old-growth forest sites. An alternative explanation might be that litterfall rates are lower on crest sites than at lower topographic positions associated with a smaller accumulation of organic matter on top of the mineral soil. However, in Sect. 12.3.4 we show for a limited dataset of 12 study sites that the decrease in litterfall rates is overcompensated by the decrease in degradation rates resulting in even higher organic layer thickness at low litterfall rates. We conclude that the relationships of altitude, topographic position, and waterlogging with organic layer thickness might have to consider the state of succession after landslide to explain and predict the spatial distribution of organic layer thickness in the study area.

12.3.3 Digital Wind Maps

Figure 12.6a shows the calculated digital map of maximum wind speed which reveals spatial structures comparable to the map of mean wind speed (not shown here). Maximum wind speed increases with altitude but is locally modified by topographic shelter effects towards the predominant wind direction. Obviously, steep and narrow valleys and ravines breaching the Cordillera exhibit the lowest wind speeds (partly close to calm) on a specific altitudinal level. It is striking that especially the east-facing slopes without any protection by upstream topographic structures exhibit severe wind speeds up to 17 m s^{-1}. The reason is the all-year dominating circulation from the east (Rollenbeck and Bendix 2011) impinging particularly the eastern slopes of the Cordillera. The high wind speeds at relatively low altitudes are a result of the Andean depression (Chap. 1) which allows the easterlies to affect the upper mountain areas nearly unbridled. By blending the land use classification of Göttlicher et al. (2009) with the digital maps of mean and maximum dynamic pressure, the interaction of wind pressure and trees can be assessed, e.g., for the tree line ecotone (Fig. 12.5b). The statistical evaluation clearly reveals that the trees at the treeline of the eastern escarpment exhibit clearly stronger mechanical exposure than on the western slopes where in the most situations, wind dynamic pressure falls into the lowest category (mean < 5 N m^{-2}; maximum < 20 N m^{-2}).

12.3.4 Chemical Interactions: Soil Nutrients and Litter

There were close positive correlations between nutrient concentrations in soil solution and annual fine litterfall as proxy of biomass productivity and close negative correlations between nutrient concentrations in soil solution and mass of

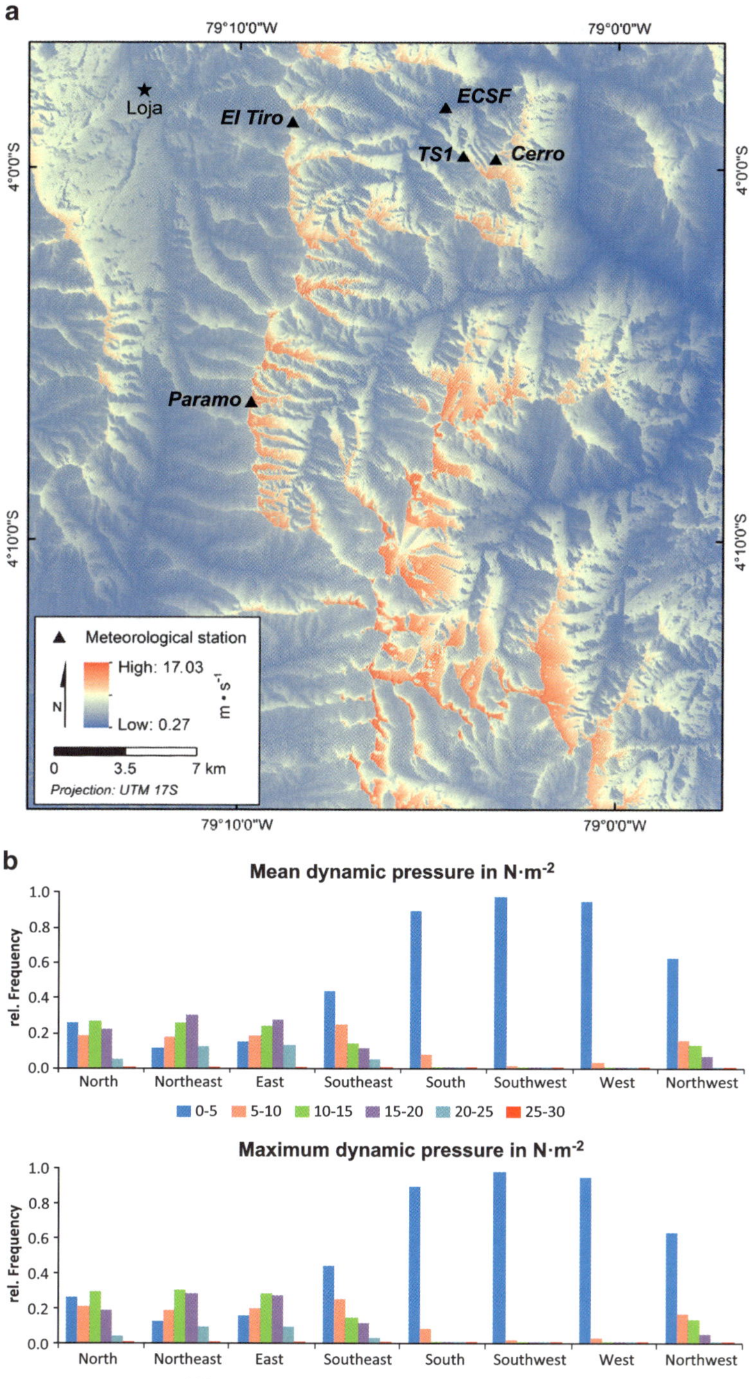

Fig. 12.5 (**a**) Digital map of maximum wind speed [95 % percentile] determined as an occurrence-weighted average of eight wind direction classes. (**b**) Mean and maximum dynamic pressure depending on aspect along the tree line ecotone

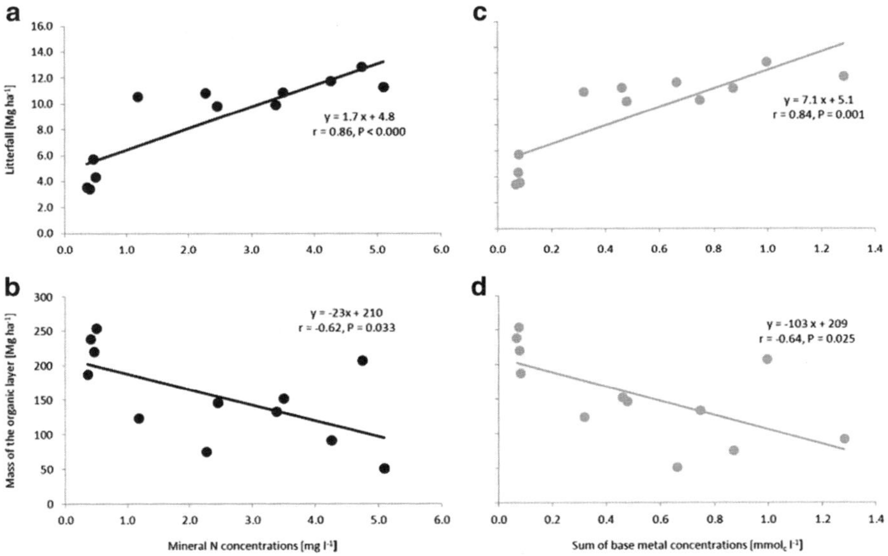

Fig. 12.6 Relationship between mean mineral N concentrations (NH_4^+–N + NO_3^-–N) in litter leachate and (**a**) mean annual litterfall and (**b**) mass of the organic layer and between mean sum of base metal concentrations (charge equivalents of K, Na, Ca, and Mg) in litter leachate and (**c**) mean annual litterfall and (**d**) mass of the organic layer. Mean nutrient concentrations in litter leachate and mean annual litterfall was determined during 1–10 years depending on the specific site

organic layer (Fig. 12.6). The effect of bases (K, Na, Ca, and Mg) is more pronounced than that of N. Consequently, increased nutrient availability resulted in increased fine litterfall production and—if litterfall is proportional to standing biomass—increased weight of vegetation. The increased weight of vegetation is counteracted by decreased weight of organic layer because of a faster turnover at higher nutrient availability. Expressed in percent of the intercept the response of litterfall is more pronounced (35 % for N and 139 % for bases) than that of mass of organic layer (−11 and −49 %) suggesting that the standing biomass will more strongly increase than the mass of the organic layer decrease if nutrient availability improves. The latter can be expected for the near future because of increased dryness enhancing release of nutrients from the organic matter by mineralization and because of rising deposition of reactive nitrogen and possibly also of base metals because of the shortening of the El Niño Southern Oscillation (ENSO) cycle (see Chap. 11). This might imply that the total weight of vegetation plus organic layer will increase in the near future in response to environmental change thereby enhancing the risk of landslides.

Nutrient availability in the study area generally decreases with increasing altitude and at the same altitude is different between valley bottom and ridge top position (Wilcke et al. 2008b, 2010). Furthermore, the frequently occurring shallow landslides in the study area remove the vegetation and the organic layer resulting in nutrient loss which is only replenished during a few decades (Wilcke et al. 2003).

The latter effect is not included in the relationship between nutrient availability and litterfall/mass of organic layer in Fig. 12.6 because all our measurements were taken at old-growth forest sites which were not impacted by landslides in the last decades. The high variability of altitude and topography in our study area results in a high spatial variability of nutrient availability and thus also a high ability of organic layer mass and standing forest biomass together determining the weight on top of the mineral soil (Wilcke et al. 2002; Moser et al. 2008).

12.4 Conclusion

The presented statistical model ensembles revealed that the occurrence of landslides is mainly controlled by factors related to the general position along a slope (i.e., ridge, open slope, or valley). However, there is a clear contradiction between the altitudinal gradient of rainfall (increasing with altitude) as an assumed major trigger and landslide probability (decreasing with altitude above 2,400 m a.s. l.). This indicates that more complex interactions control landslide activity in the study area which can be explained with a model ensemble purely forced with DEM-derived proxy predictors. Digital soil maps show a sandier soil texture and lower soil water logging probability above 2,400 m a.s.l and hence provide a good explanation. We further assume that variation in above and belowground biomass mitigating dynamic wind pressure to the forest in the higher parts are major factors causing these contradictory findings. Thus, it is necessary to provide further spatial predictor maps related to geology, vegetation biomass, and climate. By additionally considering predictors related to vegetation, soil and climate, statistical models will allow for predicting potential future changes in landslide probability patterns. Dynamic forest models like FORMIND can be used to further quantify effects on the aboveground biomass production (Chap. 24).

Regarding maps of soil conditions, statistical models based on comprehensive soil field surveys are applied to spatially predict organic layer and stagnic horizon thickness as well as stagnic horizon occurrence probability. Forcing parameters are solely derived from topographical analyses of the DEM. Even if the main influence of the relative slope position as exposed mountain ridges and flat platform-like areas on top of the ridges are the best predictors for the occurrence probability of stagnic horizons, the results point out complex interactions of different factors behind this. Particularly, the determination of the stagnic horizon thickness is less stable, most likely due to unconsidered, non geomorphologic factors. For prediction of organic layer thickness, the degree of succession after landslide might also play an important role and should be considered besides the well established relationship of waterlogging, topographic position, and altitude with organic layer thickness.

Digital maps of mean and maximum wind speeds as well as dynamic wind pressures as additional potential forcing parameters were derived by means of field observations of wind speed, data on air density, and a DEM by introducing a terrain shelter factor. It could be shown that dynamic pressure to the forest generally increases with altitude but differs with exposition to the main wind direction. Because easterly wind directions are predominant, the tree line ecotone on the eastern slopes is affected by clearly higher wind stress.

Finally, it could be shown that interactions in the biogeochemical cycles might be relevant for the risk of landslides. Nutrient availability in soil influenced litterfall production positively and organic layer thickness negatively. An increased nutrient availability in the future will most likely result in an increase of standing biomass, thus, enhancing the risk of landslides in response to future environmental change.

Regarding ecosystem services, landslide dynamics will influence different service levels. As emphasized in the introduction, landslides are most likely a precondition for the high biological diversity in the mountain forest and thus, directly related to the cultural services of the forest (Chap. 4). Because the forest structure characterized by its high species richness properly regulates abiotic processes, landslides indirectly contribute to the regulating services of the forest (Chap. 4) too. On the scale of a single landslide, regulation of abiotic parameters changes significantly. For instance, microclimate (temperature, humidity) regulation is reduced in comparison to areas sheltered by tree canopies (Fries et al. 2009, 2012; Chap. 9). On this scale also sediment and nutrient regulation are affected. While sediment and its nutrient is accumulated at the forest edges on the foot of the slide, also slope wash of matter is higher in young landslides than under natural forest (e.g., Larsen et al. 1999). On the scale of the forest as a mosaic of trees and gap areas originating, e.g., from landslides, mass and energy fluxes to the atmosphere are different than those of closed canopies (e.g., Zhang et al. 2007) which means that landslides maintain the specific regulation of forest–atmosphere interactions. Also the carbon regulation function of the mountain ecosystem is determined by the landslide occurrence. Landslides increase carbon turnover and change the forest composition towards a higher fraction of pioneer species—however overall forest productivity may be reduced compared to old growth forest without landslide disturbances due to the unfavourable environmental conditions on landslide sites (Chap. 24).

References

Benner J, Vitousek PM, Ostertag R (2010) Nutrient cycling and nutrient limitation in torpical montane cloud forests. In: Bruijnzeel LA, Scatena FN, Hamilton LS (eds) Tropical montane cloud forests. International hydrology series. Cambridge University Press, Cambridge, pp 90–100

Böhner J, McCloy KR, Strobl J (2006) SAGA – analysis and modelling application. Göttinger Geographische Abhandlungen, vol 115. Geographisches Institut der Universität Göttingen

Bräuning A, Homeier J, Cueva E, Beck E, Günter S (2008) Growth dynamics of trees in tropical mountain ecosystems. In: Beck E, Bendix J, Kottke I, Makeschin F, Mosandl R (eds) Gradients in a tropical mountain ecosystem of Ecuador (Ecological Studies 198). Springer, Berlin, pp 291–302

Breiman L (2001) Random forests. Machine Learning 45:5–32

Brenning A (2005) Spatial prediction models for landslide hazards: review, comparison and evaluation. Nat Hazards Earth Syst Sci 5:853–862

Brown DJ, Clayton MK, McSweeney K (2004) Potential terrain controls on soil color, texture contrast and grain-size deposition for the original catena landscape in Uganda. Geoderma 122:51–72

Bussmann RW, Wilcke W, Richter M (2008) Landslides as important disturbance regimes causes and regeneration. In: Beck E, Bendix J, Kottke I, Makeschin F, Mosandl R (eds) Gradients in a tropical mountain ecosystem of Ecuador. Ecological studies, vol 198. Springer, Berlin, pp 319–330

Chaplot V, Walter C, Curmi P (2000) Improving soil hydromorphy prediction according to DEM resolution and available pedological data. Geoderma 97:405–422

Connell JH (1978) Diversity in tropical rain forests and coral reefs. Science 199:1302–1310

Dislich C, Huth A (2012) Modelling the impact of shallow landslides on forest structure in tropical montane forests. Ecol Model 239:40–53

Fries A, Rollenbeck R, Göttlicher D, Nauß T, Homeier J, Peters T, Bendix J (2009) Thermal structure of a megadiverse Andean mountain ecosystem in southern Ecuador, and its regionalization. Erdkunde 63:321–335

Fries A, Rollenbeck R, Nauß T, Peters T, Bendix J (2012) Near surface air humidity in a megadiverse Andean mountain ecosystem of southern Ecuador and its regionalization. Agric For Meteorol 152:17–30

Göttlicher D, Obregón A, Homeier J, Rollenbeck R, Nauß T, Bendix J (2009) Land cover classification in the Andes of southern Ecuador using Landsat ETM+ data as a basis for SVAT modelling. Int J Remote Sens 30:1867–1886

Grieve IC, Proctor J, Cousins SA (1990) Soil variation with altitude on volcan Barva, Costa Rica. Catena 17:525–534

Jenny H (1941) Factors of soil formation. A system of quantitative pedology. Dover, New York

Larsen MC, Torres-Sánchez AJ, Concepción IM (1999) Slopewash, surface runoff and fine-litter transport in forest and landslide scars in humid-tropical steeplands, Luquillo Experimental Forest, Puerto Rico. Earth Surf Process Landforms 24:481–502

Leuschner C, Moser G, Bertsch C, Röderstein M, Hertel D (2007) Large altitudinal increase in tree root/shoot ratio in tropical mountain forests of Ecuador. Basic Appl Ecol 219–230

Ließ M (2011) Soil-landscape modelling in an Andean mountain forest region in southern Ecuador. PhD Thesis, University of Bayreuth, Bayreuth

Liess M, Glaser B, Huwe B (2009) Digital soil mapping in southern Ecuador. Erdkunde 63:309–319

Ließ M, Glaser B, Huwe B (2011) Functional soil-landscape modelling to estimate slope stability in a steep Andean mountain forest region. Geomorphology 132(3–4):287–299

Ließ M, Glaser B, Huwe B (2012) Uncertainty in the spatial prediction of soil texture – comparison of regression tree and random forest models. Geoderma 170:70–79

Marrs RH, Proctor J, Heaney A, Mountford MD (1988) Changes in soil nitrogen-mineralization and nitrification along an altitudinal transect in tropical rain forest in Costa Rica. J Ecol 76:466–482

Molino JF, Sabatier D (2001) Tree diversity in tropical rain forests: a validation of the intermediate disturbance hypothesis. Science 294:1702–1704

Moser G, Röderstein M, Soethe N, Hertel D, Leuschner C (2008) Altitudinal changes in stand structure and biomass allocaton of tropical mountain forests in relation to microclimate and soil chemistry. In: Beck E, Bendix J, Kottke I, Makeschin F, Mosandl R (eds) Gradients in a tropical mountain ecosystem of Ecuador. Ecological studies, vol 198. Springer, Berlin, pp 229–242

Park SJ, Vlek PLG (2002) Environmental correlation of three-dimensional soil spatial variability: a comparison of three adaptive techniques. Geoderma 109:117–140

Rollenbeck R (2006) Variability of precipitation in the Reserva Biologica San Francisco/Southern Ecuador. Lyonia 9:43–51

Rollenbeck R, Bendix J (2011) Rainfall distribution in the Andes of southern Ecuador derived from blending weather radar data and meteorological field observations. Atmos Res 99:277–289

Roman L, Scatena FN, Bruijnzeel LA (2010) Global and local variations in tropical montane cloud forest soils. In: Bruijnzeel LA, Scatena FN, Hamilton LS (eds) Tropical montane cloud forests. International hydrology series. Cambridge University Press, Cambridge, pp 77–89

Roxburgh SH, Shea K, Wilson B (2004) The intermediate disturbance hypothesis: patch dynamics and mechanisms of species coexistence. Ecology 85:359–371

Schrumpf M, Guggenberger G, Schubert C, Valarezo C, Zech W (2001) Tropical montane rain forest soils: development and nutrient status along an altitudinal gradient in the south Ecuadorian Andes. Die Erde 132:43–59

Schuur EAG, Matson PA (2001) Net primary productivity and nutrient cycling across a mesic to wet precipitation gradient in Hawaiian montane forest. Oecologia 128:431–442

Sheil D, Burslem DFRP (2003) Disturbing hypotheses in tropical forests. Trends Ecol Evol 18:18–26

Silver WL (1994) Is nutrient availability related to plant nutrient use in humid tropical forests? Oecologia 98:336–343

Soethe N, Lehmann J, Engels C (2006a) Root morphology and anchorage of six native tree species from a tropical montane forest and an elefin forest in Ecuador. Plant Soil 279:173–185

Soethe N, Lehmann J, Engels C (2006b) The vertical pattern of rooting and nutrient uptake at different altitudes of a south Ecuadorian montane forest. Plant Soil 286:287–299

Stoyan R (2000) Aktivität, Ursachen und Klassifikation der Rutschungen in San Francisco/ Südecuador. Diploma Thesis, Friedrich-Alexander-Unversität Erlangen-Nürnberg

Vorpahl P, Elsenbeer H, Märker M, Schröder B (2012) How can statistical models help to determine driving factors of landslides? Ecol Model 239:27–39. doi:10.1016/j.ecolmodel.2011.12.007

Weisser D (2003) A wind energy analysis of Grenada: an estimation using the 'Weibull' density function. Renew Energy 28:1803–1812

Wilcke W, Yasin S, Abramowski U, Valarezo C, Zech W (2002) Nutrient storage and turnover in organic layers under tropical montane rainforest in Ecuador. Eur J Soil Sci 53:15–27

Wilcke W, Valladarez H, Stoyan R, Yasin S, Valarezo C, Zech W (2003) Soil properties on a chronosequence of landslides in montane rain forest, Ecuador. Catena 53:79–95

Wilcke W, Oelmann Y, Schmitt A, Valarezo C, Zech W, Homeier J (2008a) Soil properties and tree growth along an altitudinal transect in Ecuadorian tropical montane forest. J Plant Nutr Soil Sci 171:220–230

Wilcke W, Yasin S, Schmitt A, Valarezo C, Zech W (2008b) Soils along the altitudinal transect and in catchments. In: Beck E, Bendix J, Kottke I, Makeschin F, Mosandl R (eds) Gradients in a tropical mountain ecosystem of Ecuador. Ecological studies, vol 198. Springer, Berlin, pp 75–85

Wilcke W, Boy J, Goller R, Fleischbein K, Valarezo C, Zech W (2010) Effect of topography on soil fertility and water flow in an Ecaudorian lower montane forest. In: Bruijnzeel LA, Scatena FN, Hamilton LS (eds) Tropical montane cloud Forests, International hydrology series. Cambridge Academic Press, Cambridge, pp 402–409

Winstral A, Marks D (2002) Simulating wind fields and snow redistribution using terrain based parameters to model snow accumulation and melt over a semi-arid mountain catchment. Hydrol Process 16:3585–3603

Zhang G, Thomas C, Leclerc MY, Karipot A, Gholz HL, Binford M, Foken T (2007) On the effect of clearcuts on turbulence structure above a forest canopy. Theor Appl Climatol 88:133–137

Chapter 13
Conservation, Management of Natural Forests and Reforestation of Pastures to Retain and Restore Current Provisioning Services

Michael Weber, Bernd Stimm, Maria Fernanda López, Andrés Gerique, Perdita Pohle, Patrick Hildebrandt, Thomas Knoke, Ximena Palomeque, Baltazar Calvas, Sven Günter, Nikolai Aguirre, and Daniel Kübler

13.1 Introduction

In 2010, forests covered 36 % (9.87 million ha) of the Ecuadorian land area (FAO 2010). 4.5 million ha are estimated to be potentially used as production forest, but only about 2 million ha are considered as permanent productive forest estate (Blaser et al. 2011). Despite the fact that the forest sector contributes only about 2 % to the gross domestic product (GDP) and 1–2 % to the total exports of Ecuador (FAO 2011), provisioning services of forests can be considered to be of high importance, especially for the rural population. As more than 50 % of the forests are under community or indigenous ownership, many products from forests are consumed by

M. Weber (✉) • B. Stimm • P. Hildebrandt • X. Palomeque • D. Kübler
Institute of Silviculture, Center of Life and Food Sciences Weihenstephan, Technische Universität München, Hans-Carl-von-Carlowitz-Platz 2, 85354 Freising, Germany
e-mail: m.weber@forst.wzw.tum.de

M.F. López
Pontificia Universidad Católica (PUCE), Apt. 17-01-2184, Quito, Ecuador

A. Gerique • P. Pohle
Institute of Geography, Friedrich-Alexander University of Erlangen-Nuremberg, 91054 Erlangen, Germany

T. Knoke • B. Calvas
Institute of Forest Management, Center of Life and Food Sciences Weihenstephan, Technische Universität München, Hans-Carl-von-Carlowitz-Platz 2, 85354 Freising, Germany

S. Günter
Tropical Agricultural Research and Higher Education Center, Latin-American Chair for Protected Areas and Biological Corridors "Kenton Miller", CATIE 7170, Turrialba 30501, Costa Rica

N. Aguirre
Center for Studies and Development of Amazonia (CEDAMAZ), National University of Loja, Loja, Ecuador

J. Bendix et al. (eds.), *Ecosystem Services, Biodiversity and Environmental Change in a Tropical Mountain Ecosystem of South Ecuador*, Ecological Studies 221,
DOI 10.1007/978-3-642-38137-9_13,

the forest owners or sold locally and thus not registered by market statistics. Consequently, reliable data on forest production are very rare. The Government of Ecuador (2009; cited in Blaser et al. 2011) estimated that about 75 % of the total forest area is used directly or indirectly by indigenous communities to contribute to their livelihoods, while 850,000 people depend directly on forest resources.

As revealed in several studies (Gerique 2010; Pohle and Gerique 2008), the importance of the provisioning services of forest ecosystems depends highly on the ethnic-cultural affiliation of the population and their socio-economic setting. In the study area of our Research Unit (see Chap. 1), three different ethnic groups have to be distinguished (Pohle 2008): The Shuar, Amazonian Indians, settling in the lower area (<1,400 m a.s.l.) of the tropical mountain rainforest range, who are typical forest dwellers practicing shifting cultivation in a subsistence economy. The Saraguros are highland Indians who live as agropastoralists in the mid-altitudes (1,700–2,800 m a.s.l.), and the Mestizos, a heterogeneous group of mixed Spanish and indigenous descents, who immigrated to the area since 1960s. They live in rural communities or scattered farms as cattle farmers and agriculturalists. The farmers of all groups highly value the provisioning services of the forests as they strengthen their livelihoods and overall economic situation (Pohle et al. 2010; Gerique 2010). A study in two northeastern provinces of Ecuador revealed that forest uses contributed 10–30 % to the family income.

The various forest goods and its consideration in official statistics are also closely linked to the type of forest they are originating from. While plantations are preferably established or managed for industrial roundwood production, products from natural forest are predominantly used to satisfy the various demands of the local population. Furthermore, the forest type also affects the sustainable yield that can be expected: that of the natural forests is estimated on 0.9 cbm ha^{-1} $year^{-1}$ while that of plantations is 10 cbm ha^{-1} $year^{-1}$ (FAO 2006).

The provisioning services comprise a multifaceted bundle of goods and services that are already or can be influenced and governed by human interventions and management. The following paragraphs provide an overview on corresponding aspects of conservation, management of natural forests and reforestation in the greater study area of the Reserva Biológica San Francisco (RBSF, see Fig. 1.1).

13.2 Material and Methods

The results of this chapter are based upon comprehensive studies conducted within the research unit RU 816 comprising ethnobotanical inventory, livelihood analyses, field monitoring as well as experiments. The details on material and methods for the several issues can be found in Pohle et al. (2010) and Gerique (2010) (ethnobotanical inventory and livelihood analyses), Günter et al. (2008) (natural forest management), Aguirre (2007) (reforestation of abandoned pastures) and Stimm et al. (2008) (forest reproductive material).

13.3 Results and Discussion

13.3.1 Conservation

People depend on trees as raw material for the production of, e.g. furniture, buildings, boats, handicrafts, fence posts or fuel. Consequently, conservation measures must be organised in a way that does not completely exclude people from the use of the forest resources. In Ecuador, a major state strategy for in situ conservation of biodiversity, ecosystems functions and services has been the protected area (PA) model. In 1976, the government established the National System of Protected Areas (*Sistema Nacional de Áreas Protegidas*-SNAP) under the notions of "protection from use" and command and control mechanisms (Gerique 2010).

Nowadays, the Ecuadorian reserve network SNAP has modified its structure, aiming to incorporate more participative management concepts. The SNAP has currently four subsystems:

(a) The State Patrimony of Natural Areas (*Patrimonio de Áreas Naturales del Estado-PANE*)
(b) The decentralised areas under administration of subnational governments, mainly municipalities (*Áreas de Gobiernos Autónomos Decentralizados*)
(c) The communal protected areas (*Áreas Protegidas Comunitarias*)
(d) The private protected areas (*Áreas Protegidas Privadas*)

Biosphere reserves are considered conservation tools (*herramientas de conservación*). The PANE has currently 45 protected areas, which cover around 19 % of Ecuador's total surface (MAE 2013). The protected areas inside this subsystem are managed and administered by the state through the Ministry of Environment, except for few decentralised experiences such as the Cajas National Park. Only the protected areas inside the PANE have officially a correspondent to the IUCN categories, e.g. the biological and ecological reserves to category I, the national parks to category II, representing the highest degree of protection. The Forestry Law of Silvestre Life regulates the legislation over protected areas since 1992 (Ulloa et al. 2007).

Particularly in the protected areas in the Andean region, a main goal of the SNAP is the conservation of water resources as a main provisioning service for civil society; four national parks are located in the eastern Cordillera and three of them serve as water reserves for important urban centres. A detailed analysis of water resources as a provisioning service and the impacts of forest conversion on water regulation and quality are provided in Sect. 9.1. The Ministry of Environment has overall responsibility for the protection of these areas, although it is beginning to share management responsibilities with municipalities and private organisations (Gerique 2010). Multistakeholder initiatives, such as "water protection funds", are implemented as finance mechanisms to support conservation by linking consumers and protected areas through trust funds to compensate for water provision

Table 13.1 Main state protected areas (IUCN categories I and II) and protective forest in the Biosphere Reserve Podocarpus-El Cóndor

Protected area	Extension (ha)	Year of establishment
Podocarpus National Park	144,993	1982
Colambo–Yacuri National Park	43,000	2010
Cerro Plateado biological reserve[a]	26,114	2010
Alto Nangaritza protective forest	128,866	2002
Corazón de Oro protective forest	54,000	2000
Podocarpus–El Cóndor biosphere reserve	1,140,080	2007

Sources: [a]Registro oficial No. 318. Quito; Gerique (2010), Com. López Sandoval, 2011

(Echavarria 2002). The *Fondo Regional del Agua* (FORAGUA) supports, among others, conservation in and around the Podocarpus National Park, PNP (FORAGUA 2009). As an example, water management by the municipality of Loja is accomplished with money raised through a city water tax, created for the protection of watersheds and other priority conservation areas (NCI 2012).

To preserve biodiversity and water resources around PNP has been the goal of the establishment of other protected areas in the region (Table 13.1). The declaration of the Biosphere Reserve Podocarpus-El Cóndor by the UNESCO in October 2007 was a major strategy to improve governance of natural resource management and provision of ecosystem services. This was supported by local institutions, universities, NGOs and the subnational governments of the provinces of Loja and Zamora Chinchipe (NCI 2009). Besides state protected areas, several private reserves under different management initiatives are being established in the region, i.e. the San Francisco Research Station (ECSF) or the Angashcola Communal Forest Reserve.

The effectiveness of protected areas as a strategy for the conservation of ecosystem services is still under discussion. In Ecuador, not only the reforms of the legal framework for land colonisation (Pohle et al. 2010 and Chap. 16) but also the establishment of protected areas have influenced deforestation rates in and around them (Mena et al. 2006). However, illegal actions, such as logging, mining (Naughton-Treves et al. 2006) or increasing hunting and fishing (Castro 2008), are ongoing facts in the region. Furthermore, direct conflicts of "local populations vs. protected areas" in the study area (Burbano 2008; Cabrera and Dumas 2008) clearly show that community-based strategies shall consider the specific cultural and social demands of local populations and reinforce protected area management. The establishment of the Biosphere Reserve offers concrete opportunities to intensify and facilitate partnerships among different stakeholders, i.e. resource users, managers, administrators and local communities at different scales of governance (Gerique 2010).

A valuable contribution to provisioning services that can be provided by protected areas may be ecotourism. Ecotourism is widely considered as a provisioning service, which can help to conserve natural resources while improving the livelihoods of local people (Kiss 2004; Page and Dowling 2001). However, to be able to substitute non-sustainable or degrading practices of forest users, they must receive significant economic benefits from the ecotourism activities. According to Moran-Cahusac (2009), 66 % of the 22,707 foreign visitors of the south of Ecuador are visiting the province of Loja. As for example birders are considered to stay for periods of up to 15 days and spending up to USD 100 per day (Moran-Cahusac 2009), the annual financial contribution to the local economies can be estimated to be in the order of about 10 million USD.

13.3.2 Management of Natural Forests

In the San Francisco valley, natural forest is still the dominating land-cover type in altitudes >2,200 m a.s.l. with a share of 85.4 % (Göttlicher et al. 2009). However, below this altitude, forest cover is only 45.2 %, while forests have been converted into anthropogenic replacement systems (predominantly pastures), which were considered to be more profitable than forest. Homeier et al. (Sect. 8.3.6) estimate that the anthropogenic transformation of natural forests in the study area reduces the average value of the ecosystem services by 10.9 million $ annually!

Most of the 120 timber species that can be found on the Ecuadorian market (Blaser et al. 2011) come from natural forests. About 80 % of the wood species utilised in the city of Loja are harvested in natural forests of the Province of Zamora-Chinchipe, with most species coming from the lowland rainforest, i.e. Almendro (*Swietenia macrophylla*), Yumbingue (*Terminalia amazonia*) and Seique (*Cedrelinga catenaeformis*) (Leischner 2000). The remaining 20 % come from Eucalypt and Pine plantations as well as from dry forests of the Province of Loja, particularly Guayacán (*Tabebuia chrysantha*) and Gualtaco (*Loxopterigum huasango*). In her study of the local wood market in the city of Loja, Leischner (2000) identified the following highly valued woods from the natural forest: Cedro (*Cedrela odorata*, *Cedrela montana* and *Cedrela lilloi*), Forastero (*Ocotea* spec., *Nectandra* spec.), Romerillo (*Podocarpus oleifolius*, *Podocarpus sprucei* and *Prumnopitys montana*). Nogal (*Juglans neotropica*) is also present in the Province of Loja. Consequently, natural forests have an important provisioning function for the local wood industry. Despite this fact there are no coordinated efforts towards a real "management" of natural forests. The normative framework of the "Sustainable Forest Management Programme" (PAFSU) for mechanised logging and the "Simplified Management Programme" (PAFSI) for non-mechanised extraction, introduced in the year 2000, are well adapted to the local reality and are a valuable tool to prevent forest destruction, but they are based on the fact that forests are not managed but harvested (FAO/World Bank 2006). The measurable criteria and indicators defined in these norms are predominantly intended to limit harvesting operations to an ecologically compatible level.

However, silvicultural management techniques, which enhance the economic value of natural forests, such as, e.g. improvement thinning or enrichment planting, are not yet scientifically investigated enough. This was the reason for the establishment of a natural forest experiment within our Research Unit in the forest of the

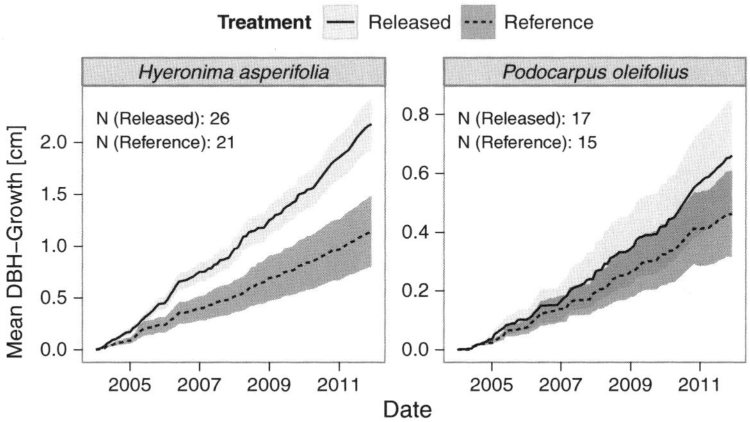

Fig. 13.1 Growth reaction (measured: diameter at breast height, DBH) of individuals of the valuable tree species *Hyeronima asperifolia* and *Podocarpus oleifolius* to an improvement thinning experiment at the RBSF ("released") compared to not treated references. Shown are means ± standard deviations

RBSF (Günter et al. 2008). First results show that improvement felling can slightly increase the number and growth of individuals of valuable timber species at the expense of very frequent but low-value species (Fig. 13.1) (Günter et al. 2004, 2008; Mosandl and Günter 2008; Weber et al. 2008).

An inventory of all trees within the area (13 ha) of the natural forest experiment (Günter et al. 2008) also allowed to calculate the sustainable utilisation potential in the natural forest (Knoke et al. 2009). In the research area, about 140 different tree species with a diameter at breast height (dbh) above 40 cm could be distinguished with mean densities of 42.9 trees per ha. Under consideration of the net mortality and the mean annual diameter increment, 1.5 trees per ha could be harvested annually without changing the actual forest structure. The harvestable trees represented an average bole volume of 1.482 m^3, of which 50 % (0.741 m^3) were considered merchantable (Leischner 2000). Based on these facts, Knoke et al. (2009) calculated yearly sustainable net revenues of US$ 31.4 per ha with an SD of ±21.80 [considering uncertainty of timber biophysical yield (30 %), timber price volatility (10 %) and logging costs (10 %)].

However, as Ecuador's forests are predominantly under the ownership of local communities and indigenous groups, non-timber forest products (NTFPs) may be of similar or even higher importance as timber. According to Blaser et al. (2011), at least 589 species are used for NTFPs in Ecuadorian forests. Although a substantial part of the products may be home consumption to cover the basic needs for food, fuel, constructive material, medicinal purposes or ornamentals and are therefore not captured by the official market statistics, the exports of NTFPs amounted to 13 million USD per year between 2006 and 2008 (Blaser et al. 2011).

From an ethnobotanical survey in the research area, it can be concluded that NTFPs play a substantial role in the livelihoods of all communities. However, the extent differs among the different ethnic groups. In their study Pohle et al. (2010) investigated the use of NTFPs in communities of the Shuar, Saraguros and Mestizos. As typical rainforest dwellers, the Shuar have the highest number of plant uses. They use 204 forest plants (43.5 % of all species used) and most of them are collected in the forest (Table 13.2). For them, the forest is not only an elementary part of their livelihoods but has also a deep cultural and spiritual meaning and provides them with their cultural identity. In fact, timber extraction has been the most profitable income activity for the Shuar over recent decades (Gerique 2010).

In the Saraguro and Mestizo communities, the use of forest plants is noticeable lower (Table 13.2). However, the forest basically supplies them with timber for their own use or to sell occasionally outside the community.

For instance, the Saraguros use 37 tree species for wood products (posts for fences, furniture, house walls and floors) from which 16 are considered wild forest species. Among them, the most important are *Podocarpus oleifolius*, *Prumnopitys montana*, *Tabebuia chrysantha* and *Cedrela* spp. Only six timber species are cultivated (*Cupressus lusitanica*, *Pinus patula*, *Eucalyptus globulus*, *Persea americana*, *Pouteria lucuma* and *Juglans neotropica*). In the Mestizo communities, the use of cultivated species is more dominant. Nevertheless, they also harvest timber from 29 non-cultivated forest species. Furthermore, the forest is used to collect Bromelia and Orchids and to transplant them to their home gardens.

Due to the clearing and overexploitation of the forest resources, timber is progressively becoming scarce. Hence, the Saraguros started to grow transplanted *Cedrela* spp., *Prumnopitys montana* and *Tabebuia chrsysantha* in home gardens or protected pastures to meet their future demand for timber. The Mestizos substitute high-quality timber by low-quality species for the same purpose.

Figure 13.2 provides a summarised overview on the importance for the resource "forest" for the different ethnic groups.

Another very important but still neglected service of natural forests which is closely linked with sustainable forest management is its relevance as a resource for reproductive material (Stimm et al. 2008). The continuous supply with tree reproductive material of desired species, i.e. seed, seedlings and/or material from vegetative propagation, builds the basis for the reestablishment of native species and their extension. Consequently, in many countries of the world the conceptual development and implementation of national tree seed management programmes are a widely accepted and recognised measure of sustainable forestry.

A conservative management of tree genetic resources should focus on the conservation of natural forests with their furnishing of sets of diverse species and—within the species level—diverse populations of trees that are well adapted to the conditions in their natural habitat. Because of exploitation or changes to agricultural land use, Ecuadorian forests have not only declined drastically in area but their diversity has also been depleted, the latter often indicating "genetic degradation" as well.

Reforestation or restoration of course needs an infrastructural network of productive nurseries, whether public or private, to provide the urgently needed reproductive material (Kindt et al. 2006).

Table 13.2 Number of different applications of forest plants used by the Shuar (Shaime, Chumpias, Napints), the Saraguros (El Tibio, El Cristal) and the Mestizos (Los Guabos, Sabanilla, El Retorno, La Fragancia)

	Number of uses of forest plants										
Group	Medicine	Food	Construction	Tools	Fodder	Fuel	Ornamentals	Ritual	Veterinary	Other	Total
Shuar	37	34	43	9	21	18	4	6	9	23	204
Saraguro	1	9	16	9	0	4	1	2	0	1	43
Mestizo	2	10	14	4	0	4	10	1	0	6	51

Note: One plant species can be found in more than one use category (compiled after Gerique 2010)

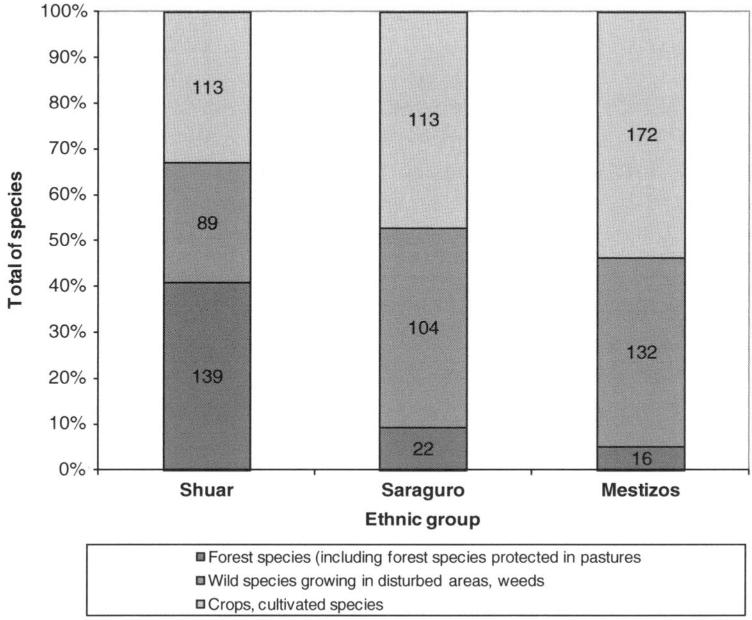

Fig. 13.2 Plant species used by Shuar, Saraguros and Mestizos according to their gathering places. *Note*: The numbers inside the *columns* indicate total values and not percentages (Source: Gerique 2010)

The monitoring, documentation and conservation of seed sources with broad genetic variability is of high importance. Appropriate seed collection for plant propagation is an opportunity to reverse the trends of genetic degradation and species loss, where nurseries play a key role in conserving the gene pool of native trees (Luna and Wilkinson 2009).

Natural forests are usually the major sources for the collection of seeds. In an attempt to evaluate seed sources for seed collection in Loja Province, Samaniego et al. (2005) identified 17 seed sources for *Clethra fimbriata*, *Cedrela montana*, *Myrica pubescens*, *Tecoma stans*, *Eugenia* sp., *Ilex* sp., *Juglans neotropica*, *Ocotea* sp., *Hyeronima macrocarpa*, *Clusia* sp., *Clethra revoluta* and *Lafoensia acuminata*.

Loja Province has an area of 11,000 km^2 and shows an altitudinal variation from 450 m a.s.l. (Macará) to 2,850 m a.s.l. (El Tiro). The larger part of the province is covered with deciduous dry forest (*bosque semideciduo y deciduo piemontano*) characterised by drought tolerant tree species like *Acacia pelyacantha*. Along the western cordillera in an elevation between 1,100 and 1,500 m a.s.l., there is a transition zone between dry to humid forests, so-called sub-montane semideciduous forest (*bosque semideciduo montano bajo*), characterised by the presence of *Tabebuia chrysantha*, *Cecropia litoralis*, *Pleurothyrium obovatum* and *Miconia denticulate* (Leischner 2000), which is followed in elevations up to 2,400 m a.s.l., particularly in the eastern part of the province by sub-montane humid rainforest zone (*bosque húmedo montano bajo*) and subsequently by the montane rainforest ecotone (*bosque húmedo montano*). As a consequence of this variation, Günter et al. (2004) revealed a total of 134 potential gene-ecological zones for the Loja Province, which need to be considered in reforestation planning. Furthermore, because of high interzonal variability, seed collection calendars must be elaborated for each species and separately for different eco-zones of the province. To receive sufficient genetic variety for large-scale reforestation, it is suggested to harvest a minimum of 50 seed trees of one provenance and species (Stimm et al. 2008).

13.3.3 Reforestation

The actual area of planted forest in Ecuador is estimated at 175,000 ha (Blaser et al. 2011). Between 2005 and 2010, the increase of planted forests for production was 11,000 ha, which corresponds to an annual afforestation rate of 2,200 ha. This has to be seen alongside the average annual loss of forest cover of 198,000 ha in the same period. Furthermore, there exists an area of degraded forest land of 3.8 million ha (Blaser et al. 2011). As abandoned and degraded land is barely used and thus contributes only little to the livelihoods of the people, it is important to put it back into production as fast as possible. For instance, for our study area Göttlicher et al. (2009) revealed that pasture management is obviously not be applied in a sustainable manner as only 15.4 % of the converted land is still in use. 32.3 % of the land resource consists of areas that are already abandoned and does neither contribute to the livelihoods of the local people nor to the conservation of biodiversity.

Many studies have shown that the establishment of plantations is an adequate means to restore forest cover on abandoned and degraded lands and to rehabilitate its provisioning functions in a reasonable time (Lamb 1998; Parrotta 1992). Mostly, exotic tree species are used for this purpose because they usually grow well, seedlings are easily accessible, and sufficient knowledge exists on their biological and silvicultural characteristics (Weber et al. 2008, 2011; Aguirre et al. 2011). Correspondingly, 80 % of the above-mentioned afforestations are Eucalypt and Pine plantations in the Andes.

However, our reforestation trials at the RBSF revealed that the growth of the native species *Alnus acuminata* is competitive to that of the exotics *Eucalyptus saligna* and *Pinus patula* (Table 13.3). Moreover, the native *Tabebuia chrysantha* showed excellent survival although the growth was very slow during the first 5 years after establishment. But as Weber et al. (2008) stated, in the initial phase of growth this species invests especially into the development of the root system. Thus, on the long run, the initial inferiority may be compensated by improved growth, superior wood quality and higher timber price at the market.

Aguirre et al. (2006) revealed that many native species can better establish under the shelter or in gaps of *Pinus patula* than at a neighbouring pasture site without any tree cover (Fig. 13.3). This offers the opportunity to use exotic species as pioneer crop to achieve a fast forest cover and to subsequently convert them into more

Table 13.3 Mean height, basal diameter and survival of three native and two exotic species 60 months (*Alnus* 48 months) after planting on three different sites: pasture, bracken and shrub, respectively

	Native species			Exotic species	
	Aa	Mp	Tc	Es	Pp
Height [cm ±SD]					
Pasture	409.1 (±245.4)	146.2 (±72.2)	25.8 (±10.2)	313.8 (±244.8)	612.4 (±140.9)
Bracken	191.8 (±104.7)	224.2 (±70.9)	44.1 (±35.9)	169.6 (±131.2)	425.9 (±124.1)
Shrub	161.0 (±82.0)	100.1 (±68.9)	48.4 (±26.4)	221.9 (±105.2)	401.8 (±132.4)
Basal diameter [cm ± SD]					
Pasture	7.1 (±4.7)	2.5 (±1.5)	1.2 (±0.2)	5.2 (±4.9)	14.8 (±4.3)
Bracken	4.5 (±2.1)	6.7 (±2.2)	1.6 (±0.8)	2.5 (±2.2)	10.8 (±3.7)
Shrub	3.0 (±1.2)	3.8 (±1.9)	1.8 (±0.6)	2.5 (±1.7)	7.5 (±3.2)
Survival [% ± SD]					
Pasture	49 (±14.9)	44 (±24.0)	80 (±13.8)	82 (±11.1)	95 (±5.8)
Bracken	38 (±24.3)	79 (±10.9)	82 (±13.2)	88 (±16.7)	88 (±12.8)
Shrub	39 (±37.7)	37 (±14.6)	97 (± 7.0)	76 (±27.0)	92 (±7.3)

Aa, *Alnus acuminata*; Mp, *Morella pubescens*; Tc, *Tabebuia chrysantha*; Es, *Eucalyptus saligna*; Pp, *Pinus patula*. *N* = 25 per species (*SD* standard deviation)

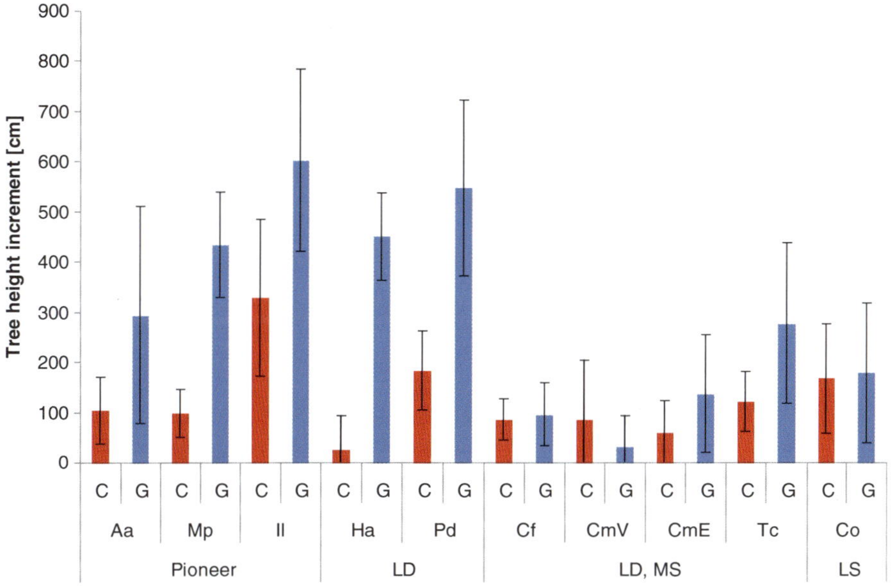

Fig. 13.3 Growth of individuals of ten native species planted in an enrichment planting experiment under the shelter and in gaps of a Pine plantation: shown are means and standard deviations of tree height increments after 85 months; Aa, *Alnus acuminata*; Mp, *Morella pubescens*; Il, *Isertia laevis*; Ha, *Heliocarpus americanus*; Pd, *Piptocoma discolor*; Cf, *Cupania* sp.; CmV, *Cedrela montana* (seed source Vilcabamba); CmE, *Cedrela montana* (seed source RBSF); Tc, *Tabebuia chrysantha*; Co, *Cinchona officinalis*. *C* canopy, *G* gaps, *LD* light demanding, *MS* mid successional, *LS* late successional

natural stands via underplanting. Consequently, the reforestation of unproductive land is an important element of the efforts to reduce the pressure on natural forests and to rehabilitate the provisioning services of forests.

In 2006, the Ecuadorian government approved a National forest and reforestation programme in order to restore a higher forest cover and to increase forest production. According to this programme, 750,000 ha of industrial forest plantations, 150,000 ha of agroforestry practices and 100,000 ha of protective plantations shall be established in Ecuador within 20 years. To fulfil the demand for the establishment of those plantations, about 56 million seedlings are needed on an annual basis. Based on information provided by UNDP/ILO (1989) on work organisation in nurseries, a permanent workforce of 7–8 workers (incl. foreman and watchman) is needed for the operation of a nursery producing about 100,000 containerised seedlings with only one planting season (begin of rainy season) per year. For producing annually 56 million seedlings this means a need of 560 nurseries with a total of nearly 4,500 permanent jobs and a high number of additional temporary workforces. Annual sales volume of 56 million seedlings may be calculated with 28 million US$ (average of 0.5 US$ per seedling).

Under the assumption that the Province of Loja is planning to reforest 40,000 ha of abandoned land in the next 20 years, this would result in an average annual establishment of 2,000 ha and a calculated number of 2.2 million tree seedlings per year. If 80 % of the area (1,600 ha) should be established with native tree seedlings, about 1.8 million of native seedlings have to be produced yearly. Let us further assume that 10 % of the seedlings (180,000) should be from *Cedrela*, another 10 % of *Tabebuia*, 10 % from *Alnus* and another 10 % of *Juglans* the required seed mass is 4.3 kg for *Cedrela*, 4.13 kg for *Tabebuia*, 0.23 kg for *Alnus* and 7,650 kg for *Juglans*. Our studies of *Cedrela montana* fruiting phenology showed that in a good seed year, one individual tree is producing up to 11,000 seeds (Stimm et al. 2008), which means a population of 32 seed trees will be needed to produce the required amount.

Of course such an estimate is theoretical because it does for instance not take into account the necessity of seed stands from different gene-ecological zones. However, it makes clear that the establishment of competent national and regional tree seed centres in Ecuador is a prerequisite to meet the ambitious goals of the national reforestation programme. The seed centres task is the appropriate management of a network of seed production stands, which is a prerequisite of sustainable seed and seedling production.

13.4 Conclusions

Forests do still have an important provisioning function for the people in the study region. As many of the products taken from the forests are used for home consumption, it is very difficult to quantify their total amount and market value exactly. However, there is no doubt that there exists a high potential to enhance respectively rehabilitate their role in the provision of goods and services for the local communities. Sustainable management of natural forest considering low impact harvesting, improvement thinning and enrichment planting is considered as one important option. The most important need is to reintegrate the increasing amount of unproductive and degraded land into the production area. Plantations either with native or exotic species are an adequate tool for this. To meet the goals of the national afforestation programme, the establishment of national and regional tree seed centres is seen as an indispensable prerequisite. However, this would provide thousands of job opportunities and generate a huge market for forest reproductive material. Finally, agricultural and forestry activities must be integrated to come to more sustainable land use. Knoke et al. (2009) have already presented a model, which combines agriculture, plantations and selective management of natural forests leading to more stable and higher income while halting deforestation. Further examples are presented by Knoke et al. in Chap. 25.

References

Aguirre N, Günter S, Weber M, Stimm B (2006) Enrichment of Pinus patula plantations with native species in southern Ecuador. Lyonia 10(1):17–29

Aguirre N (2007) Silvicultural contributions to the reforestation with native species in the tropical mountain rainforest region of South Ecuador. Dissertation Fakultät Wissenschaftszentrum Weihenstephan. http://mediatum2.ub.tum.de/doc/623642/623642.pdf

Aguirre N, Palomeque X, Weber M, Stimm B, Günter S (2011) Reforestation and natural succession as tools for restoration on abandoned pastures in the Andes of South Ecuador. In: Günter S, Weber M, Stimm B, Mosandl R (eds) Silviculture in the tropics. Springer Verlag, Heidelberg, pp 513–524

Blaser J, Sarre A, Poore D, Johnson S (2011) Status of tropical forest management 2011. ITTO Technical Series No 38. International Tropical Timber Organization, Yokohama, Japan

Burbano M (2008) La conservación y el desarrollo en zonas rurales habitadas: implicaciones entre el Bosque Protector Corazón de Oro y la comunidad de El Tibio, Parroquia Imbana, Cantón Zamora. Bachelor Thesis, Institute of Geography, PUCE, Quito

Cabrera P, Dumas J (2008) Diálogo en Nangaritza: caminando de la violencia a la colaboración. Fundación Futuro Latinoamericano, Quito

Castro P (2008) Diagnóstico socioambiental del uso de la fauna Silvestre en el Bosque Protector Alto Nangaritza – Región Sur del Ecuador. Thesis, Universidad Técnica Particular de Loja, Loja, Ecuador

Echavarria M (2002) Financing watershed conservation: the FONAG waterfund in Quito, Ecuador. In: Pagiola S, Bishop J, Landell-Mills N (eds) Selling forest environmental services: market-based mechanisms for conservation and development. Earthscan, London, pp 91–101

FAO (2010) Global forest resources assessment 2010 country report: Ecuador. http://www.fao.org/forestry/fra/67090/en/

FAO (Food and Agriculture Organization) (2011) State of the World's Forests. FAO, Rome

FAO/World Bank (2006) Andean countries: a strategy for forestry – case studies, vol III of V. FAO/World Bank Cooperative Programme. Latin America and the Caribbean Service, Investment Centre Division, 06/028 CP-LAC. FAO/World Bank, Rome/Geneva

FORAGUA (Fondo Regional del Agua) (2009) Fideicomiso. http://www.foragua.org/fideicomiso.html. Accessed 10 Jan 2012

Gerique A (2010) Biodiversity as a resource: plant use and land use among the Shuar, Saraguros, and Mestizos in tropical rainforest areas of southern Ecuador. Dissertation, Friedrich Alexander Universitaet Erlangen-Nürnberg, Erlangen

Göttlicher D, Obregón A, Homeier J, Rollenbeck R, Nauss T, Bendix J (2009) Land-cover classification in the Andes of southern Ecuador using Landsat ETM+ data as a basis for SVAT modelling. Int J Remote Sens 30(8):1867–1886

Günter S, Stimm B, Weber M (2004) Silvicultural contributions towards sustainable management and conservation of forest genetic resources in Southern Ecuador. Lyonia 6(1):75–91

Günter S, Cabrera O, Weber M, Stimm B, Zimmermann R, Fiedler K, Knuth J, Boy J, Wilcke W, Iost S, Makeschin F, Werner FA, Gradstein SR, Mosandl R (2008) Natural forest management in neotropical mountain rain forests: an ecological experiment. In: Beck E, Bendix J, Kottke I, Makeschin F, Mosandl R (eds) Gradients in a tropical mountain ecosystem of Ecuador. Ecological studies, vol 198. Springer, Berlin, pp 363–376

Kindt R, Lillesø JPB, Mbora A, Muriuki J, Wambugu C, Frost W, Beniest J, Aithal A, Awimbo J, Rao S, Holding-Anyonge C (2006) Tree seeds for farmers: a toolkit and reference source. World Agroforestry Centre, Nairobi

Kiss A (2004) Is community-based ecotourism a good use of biodiversity conservation funds? Trends Ecol Evol 19(54):232–237

Knoke T, Weber M, Barkmann J, Pohle P, Calvas B, Medina C, Aguirre N, Günter S, Stimm B, Mosandl R, Walter F, Maza B, Gerique A (2009) Effectiveness and distributional impacts of payments for reduced carbon emissions from deforestation. Erdkunde 63:365–384

Lamb D (1998) Large-scale ecological restoration of degraded tropical lands: the potential role of timber plantations. Restor Ecol 6:271–279

Leischner B (2000) Holznutzung und Holzmarkt in den Provinzen Loja und Zamora-Chinchipe, Süd-Ecuador. Unpublished Diploma Thesis, School of Forest Science and Resource Management, Technische Universität München, Munich

López Sandoval (2011) personal communication

Luna T, Wilkinson KM (2009) Collecting, processing, and storing seeds. In: Dumroese R, Luna T, Landis TD (eds) Nursery manual for native plants: a guide for tribal nurseries, vol 1: nursery management. Agriculture handbook 730. US Department of Agriculture, Forest Service, Washington, DC, 302 p

Mena C, Barbieri A, Walsh S, Erlien C, Holt F, Bilsborrow R (2006) Pressure on the Cuyabeno wildlife reserve: development and land use/cover change in the northern Ecuadorian Amazon. World Dev 34(10):1831–1849

Ministerio del ambiente (2013) Subsistema Patrimonio de Áreas Naturales del Estado (PANE). http://web.ambiente.gob.ec/sites/default/files/users/jloartefls/CUADRO%20PANE.pdf. Accessed 13 May 2013

Moran-Cahusac C (2009) Ecotourism in Podocarpus National Park. J Sustain For 28 (6–7):777–798

Mosandl R, Günter S (2008) Sustainable management of tropical mountain forests in Ecuador. In: Gradstein SR, Homeier J, Gansert D (eds) The tropical mountain forest. Biodiversity and ecology series 2. Universitätsverlag Göttingen, pp 179–195

Naughton-Treves L, Nora Alvarez-Berríos N, Brandon K, Bruner A, Buck Holland M, Ponce C, Sáenz M, Suárez L, Treves A (2006) Expanding protected areas and incorporating human resource use: a study of 15 forest parks in Ecuador and Perú. Sustain Sci Pract Policy 2 (2):32–44

NCI (Nature and Culture International) (2009) Podocarpus–El Cóndor Biosphere Reserve. http://www.natureandculture.org/htm/ecuador/areas-andes-podocarpus.htm. Accessed 10 Jan 2012

NCI (2012) http://www.naturalezaycultura.org/concept/htm/news/2008-08-ElCarmen.htm. Accessed 29 Feb 2012

Page SJ, Dowling RK (2001) Ecotourism. Longman Group United Kingdom, 338 p

Parrotta JA (1992) The role of plantation forests in rehabilitating degraded tropical ecosystems. Agric Ecosyst Environ 41:115–133

Pohle P (2008) The people settled around Podocarpus National Park. In: Beck E, Bendix J, Kottke I, Makeschin F, Mosandl R (eds) Gradients in a tropical mountain ecosystem of Ecuador. Ecological studies, vol 198. Springer, Berlin, pp 25–36

Pohle P, Gerique A (2008) Sustainable and non sustainable use of natural resources by indigenous and local communities. In: Beck E, Bendix J, Kottke I, Makeschin F, Mosandl R (eds) Gradients in a tropical mountain ecosystem of Ecuador, vol 198, Ecological studies. Springer, Berlin, pp 347–362

Pohle P, Gerique A, Park M, López Sandoval MF (2010) Human ecological dimensions in sustainable utilization and conservation of tropical mountain forests under global change in southern Ecuador. In: Tscharntke T, Leuschner C, Veldkamp E, Faust H, Guhardja E, Bidin A (eds) Tropical rainforests and agroforests under global change. Environmental science and engineering. Springer, Berlin, pp 477–509

Samaniego C, Ordóñez O, Prado L, Morocho M (2005) Las fuentes semilleras y semillas forestales nativas de Loja y Cañar: participación social en el manejo. Fundación Ecológica Arcoiris, Loja, Ecuador

Stimm B, Beck E, Günter S, Aguirre N, Cueva E, Mosandl R, Weber M (2008) Reforestation of abandoned pastures: seed ecology of native species and production of indigenous plant material. In: Beck E, Bendix J, Kottke I, Makeschin F, Mosandl R (eds) Gradients in a tropical mountain ecosystem of Ecuador, vol 198, Ecological studies. Springer, Berlin, pp 417–429

Ulloa R, Aguirre M, Camacho J, Cracco M, Dahik A, Factos M, Gutiérrez I, Kigman S, Quishpe MD, Moreno F, Ribadeneira-Sarmiento M, Rivas J, Rodríguez A, Samaniego J, Suárez L, Tobar M, Viteri X, von Horstman E (2007) Situación Actual del Sistema Nacional de Áreas Protegidas del Ecuador. Informe Nacional – Ecuador 2007. II Congreso Latinoamericano de Áreas Protegidas, Sept 30 a Oct 6. Bariloche, Argentina. Ministerio del Ambiente, Comité Ecuatoriano de la UICN, Comisión Mundial de Áreas Protegidas CMAP, Quito, Ecuador

UNDP, ILO (1989) Tree nurseries – an illustrated technical guide and training manual. Special public works programmes. Booklet No. 6

Weber M, Günter S, Aguirre N, Stimm B, Mosandl R (2008) Reforestation of abandoned pastures: silvicultural means to accelerate forest recovery and biodiversity. In: Beck E, Bendix J, Kottke I, Makeschin F, Mosandl R (eds) Gradients in a tropical mountain ecosystem of Ecuador, vol 198, Ecological studies. Springer, Berlin, pp 447–458

Weber M, Stimm B, Mosandl R (2011) Plantations for protective purposes and rehabilitation. In: Günter S, Weber M, Stimm B, Mosandl R (eds) Silviculture in the tropics. Springer Verlag, Heidelberg, pp 475–490

Chapter 14
Mycorrhiza Networks Promote Biodiversity and Stabilize the Tropical Mountain Rain Forest Ecosystem: Perspectives for Understanding Complex Communities

Ingrid Kottke, Sabrina Setaro, Ingeborg Haug, Paulo Herrera, Dario Cruz, Andreas Fries, Julia Gawlik, Jürgen Homeier, Florian A. Werner, Andrés Gerique, and Juan Pablo Suárez

I. Kottke (✉)
Plant Evolutionary Ecology, Institute of Evolution and Ecology, Eberhard-Karls-University Tübingen, Auf der Morgenstelle 1, 72076 Tübingen, Germany

Universidad Técnica Particular de Loja, Centro de Biologia Celular y Molecular, San Cayetano Alto s/n C. P., 1101608 Loja, Ecuador
e-mail: ingrid.kottke@unibund.uni-tuebingen.de

S. Setaro
Department of Biology, Wake Forest University, 1834 Wake Forest Road, Winston-Salem, NC 27106, USA

I. Haug
Plant Evolutionary Ecology, Institute of Evolution and Ecology, Eberhard-Karls-University Tübingen, Auf der Morgenstelle 1, 72076 Tübingen, Germany

P. Herrera • D. Cruz • J.P. Suárez
Universidad Técnica Particular de Loja, Centro de Biologia Celular y Molecular, San Cayetano Alto s/n C. P., 1101608 Loja, Ecuador

A. Fries
Department of Geography, Philipps-University Marburg, Deutschhausstrasse 10, 33037 Marburg, Germany

J. Gawlik • A. Gerique
Institute of Geography, University of Erlangen, Kochstrasse 4/4, 91054 Erlangen, Germany

J. Homeier
Plant Ecology, University of Göttingen, Untere Karspüle 2, 37075 Göttingen, Germany

F.A. Werner
Institute of Biology and Environmental Sciences, Carl von Ossietzky University Oldenburg, PO Box 2503, 26111 Oldenburg, Germany

J. Bendix et al. (eds.), *Ecosystem Services, Biodiversity and Environmental Change in a Tropical Mountain Ecosystem of South Ecuador*, Ecological Studies 221,
DOI 10.1007/978-3-642-38137-9_14,

14.1 Introduction

Mutualism in fungal–plant root associations was predicted 130 years ago by Frank (1885) after studying absorptive roots of Fagaceae morphologically. Frank found all the absorptive roots ensheathed by fungal mycelium not only spreading in the soil but also penetrating among the root cortical cells. He concluded from the observations that these trees depend on fungi for mineral and water uptake while supplying carbohydrates to the fungi promoting their growth and propagation. Comprehensive field studies and experimental investigations of physiological and molecular mechanisms during the last decades did not only corroborate Frank's findings and farsighted suggestions but also revealed further, differently structured, mutually beneficial (mutualistic) fungal–root associations (mycorrhizae) among the majority of land plants and diverse fungal groups (mycobionts). Thus, mycorrhizae provide an essential ecosystem service and by this a provisioning ecosystem service to humans.

Theories on biodiversity and ecosystem stability of multi-species assemblages only recently integrated mutualistic interactions (Thompson 2005), and mycorrhizae were not in the focus of the theoretical research (Ings et al. 2009). To compensate for this lack of knowledge, we carried out a joined inventory of plants and mycorrhizal fungi in the tropical mountain rain forest area of Southern Ecuador comparing pristine forest and open landscape. The floristic inventory revealed high plant diversity, details given in Chap. 8. By sampling mycorrhizae of identified species of trees, Orchidaceae and Ericaceae we compiled presumably the broadest, culture-independent molecular data set on mycobionts in a tropical area so far. We found an unexpected richness in each of the associated fungal groups Glomeromycota, Sebacinales, Tulasnellales and Atractiellales (Setaro et al. 2006a, b; Suárez et al. 2006, 2008; Kottke et al. 2010; Haug et al. 2010). We rely on the ultrastructural features as documented by us (references cited above) indicating mycorrhizal interactions. It is just prohibitive to isolate all the mycobionts and investigate their mycorrhizal potential and the cost and win situation for plants with the large number of species in the tropical area.

While Chap. 8 focuses on the environment-driven species turnover, we apply network theories to mycorrhizal assemblages to interpret the observed "long-tailed" link distributions among few frequent and many rare species of plants and mycobionts, respectively, and the "nested" structure of the networks as stabilizing and promoting the multispecies ecosystem. During the recent decade, these key tools based on appropriate non-linear functions were applied to analyse species-rich, mutualistic plant–animal assemblages (pollinator and seed disperser webs). Analyses revealed general features in network architecture of fundamental importance for community formation, biodiversity persistence and stability of ecosystems (Bastolla et al. 2009; Mello et al. 2011 and literature therein). So far, studies on network architecture of mycorrhizae have only addressed specialization and phylogenetic signals in one orchid genus in Europe (Jacquemyn et al. 2011), epiphytic versus terrestrial orchids on La Réunion island (Martos et al. 2012) and community structure of plant–arbuscular mycorrhizal fungi (Montesinos-Navarro et al. 2012; Chagnon et al. 2012). We discuss biological arguments for establishment of the observed mycorrhizal network architecture. We tentatively relate distinct climate situations of the investigation sites to potential modularity in the mycorrhizal networks. Concluding, we point to the outcome of our studies in respect to provisioning ecosystem services for human society.

14.2 Floristic and Mycobiont Inventories

14.2.1 Sampling Methods

Floristic and mycobiont inventories were carried out in the natural forest and on human changed sites in the area of Reserva Biológica San Francisco (RBSF, see Fig. 14.1) on the eastern slope of the Cordillera El Consuelo bordering the Podocarpus National Park, Zamora-Chinchipe Province (3°58′ S and 79°04′ W), Southern Ecuador. Detailed description of the study area is given in Chap. 1. Tree inventory in the pristine forest was based on 17 permanent plots (0.68 ha), along an altitudinal gradient between 1,850 and 2,570 m on the steep, north facing slope of the Cordillera (Homeier et al. 2010). In addition, 1,191 tree individuals were identified on 18 plots (0.72 ha) located on different slope positions between 1,900 and 2,100 m a.s.l. Floristic inventories on herbs and shrubs were carried out on the anthropogenic influenced south facing slope comprising used and abandoned pastures, bracken fern areas, forest relicts and a 30-year-old pine afforestation along transects within 1,900–2,200 m a.s.l. according to Peters et al. (2010). Epiphytic and hemiepiphytic Ericaceae and epiphytic Orchidaceae were recorded in the pristine forest in 16 plots of 100 m^2 at 1,810–2,090 m a.s.l. and a regenerating forest site of 0.1 ha at 2,140 m. Terrestrial Orchidaceae and Ericaceae were recorded on a human caused, recovering road cut on the north facing slope. More details on methods are given in Chap. 8.

Mycorrhizae formed by Glomeromycota with tree roots were sampled from 28 tree species, 63 individuals on total, on some of the tree inventory plots, some additional plots of the pristine forest and from tree seedlings of four native species planted 1–3 years in advance of sampling on the abandoned pastures, bracken fern and shrub areas mentioned above (Günter et al. 2009; Haug et al. 2010; intention of reforestation is explained in Sect. 13.3). Our pioneer sampling intended to obtain mycorrhizae from as many forest tree species as possible and did not respect abundance of individual tree species. Only the seedlings on the plantation site were repetitively sampled and these species will, thus, appear as core tree species in the networks. Mycorrhizae formed by terrestrial and hemiepiphytic Ericaceae were sampled by random from flowering individuals in the pristine and the regenerating forest, on the road cut and in the bushy páramo. Mycorrhizae of 59 Orchidaceae species, identified by barcoding, were sampled on terrestrial and

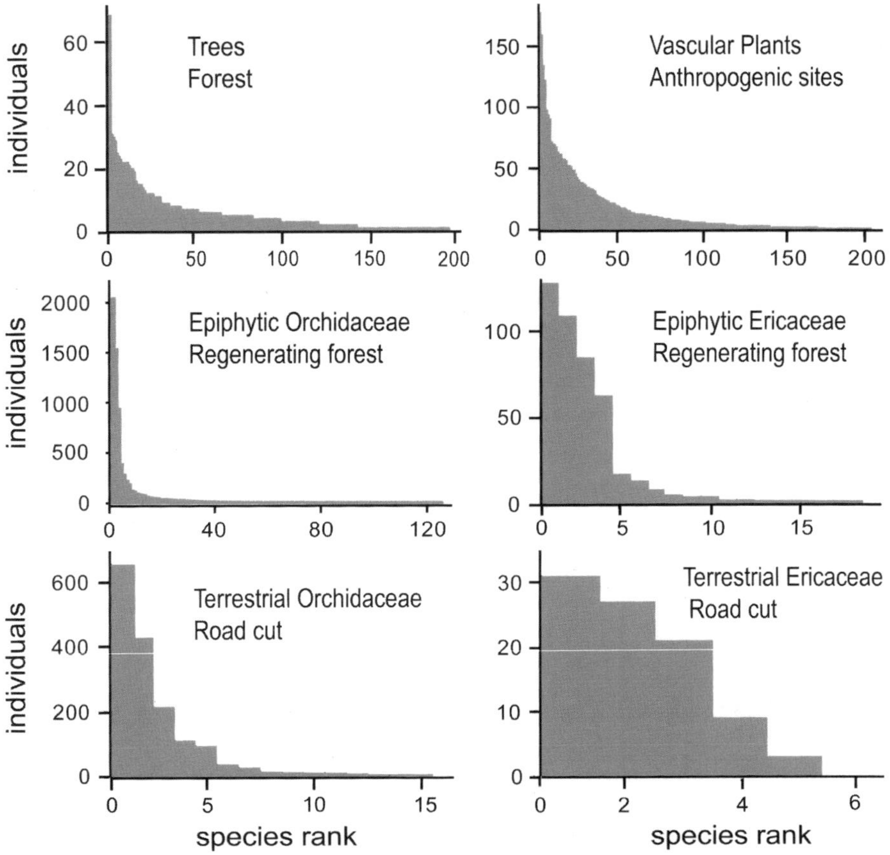

Fig. 14.1 Examples of plant frequency distributions obtained by the inventories display long-tailed distributions with few frequent and many rare species, except for terrestrial Ericaceae. Respective species numbers are displayed by the species ranks: frequency of 197 tree species based on 1,191 individuals within 18 plots; frequency of vascular plants based on 204 species and 3,392 individuals along a 50 × 2 m transect; frequency of 126 epiphytic Orchidaceae species based on 7,251 individuals within 0.1 ha; frequency of 18 epiphytic Ericaceae species based on 457 individuals within 0.1 ha; frequency of 15 terrestrial orchid species based on 1,663 individuals; frequency of 5 terrestrial Ericaceae species based on 91 individuals

epiphytic plots on two sites in the pristine forest, one in the regenerating forest site and one on the human-caused road cut (Kottke et al. 2010).

Mycobionts were accessed by molecular tools as given in Haug et al. (2010), Setaro and Kron (2012) and Kottke et al. (2010). Operational Taxonomic Units (OTUs) were defined as surrogates for species on the basis of sequence similarity by use of OPTSIL (Göker et al. 2009) at <3 % divergence of ITS/5.8S sequences for Tulasnellales and Sebacinales and <1 % divergence of 18S for Glomeromycota. As delimitation of OTUs is still discussed (Hibbett et al. 2011; Cruz et al. 2011), we explored also delimitation of Tulasnellales at <5 % and Glomeromycota at <2 % divergence.

14.2.2 Diversity of Plants and Mycobionts

More than 280 tree species (diameter at breast height, dbh > 5 cm), presenting 53 families were so far identified. The most species-rich families are Lauraceae (39 species, *Ocotea* with 12 and *Persea* with 10 species), Rubiaceae (24 species) and Melastomataceae (20 species) with the genus *Miconia* as the most speciose (13 species); 21 families are represented by just one species. Epiphytic Ericaceae covered 31 species in 14 genera. The most speciose vascular plant group are Orchidaceae, with 337 species within 56 genera in the natural forest, the epiphytic genera *Pleurothallis* (52 species), *Stelis* (46 species) and *Maxillaria* (35) representing the most speciose (Homeier and Werner 2007). In the regenerating forest (0.1 ha), 126 epiphytic Orchidaceae and 18 species of Ericaceae were found. Five species of Ericaceae and 12 species of Orchidaceae were found on the human-caused road cut. In total 245 species of 169 genera and 73 families, mostly native plants, were recorded on the human-influenced sites. The most speciose families were Asteraceae (27 genera and 45 species), followed by Poaceae (16 genera and 24 species) and Melastomataceae (6 genera and 15 species); Orchidaceae comprised 11 species. The highest species richness was found in the pine afforestation and the lowest on the abandoned *Setaria* pastures. Species frequency distributions appeared as strongly skewed, with few frequent and many rare species (Fig. 14.1). Only abundance of the few Ericaceae species on the road cut was more even (Fig. 14.1).

The 28 tree species were found associated with 60 Glomeromycota OTUs (<1 % divergence; 28 at <2 % divergence) covering Glomus A (44), Glomus B (1), Acaulospora (6), Archaeosporales (6), Gigasporaceae (2) and Paraglomeraceae (1). The 11 Ericaceae species were found associated with 20 Sebacinales. The 59 Orchidaceae species were found associated with Tulasnellales (33 OTUs at <3 % divergence; 29 at <5 % divergence), Sebacinales (12 OTUs at <3 % divergence) and Atractiellales (3).

14.3 Mycorrhiza Networks

14.3.1 Networks Analysis

Networks as addressed here are by definition a set of nodes (species) connected through links (interactions). Mutualistic networks, such as mycorrhizae, are two-mode networks since there are two well-defined types of nodes, the fungi and the plant species, and mutualistic interactions occur between but not within these node types. We treat trees associated with Glomeromycota, Ericaceae associated with Sebacinales and Orchidaceae associated with Tulasnellales, Sebacinales and Atractiellales as three distinct networks. No overlap among mycobionts and plant

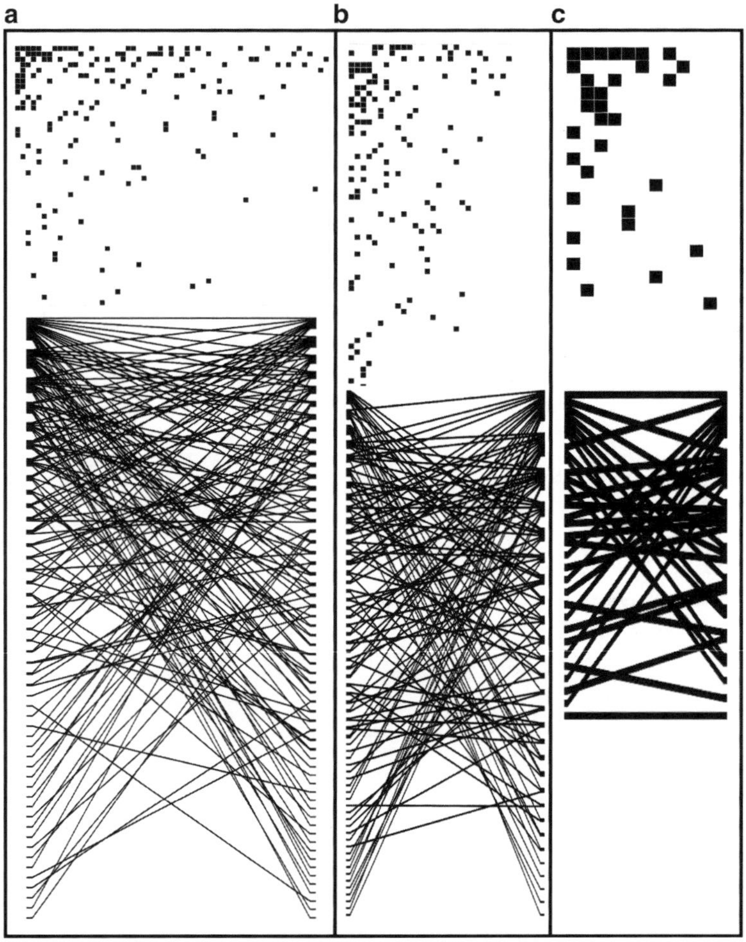

Fig. 14.2 Grid and network graphs of Orchidaceae (**a**), trees (**b**), Ericaceae (**c**) and their respective mycobionts. Grids show plants in columns and mycobionts in rows ordered from the original matrix according to declining frequency of links from *left* to *right* and *top* to *bottom*. Networks are based on the ordered matrices (grids) and display mycobionts on the *left* and plants on the *right* side. The nested associations are clearly visible in all the examples: species with few links are predominantly associated with subsets of frequently linked species

groups was found by our investigations so far (Kottke et al. 2008; Setaro et al. 2013). Mycorrhiza network topology was based on qualitative links (presence/absence data). The interactions were assembled in binary matrices. Species degrees as number of links of each species were obtained as marginal totals from the matrices (Fig. 14.2). The cumulated probabilities of the degree distributions were displayed for plants and mycobionts in log–log plots and were tested in respect to linear or exponential fit by the use of Excel in Open Office (not shown). Interactions are displayed as bipartite graphs sorted by decreasing number of interactions (Fig. 14.2; Bascompte and Jordano 2007). Nestedness was calculated as *NODF*

(N = paired Overlap, Decreasing Fill) and significance tested against two null models, ER model assigning interactions completely at random and CE model randomizing each species similar to observed data matrix, implemented in ANINHADO vers 3.0 Bangu (Guimaraes and Guimaraes 2006; Almeida-Neto et al. 2008).

14.3.2 Network Topologies

A nested distribution of links among individual species with a core of frequent interactions and rare species linked to subsets of the higher linked species are apparent in all three networks (Fig. 14.2a–c). We found the mycorrhiza network of trees and Orchidaceae as significantly nested (Table 14.1). Trees and Glomeromycota are almost similarly nested, but mycobionts of Orchidaceae are less nested than the plants (Table 14.1). In the Ericaceae–Sebacinales network, nestedness was not significant (Table 14.1). The matrix size may be just too small to obtain significance (Bascompte et al. 2003). Future inclusion of Ascomycete mycobionts associated with the Ericaceae may improve the database. Values for *NODF* are low in all our networks when compared with published data (Jacquemyn et al. 2011; Mello et al. 2011). Almeida-Neto et al. (2008) showed that *NODF* is independent of network size, but values are related to matrix fill, decreasing strongly with decreasing matrix fill. Due to tropical species richness, thus largeness of the matrices, and a still strong constrain for molecular sampling of mycobionts (DNA product success rate, selective primers and costs), up to now we obtained only low fills of the matrices (Table 14.1), means that we have probably a disproportional high number of singular links. Matrix fill and *NODF* values in our networks increase when we define the OTUs at lower resolution level as the number of singular links becomes lower (Table 14.1). It is, however, biologically more appropriate to improve the database on the conventional 3 % or 1 % OTU-level, respectively, by further sampling of mycobionts. Ongoing sampling of mycobionts should probably respect the now known frequencies of the plant species. Higher fills of the matrices and quantitative data on repetitive pairwise interactions may be obtained by more appropriate sampling of mycorrhizae from frequent plant species as these are linked with frequent and with rare mycobionts. Because of the lack of quantitative data for interaction strength, we have not addressed the question of potential complementary specialization in the mycorrhizal associations (Blüthgen et al. 2006).

Degree distributions of plants and mycobionts in our three networks are strongly skewed by few species with many links and many species with few links (Fig. 14.3). Degree distributions thus resemble those found for pollinator and frugivore webs

Table 14.1 Network metrics

Mycorrhizal association	Richness [matrix size $S = p + m$]	NODF [significance against ER and CE models respectively]	Matrix fill [%]	Number of links [degree k and k_i]
Trees—Glomeromycota (OTUs < 1 % divergence)	*S* 88 *p* 28 *m* 60	*t* 15.21 ($p = 0.001$/ 0.01) *p* 15.84 *m* 12.28	7.74	*k* 130; mean 1.5 mean k_i *p* 4.64 k_i *m* 2.16
Trees—Glomeromycota (OTUs < 2 % divergence)	*S* 56 *p* 28 *m* 28	*t* 28.0 ($p = 0.001$) *p* 26.71 *m* 29.3	13.01	*k* 102; mean 1.8 mean k_i *p* 3.6 k_i *m* 3.6
Ericaceae—Sebacinales (OTUs < 3 % divergence)	*S* 31 *p* 11 *m* 20	*t* 17.27 (ns) *p* 22.06 *m* 15.88	15.00	*k* 33; mean 1.1 mean k_i *p* 3.0 k_i *m* 1.6
Orchidaceae—Tulasnellales/ Sebacinales/Atractiellales (Tul. and Seb. OTUs < 3 % divergence)	*S* 107 *p* 59 *m* 48	*t* 14.10 ($p = 0.001$) *p* 17.18 *m* 9.42	5.16	*k* 159; mean 1.5 mean k_i *p* 2.7 k_i *m* 3.3
Orchidaceae—Tulasnellales/ Sebacinales/Atractiellales (Tul. OTUs < 5 %, Seb. OTUs < 3 % divergence)	*S* 103 *p* 59 *m* 44	*t* 16.47 ($p = 0.001$) *p* 20.07 *m* 9.95	5.93	*k* 154; mean 1.5 mean k_i *p* 2.61 k_i *m* 3.5

t total, *p* plants, *m* mycobionts, *OTU* operational taxonomic units, *ER* model assigning interactions completely at random, *CE* model randomizing each species similar to observed data matrix, *Tul.* Tulasnellales, *Seb.* Sebacinales

shown to following a power law or truncated power law function (Jordano et al. 2003; Medan et al. 2007). Power law distribution of links in networks was found to be a consequence of (1) continuous integration of new species and (2) new species linking preferentially to already well-associated species (Barabási and Albert 1999; Zhang et al. 2011). This way of assemblage building leads to an increasing "nested architecture" of the mutualistic networks. Preferential attachment and nestedness were mathematically shown to be "two sides of the same coin" (Medan et al. 2007). Because of still low numbers of data points, we could not find definite support for truncated power law fit for cumulated probabilities of degree distributions. Significant nestedness indicates, however, non-randomly structured assemblages based on preferential attachment at least for the arbuscular mycorrhizal fungi (AMF) and orchid mycorrhiza networks.

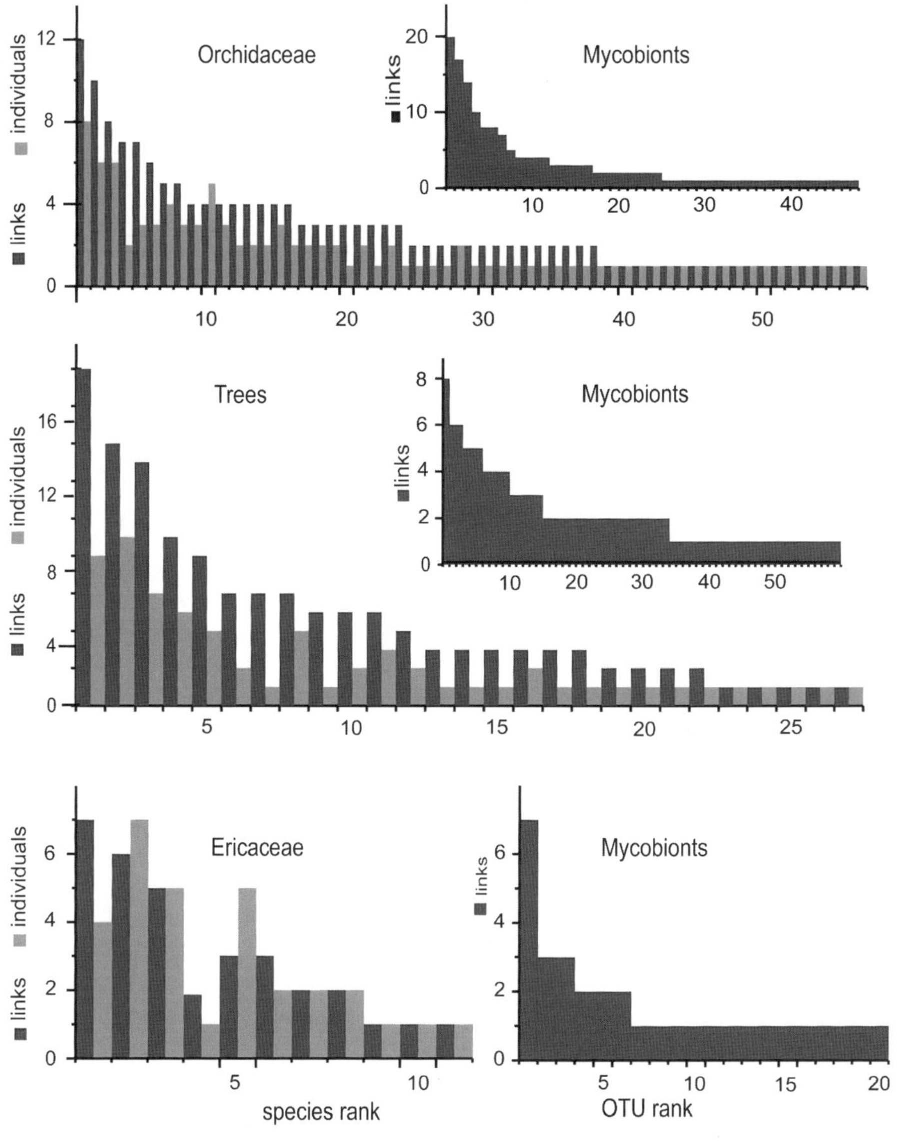

Fig. 14.3 Long-tailed degree distributions (frequency distribution of number of links per species) of plants and their respective mycobionts. Inserted number of plant individuals from which the mycobionts were generated are generally, but not individually congruent with degree distributions

14.3.3 Underlying Biological Mechanisms Maintaining High Biodiversity

The independently assessed abundances of the plant species in the investigation area show long-tailed distributions very similar to degree distributions (Figs. 14.1 and 14.3). No strict parallel is, however, apparent when comparing the numbers of sampled plant individuals and the degree distribution of the individual plant species (Fig. 14.3). The interpretation that frequent species just associate with more partners than rare species is thus probably too simple. Instead we may expect important biological mechanisms behind the striking parallel of species abundances and degree distributions. We follow Medan et al. (2007) in hypothesizing that the long-tailed species abundances are promoted by species integration into mutualistic networks through preferential attachment to higher linked species. Mycorrhiza-, pollinator- and seed disperser-webs of similar network structure may simultaneously operate in the tropical mountain rainforest, all fostering maintenance of rare species by associating them mutualistically with well-integrated species. Rare species interacting only with rare species would most likely be prone to extinction. Extinction of rare species is best minimized by mutual links with "generalists", higher linked and thus more frequently available partners (Gómez et al. 2011). Higher propagation or growth rate of highly linked partners in mycorrhiza networks is difficult to prove during field studies. However, Kiers et al. (2011) showed experimentally that plants detect, discriminate and reward the best fungal partners, those which supply more phosphate, with more carbohydrates. Thus, plants and fungi of most efficient associations may finally become the more frequent ones.

Preferential attachment was, e.g. observed when local mycobionts associated faster with seedlings inoculated in the nursery than with non-inoculated seedlings (Kottke and Hönig 1998). Biologically, "preferential attachment" is based on trait matching through complementarity and convergence in evolution (Thompson 2005). Complementarity in mycorrhizae is evident by the intimate interaction on the cellular level where bidirectional nutrient exchange is executed and complex recognition processes are required for mycorrhiza establishment (e.g. Bonfante and Genre 2008; Parniske 2008). Convergence in evolution of the two processes, nutrient exchange and partner recognition, is presumably the strongest selection mechanisms in mycorrhizal networks favouring integration of close relatives among nodes as these are functionally the most similar (Thompson 2005). Our inventories revealed closely related species in diverse plant groups (see above) and molecular phylogeny showed close relationships in the mycobiont groups (Haug et al. 2010; Setaro et al. 2006b; Suárez et al. 2006 and unpublished data). Bastolla et al. (2009) analytically combined interspecies competition with mutualism among the groups and found nested networks reducing competition and enhancing the number of potentially coexisting species when compared with random networks. Some rather simple experiments using Glomeromycota and herbaceous plants had shown influence of mycorrhizae on competition and species diversity resembling the simulations of Bastolla et al. (2009), but results depended on individual plant and fungal species (Grime et al. 1987; Scheublin et al. 2007 and literature therein). In case of the tropical rainforest, we are currently restricted to hypothetically corresponding conclusions.

14.3.4 Perspectives for Ecosystem Stability

Approach by network theory takes the importance of rare species into account and makes this approach especially suited for understanding highly diverse, tropical mutualistic assemblages. Numerical simulations revealed that the nested pattern with many weak, asymmetric dependences was crucial for stabilizing complex mutualistic networks (Thébault and Fontaine 2010 and literature therein), made the communities more robust to random extinctions or loss of keystone species (Memmot et al. 2004; Burgos et al. 2007; Mello et al. 2011) and to habitat destruction (Fortuna and Bascompte 2006; Abramson et al. 2011). Higher species richness, symmetric interaction strength and increased nestedness improved stability of mutually interacting assemblages when simultaneously considered in simulations based on non-linear functional responses (Okuyama and Holland 2008). Natural evolution obviously tends for long-lasting nestedness architecture most likely by a kind of feed-back mechanism, where links increase with abundance and abundance is promoted by increasing number of links, means that we deal with dynamic, adaptive networks (Zhang et al. 2011). This evolutionary scenario protects against environmental changes because rare species may become abundant and dominating declining while mutualistic interactions will be pertained by switching to other partners. The small-scale fragmentation presently found in the RBSF region has not destroyed the plant mycobiont networks yet. Chances for regeneration of natural forest are obviously not limited by lack of appropriate mycobionts. The highly diverse flora on the human-influenced sites most likely serves as provider of the diverse generalist mycobionts. Forest management by planting native tree seedlings can foster the ongoing process of rehabilitation. Further research is on the way to specify and corroborate these hypotheses.

14.4 Potential Network Modularity by Influence of Climate

Plant inventories and mycobiont samplings in the area under investigation cover a steep altitudinal gradient and also forest fragmentations caused by human activities. Differences among sampling sites are fairly well visible from climate data and climate may be a main driver of potential plant and mycobiont turnover among sites. Network theories were first applied to explain floristic turnover among islands of a fragmented Archipelago (Atmar and Patterson 1993) and situation in RBSF may be considered to be similar. Currently, our mycobiont datasets are incomplete and not appropriately large for calculation of modularity in the networks

Table 14.2 Climate situation on mycorrhiza sampling sites 2 m above ground, generated by integrated thermal and near-surface air humidity maps; yearly means of air temperature (*T*) and relative humidity (RH); highest and lowest temperature (T_{min}, T_{max}) and lowest air humidity (RH_{min}) as indicating potential stress for plants and mycobionts

Altitude of sampling sites [m a.s.l.]	Site type	Forest cover [%]	T_{mean} [°C]	T_{min} [°C], August	T_{max} [°C], Nov.	RH_{mean} [%]	RH_{min} [%], Sep., Oct. or Nov.
Arbuscular mycorrhizae							
1,866–2,002	Pristine forest	100	15.0–15.5	12.0	19.6	96.7–98.6	83.7
2,105–2,142	Pristine forest	100	14.5–14.6	11.5	19.5	94.8–95.3	77.9
1,900–1,950	Grasses, bracken fern, few shrubs, planted seedlings	0	16.0–16.5	11.1	23.9	85.5–86.3	55.1
2,130–2,210	Shrubs, bracken fern, herbs, planted seedlings	0	13.7–14.4	10.1	21.6	82.0–83.1	45.8
Orchid mycorrhizae							
1,885	Recovering landslide	0	15.3	11.2	24.1	85.9	54.8
1,980	Pristine forest	82	14.9	11.4	21.1	91.7	71.1
2,140	Regenerating forest	73	14.2	10.8	20.3	90.1	66.3
2,150	Pristine forest	80	14.1	10.9	20.0	90.8	68.4

(Olesen et al. 2007; Fortuna et al. 2010; Mello et al. 2011). Instead, we present a first, descriptive indication for influence of climate on plants and their associated mycobionts.

Climate data for the study sites were obtained from integrated thermal and near-surface air humidity maps (Fries et al. 2009, 2012). The selected site data (Table 14.2) show that the long-term mean values depend on elevation and forest density. Generally, temperature (*T*) decreases with altitude, but the canopy provides a shelter effect that dampens the annual and daily amplitudes (Körner and Paulsen 2004). Maximum temperatures occur during the relatively dry season (November to February) and minimum temperatures during the main rainy season (May to August). In contrast to the free atmosphere, mean relative humidity decreases with altitude inside the forest stands due to forest structure becoming more open at higher elevations. At the anthropogenic sites, relative humidity decreases with altitude because of humidity transport from the main river. This benefits especially the lower parts of the pastures where no dense tree vegetation hampers the moisture transport. However, minimum relative humidity (RH) values inside the forest do not fall below 72 %, while open sites display values down to 60 % on average when *T* is highest; extreme values are even lower (Table 14.2).

High turnover of tree species was observed previously and related to climate at different elevation (Homeier et al. 2010). Tree species of the reforestation experiment were selected according to climate site resistance (Günter et al. 2009), but the surrounding bushy and herbaceous flora is significantly distinct from the forest (see floristic inventories). No Orchidaceae were sampled as common to the road cut and the forest. Inventories showed that the great majority of local Orchidaceae have very specific habitat requirements, moisture and light availability appearing as strongest predictors (Werner et al. 2005; Werner 2011). The vast majority of the sampled hemiepiphytic Ericaceae occur only in the forest, some of the terrestrials are adapted to the open sites. In contrast, occurrence of the 28 Glomeromycota encountered more than one time in the forest plots so far could not be related to different altitudes. Nine associated Glomeromycota were, however, only encountered on the reforestation plots and 16 were only found on the forest plots, while 12 were shared among the reforestation and the pristine forest. The planted seedlings, obviously, trapped their mycobionts from the surrounding species rich flora. We tentatively conclude that the more stressful situation on the reforestation plots than in the forest (Table 14.2) is a driving force for establishment of different Glomeromycota communities. On contrast, mycobionts of Orchidaceae widely overlapped among epiphytic and terrestrial plants as well as among the road cut and the forest sites. Six of the eight mycobionts encountered on the road cut were also found in the forest. The orchid mycobionts may potentially profit from the shelter against drought in the orchid root velamen. Orchid mycobionts can, thus, promote orchid seed germination and protocorm growth independent of site situation. In contrast, Martos et al. (2012) found significant differences among the mycobionts of terrestrial and epiphytic orchids on La Réunion Island. Sebacinales that associate with Ericaceae show some overlap among different habitats, but most mycobionts from the road cut are unique so far. Thus, situation appears as quite different among plants and mycobionts and again among the mycobiont groups. Improved datasets, provided by ongoing work, will allow for numerical approach on impact of modularity in network architecture caused by environment or potential preferences among partners (Donatti et al. 2011; Mello et al. 2011; Schleuning et al. 2011).

14.5 Conclusions

The conceptual framework of network theory explains for the first time convincingly the biologically well-documented, long-lasting stability of mutualistic interactions, here mycorrhizae. This theory also considers the large numbers of rare species as found in the tropics as an important fact. Our data analysis shows that diversity of plants is intimately linked with diversity of mycorrhizal fungi maintaining rare species by mutualistic interaction with frequent partners. Rare species enhance stability of the complex community as they may become frequent partners under changing environment. Complementarity and convergence in evolution of bidirectional nutrient exchange and recognition processes favour generalist partnerships in mycorrhizae and maintain phylogenetically closely related species in the networks.

We propose the use of generalists Glomeromycota, adapted to planting sites, for inoculation of native tree seedlings in the nurseries as good option for reforestation. Greenhouse cultivation of orchids is already well established in Ecuador, but application of mycobionts needs to be improved by cooperative research among scientists from local universities and commercial companies. According to our results, stimulation of seed germination and further growth of orchids can be achieved by fungi from frequent orchid species, best from terrestrial, well-colonized roots. Further scientific studies are recommended for domestication and off-site production of Ericaceae. As further non-timber product, honey production has probably the highest potential. A high number of species are of melliferous interest and beekeeping has environmental benefits as well, reducing fire hazard and making reforestation with native species attractive (Ordoñez and Lalama 2006).

The main outcome in respect to provisioning ecosystem services for local people is, however, an intellectual one. The models, saying that supportive bipartite interactions are playing a fundamental role in complex ecosystems, need to be realized also for human society and are promising tools in further research there (e.g. Saavedra et al. 2009). While educational capacity building is generally outlined in Sect. 29.4, we claim here that our three Ecuadorian co-authors were not only introduced into technical scientific tools but also in understanding of mutualistic network models and their overall implication in most recent ecological research. Further education through self-responsible scientific research, the leading idea of university education since von Humboldt, will provide fundamental cultural ecosystem services to the entire local society as to reflect on the obligate future sustainability. Such education will by far outweigh short-term improvement of pastures acquired by slash and burn. Our conclusion is further supported and exemplified in Sect. 17.2 and modelled in Sect. 25.3 by socio-economical tools.

Acknowledgements The authors are grateful for fruitful discussions with Pedro Jordano on network analysis.

References

Abramson G, Trejo Soto CA, Leonardo O (2011) The role of asymmetric interactions on the effect of habitat destruction in mutualistic networks. PLoS One 6:e21028

Almeida-Neto M, Guimaraes P, Guimaraes PR Jr, Loyola RD, Ulrich W (2008) A consistent metric for nestedness analysis in ecological systems: reconciling concept and measurements. Oikos 117:1227–1239

Atmar W, Patterson BD (1993) The measure of order and disorder in the distribution of species in fragmented habitat. Oecologia 96:373–382

Barabási AL, Albert R (1999) Emergence of scaling in random networks. Science 286:509–512

Bascompte J, Jordano P (2007) Plant-animal mutualistic networks: the architecture of biodiversity. Annu Rev Ecol Syst 38:567–593

Bascompte J, Jordano P, Melián CL, Olesen JM (2003) The nested assembly of plant-animal mutualistic networks. Proc Natl Acad Sci 5:9383–9387

Bastolla U, Fortuna MA, Pascual-Garcia A, Ferrera A, Luque B, Bascompte J (2009) The architecture of mutualistic networks minimizes competition and increases biodiversity. Nature 458:1018–1020

Blüthgen N, Menzel F, Blüthgen N (2006) Measuring specialization in species interaction networks. BMC Ecol 6:9

Bonfante P, Genre A (2008) Plants and arbuscular mycorrhizal fungi: an evolutionary-developmental perspective. Trends Plant Sci 13:492–498

Burgos E, Ceva H, Perazzo RP, Devoto M, Medan D, Zimmermann M, Delbue M (2007) Why nestedness in mutualistic networks? J Theor Biol 249:307–313

Chagnon PL, Bradley RL, Klironomos JN (2012) Using ecological network theory to evaluate the causes and consequences of arbuscular mycorrhizal community structure. New Phytol 194:307–312

Cruz D, Suárez JP, Kottke I, Piepenbring M, Oberwinkler F (2011) Defining species in *Tulasnella* by correlating morphology and nrDNA ITS-5.8S sequence data of basidiomata from a tropical Andean forest. Mycol Prog 10:229–238

Donatti CI, Guimaraes PR, Galetti M, Pizo MA, Marquitti FM, Dirzo R (2011) Analysis of a hyper-diverse seed dispersal network: modularity and underlying mechanisms. Ecol Lett 14:773–781

Fortuna MA, Bascompte J (2006) Habitat loss and the structure of plant-animal mutualistic networks. Ecol Lett 9:281–286

Fortuna MA, Stouffer DB, Olesen JM, Jordano P, Mouillot D, Krasnov BR, Poulin R, Bascompte J (2010) Nestedness versus modularity in ecological networks: two sides of the same coin? J Anim Ecol 79:811–817

Frank B (1885) Über die auf Wurzelsymbiose beruhende Ernährung gewisser Bäume durch unterirdische Pilze. Ber Deutsch Bot Ges 3:128–145; English translation in Mycorrhiza (2005) 15:267–275

Fries A, Rollenbeck R, Göttlicher D, Nauß T, Homeier J, Peters T, Bendix J (2009) Thermal structure of a megadiverse mountain ecosystem in Southern Ecuador and its regionalization. Erdkunde 63:321–335

Fries A, Rollenbeck R, Nauß T, Peters T, Bendix J (2012) Near surface air humidity in a megadiverse mountain ecosystem in Southern Ecuador and its regionalization. Agric For Meteorol 152:17–30

Göker M, García-Blázquez G, Voglmayr H, Tellería MT, Martín MP (2009) Molecular taxonomy of phytopathogenic fungi: a case study in *Peronospora*. PLoS One 29:e6319

Gómez JM, Perfectti F, Jordano P (2011) The functional consequences of mutualistic network architecture. PLoS One 6:e16143

Grime JP, Mackey JM, Hillier SH, Read DJ (1987) Floristic diversity in a model system using experimental microcosms. Nature 328:420–422

Guimaraes PR Jr, Guimaraes P (2006) Improving the analysis of nestedness for large sets of matrices. Environ Model Softw 21:1512–1513

Günter S, Gonzalez P, Alvarez G, Aguirre N, Palomeque X, Haubrich F, Weber M (2009) Determinants for successful reforestation of abandoned pastures in the Andes: soil conditions and vegetation cover. For Ecol Manage 258:81–91

Haug I, Wubet T, Weiß M, Aguirre N, Weber M, Günter S, Kottke I (2010) Species-rich but distinct arbuscular mycorrhizal communities in reforestation plots on degraded pastures and in neighboring pristine tropical mountain rain forest. J Trop Ecol 51:125–148

Hibbett DS, Ohman A, Glotzer D, Nuhn M, Kitk P, Nilsson RH (2011) Progress in molecular and morphological taxon discovery in *Fungi* and options for formal classification of environmental sequences. Fungal Biol Rev 25:38–47

Homeier J, Werner FA (2007) Spermatophyta checklist – Reserva Biológica San Francisco (Prov. Zamora-Chinchipe, S. Ecuador). In: Liede-Schumann S, Breckle S-W (eds) Provisional checklist of flora and fauna of San Francisco Valley and its surroundings. Ecotropical monographs, vol 4. Society for Tropical Ecology, pp 15–58

Homeier J, Breckle S-W, Günter S, Rollenbeck RT, Leuschner C (2010) Tree diversity, forest structure and productivity along altitudinal and topographical gradients in a species-rich Ecuadorian montane rain forest. Biotropica 42:140–148

Ings TC, Montoya JM, Bascompte J, Blüthgen N, Brown L, Dormann CF, Edwards F, Figueroa D, Jacob U, Jones JI, Lauridsen RB, Ledger ME, Lewis HM, Olesen JM, van Veen FJ, Woodward G (2009) Ecological networks – beyond food webs. J Anim Ecol 78:253–269

Jacquemyn H, Merckx V, Brys R, Tyteca D, Cammue BP, Honnay O, Lievens B (2011) Analysis of network architecture reveals phylogenetic constraints on mycorrhizal specificity in the genus *Orchis* (Orchidaceae). New Phytol 192:518–528

Jordano P, Bascompte J, Olesen JM (2003) Invariant properties in coevolution networks of plant-animal interactions. Ecol Lett 6:69–81

Kiers ET, Duhamel M, Beesetty Y, Mensah JA, Franken O, Verbruggen E, Fellbaum CR, Kowalchuk GA, Hart MM, Bago A, Palmer TM, West SA, Vandenkoornhuyse P, Jansa J, Bücking H (2011) Reciprocal rewards stabilize cooperation in the mycorrhizal symbiosis. Science 333(6044):880–882

Körner C, Paulsen J (2004) A world-wide study of high altitude treeline temperatures. J Biogeogr 31:713–732

Kottke I, Hönig K (1998) Improvement of maintenance and autochthones mycorrhization of beech (*Fagus sylvatica* L.) and oak (*Quercus robur* L.) plantlets by pre-mycorrhization with *Paxillus involutus* (Batsch) Fr. In: Misra A (ed) Problems of wasteland development and role of microbes. AMIFM Publications, Bhubaneswar, pp 187–218

Kottke I, Beck A, Haug I, Setaro S, Jeske V, Suárez JP, Pazmiño L, Preußing M, Nebel M, Oberwinkler F (2008) Mycorrhizal state and new and special features of mycorrhizae of trees, ericads, orchids, ferns and liverworts. In: Beck E, Bendix J, Kottke I, Makeschin F, Mosandl R (eds) Gradients in a tropical mountain ecosystem of Ecuador. Ecological studies, vol 198. Springer Verlag, Berlin, pp 137–148

Kottke I, Suárez JP, Herrera P, Cruz D, Bauer R, Haug I, Garnica S (2010) Atractiellomycetes belonging to the 'rust' lineage (Pucciniomycotina) form mycorrhizae with terrestrial and epiphytic neotropical orchids. Proc R Soc London B 277:1289–1296

Martos F, Munoz F, Pailler T, Kottke I, Gonneau C, Selosse MA (2012) The role of epiphytism in architecture and evolutionary constraint within mycorrhizal networks of tropical orchids. Mol Ecol 21:5098–5109

Medan D, Perazzo RP, Devoto M, Burgos E, Zimmermann MG, Ceva H, Delbue AM (2007) Analysis and assembling of network structure in mutualistic systems. J Theor Biol 246:510–521

Mello MA, Marquitti FM, Guimaraes PR Jr, Kalko EK, Jordano P, de Aguiar MA (2011) The missing part of seed dispersal networks: structure and robustness of bat-fruit interactions. PLoS One 6:e17395

Memmot J, Waser NM, Price MV (2004) Tolerance of pollination networks to species extinctions. Proc R Soc London B 271:2605–2611

Montesinos-Navarro A, Segarra-Moragues JG, Valiente-Banuet A, Verdú M (2012) The network structure of plant-arbuscular mycorrhizal fungi. New Phytol 194:536–547

Okuyama T, Holland JN (2008) Network structural properties mediate the stability of mutualistic communities. Ecol Lett 11:208–216

Olesen JM, Bascompte J, Dupont YL, Jordano P (2007) The modularity of pollination networks. Proc Natl Acad Sci 104:19891–19896

Ordoñez O, Lalama K (2006) Experiencias del manejo apícola en Uritusinga, Loja, Ecuador. Fundación Ecológica Arco Iris, PROBONA, Loja

Parniske M (2008) Arbuscular mycorrhiza: the mother of plant root endosymbioses. Nat Rev Microbiol 6:763–775
Peters T, Diertl KH, Gawlik J, Rankl M, Richter M (2010) Vascular plant diversity in natural and anthropogenic ecosystems in the Andes of Southern Ecuador. Mt Res Dev 30:344–352
Saavedra S, Reed-Tsochas F, Uzzi B (2009) A simple model of bipartite cooperation for ecological and organizational networks. Nature 457:463–466
Scheublin TR, van Logtestijn RS, van der Heijden MG (2007) Presence and identity of arbuscular mycorrhizal fungi influence interactions between plant species. J Ecol 95:631–638
Schleuning M, Blüthgen N, Flörchinger M, Braun J, Schaefer HM, Böhning-Gaese K (2011) Specialization and interaction strength in a tropical plant–frugivore network differ among forest strata. Ecology 92:26–36
Setaro S, Kron K (2012) Neotropical and North American Vaccinioideae (Ericaceae) share their mycorrhizal Sebacinales – an indication for concerted migration? PLoS Currents: Tree of Life. doi:10.1371/currents.RRN1227
Setaro S, Oberwinkler F, Kottke I (2006a) Anatomy and ultrastructure of mycorrhizal associations of Neotropical Ericaceae. Mycol Prog 5:243–254
Setaro S, Weiß M, Oberwinkler F, Kottke I (2006b) Sebacinales form ectendomycorrhizas with *Cavendishia nobilis*, a member of the Andean clade of Ericaceae, in the mountain rain forest of southern Ecuador. New Phytol 169:355–365
Setaro S, Suarez JP, Herrera P, Cruz D, Kottke I (2013) Distinct but closely related Sebacinales from mycorrhizae with co-existing Ericaceae and Orchidaceae in a neotropical mountain area. In: Varma A, Kost G, Oelmüller R (eds) Sebacinales. Forms, functions and biotechnological application. Springer, Berlin, pp 81–105
Smith S, Read D (2008) Mycorrhizal symbiosis, 3rd edn. Academic, San Diego
Suárez JP, Weiß M, Abele A, Garnica S, Oberwinkler F, Kottke I (2006) Diverse tulasnelloid fungi form mycorrhizas with epiphytic orchids in an Andean cloud forest. Mycol Res 110:1257–1270
Suárez JP, Weiß M, Abele A, Oberwinkler F, Kottke I (2008) Members of Sebacinales subgroup B form mycorrhizae with epiphytic orchids in a neotropical mountain rain forest. Mycol Prog 7:75–85
Thébault E, Fontaine C (2010) Stability of ecological communities and the architecture of mutualistic and trophic networks. Science 329:853–856
Thompson JN (2005) The geographic mosaic of coevolution. University of Chicago Press, Chicago
Werner FA (2011) Reduced growth and survival of vascular epiphytes on isolated remnant trees in a recent tropical montane forest clear-cut. Basic Appl Ecol 12:172–181
Werner FA, Homeier J, Gradstein SR (2005) Diversity of vascular epiphytes on isolated remnant trees in the montane forest belt of Southern Ecuador. Ecotropica 11:21–40
Zhang F, Hui C, Terblanche JS (2011) An interaction switch predicts the nested architecture of mutualistic networks. Ecol Lett 14:797–803

Chapter 15
Current Provisioning Services: Pasture Development and Use, Weeds (Bracken) and Management

Kristin Roos, Jörg Bendix, Giulia F. Curatola, Julia Gawlik, Andrés Gerique, Ute Hamer, Patrick Hildebrandt, Thomas Knoke, Hanna Meyer, Perdita Pohle, Karin Potthast, Boris Thies, Alexander Tischer, and Erwin Beck

15.1 Introduction

Pasture farming for milk and meat is the prevalent form of agriculture in the humid and semi-humid Andes of Southern Ecuador, in which large areas of primary mountain rain forest have been and still are being cleared, mostly by slash-and-burn practices (Mosandl et al. 2008). Ortho-rectified satellite scenes (Landsat TM/ETM+) of the research area, the Rio San Francisco valley, show that 4.1 % of the primary mountain rain forest, which originally covered approximately

K. Roos (✉) • E. Beck
Department of Plant Physiology and Bayreuth Centre of Ecology and Environmental Research, University of Bayreuth, Universitätsstr. 30, 95440 Bayreuth, Germany
e-mail: kristin.roos@uni-bayreuth.de

J. Bendix • G.F. Curatola • H. Meyer • B. Thies
Laboratory of Climatology and Remote Sensing, University of Marburg, Deutschhausstr. 10, 35032 Marburg, Germany

J. Gawlik • A. Gerique • P. Pohle
Institute of Geography, Friedrich-Alexander University of Erlangen, Kochstr. 4/4, 91054 Erlangen, Germany

U. Hamer
Institute of Landscape Ecology, University of Muenster, Robert-Koch-Straße 28, 48149 Muenster, Germany

P. Hildebrandt
Institute of Forest Management, Center of Life and Food Sciences Weihenstephan, Technische Universität München, Hans-Carl-von-Carlowitz-Platz 2, 85354 Freising, Germany

T. Knoke
Institute of Silviculture, Center of Life and Food Sciences Weihenstephan, Technische Universität München, Hans-Carl-von-Carlowitz-Platz 2, 85354 Freising, Germany

K. Potthast • A. Tischer
Institute of Soil Science and Site Ecology, Dresden University of Technology, Pienner Str. 19, 01737 Tharandt, Germany

J. Bendix et al. (eds.), *Ecosystem Services, Biodiversity and Environmental Change in a Tropical Mountain Ecosystem of South Ecuador*, Ecological Studies 221,
DOI 10.1007/978-3-642-38137-9_15,

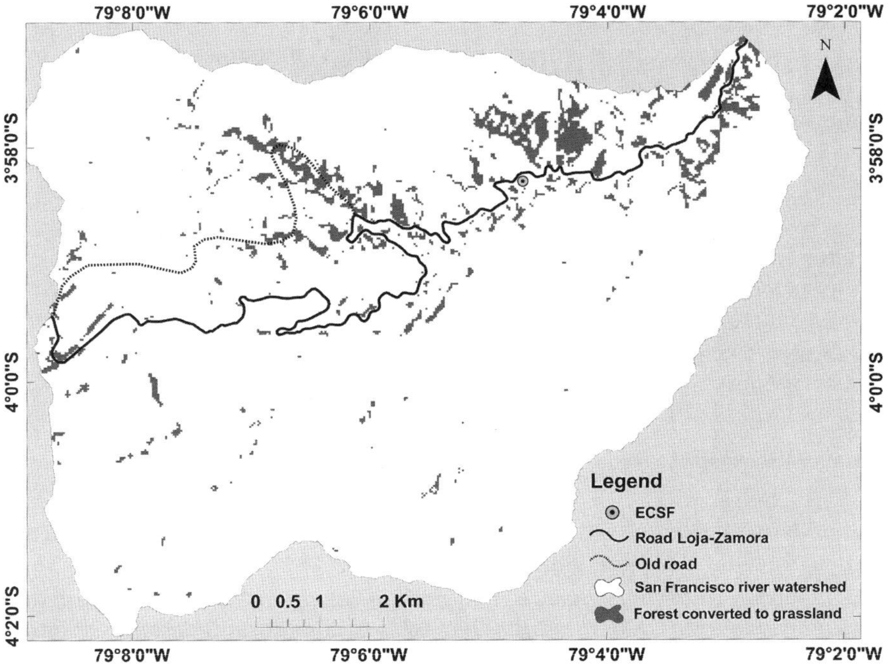

Fig. 15.1 Conversion of natural forest to grassland in the catchment area of the Rio San Francisco between 1987 and 2001. Based on Landsat TM change detection (from Thies et al. 2012, modified by G. Curatola). *ECSF* Estación Científica San Francisco

120 km^2 of the area, have been cleared and converted to agricultural land (mainly pastures) between 1987 and 2001 (Fig. 15.1). This chapter presents a review of the current state of pastures and cattle ranching in the core research area between 1,600 and 2,400 m a.s.l. The natural vegetation in the humid eastern ranges of the Andes is a dense mountain rain forest, and therefore indigenous tropical grass species are rare, except bamboo (*Neurolepis* spec.) which is not useful for agriculture. At least the dominant pasture grasses are thus introduced grass species like the manually planted *Setaria sphacelata* from South and Central Africa, or *Melinis minutiflora* probably from the same provenience. The nutritive value of the pasture vegetation is rather low, and the majority of these pastures are overgrazed. As everywhere overgrazing ends in a vicious cycle as pasture weeds successively overgrow the pastures but a return of the natural forest on the vast pasture areas, even after abandonment has never been observed.

Most of the farmers in the area belong to the so-called Mestizos whereas neighboring valleys are inhabited predominantly by the Saraguros. Both ethnic groups are agropastoralists who combine market economy (cattle ranching) and subsistence economy (crop production and horticulture) while considering the forest as a reserve for new agricultural areas (Pohle and Gerique 2006).

15.2 Results and Discussion

15.2.1 The Pasture Types

Flat parts of the terrain accommodate the so-called *pastos azules* or *pastos naturales*. These pastures are dominated by the cosmopolitan grass *Holcus lanatus* L. or, in the lower San Francisco valley, by *Pennisetum clandestinum* Hochst. Ex Chiov. (Gerique 2010) and harbour a great variety of other grasses and herbs. In spite of its attractive appearance, *Holcus lanatus* (common velvet grass, Yorkshire fog) is a low-value forage grass (Watt 1978; Merrill et al. 1995), and its dominance on the *pastos azules* results from spurning by the cattle. It is an undemanding species with regard to soil quality, structure and pH value. The *pastos azules* are nevertheless considered to be valuable pastures, partly because they are reasonably productive and evidence considerable herbal variety, and partly due to the fact that bracken cannot invade the compacted and frequently water-logged soils of these pastures. In addition, cattle tend to gather on the flat terrain, fertilizing it with their dung and allowing easy access by the farmers for milking. Maintenance of these pastures requires almost no management efforts and financial input. *Pastos azules* are usually not burned, as the abundance of herbaceous and woody pastoral weeds growing on them has decreased during the last decade, and their cover abundance is now in the range of only about 10 % (Gawlik 2010). Sorrel, dock and some other weeds may again increase their cover abundance when grazing intensity increases, but in principle the *Holcus lanatus* pastures can be expected to preserve their moderate pastoral value.

Melinis pastures that cover the steepest parts of the slopes are particularly rich in species. The character grass *Melinis minutiflora* P. Beauv (Molasses grass, "yaragua") is an invasive African species (D'Antonio and Vitousek 1992) which propagates via stolons and seeds. Its strong smell and hairiness requires habituation by the cattle. The biomass production of this C3-grass is rather low (2–16 t ha^{-1} a^{-1}), and its nutritive value has also been rated as only "low to acceptable" (Duke 1979). The organic matter digestibility of the grass is approximately 50 % (Axtmayer et al. 1938; Göhl 1982). The metabolizable energy value lies between 5.9 and 8.6 MJ kg^{-1} dry matter, and its digestible crude protein content has been quoted as being as low as 0.58–2.18 % of the dry matter (da Costa et al. 1981). *Melinis* nevertheless provides a proper balance in diet for cattle and sheep. Local farmers consider it to be the best grass for milk production, as the milk is very creamy. The distribution of *Melinis* pastures, that interfere here and there with the *Setaria* pastures, gives the impression of randomness. Frequent grazing results in horizontal trampling contours that give the steeper slopes a terrace-like appearance. *Melinis* is a pioneer grass that tolerates drought, poor soils and steep slopes, where it forms closed swards. It also tolerates burning, because the rather low-standing crop does not give rise to very hot fires. The presence of fire-resistant weeds such as bracken and shrubs evidences repeated burning (Hartig and Beck 2003). *Melinis* is rarely planted or sown. However, as an invasive species it is part of the anthropogenic flora of the tropical Andes, where it not only forms pastures but also constitutes a component of the way-side flora.

"*Pastos mieles*" are the most common and wide-spread pastures found in the humid valleys of South Ecuador, and are dominated by the manually planted grass *Setaria sphacelata* (Schumach.) Stapf & C.E. Hubb. ("mequerón"). *Setaria* represents a polyploid complex of 4n-, 6n-, 8n- and 10n-races, all of which are cultivated (Hacker and Jones 1969). Which of the races are used in southern Ecuador is not known, as the farmers do not buy the grass when planting a new pasture, but rather obtain it as bunches from other pastures. Its nutritive value is generally rated to be very low, but this estimation depends greatly on the age of the leaves. Young leaves have a reasonable matter digestibility of 52–56 % (Hacker 1974), which declines rapidly due to lignification as the leaves mature. The fibre content of this grass is lower than that of *Melinis* (~30 vs. 36 %), whereas its total protein, ash and potassium contents are much higher (protein ~16 vs. ~9 %, ash ~15 vs. ~8 % and potassium ~4 vs. ~1.4 %; Hacker and Jones 1969). *Setaria* has an extremely high content of soluble oxalate (about 10-fold higher than that of other grasses; Jones and Ford 1972). This can trigger calcium deficiency in livestock ("hypocalcaemia"), which in rare cases leads to the death of the animals (Seawright et al. 1970). A high nitrogen supply (and uptake) favours the accumulation of soluble oxalate; but the concentration of this ion substantially decreases with the aging of the leaves (Middleton and Barry 1978). Although it has not been examined, this difference in the oxalate content could be the reason why the cows prefer the terminal parts of the leaves, although these are the oldest, least palatable and nutritionally poorest parts of the foliage. Farmers recognize the poor nutritive value of a *Setaria* diet in the rather "watery" milk produced by the cows. In summary, only a minor part of the biomass of the large *Setaria* tussocks can be considered to be good fodder. This disadvantage is, however, balanced by the comparatively high productivity of the C4-grass that makes it a strong competitor for other species. It propagates by stolons and rhizomes, and thus extends its tussocks mostly along the trampling contours.

Like *Melinis, Setaria sphacelata* was introduced from Africa and therefore has only few natural enemies in Ecuador. As typical of C4-plants it is, however, a light-demanding species which can be suppressed by shading, e.g. by shrubs and bracken fern. *Setaria* pastures prevail in the humid montane areas of South Ecuador. The tussocks grow up to 60 cm high and the flowering culms can approach 3 m in height. Small tufts of leaves are produced on the nodes of the culms, which frequently fall over under their own weight. Upon contact with the soil the small tufts produce adventitious roots, which thus initiate the formation of a sort of thicket. Neither the culms nor the seeds of the grass are eaten by cattle. As mowing is not practised in Ecuador, the farmers set fire to *Setaria* pastures in order to rejuvenate them, especially when the grass has started to flower (Hartig and Beck 2003; Beck et al. 2008; Gerique 2010). The lack of combustible material on these pastures requires the use of petrol for burning. Since the use of fire is illegal, reliable data on burning practices—apart from own observations—are difficult to obtain.

Setaria pastures on sloping ground are readily infested with bracken fern and later on with a variety of fire-resistant shrubs. Setting fire on the pastures also serves the suppression of these weeds. However, the fire stimulates not only the rejuvenation of the *Setaria* tussocks but also the regrowth of the weeds (Roos et al. 2010).

15.2.2 Bracken Vegetation

In an altitudinal range from the lowlands up to about 3,000 m a.s.l., bracken fern (two species, *Pteridium arachnoideum* (Kaulf.) Maxon and *Pt. caudatum* (L.) Maxon) spreads readily into areas where the forest has been cleared by the use of fire. Its vegetative propagation from rhizomes is stimulated by heat, even when its aerial parts—the fronds—are completely destroyed by fire (Roos et al. 2010). In addition, there is reason to assume that a wave of heat that does not kill the fronds triggers the formation of spores which readily germinate on the burned, ash-covered and vegetation-free soil. This leads to the formation of mats of livermoss-like prothallia, and 4–6 weeks later a lawn of young sporophytes can be observed. Other weeds that have proved to be fire resistant are shrubs that belong mainly to the Ericaceae, Asteraceae, Melastomataceae and Escalloniaceae (Hartig and Beck 2003). After the aerial shoots have been burned, new shoots sprout readily from buds at or beneath the soil surface and the shrub revives. *Setaria,* as a light-demanding species, becomes more and more out-competed by the weeds, and without management disappears almost completely from the area after a few years. The indestructibility of the bracken rhizomes on the one hand and the exuberant seed production of the shrubs on the other appear to lead to a balance between the different life-forms and to the formation of a stable weed vegetation (Hartig and Beck 2003) that has been termed "succession fallow" (Makeschin et al. 2008).

15.2.3 Properties and Stability of the Pasture Soils

Conversion of the mountain rainforest into pastures by slash-and-burn practices, as described in more detail in Chap. 11, has resulted in an augmentation of the carbon and nitrogen stocks in the mineral topsoil of old pastures (Makeschin et al. 2008; Hamer et al. 2012b). This compensates for the loss of organic carbon caused by the destruction of the thick organic layer upon the clearing of the forest. Pastoral use leads to the accumulation of soil organic matter (SOM) in the mineral topsoil due to the high inputs of root biomass from the C4-grass *Setaria sphacelata*. However, the grass-derived SOM is lost rapidly after bracken invasion as a result of the crowding out of the grass and the abandonment of the pastures (Hamer et al. 2012a).

The properties of the pasture soils depend strongly on the age and the fire history of the pasture and on the degree of infestation by bracken or shrub vegetation. The soil of a 17-year-old pasture harboured a very active microbial community, the biomass of which was about three times as high as that determined for the forest soil (Table 15.1). The enhanced microbial growth was related to three major factors: (1) an increase in the pH-value of the soil due to the input of alkaline ashes (Makeschin et al. 2008), (2) a better availability of the nutrients, and (3) a narrower soil C:N ratio. Microbial activity, as indicated by the mineralization rates of soil organic carbon (SOC), was up to four times greater than that evidenced by the forest soil (Potthast et al. 2011). While microbial biomass and activity remained high in the soil of long-standing active pastures, the contents of inorganic and organic phosphates decreased with time to values comparable to those found in the forest soils (Table 15.1). Ten years after bracken invasion and abandonment of the pastures, also microbial biomass and activity had declined to the forest level and the contents of total N and both inorganic and organic NH_4F-extractable P had decreased to values even below those found in forests (Table 15.1). Extreme bracken infestation leads to a further decrease of inorganic plant-available P to values of below 0.6 mg PO_4–P kg^{-1} soil. Contents below 15 mg kg^{-1} are known to limit plant productivity (Landon 1991). An increase in the readily available phosphate pool above this threshold of 15 mg kg^{-1} was detected shortly after the burning of an abandoned pasture site. Higher P availability lasted for up to 5 months subsequent to the fire, while the dynamics of enhanced C and N availability were much more rapid and became evident within a timescale of days. Burning of the pastures may on the one hand trigger regrowth of the grass due to the heat shock and on the other hand may also stimulate the growth of new leaves by increasing the supply of phosphate (if P is the limiting nutrient for grass growth). In the long run, however, repeated burning leads to an enormous loss of nutrients from the ecosystem (mainly due to volatilization and by leaching upon rainfall) and to an accumulation of black carbon. Black carbon is very resistant to microbial decomposition and can sequester ions by sorption (Knicker 2007). This results in lower nutrient cycling rates and nutrient availability. In the long run the soil properties accruing from burning favour the growth of bracken, the roots and rhizomes of which have a high capacity for P (and N) acquisition. This gives the fern a considerable competitive advantage over *Setaria* (Chap. 26). The invasion of former pasture land by successional shrub (succession fallow, Table 15.1) leads to a further decrease in nutrient availability, in particular respective of nitrogen (and P) in the mineral soil as indicated by the wide C:N ratio.

15.2.4 Vegetation Development During the Past Decade

The vegetation of the three types of pastures found in the upper San Francisco valley was first analyzed in 1999 (Hartig and Beck 2003). This analysis was repeated after 10 further years of customary cattle grazing (Gawlik 2010; Peters et al. 2010), during which at least one fire occurred on the *Setaria* pastures (1 year

Table 15.1 Comparison of biochemical soil properties along a land use gradient

						NH_4-F-extraction	
		C:N ratio	SOC mineralization [%]	MBC [mg kg^{-1}]	MBP [mg kg^{-1}]	PO_4–P [mg kg^{-1}]	P_{org} [mg kg^{-1}]
Forest	Oi	26.6 (2.2)[a]	2.46 (0.43)[a]	3,866 (709)[a]	388 (83)	34.3 (27.0)	26.4 (2.8)
	OeOa	19.2 (1.6)[a]	0.86 (0.44)[a]	3,075 (678)[a]	366 (30)	16.3 (9.2)	17.5 (7.0)
	0–5 cm	16.6 (0.7)[a]	0.41 (0.06)[a]	1,048 (205)[a]	84 (27)	1.8 (0.3)	5.4 (0.8)
	5–10 cm	17.1 (1.5)[a]	0.27 (0.06)[a]	801 (240)[a]	74 (29)	1.4 (0.3)	6.9 (2.0)
	10–20 cm	18.7 (1.3)[a]	0.17 (0.04)[a]	458 (114)[a]	52 (45)	0.9 (0.6)	8.9 (3.4)
Pasture 17a (FERPAST)	0–5 cm	13.3 (0.3)[a]	1.56 (0.30)[a]	3,445 (761)[a]	366 (66)	4.5 (2.5)	12.8 (5.6)
	5–10 cm	12.5 (0.2)[a]	0.76 (0.30)[a]	1,631 (601)[a]	128 (11)	3.0 (1.4)	22.6 (11.6)
	10–20 cm	13.0 (0.2)[a]	0.41 (0.08)[a]	846 (180)[a]	67 (15)	2.2 (0.5)	17.7 (7.6)
Pasture 50a	0–5 cm	15.5 (1.4)[b]	0.73 (0.17)	1,408 (345)[b]	116 (37)[b]	1.6 (1.9)[b]	3.5 (1.2)[b]
	5–10 cm	16.4 (1.3)[b]	0.36 (0.04)	850 (231)[b]	45 (21)[b]	1.2 (0.5)[b]	3.6 (1.7)[b]
	10–20 cm	18.6 (2.6)[b]	0.25 (0.07)	566 (271)[b]	29 (13)[b]	0.7 (0.3)[b]	7.0 (3.1)[b]
Abandoned pasture	0–5 cm	21.1 (1.1)	0.57 (0.07)	910 (107)	103 (57)	1.3 (0.9)	1.5 (1.0)
	5–10 cm	23.8 (2.3)	0.29 (0.12)	749 (162)	57 (24)	0.4 (0.5)	0.9 (0.5)
	10–20 cm	26.7 (3.4)	0.18 (0.04)	678 (201)	21 (11)	0.2 (0.2)	1.5 (1.8)
Abandoned pasture (Llashipa)	0–5 cm	21.1 (2.4)	0.45 (0.06)	1,125 (73)	65 (39)	0.5 (0.4)	3.7 (0.8)
	5–10 cm	21.2 (1.7)	0.32 (0.02)	1,050 (366)	56 (22)	0.3 (0.3)	4.6 (0.5)
	10–20 cm	24.6 (0.5)	0.19 (0.02)	596 (49)	27 (5)	0.2 (0.0)	5.3 (0.6)
Succession	Oi	45.0 (5.2)[b]	0.94 (0.16)	2,534 (524)[b]	90 (47)[b]	26.8 (15.8)[b]	48.8 (38.0)[b]
	OeOa	23.0 (1.7)[b]	0.11 (0.03)	1,376 (396)[b]	260 (73)[b]	16.7 (12.4)[b]	16.5 (12.5)[b]
	0–5 cm	24.4 (1.2)[b]	0.29 (0.09)	1,024 (339)[b]	101 (52)[b]	3.2 (1.9)[b]	3.6 (0.6)[b]
	5–10 cm	25.3 (1.8)[b]	0.18 (0.02)	581 (276)[b]	35 (19)[b]	1.0 (0.4)[b]	3.3 (0.6)[b]
	10–20 cm	25.4 (1.9)[b]	0.17 (0.02)	401 (270)[b]	13 (5)[b]	0.7 (0.2)[b]	3.4 (0.7)[b]

The C:N ratio, soil organic carbon (SOC) mineralization (% during 2-weeks of laboratory incubation), microbial biomass carbon (MBC), microbial biomass phosphorus (MBP) and NH_4F-extractable inorganic and organic phosphorus (P) are recorded for the mineral topsoil (at depths of 0–5, 5–10 and 10–20 cm) and the organic layer (Oi and OeOa horizons) of an old-growth forest, 17- and 50-year-old pastures, a pasture having been abandoned for about 10 years, and successional shrub vegetation occurring on a pasture having been abandoned for about 20 years ("Succession fallow"). Mean values are quoted, with SDs in parentheses

[a]Data from Potthast et al. (2011)

[b]Data from Hamer et al. (2012b)

Table 15.2 Plant species richness of the three types of pastures found in the upper Rio San Francisco valley in the years 2000 and 2010

Pasture type	*Pastos azules*	*Melinis* pastures	*Setaria* pastures
	Number of species		
Year 2000	30	31	21
Year 2010	43	59	34
Common species in both years	16	21	8
Number of species in all examined plots of the respective pasture type	57	69	47

The numbers are mean values of six homogeneous plots of 4 m^2 each of each pasture type

after the first vegetation record). The changes taking place in the pattern of functional plant types are shown in Table 15.2 and Fig. 15.2, respectively.

Several questions having been posed about tendencies that become apparent after the first decade of this century can be now be addressed on the basis of the currently available data:

- *Has phytodiversity increased, decreased or remained constant?*
  All three types of pastures showed substantial increases in the numbers of the species that they harbour (Table 15.2). This increase was most impressively on the *Melinis* pastures, where the number of species almost doubled while the abundance of *Melinis* remained constant (75–72 %; Fig. 15.2). The increases in species diversity on the other two types of pasture were less spectacular, but the cover abundance of the dominant grasses also increased dramatically. The cover abundance of *Holcus* more than doubled on the "*pastos azules*", and that of *Setaria* increased from 75 to 92 % on the "*pastos mieles*".
- *Has the plant species composition recorded in 1999 changed over the decade?*
  The turnover of species composition was higher than the increase in species numbers. Only 8 of the 21 species recorded on the *Setaria* pastures in 1999 still contributed to the set of 34 species present in 2010. Thirteen of the original 21 species vanished, to be replaced by 26 new ones. The situation respective of the other two pasture types was similar, although less pronounced.
- *Did the spectrum of the species shift in favour of useful pasture plants or did the share of weeds increase?*
  The contribution of herbaceous as well as of woody weeds decreased significantly, but potentially N-fixing species such as trefoil also declined in importance. In principle, herbs were replaced by the dominant grasses on all three pasture types.
- *Was the abundance of bracken progressive or regressive?*
  In contrast to a general trend of progressive invasion by bracken, the share of this fern decreased considerably on the active pastures, most prominently on the *Melinis* and the *Setaria* pastures.
- *Did the area of total pastures and of abandoned pastures in the Rio San Francisco valley increase or decrease?*

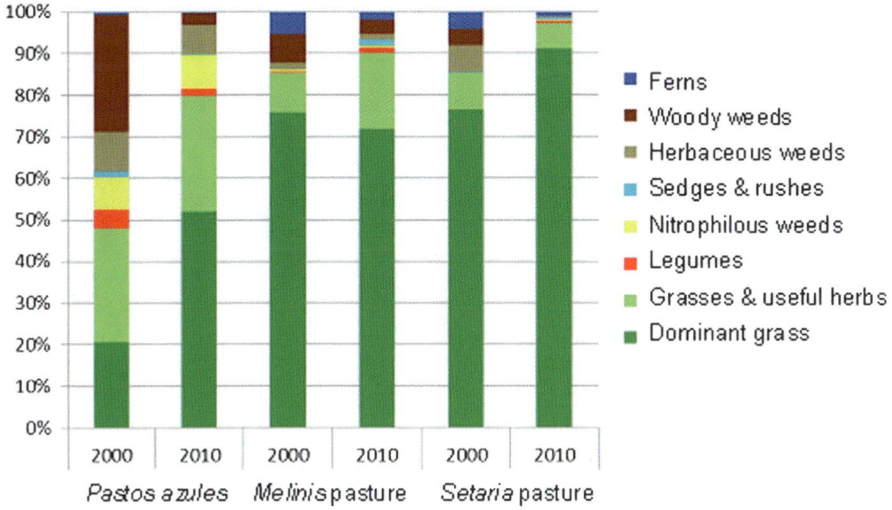

Fig. 15.2 Changes of the composition of plant functional types on the three pasture types from 2000 to 2010. Plant species were grouped to represent functional types. For comparison the abundance of species on each pasture type was set 100 %

As shown in Chap. 2 (e.g. Fig. 2.2), the total area of pasture land in the Rio San Francisco valley has increased dramatically during the past 50 years, but a considerable proportion of the established pastoral area has also been abandoned (approximately 27 % until 2003). A recent estimate derived from high-resolution satellite image (QuickBird) shows that 14.3 % of the study area (shown in Fig. 15.3) is covered by pasture land (5.3 % *Holcus*-, 0.5 % *Melinis*- and 8.5 % *Setaria* pastures) and 10.7 % is not in use due to infestation by weeds (mainly bracken; Fig. 15.3).

15.2.5 Pasture Farming

Rotational grazing between the three different pasture types appears to be crucial to avoiding the potential problems associated with dominant pasture grasses. This will not only prevent overgrazing but will also insure a balanced diet for the cattle that will help to prevent health problems. Fertilization of the pastures is not practised in the research area except fertilization by cow dung during the grazing period. Preliminary studies have shown a dramatic response of *Setaria* to fertilization with macronutrients (see Chap. 26). The pasture management commonly practised by at least the Mestizos in the San Francisco valley is not in ecological balance with the environment and is therefore neither sustainable nor economically viable. This is revealed by the following rough estimate (see also Chap. 26).

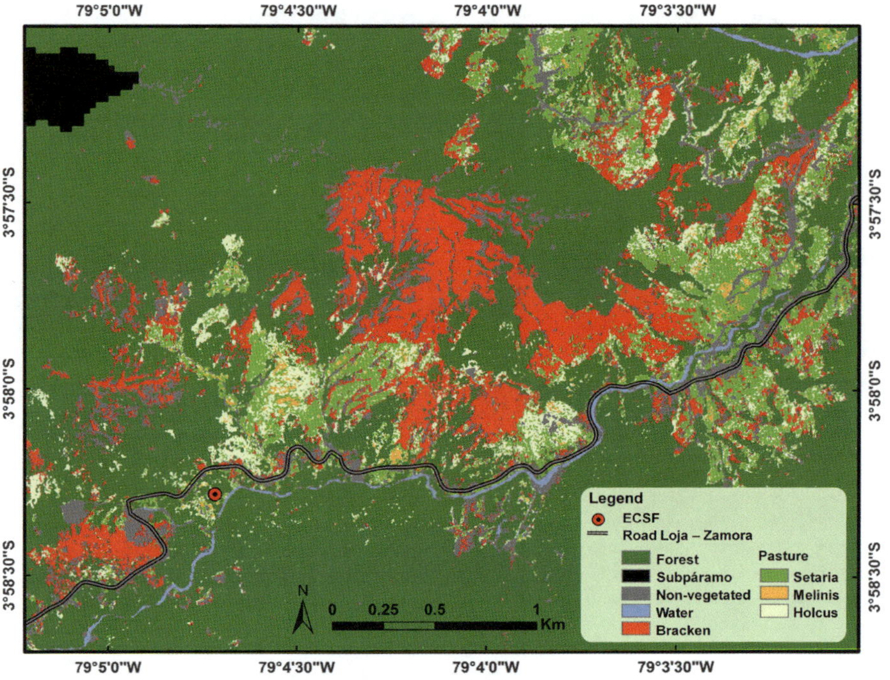

Fig. 15.3 Land use in the study area in the Rio San Francisco valley in October 2010. The spatial distribution of bracken and of the different pasture types is based on a QuickBird satellite image. Spectral signatures determined by field spectrometry were used for the classification (Göttlicher et al. 2011). *ECSF* Estación Científica San Francisco

Cattle numbers per hectare of pasture are low due to the poor nutritive value of the major grass species. The livestock density varies between 0.5 and 1.5 head per ha with an average value of 0.7 head per ha (Gerique 2010), resulting in a low income from pasture farming. Empirical data from a household survey showed that average net revenues from milk and meat production amounted to 137 US$ per ha and year with a large standard deviation of ±141 US$ (Knoke et al. 2009). Another recent study calculated net pasture revenues from milk production as 131 ± 18 US $ per ha and year (Knoke et al. 2011). A rough calculation can be performed on the assumption that there is an average of 0.8 cows per ha, half of which produce 4.5 l of milk per day over a lactation period of 200 days. At a milk selling price of 0.34 US$ per litre, this production would result in a gross annual revenue of 122 US$ per ha and year. The meat production of the other 50 % of the cattle can be calculated at 88 kg meat per ha. At a meat price of 1.9 US$ per kg, this meat would bring a revenue of 167 US$ per ha and year. Maintenance and labour costs must, however, be deducted from the gross annual revenue. These costs may amount to up to 135 US$ per ha and year according to farmer's estimates, and so a yearly net revenue of only 154 US$ per ha of pasture would result.

An average per capita GNP of approximately 4,000 US$ was quoted for Ecuador for 2010 (US State Department 2011. Assuming the presence of four adults in each family, the average annual household income would amount to 16,000 US$. A farm in the San Francisco valley would require a pasture area of at least 100 ha to be able to achieve this average income. This pasture size is, however, not realistic for the Rio San Francisco valley. From a survey of 130 households carried out by Barkmann (Knoke et al. 2009), only one farm comprised more than 100 ha, the median being about 13 ha of pasture area. Even upon consideration of the products grown in the small obligatory home gardens, subsistence farming is unable to provide sufficient means for securing a livelihood corresponding to the national average for the farmers in the research area. This also holds even for the better organized communities of the Saraguros in the neighbouring valleys, whose annual family assets may be four times as high as those of the Rio San Francisco valley (Gerique 2010). Diversification of the income portfolio, including the utilization of non-farm sources, thus acquires great importance for the farmers. Members of the farm families accordingly aspire to non-farm wage income and invest their urban earnings in the maintenance of their farms.

15.3 Conclusions

This chapter presents selected aspects of the pastures and pasture farming in the Rio San Francisco valley from the viewpoint of provisioning ecosystem services subsequent to clearing of the forest. Three types of pastures could be differentiated according to the dominant pasture grasses, two of which are threatened by highly invasive weeds such as the tropical bracken fern. Propagation of the weeds is fostered by an ecologically insufficient pasture management that eventually results in the abandonment of the pastures following depletion of essential macronutrients. The species diversity of the pasture vegetation has increased considerably in the course of the last decade, but the income from livestock farming is still too low for an adequate livelihood of the local farmers. A rough estimate shows that subsistence farming is practically not possible under the current regime of pasture management and that diversification of the income portfolio is imperative.

References

Axtmayer JH, Asenjo CJ, Cook DH (1938) The nutritive value of some forage crops of Puerto Rico. I. Grasses. J Agric Univ P Rico 22:95–121

Beck E, Hartig K, Roos K (2008) Forest clearing by slash and burn. In: Beck E, Bendix J, Kottke I, Makeschin F, Mosandl R (eds) Gradients in a tropical Mountain Ecosystem of Ecuador. Ecological studies, vol 198. Springer, Berlinm, pp 371–374

D'Antonio CM, Vitousek PM (1992) Biological invasions by exotic grasses, the grass/fire cycle, and global change. Annu Rev Ecol Syst 23:63–87 da Costa JL, Campos J, Garcia R, do Nascimento Junior D (1981) Effect of closing date on the nutritive value of molasses grass (*Melinis minutiflora* Pal de Beauv) with reserve pasture for the dry season. Rev Bras Zool 10:765–784

Duke JA (1979) Ecosystematic data on economic plants. Quart J Crude Drug Res 17:91–110

Gawlik J (2010) Phytodiversität auf anthropogen veränderten Standorten im San Francisco Tal – Südecuador. Diploma Thesis, Universität Erlangen-Nürnberg

Gerique A (2010) Biodiversity as a resource: plant use and land use among the Shuar, Saraguros, and Mestizos in tropical rainforest areas of southern Ecuador. PhD Thesis, University of Erlangen-Nürnberg

Göhl B (1982) Les aliments du bétail sous les tropiques. FAO, Division de Production et Santé Animale, Roma

Göttlicher D, Albert J, Nauß T, Bendix J (2011) Optical properties of selected plants from a tropical mountain ecosystem – traits for plant functional types to parameterize a land surface model. Ecol Model 222:493–502

Hacker JB (1974) Variation in oxalate, major cations and dry matter digestibility of 47 introduction of the tropical grass *Setaria*. Trop Grasslands 8:145–154

Hacker JB, Jones RJ (1969) The *Setaria sphacelata* complex – a review. Trop Grasslands 3:13–34

Hamer U, Rumpel C, Dignac MF (2012a) Cutin and suberin biomarkers as tracer for the turnover of shoot and root derived organic matter along a chronosequence of Ecuadorian pasture soils. Eur J Soil Sci 63:808–819. doi:10.1111/j.1365-2389.2012.01476.x

Hamer U, Potthast K, Burneo JI, Makeschin F (2012b) Nutrient stocks and phosphorus fractions in mountain soils of Southern Ecuador after forest to pasture conversion. Biogeochemistry 112:495–510. doi:10.1007/s10533-012-9742-z

Hartig K, Beck E (2003) The bracken fern (*Pteridium arachnoideum* (Kaulf.) Macon dilemma in the Andes of Southern Ecuador. Ecotropica 9:3–13

Jones RJ, Ford CW (1972) The soluble oxalate content of some tropical pasture grasses grown in South-East Queensland. Trop Grasslands 6:201–204

Knicker H (2007) How does fire affect the nature and stability of soil organic nitrogen and carbon? A review. Biogeochemistry 85:91–118

Knoke T, Weber M, Barkmann J, Pohle P, Calvas B, Medina C, Aguirre N, Günter S, Stimm B, Mosandl R, von Walter F, Maza B, Gerique A (2009) Effectiveness and distributional impacts of payments for reduced carbon emissions from deforestation. Erdkunde 63:365–384

Knoke T, Steinbeis OE, Bösch M, Román-Cuesta RM, Burkhardt T (2011) Cost-effective compensation to avoid carbon emissions from forest loss: an approach to consider price–quantity effects and risk-aversion. Ecol Econ 70:1139–1153

Landon JR (1991) Booker tropical soil manual. Longman, London, 450 p

Makeschin F, Haubrich F, Abiy M, Burneo JI, Klinger T (2008) Pasture management and natural soil regeneration. In: Beck E, Bendix J, Kottke I, Makeschin F, Mosandl R (eds) Gradients in a tropical mountain ecosystem of Ecuador. Ecological studies, vol 198. Springer, Berlin, pp 397–408

Merrill EH, Callahan-Olson A, Raedeke KJ, Taber RD, Anderson RJ (1995) Elk (*Cervus elaphus roosevelti*) dietary composition and quality in the Mount St. Helens blast zone. Northwest Sci 69:9–18

Middleton CH, Barry GA (1978) A study of oxalate concentration in fine grasses in the wet tropics of Queensland. Trop Grasslands 12:28–35

Mosandl R, Günter S, Stimm B, Weber M (2008) Ecuador suffers the highest deforestation rate in South America. In: Beck E, Bendix J, Kottke I, Makeschin F, Mosandl R (eds) Gradients in a tropical mountain ecosystem of Ecuador. Ecological studies, vol 198. Springer, Berlin, pp 37–40

Peters T, Diertl K-H, Gawlik J, Rankl M, Richter M (2010) Vascular plant diversity in natural and anthropogenic ecosystems in the Andes of southern Ecuador. Mt Res Dev 30:344–352

Pohle P, Gerique A (2006) Traditional ecological knowledge and biodiversity management in the Andes of southern Ecuador. Swiss J Geogr 4:275–285

Potthast K, Hamer U, Makeschin F (2011) Land-use change in a tropical mountain rainforest region of Southern Ecuador affects soil microorganisms and nutrient cycling. Biogeochemistry. doi:10.1007/s10533-10011-19626-10537

Roos K, Rollenbeck R, Peters T, Bendix J, Beck E (2010) Growth of tropical bracken (*Pteridium arachnoideum*): response to weather variations and burning. Inv Plant Sci Manage 3:402–411. doi:10.1614/IPSM-D-09-00031.1

Seawright AA, Groenendyk S, Silva KING (1970) An outbreak of oxalate poisoning in cattle grazing Setaria sphacelata. Austral Vet J 46:293–296

Thies B, Meyer H, Nauss T, Bendix J (2012) Projecting land use/land cover changes in a tropical mountain forest of southern Ecuador. J Land Use Sci. doi:10.1080/1747423X.2012.718378

US State Department (2011) http://www.state.gov/r/pa/ei/bgn/35761.htm. Accessed 21 Jan 2012

Watt TA (1978) The biology of *Holcus lanatus* (Yorkshire fog) and its significance in grassland. Herbage Abs 48(6):195–204

Chapter 16
Current Provisioning Ecosystem Services for the Local Population: Landscape Transformation, Land Use, and Plant Use

Perdita Pohle, Andrés Gerique, Maria Fernanda López, and Regine Spohner

16.1 Introduction

Provisioning ecosystem services are defined as products obtained directly from nature (e.g., freshwater, hunting and gathering of a range of species, fisheries, harvesting of plants for timber, fuel, fiber, and medicines). These services also come from domesticated species through pastoralism, agriculture, and aquaculture (Daily 1997; Millennium Ecosystem Assessment 2005). In rural areas of tropical developing countries, agricultural biodiversity plays a major role in the production of goods and provides local people with a wider range of responses to environmental or market risks (Coomes and Burt 1997). However, even if farmland in forest areas can support biodiversity through preservation of important forest ecosystem elements on a small scale (Mendenhall et al. 2011), the long-term sustainability of forests and the array of ecosystem services they provide may be under threat from the expansion of farming activities. Given the fragmented state of most tropical ecosystems, agricultural landscapes should be a crucial concern of any conservation strategy.

P. Pohle (✉) • A. Gerique
Institute of Geography, Friedrich-Alexander University of Erlangen-Nuremberg, 91054 Erlangen, Germany
e-mail: perdita.pohle@geographie.uni-erlangen.de

M.F. López
Pontificia Universidad Católica del Ecuador (PUCE), Apt. 17-01-2184, Quito, Ecuador

R. Spohner
Department of Geography, University of Cologne, 50923 Cologne, Germany

J. Bendix et al. (eds.), *Ecosystem Services, Biodiversity and Environmental Change in a Tropical Mountain Ecosystem of South Ecuador*, Ecological Studies 221,
DOI 10.1007/978-3-642-38137-9_16, © Springer-Verlag Berlin Heidelberg 2013

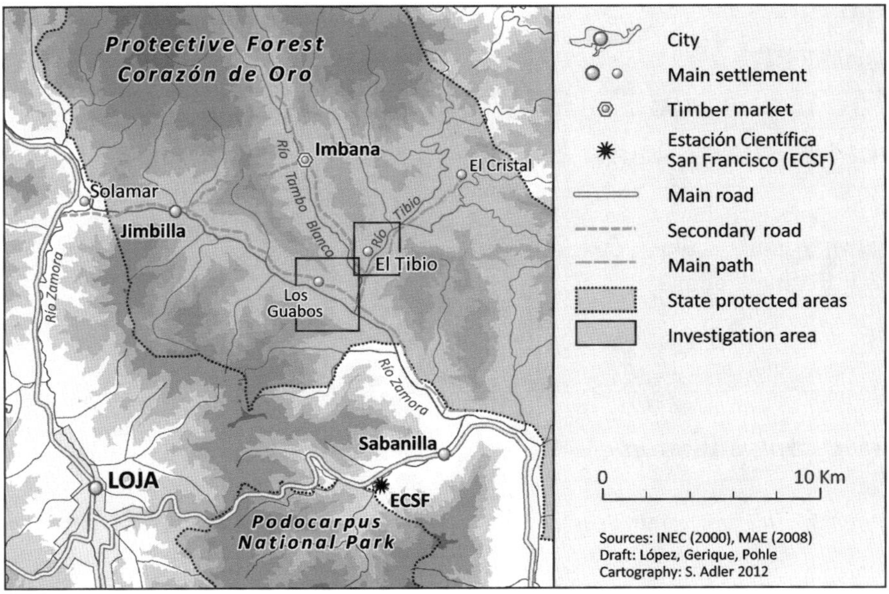

Fig. 16.1 Location of the study sites

16.2 Aims, Materials, and Methods

The megadiverse ecosystem of the tropical mountain forests of the eastern Andean Cordillera of southern Ecuador has been under severe pressure since the first colonists arrived during the first half of the twentieth century. Given the current fragmented stage of the former forest landscape which has been replaced by a mixed matrix of forest remnants, pastures, fields/gardens, and wasteland (*matorral*), the aims of the study are as follows:

1. To document and to analyze the landscape transformation process and its implications for provisioning ecosystem services for the local population,
2. To determine current ecosystem provisioning services for food production of smallholder farmers and to identify the services provided by local plant diversity for specific ethno-cultural communities.

The outcome of this study aims to help experts to develop strategies for biodiversity conservation and a sustainable use of ecosystem services.

Research was undertaken north of Podocarpus National Park in the Upper Zamora Valley (Fig. 16.1). The main study sites were the settlements of El Tibio (1,770 m a.s.l.) with an indigenous Saraguro population and Los Guabos (1,900 m a.s.l.) mainly inhabited by *mestizos*.[1] Other study sites were El Cristal

[1] The Saraguros are Quechua-speaking highland Indians who traditionally inhabit the northern Andean area of Loja Province. The mestizos are of Spanish and indigenous descent and represent at more than 80 % the major population group of southern Ecuador (Pohle 2008).

(1,980 m a.s.l.), a small Saraguro settlement north of El Tibio, and 12 *mestizo* farms along the road Loja–Zamora in areas known as Sabanilla, El Retorno, and La Fragancia (2,200–1,100 m a.s.l.).

The study makes use of data generated by 6 years of ethnobotanical/ ethnoecological and agrogeographical research among the ethnic groups of southern Ecuador (Pohle and Gerique 2006, 2008; Gerique and Veintimilla 2008; Pohle et al. 2010). Three methodological approaches were used: (a) analysis of qualitative and quantitative data from field surveys concerning settlement history and livelihood strategies, (b) analysis of land use/land cover (LULC) changes, and (c) compilation of an ethnobotanical inventory of useful wild and cultivated plants.

16.3 Results and Discussion

16.3.1 Colonization, Access to Land, and Extraction of Plant Resources

Historical insights into the colonization process, access to land, and resource extraction are necessary preconditions to assess current ecosystem services. The conversion of land from natural ecosystems to agriculture by small-scale farming colonists has been a main factor of environmental change in Ecuador (Bromley 1981; Pichón 1996).

Data provided by local informants and secondary sources (Arias Benavides 2004) mention the existence of large land holdings (*haciendas*) in the Upper Zamora Valley, such as the *hacienda* Los Guabos, which developed through land accumulation by absentee landowners (*terratenientes*) from the city of Loja since the late nineteenth century. Settlers who moved to the area from the beginning of the twentieth century were laborers of the large holdings or landless colonists in search of vacant public lands (*tierras baldías*). The colonization route mainly used by *mestizo* settlers was the ancient trail between the cities of Loja and Zamora (Fig. 16.1). Sabanilla originated as a resting place (*tambo*) along this trail. A second route followed the Río Tambo Blanco and was mainly used by Saraguro colonists from the high Andean town of San Lucas. They founded El Tibio in the 1950s. *Mestizos* established various settlements such as Los Guabos in the early twentieth century.

Formalization of land rights for peasants was only possible after the first Agrarian Reform and Colonization Law came into force in 1964. The law abolished the traditional forms of labor compensation inherited from the colonial period and encouraged the expropriation and adjudication of *haciendas* which were not productive (Barsky 1984). Landowners undertook indirect measures to avoid expropriation, such as selling parts of their property to the workers (Pohle 2008), as was the case in Los Guabos. The second Law of Agrarian Reform and Colonization in 1973 strengthened the opportunities for peasants to acquire land, both through means of agrarian reform procedures (e.g., expropriation of private *haciendas*) or through colonization (e.g., adjudication of public land). Both processes occurred in the study site; according to local informants, many of the first settlers bought land from former *hacienda* owners or received it as compensation for their work (Tutillo 2005; Gerique 2010). The arrival of settlers was intense until the 1970s, and active pioneer fronts accompanied the development of settlements. Later pioneer activities continued mainly to establish pastures and to extract timber. Deforestation became the main condition to guarantee access to land in frontier areas as a response to the legal demands of the colonization laws requiring land clearance of between 25 and 50 % of the claimed land to get land titles (Southgate et al. 2009). Between 1975 and 1980 the greatest amount of land was granted by the state to colonists in the country (Gondard and Mazurek 2001). The Agrarian Development Law of 1994 derogated the Agrarian Reform Law and eliminated the condition of forest clearing for adjudication (Pohle et al. 2010).

As in other frontier areas (Rudel and Horowitz 1993; Marquette 2006), the welfare of the first settlers of the Upper Zamora Valley relied greatly on the provisioning services of the ecosystem. Important non-timber forest products (NTFPs) for medicinal purposes were overexploited, like the bark of *cascarilla* (*Cinchona* spp.) in the eighteenth and nineteenth centuries and again in the 1940s, and the latex of *sangre de drago* (*Croton lechleri* and *Croton mutisianus*) in the 1990s (Gerique 2010). Logging became the main source of income for the colonists, starting in the 1950s with intense extraction of the high quality timber of *romerillo* (*Podocarpus oleifolius* and *Prumnopitys montana*) in Sabanilla. These species were very abundant at that time and colonists perceived their existence as "inexhaustible." The road Loja–Zamora (1950–1960) also favored timber extraction. In the 1990s *romerillo* was scarce, and extraction was undertaken in remote areas inside or near the Podocarpus National Park (Romerillos, Tunantza Alto). Other species such as *cedro* (*Cedrela* spp.), *sanón* (*Hyeronima* spp.), *canelo* (*Nectandra* spp.), and *guayacán* (*Tabebuia chrysantha*) also became valuable and were heavily sought after (Gerique 2010).

During the 1990s a turnover of resource utilization took place. The pace of colonization decreased and, although pasture expansion continued to more remote sites, few new immigrants arrived in the area. Profitable timber species became overexploited, while cattle ranching became the main economic activity (see Sect. 16.3.3). The declaration of the Podocarpus National Park (1982) and the Bosque Protector Corazón de Oro (2000, see Table 13.1) notably influenced the allocation of land to conservation purposes and set legal barriers to the conversion of forests to agricultural land and to the acquisition of property titles. The existence of the Podocarpus National Park currently opens up opportunities for the implementation of conservation projects with the participation of local stakeholders. One example is the declaration of the Biosphere Reserve Podocarpus—El Condor in 2007 which aims to be an important instrument for further negotiations towards conservation and sustainable development.

16.3.2 Land Use/Land Cover Change Analysis at Local Scale

In the research area spatiotemporal landscape transformations are linked to the political and land use history, especially to the colonization process and the allocation of land. Additionally, land use/land cover changes largely depend on the decisions of individual farming households, especially at local scales.

The land use/land cover change maps of Los Guabos and El Tibio (Figs. 16.2 and 16.3)[2] give insights into the spatial distribution of three land use/land cover classes—forest, *matorral*, pasture[3]—and their spatiotemporal development in the period 1969–2001. Adjacent to the maps, change detection graphs are presented, in the case of El Tibio with three intermediate change periods.

The LULC change analysis shows two dynamics: (a) a main process of forest loss due to pasture expansion and (b) a secondary process of vegetation succession (*matorral* and forest). In both study sites a substantial loss of forest cover in favor of pastures has taken place: in 2001 the forest coverage in both areas was below 50 %.

Regarding the spatial distribution of land use/land cover classes, similar features can be observed for Los Guabos and El Tibio. The maps show a clear prevalence of pastures at lower altitudes, along riversides, and on the valley sides where settlements were established in order to keep distances short to control livestock. Accordingly, forest remnants are restricted to the upper parts of the slopes, the more inclined slopes, steep *quebradas*, and the valley sides opposite the villages. Thus, the deforestation process in both areas seems to have followed a general pattern: (a) from lower to higher altitude, (b) along rivers, and (c) from center to periphery (Lambin and Geist 2006).

Regarding the spatiotemporal development of specific land use/land cover classes, differences between both villages are obvious. Whereas in Los Guabos 33 % of the land use/land cover was classified as pasture in 1969 and 2001, respectively, in El Tibio the proportion of pastures increased considerably from 25 % in 1969 to 39 % in 2001, while forests declined dramatically from 68 % to 42 % (Figs. 16.2 and 16.3, bare graphs in black). Accordingly, in El Tibio the highest proportion of land use/land cover change can be attributed to the change category "forest to pasture" (44 %) compared to Los Guabos with 20 % (Figs. 16.2 and 16.3, bare graphs in color). On the valley side of El Tibio many acres of pasture were established between 1969 and 2001, whereas on the valley side of Los Guabos pastures were predominantly established before 1969. The differences in pasture

[2] The LULC change analysis is based on a visual interpretation of a sequence of orthorectified aerial photographs of Los Guabos (1969, 2001) and El Tibio (1969, 1976, 1989, 2001) with ArcGis. Field work for ground-truthing and qualitative data assessment was carried out between 2003 and 2007. The LULC change maps cover an area of about 2,000 ha (Los Guabos) and 500 ha (El Tibio).

[3] The forest category comprises tropical mountain forest, either as primary forest or in a successional stage. The category *matorral* comprises shrub (*lusara*) and bracken (*llashipa*) vegetation. Pastures in the research area are either *pastos naturales* (Sect. 15.2.1, prevalent in Los Guabos) or cultivated *mequerón* (*Setaria sphacelata*) pastures (dominant in El Tibio).

Los Guabos - Quality, quantity and spatial distribution of land use/land cover change (1969 - 2001)

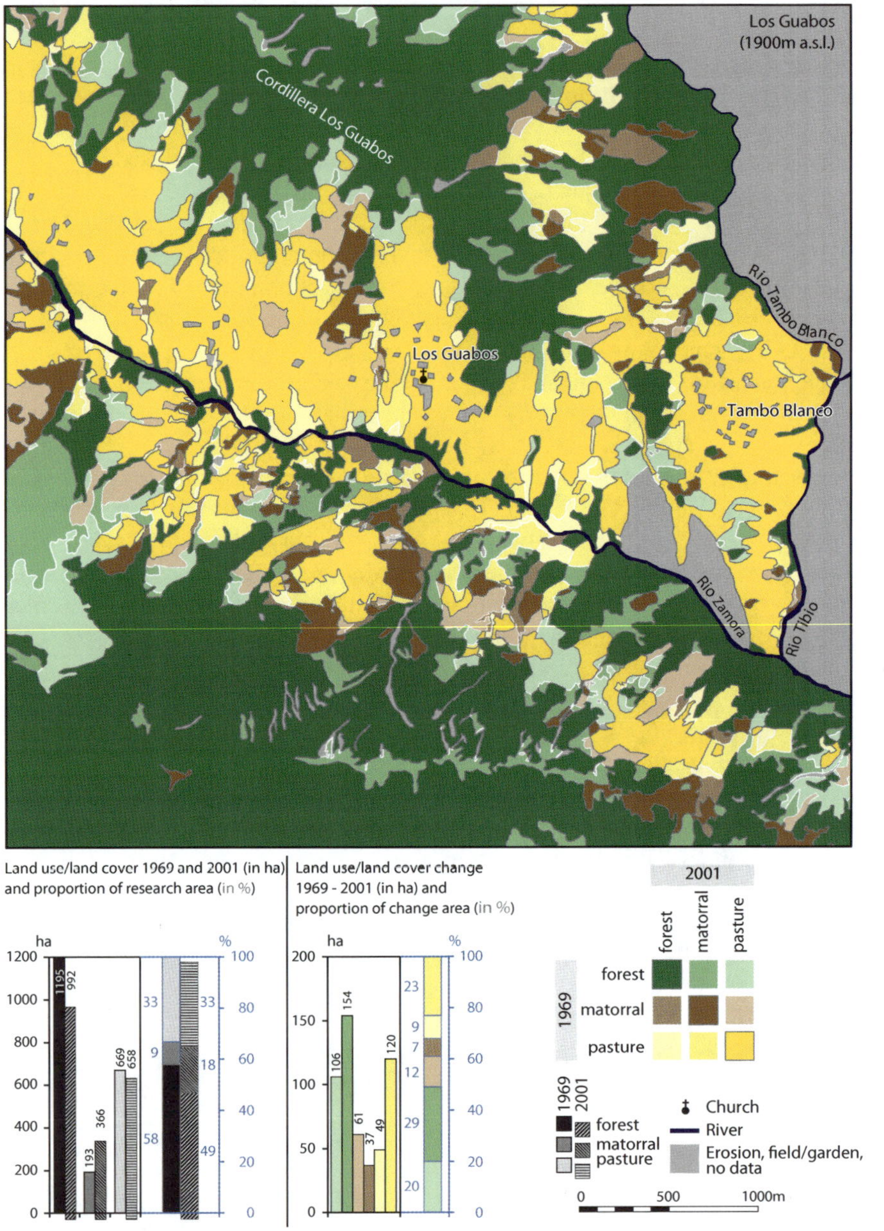

Source: Orthoimage 1969, 2001, Instituto Geográfico Militar, Quito; Fieldwork, Supervision: P. Pohle; Evaluation, Cartography: R. Spohner (2011)

Fig. 16.2 Spatiotemporal land use/land cover change detection at Los Guabos between 1969 and 2001

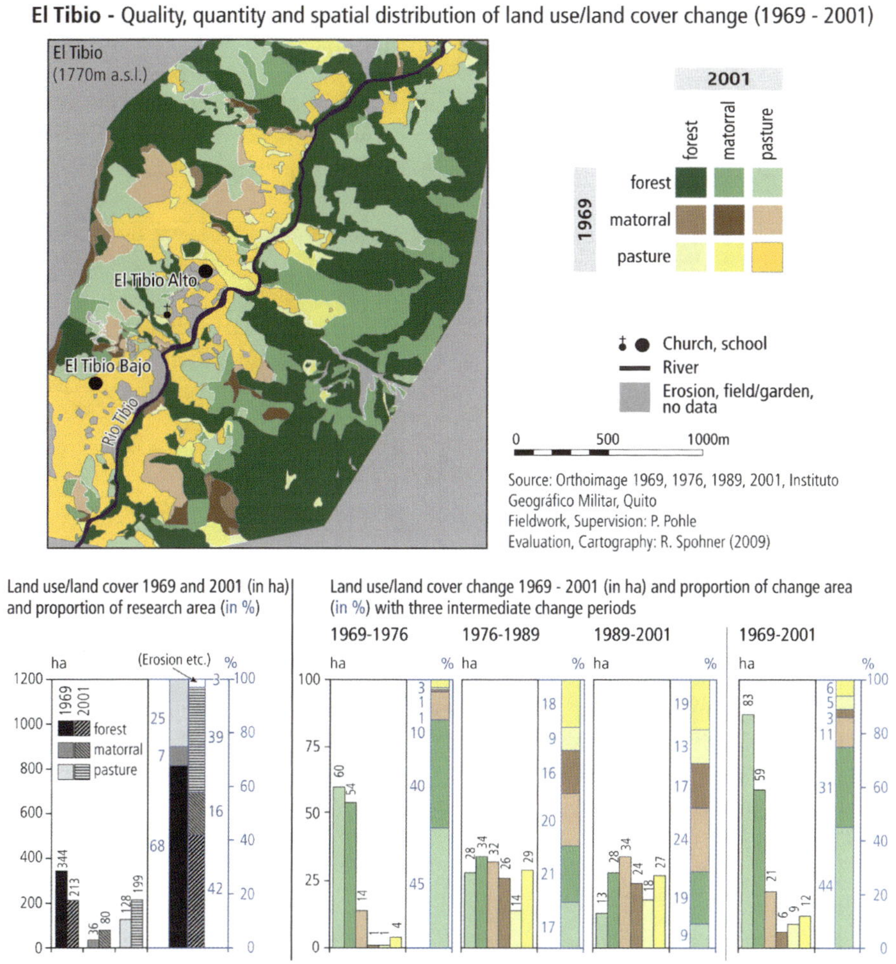

Fig. 16.3 Spatiotemporal land use/land cover change detection at El Tibio between 1969 and 2001

expansion between the two communities can be related to their history of settlement and colonization. As reported by the villagers, the area of Los Guabos was colonized more than 100 years ago, whereas El Tibio was founded in the 1950s. Thus it appears that Los Guabos with its stable or decreasing deforestation rate is in a more advanced phase of the landscape transformation process.

Concerning the process of vegetation succession, similar features could be observed in Los Guabos and El Tibio. From 1969 to 2001 in both research areas the proportion of the land cover class *matorral* at least doubled: from 9 to 18 % in Los Guabos and 7 to 16 % in El Tibio (Figs. 16.2 and 16.3, bare graphs in black). According to the transformation matrix (Figs. 16.2 and 16.3, bare graphs in color) this doubling can be attributed mainly to the change category "forest to *matorral*" comprising 29 % of the change area in Los Guabos and 31 % in El Tibio, and to a lesser degree to the change category "pasture to *matorral*," in El Tibio with 6 %, whereas in Los Guabos this category is more pronounced with 23 %. While the changes from forest to *matorral* suggest an initial stage in postfire vegetation regeneration, changes of pasture to *matorral* indicate a degradation or abandonment of pastures to successional vegetation.

The relatively high rates of change from forest to *matorral* can be understood in view of the legal demands for land adjudications given by the two Laws of Agrarian Reform and Colonization in 1964 and 1973, which encouraged land clearing for obtaining official land titles (Sect. 16.3.1). As illustrated by the high proportion of the change category "forest to *matorral*" (40 %) in the intermediate change period 1969–1976 of the LULC change graphs of El Tibio (Figs. 16.3), obviously more forest was cleared than was needed for pastures. With the Law of Agrarian Development of 1994 forest clearing as a pre-condition for land adjudication was eliminated. Consequently, in El Tibio the proportion of the change category "forest to *matorral*" decreased from 40 % (1969–1976) to 28 % (1989–2001). As reported in the interviews, the cleared land was often too large or located too far from the village for effective maintenance. These areas were therefore left abandoned and secondary vegetation developed.

Another reason for the high rates of change from forest to *matorral* can be seen in the slash and burn practice to establish pastures among the *mestizos* and Saraguros where fire often gets out of control. The unintentionally burned forest areas just give way to the development of a secondary bracken and shrub vegetation (see Sect. 15.2). These plots—sometimes extensive—are mainly located next to the recently established pastures.

The higher percentage of change from pasture to *matorral* in Los Guabos (23 %) can partly be related to the emigration of landowners to Loja and the scarcity of labor for the maintenance of pastures as stated by the interviewed farmers. These plots are found in favorable locations close to the village or close to previously (before 1969) established pastures (Fig. 16.2).

From the LULC change analysis it can be concluded that due to the substantial loss of forest cover in favor of pastures, forest products play only a marginal role in food and income (from timber) supply for the local population who are becoming increasingly dependent on cattle ranching and products derived from that source. Although *matorral* areas in general are of limited use for the regulating and provisioning ecosystem services, their potential towards sustainable land use options—either for forest recovery by succession, for reforestation with native tree species (Chap. 13), or for pasture rehabilitation (Chap. 15)—might be rated as promising, with complementary financial incentives, as suggested by Knoke et al. (2011, see also Chap. 25).

16.3.3 *Food Production of Small-Scale Farming Households: Livelihood Strategies, Cattle Ranching, Field and Garden Cropping*

16.3.3.1 Livelihood Strategies

The Saraguros and the *mestizos* of the research area are mainly engaged in agro-pastoral activities that combine both a market economy (cattle ranching for cheese, milk, and meat production) and a subsistence economy (crop production, horticulture, and cattle ranching for subsistence needs). Whereas corn and beans are cropped in shifting fields (*chacras*), vegetables, fruits, spices, and other useful plants are cultivated in permanent home gardens (*huertas*). The main product drawn from cattle ranching is cheese, which is sold weekly in the markets of Loja.

According to the livelihood survey (Pohle et al. 2010, 2012) in the communities of El Tibio and Los Guabos revenues from employment and cattle ranching were the most important sources of household income, comprising in 2007 83 % (El Tibio, $n = 28$ households) and 80 % (Los Guabos, $n = 18$ households). Revenues from cattle ranching (mainly sales of cheese) are far higher in Saraguro households (41.2 %) than in m*estizo* households (25.5 %). The contribution of employment (mainly in the form of irregular work, day labor) is higher in the *mestizos* households (54.5 %) than in those of the Saraguros (41.8 %).

The stronger engagement of Saraguros in cattle ranching becomes obvious also in the share of land per land use category and the number of cattle per household: the Saraguros of El Tibio maintain more pasture (11.0 ha per household, $n = 29$) than the *mestizos* of Los Guabos (8.4 ha per household, $n = 18$) and own more cattle (11.4 head compared to 9.4 head). In contrast, the *mestizos* of Los Guabos show a stronger engagement in cropping than the Saraguros (4.2 ha crop fields (*chacras*) per household compared to 2.1 ha).

16.3.3.2 Cattle Ranching

In the study area cattle ranching is the most relevant productive activity and from the farmers' point of view the most profitable. *Quesillo*, an unsalted fresh white cheese, is the main product drawn from raising cattle. Production usually takes place in the pastures, since cheese is easier to transport and more durable than milk. Only farmers who have good access to roads can sell milk to regional producers of dairy products. Although cattle farming for meat production requires less labor input and produces high benefits, it is only practiced by ranchers who do not depend on daily or weekly revenues, as cows must be raised for 1 year before they can be sold.

Slash and burn is the traditional way of establishing pastures (see Chap. 15). The Saraguros and *mestizos* follow a system of rotations on pasture paddocks (*potreros*) of about 4–10 ha. Giving cattle enough access to fodder, farmers move livestock between paddocks at intervals of 8–60 days, preferably every 15 days. Decisions about when to move the cattle depend on the grass species, the number and type of animals grazing (dairy cows, heifers, bulls, steers, calves), and the weather conditions. The *potreros* are used again when fodder grasses have recovered, which takes 45–90 days, again depending on the grass species. Rotations keep grasses from going to seed which preserves forage quality and avoids damage to the udders of cows from the rigid culms. Pregnant dairy cows and cows with calves are kept in *potreros* close to the houses, as they need milking every day. Bulls which do not require daily care are usually raised on distant pastures. One farmer with enough pasture land can manage up to 25 dairy cows and many more bulls, especially if there is assistance for seasonal activities. Another method of pasture management, especially applied by the Saraguros, is the tethering of cattle with a rope. Thus, 4–6 animals can be kept in 1 ha for 15 days. Although this method requires more labor, it allows for higher stocking density and has a second advantage: between the grazing periods, the farmers can take care of the pastures, in particular removing nasty weeds like bracken (Gerique 2010).

Stock density in the research area varies between 0.5 and 1.5 head/ha with an average of 0.7, a density which has also been reported for the Province Zamora Chinchipe (Aguirre and Maldonado 2004). Most ranchers in El Tibio and Los Guabos are smallholders who keep five or less cows per household (24 of a total of 47 households) and consider people with 20 or more cows (6 of 47 households) wealthy. The herds of the farmers in the *fincas* along the road Loja–Zamora are clearly larger: in 2005/06 the number of cattle varied between 9 and 65 head; 7 of a total of 12 households had more than 20; the mean was 25.7 head of cattle per household (Gerique 2010).

Many paddocks in the studied communities are more than 50 years old. Since ranchers do not use artificial fertilizers nor improve the soil quality by a targeted cultivation of nitrogen fixing plants, probably an optimized use of pastures has made such a long use possible, at least in favorable sites with good soils and a low risk of erosion (Gerique 2010).

16.3.3.3 Field Cropping

Maize (*Zea mays*) as the basis of the traditional diet is the most common crop in both ethnic groups. However, it is not produced for the market. Maize is intercropped with beans and squash is commonly planted at the edge of the fields. The average size of the crop fields is 0.5–1 ha. October is the main sowing period and after the harvest in February the remnants are used as fodder for the cattle which fertilize the fields with their dung. Maize is commonly not grown for more than 2 years in the same field. However, the *mestizos* of Los Guabos apparently have optimized maize cultivation. According to local informants, they have been cultivating maize by intercropping with beans (which are known to fix atmospheric nitrogen and in turn fertilize the soil) in the same plots without interruption for more than 20 years, allegedly without using any chemical fertilizer (Gerique 2010).

Farmers also cultivate maize after slashing and burning (Beck et al. 2008). This occurs mainly on marginal steep slopes and entails further clearing of forest (around 0.5 ha each time) adjacent to the pastures. Once the forest parcel has been cleared and the soil has cooled down, farmers use planting sticks for sowing the maize. On steep slopes soils are poor and crop production is limited to 1 or 2 years; afterwards the plot is abandoned or pasture grass is planted (Gerique 2010).

16.3.3.4 Gardening

Home gardens (*huertas*) of the Saraguros and *mestizos* of El Tibio and Los Guabos host a high agrobiodiversity (Pohle and Gerique 2006, 2008) and play an essential role not only in supplying food, in particular, fruits, vegetables, spices, and teas but also in the growing of medicinal and ornamental plants. As in the case of field cropping, Saraguros and *mestizos* cultivate similar plants in their *huertas* and apply analogous cultivation methods. Differences result from the altitude, the age of the gardens, and personal preferences.

Huertas are laid out adjacent to the homes in rectangular shapes (Fig. 16.4). Inside the home garden, vegetables are often cultivated in fenced plots to keep plants safe from poultry. Obligatory is an area with banana (*Musa* × *paradisiaca*) which is used as pigsty and also as latrine. Sugar cane (*Saccharum officinarum*) is either cultivated in plots inside the home gardens or in small fields outside the garden. The plant composition of a home garden in Los Guabos is given in Fig. 16.4.

As the ethnobotanical and agrogeographical surveys indicate, there is a potential for the production of plant species in demand for regional markets in home gardens (Pohle et al. 2010). Since some *mestizos* and Saraguros already have experience in market oriented gardening—they sell ornamental plants (e.g., *Lilium candidum*, *Zantedeschia aethiopica*) or ingredients for the famous *horchata* tea in the markets of Loja—the further promotion of horticultural products could generate additional income on a low investment basis.

16.3.4 Plant Use

The central role of plants in the everyday life has obvious significance in developing countries, where all kinds of daily living activity include the direct use of plant resources (FAO 2007). In this way, indigenous and local people have generated vast bodies of knowledge about the use of plant resources.

As shown in Table 16.1, plant resources provide the Saraguros and *mestizos* of the research area with a wide array of services, mainly for food, medicine, decoration, and construction. The plant inventory of the Saraguros comprises 230 plant species with a total of 310 uses; among the *mestizos* 312 useful species with a total of 409 uses have been recorded. The differences in number might not necessarily

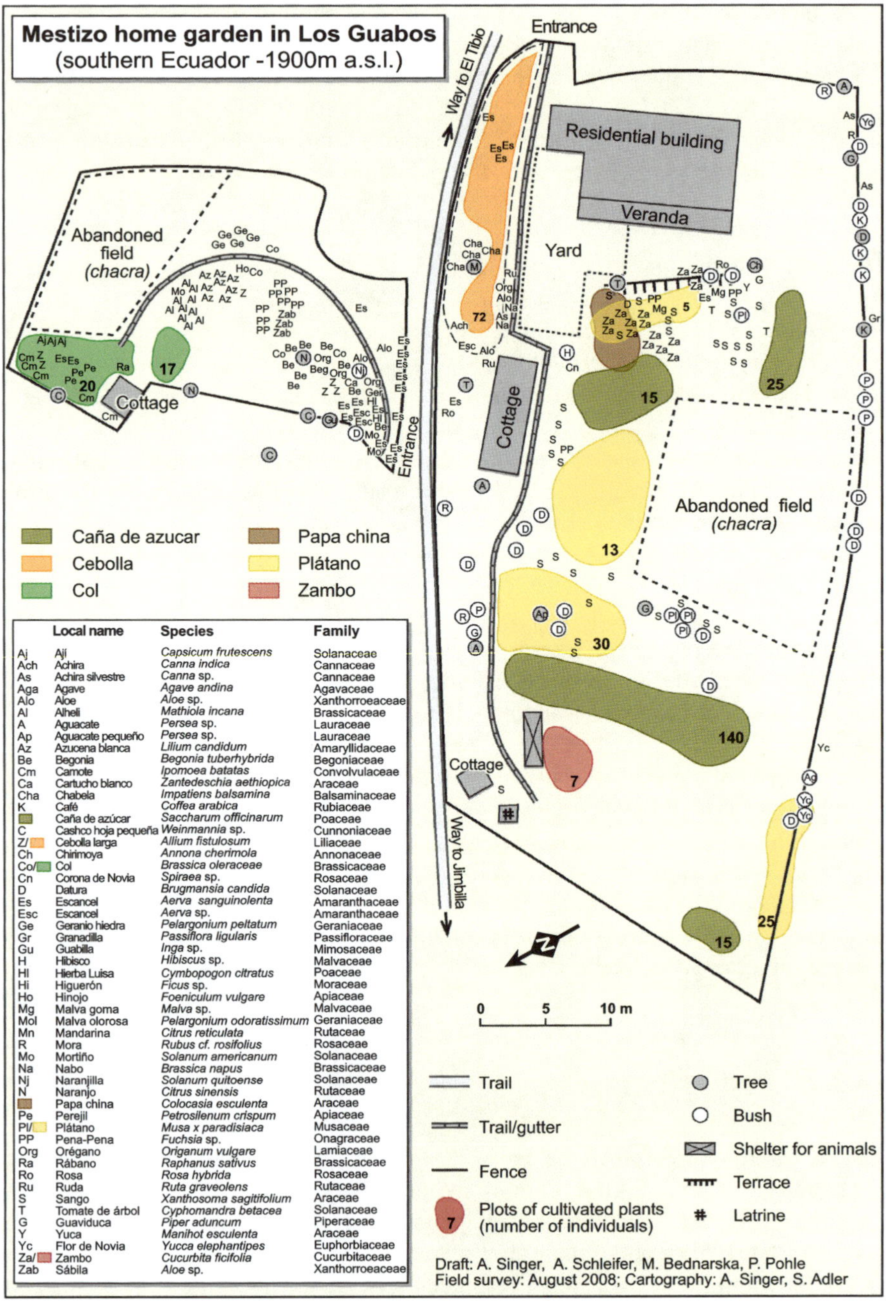

Fig. 16.4 *Mestizo* home garden in Los Guabos

Table 16.1 Plant uses among Saraguros (Sarag.) and *mestizos* (*mest.*) according to use categories and area of collection

	Uses from forest plants[a]		Uses from wild plants growing in disturbed areas[b]		Uses from cultivated plant species		Total of uses in each category	
Use categories	Sarag.	*mest.*	Sarag.	*mest.*	Sarag.	*mest.*	Sarag.	*mest.*
Food	9	10	15	25	60	68	84	103
Medicinal uses	1	2	43	49	31	48	75	99
Ornamental uses	1	10	4	2	34	66	39	78
Construction	16	14	15	16	6	7	37	37
Fodder	0	0	11	14	11	15	22	29
Living fences	0	0	3	2	10	16	13	18
Fuel	4	4	8	8	1	2	13	14
Shade trees	1	5	5	6	0	0	6	11
Other uses[c]	11	6	7	7	3	7	21	20
Total uses[d]	43	51	111	129	156	229	310	409

[a]This category includes wild species growing in home gardens, pastures, and fields. Forest species protected during forest clearing have been included in this category, even if the forest has disappeared

[b]Native pioneer species that develop on disturbed areas after clearing have been included in this category

[c]This category includes plants used to make or prepare tools, baskets, remedies for domestic animals, fibers, soaps, perfumes, dyes, amulets, incense, pesticides, and soil indicators

[d]All categories are nonexclusive categories; thus, one plant species can be represented in more than one use category if it provides more than one service. *Source*: Gerique (2010, modified)

reflect differences of ethnospecific knowledge but they can be attributed to the higher number of *mestizo* settlements studied and to the wider altitudinal gradient they cover.

More than half of the useful plants are cultivated plant species growing in fields, home gardens, and pastures. Disturbed areas and pastures are important collection sites of useful wild plants; more than 40 % of them are gathered here. Less than 10 % of the species used are from the forests, most of them are timber species (Sect. 13.3.2).

Despite the difference in numbers, the use of plants is similar in both ethnic groups. Food (including spices and infusions) and medicine (especially plants to treat gastrointestinal ailments, respiratory diseases, and skin problems) are the principal provisioning services. Both ethnic groups cultivate and collect a great variety of ornamental plants for the decoration of houses, gardens, and chapels. The *mestizos* use 78 species for such purposes which is remarkably high. Timber as construction material is another basic need, especially for house construction and furniture. The importance of pasture economy among both ethnic groups explains the relevance of fodder plants, shade trees, and species used for living fences. However, the latter are being gradually substituted by barbed wire fences. Moreover, due to the widespread availability of subsidized gas cylinders, the use of forest plants as fuel is substantially decreasing. Other plant species used to make or prepare tools and baskets, soaps, perfumes, fibers, or used as remedies for domestic animals, and as ritual or mythical plants have considerably lost importance and are often only known by older people. In general, the use of wild plants has lost importance today, as monetary income from cattle ranching and increasing supply on the market allow replacing traditional self-made products with manufactured goods.

16.4 Conclusions

Similar to other tropical frontier areas the land use/land cover changes in the research area are characterized by a substantial loss of forests and a concomitant loss of biodiversity due to pasture extension. However, the trajectories of change are nonlinear, showing high deforestation rates in the 1960s and 1970s as a result of national colonization policy and land reforms, but also various stages of vegetation succession and even some forest recovery.

The study shows that current land use does not necessarily lead to a spiral of deforestation. Local examples demonstrate that with a proper management pastures and fields can be used for a long time reducing the need for more land to convert. Furthermore, the home gardens of the Saraguros and *mestizos* conserve high levels of agrobiodiversity (Pohle and Gerique 2006). These traditional and ecologically sound agricultural practices should be promoted and conserved (Harvey et al. 2008).

Regarding cultural preferences, Rudel and Horowitz (1993) pointed out that generic factors such as land-titling requirements, year of settlement, population growth, and improvement of infrastructure homogenize the land use patterns of different ethnic groups. Indeed, the Saraguros and the *mestizos* share modes of land and plant use and are engaged in similar agro-pastoral activities. However, the Saraguros of El Tibio have a stronger engagement in cattle ranching and seem to be more successful from an economic perspective.

In order to protect the remaining biodiversity and the ecological services it is necessary to develop land use systems that conserve the existing forest patches and offer attractive and affordable alternatives for cattle ranching (Marquette 2006, Sects. 17.3 and 25.4). But due to the multiple objectives that this activity fulfills, a complete substitution will be almost impossible. As one promising approach, the use of wasteland (matorral)—either for reforestation with useful native tree species in demand or for pasture rehabilitation—could be discussed (Stimm et al. 2008; Roos et al. 2011, Sect. 13.3.3 and Chap. 26). An additional option, among others (cf. Pohle et al. 2010; Gerique 2010), might be the cultivation of useful plants (e.g., medicinal herbs, fruits, vegetables, and ornamental flowers) in home gardens for a regional market. In any case, alternative land use systems should incorporate existing sustainable practices, should be based on local knowledge and experience and should take into account cultural preferences in order to be socially accepted.

Acknowledgments We wish to thank the inhabitants of the communities of El Tibio, Los Guabos, El Cristal, Sabanilla, El Retorno, and La Fragancia for their hospitality and generous participation in this study. Our gratitude also goes to Eduardo Tapia, Mónica Burbano, Tatiana Ramón, and to the staff of the Reinaldo Espinosa Herbarium of the National University of Loja for their contribution to this work.

References

Aguirre Z, Maldonado N (2004) Ecosistemas, biodiversidad, etnias y culturas de la Región Amazónica Ecuatoriana. Universidad Nacional de Loja UNL-CEDAMAZ-PROMSA, Loja

Arias Benavides H (2004) Zamora de Ayer y Hoy. Honorable Consejo Provincial de Zamora. Chinchipe, Zamora

Barsky O (1984) La Reforma Agraria Ecuatoriana. CEN, Quito

Beck E, Hartig K, Roos K (2008) Forest clearing by slash and burn. In: Beck E, Bendix J, Kottke I, Makeschin F, Mosandl R (eds) Gradients in a tropical mountain ecosystem of Ecuador, vol 198, Ecological studies . Springer, Berlin, pp 371–374

Bromley R (1981) The colonization of humid tropical areas in Ecuador. Singap J Trop Geogr 2 (1):15–26

Coomes OT, Burt GJ (1997) Indigenous market-oriented agroforestry: dissecting local diversity in western Amazonia. Agrofor Syst 37:27–44

Daily G (1997) What are ecosystems services? In: Daily G (ed) Nature's services: societal dependence on natural ecosystems. Island Press, Washington, DC, pp 2–10

FAO (Food and Agriculture Organization) (2007) The state of food and agriculture 2007, vol 38, FAO agriculture series. FAO, Rome

Gerique A (2010) Biodiversity as a resource: plant use and land use among the Shuar, Saraguros, and Mestizos in tropical rainforest areas of southern Ecuador. Dissertation, University of Erlangen-Nuremberg

Gerique A, Veintimilla D (2008) Useful plants and weeds occurring in Shuar, Saraguro and Mestizo communities. In: Checklist – Reserva Biológica San Francisco. Ecotropical monographs, vol 4. Prov. Zamora-Chinchipe, S. Ecuador, pp 237–256

Gondard P and Mazurek H (2001) 30 Años de Reforma Agraria y Colonización en el Ecuador (1964–1994). In: Gondard P, León J (eds) Dinámicas Territoriales. Estudios de Geografía, vol 10, Quito, pp 15–40

Harvey CA, Komar O, Chazdon R, Ferguson BG, Finegan B, Griffith DM, Martínez-Ramos M, Morales H, Nigh R, Soto-Pinto L, Van Breugel M, Wishnie M (2008) Integrating agricultural landscapes with biodiversity conservation in the Mesoamerican hotspot. Conserv Biol 22 (1):8–15

Knoke T, Steinbeis O-E, Bösch M, Román-Cuesta RM, Burkhardt T (2011) Cost-effective compensation to avoid carbon emissions from forest loss: an approach to consider price-quantity effects and risk-aversion. Ecol Econ 70:1139–1153

Lambin EF, Geist HJ (2006) Land-use and land-cover change. Local processes and global impacts. Springer, Berlin

Marquette CM (2006) Settler welfare on tropical forest frontiers in Latin America. Popul Environ 27:397–444. doi:10.1007/s11111-006-0029-y

Mendenhall CD, Sekercioglu CH, Brenes FO, Ehrlich PR, Daily GC (2011) Predictive model for sustaining biodiversity in tropical countryside. Proc Natl Acad Sci USA 108(39):16313–16316

Millennium Ecosystem Assessment (2005) Ecosystems and human well-being: synthesis. Island Press, Washington DC

Pichón FJ (1996) Settler agriculture and the dynamics of resource allocation in frontier environments. Hum Ecol 24(3):341–371

Pohle P (2008) The people settled around podocarpus national park. In: Beck E, Bendix J, Kottke I, Makeschin F, Mosandl R (eds) Gradients in a tropical mountain ecosystem of Ecuador, vol 198, Ecological studies. Springer, Berlin, pp 25–36

Pohle P, Gerique A (2006) Traditional ecological knowledge and biodiversity management in the Andes of southern Ecuador. Geogr Helv 4:275–285

Pohle P, Gerique A (2008) Sustainable and Non-sustainable Use of natural resources by indigenous and local communities. In: Beck E, Bendix J, Kottke I, Makeschin F, Mosandl R (eds) Gradients in a tropical mountain ecosystem of Ecuador, vol 198, Ecological studies. Springer, Berlin, pp 331–346

Pohle P, Gerique A, Park M, López MF (2010) Human ecological dimensions in sustainable utilization and conservation of tropical mountain rain forests under global change in southern Ecuador. In: Tscharntke T, Leuschner C, Veldkamp E, Faust H, Guhardja E, Bidin A (eds) Tropical rainforests and agroforests under global change. Environmental science and engineering. Springer, Berlin, pp 477–509

Pohle P, Park M, Hefter T (2012) Livelihood analysis of small-scale farming households in southern Ecuador. Tropical Mountain Forest Newsletter 16, DFG Research Unit 816, pp 10–11

Roos K, Rödel HG, Beck E (2011) Short- and long-term effects of weed control on pastures infested with *Pteridium arachnoideum* and an attempt to regenerate abandoned pastures in South Ecuador. Weed Res 51:165–176

Rudel T, Horowitz B (1993) Tropical deforestation. Small farmers and land clearing in the Ecuadorian Amazon. Columbia University Press, New York

Southgate D, Wasserstrom R, Reider S (2009) Oil development, deforestation, and indigenous populations in the Ecuadorian Amazon. Conference presented to the Latin American Studies Association in Rio de Janeiro, Brazil, 11–14 June 2009. http://lasa.international.pitt.edu/members/congresspapers/lasa2009/files/SouthgateDouglas.pdf. Cited 12 Jan 2012

Stimm B, Beck E, Günter S, Aguirre N, Cueva E, Mosandl R, Weber M (2008) Reforestation of abandoned pastures: seed ecology of native species and production of indigenous plant material. In: Beck E, Bendix J, Kottke I, Makeschin F, Mosandl R (eds) Gradients in a tropical mountain ecosystem of Ecuador. Springer, Berlin, pp 417–429

Tutillo A (2005) La actividad agropecuaria en la Comunidad Saraguro de El Tibio: El uso del suelo y su relacióncon el ambiente. Bachelor thesis, Pontificia Universidad Católica del Ecuador (PUCE), Quito

Chapter 17
Sustainable Agriculture and Conservation Payments Are Key Factors in Mitigating Tropical Forest Loss

Baltazar Calvas, Thomas Knoke, Luz Maria Castro, Patrick Hildebrandt, Michael Weber, Bernd Stimm, Reinhard Mosandl, Sven Günter, and Nikolai Aguirre

17.1 Introduction

Southern Ecuador is one of the world's biodiversity "hotspots" and has one of the highest deforestation rates in South America (Mosandl et al. 2008). A land-use conflict is evident for this country: slash-and-burn practices are employed to obtain current provisioning services from pasturing (milk, cheese, meat) while destroying valuable forest ecosystems and their ecological services such as supporting (e.g., soil formation), regulating (e.g., climate regulation, avoidance of erosion), and cultural ecosystem services, moreover, to affect the plant composition (see Chap. 8). On account of this the land cover of tropical forests in Ecuador decreased from 13.8 million ha in 1990 to 10.8 million ha in 2005 (FRA 2010). The current land-use practice results not only in the loss of important ecological services but also in a dramatic decline of biodiversity (Günter et al. 2009).

To resolve this land-use conflict, compensation payments to farmers to motivate them to keep their forests rather than to cut them down have been discussed (Fearnside 2001; Eliasch 2008). However, such payments alone are not enough.

B. Calvas (✉) • T. Knoke • L.M. Castro
Institute of Forest Management, Center of Life and Food Sciences Weihenstephan, Technische Universität München, Hans-Carl-von-Carlowitz-Platz 2, 85354 Freising, Germany
e-mail: calvas@forst.wzw.tum.de

P. Hildebrandt • M. Weber • B. Stimm • R. Mosandl
Institute of Silviculture, Center of Life and Food Sciences Weihenstephan, Technische Universität München, Hans-Carl-von-Carlowitz-Platz 2, 85354 Freising, Germany

S. Günter
Tropical Agricultural Research and Higher Education Center, Latin-American Chair for Protected Areas and Biological Corridors "Kenton Miller", CATIE 7170, Turrialba 30501, Costa Rica

N. Aguirre
Center for Studies and Development of Amazonia (CEDAMAZ), National University of Loja, Loja, Ecuador

J. Bendix et al. (eds.), *Ecosystem Services, Biodiversity and Environmental Change in a Tropical Mountain Ecosystem of South Ecuador*, Ecological Studies 221,
DOI 10.1007/978-3-642-38137-9_17,

Improving agriculture and providing productive land-use alternatives (Knoke et al. 2008) will be important issues as well.

To sketch strategies for achieving sustainable land use, we start with general considerations focusing on agriculture under discussion of intensification, compensation payments and diversification. We then formulate conceptual requirements of sustainable agriculture and conclude with an outlook regarding future research needs.

17.2 Conceptual Considerations

17.2.1 Land

In classical economics, land was the principal source of wealth. In neoclassical economics land lost its central importance due to the premise that the products of human endeavor can substitute for natural capital. Land was regarded as any other commodity or factor of production, without giving due to the unique services that land provides which cannot be traded in markets. Due to environmental crises and increased environmental awareness in the late twentieth century, various aspects of land such as support of biodiversity and the provision of nonrenewable resources have found their way back into the economic discourse. In times of "land grabbing" (Bommert 2012), no one would deny that land is actually a scarce resource and that opportunities for its substitution are very limited. Ecological and spatial economics attempt to define and measure meaningful social and ecological scales. The notion of scale is directly related to land and space use and Daly (1992) suggests that an activity is sustainable if it does not erode the environmental carrying capacity of the landscape.

17.2.2 Land Use

The use of land can be regarded in different ways depending upon the category under which land is analyzed. Such categories are those of the environment, economics, and society. In terms of environmental studies land is viewed as soil, the interface between hydrological, atmospheric, and climatic systems. In economics land is a resource required for setting up infrastructure and dwellings and providing the productive soil that supports agriculture or other uses such as horticulture or forestry. Furthermore, land ownership provides status and a basis for socioeconomic relationships within societies. Although modern scientific disciplines artificially separate the various aspects of land in their studies, all these aspects are interwoven and must be analyzed together to better understand and design sustainable land-use patterns. All resource allocation takes place on land

and thus inappropriately allocated land inevitably leads to the deterioration or even loss of numerous valuable ecosystem functions (Hubacek and van den Bergh 2006).

17.2.3 Agriculture as a Key Factor in Land Use

Agriculture is one of the human activities most intimately associated with land, and numerous indicators for sustainable agriculture and sustainable land management have been proposed in past years (Walter and Stützel 2009). Sustainable agriculture is crucial to the well-being of people, especially in tropical countries, and it must be reconciled with the conservation of biodiversity. Intensification of agriculture focusing on short-term economic gain accelerates the loss of biodiversity and is also not conducive to human well-being in the long run. A mainstream approach to intensifying land use while conserving biodiversity will only be successful if correct policies are set down and appropriate technologies are applied.

17.2.3.1 Agricultural Intensification

Increasing demands for food from a growing population has always stimulated more intensive agriculture practices and will continue do so. The second half of the twentieth century saw a revolution in agricultural practice. Economic and technological incentives to increase agricultural productivity have resulted in unprecedented rapid agricultural intensification over the past 60 years. This has, however, been accompanied by a widespread decline in farmland biodiversity in recent decades (Benton et al. 2003).

Biodiversity is not only important for nature conservation but is also an essential resource for agriculture itself. It is an important factor in incorporating traits for resistance to diseases and pests, in improving nutritional quality, in promoting more effective soil nutrient uptake, and in applying more environmentally friendly methods to control pests. Detrimental effects due to the loss of biodiversity become evident in the process of homogenization of the matrix surrounding National Parks and other conservation areas (Perfecto and Vandermeer 2010). Heterogeneity within agricultural landscapes is associated with higher species richness, whether this is measured on a small or large scale. On a local scale, specific agricultural practices have been identified as plausible explanations for the decline in farmland biodiversity. On the landscape scale, the structure and composition of the landscape matrix affects population dynamics. Providing a diversity of habitat ensures organism diversity. For example, wildlife-friendly farming stimulates heterogeneity of habitat structure and facilitates biodiversity flow (Fischer et al. 2011). However, agricultural land use must always produce food that is sufficient in quantity and not too expensive for everyone concerned.

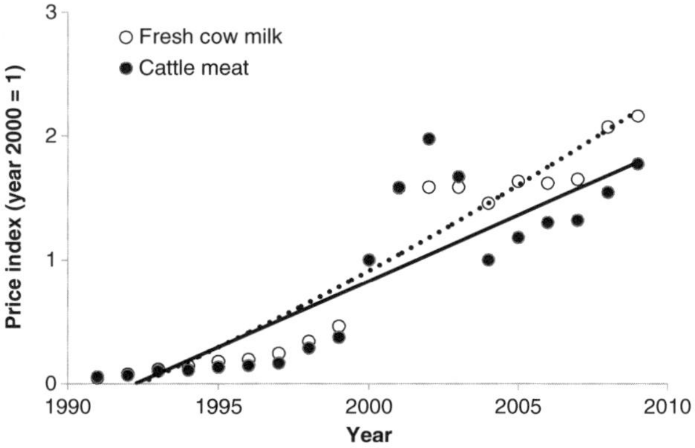

Fig. 17.1 Price indices for milk and meat in Ecuador (prices adopted from FAO 2012). The *line* and *curve* reflect the overall trends; *dotted line* stands for fresh cow milk, *solid line* for cattle meat

17.2.3.2 Conservation Payments

In the current political and scientific debate, market instruments are increasingly being advocated to implement optimized ecosystem services provision. These include making direct payments for maintaining biodiversity or for ensuring the delivery of specific environmental services (Ferraro and Kiss 2002; Wunder et al. 2008). However, focusing on compensation payments alone and disregarding the immediate needs of the people cannot be regarded as being satisfactory. This is evident from recent increases in food prices: for example, Ecuador's milk prices have increased by 116 % and meat prices by 78 % since the year 2000 (Fig. 17.1).

Such rising prices indicate increasing land scarcity. This is not peculiar to Ecuador: food prices are expected to increase worldwide in a continuing fashion in conjunction with policies to implement climate-friendly land-use practices (Wise et al. 2009). Conservation policies should thus be designed with the binding constraint of avoiding land scarcity.

This constraint is also important in that increasing food prices will severely feed back onto the effectiveness of conservation policies. In Ecuador, the program "Socio Bosque" offers compensation payments of a maximum of US$ 30 per ha per year to farmers who agree to not convert their forests into agricultural lands (e.g., Knoke et al. 2009b). Whereas a mathematically derived compensation payment (US$ 47 per ha per year) according to Knoke et al. (2011; see also Chap. 25) would only be slightly higher than the payment offered by "Socio Bosque" at food prices up till the present, a price increase of 100 % would require compensation payments to increase by 170 % according to the same model (Fig. 17.2). This massive increase in financial expenditure together with strong ethical concerns regarding high food prices would certainly call the concept of compensation payments into question.

Improved land-use concepts that avoid deforestation practices are thus needed to increase food production. At the same time the prices for agricultural products should be kept at an acceptable level without neglecting the fact that land-use concepts must be profitable and sustainable in the long run. It is thus important to

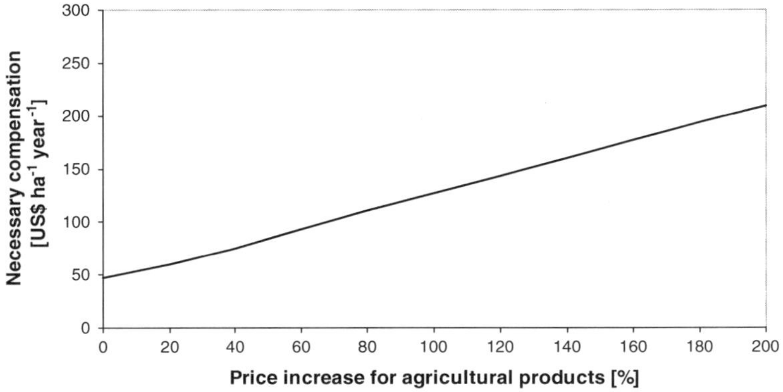

Fig. 17.2 Required compensation from the perspective of a risk-averting farmer with increasing food prices (adopted from Knoke et al. 2011, with alterations)

explore scenarios to assess the applicability and the implications of maintaining biodiversity in combination with the implementation of productive land-use alternatives.

17.2.3.3 Diversification

By relying on fewer crops, modern farms do not come close to matching the levels of species diversity and genetic variation found on traditional farms. In areas of monoculture farming, yields are increased by means of crop specialization, the inputs of fertilizers and pesticides, and mechanization. Farms in this category utilize the economics of large scale production and are thus highly specialized. Disadvantages associated with monoculture farming include reduced resistance to biological pests (plagues and diseases) and economic risks (volatility of prices).

Whereas farmers specialized in a single asset are quite vulnerable to risks, diversification of assets reduces the impacts of risks and uncertainty. The impact of uncertainty on decision making in biological and agricultural sciences has been studied only recently (e.g., Koellner and Schmitz 2006; Hildebrandt and Knoke 2011). Two major strategies for risk-averse farmers to hedge their economic risks are (1) to grow a diverse portfolio of crops and to enhance non-crop components of biodiversity on the farm as a form of natural insurance and (2) to buy financial insurance (Baumgärtner and Quaas 2010). However, strategy (2) is often not affordable for farmers and would also foster homogenized land use.

17.3 Sustainable Agriculture

In South Ecuador, pasturing is the main land-use practice and provides the main provisioning service. Its financial returns have been estimated from various data sources to range from 72 (Knoke et al. 2009a) over 99 (Knoke et al. 2009b) and 131 (Knoke et al. 2011) to 158 (Chap. 15) US$ per ha per year. These returns are superior to those obtainable from immediately available forestry options, but they are very small in comparison to the total estimated value of ecosystem goods and services inherent in a tropical forest (Torras 2000; Knoke et al. 2008). Despite their great ecosystem value, tropical forests are still being converted to farmland to secure small but readily attainable financial returns. The current type of agriculture, however understandable it may be from the farmer's point of view, can thus not be regarded as sustainable.

In contrast to the common land-use practice, sustainable agriculture can be described as an "ecosystem approach" to solve land-use conflicts. It encourages the judicious use of limited external inputs at a suitable time and to an appropriate extent during the production process. In order to achieve sustainable agriculture, all of the three already mentioned instruments of agricultural intensification, compensation payments, and diversification must be combined (Knoke et al. accepted). In developing countries it is a challenge to enhance the productivity, profitability, and stability of the major farming systems without depleting the natural resource base. However, it is inevitable that trade-offs between biodiversity and economic development will have to be made.

Future land-use concepts for South Ecuador must be focused on sustainable intensification of agriculture, which could be achieved by improved pasture management. This includes a more professional establishment and maintenance of pastures. An important prerequisite for sustainable intensification is the avoidance of the use of fire, which often gets out of control and destroys more area than necessary (Gerique 2010). Appropriate fertilization could also help to improve pasture yields. The implementation of all these measures requires, however, that sufficient working capacity is available—a condition which is often not fulfilled. It will be a future task to develop land-use concepts that integrate intensification into compartmental land-use approaches (see Knoke et al. accepted), with the aim of improving farm yields to enable farmers to employ a greater number of people.

When seeking solutions for current land-use problems, it should be considered that farmers are usually risk-averters—at least in principle (Knoke et al. 2011). However, the virtues of on-farm diversification are often not fully utilized, as, for example, when farmers practice single pasturing to supplement off-farm income. The transfer of appropriate knowledge as to on-farm agro-biodiversity could enhance the utility of general farmers compared to those operating specialized farms. On-farm agro-biodiversity would provide benefits not just at the farm level but also outside farming. A strategy to compensate farmers for preventing future land-use conversion is to internalize the benefits generated by environment-friendly land use by paying for ecosystem services. The compensation payments that have to

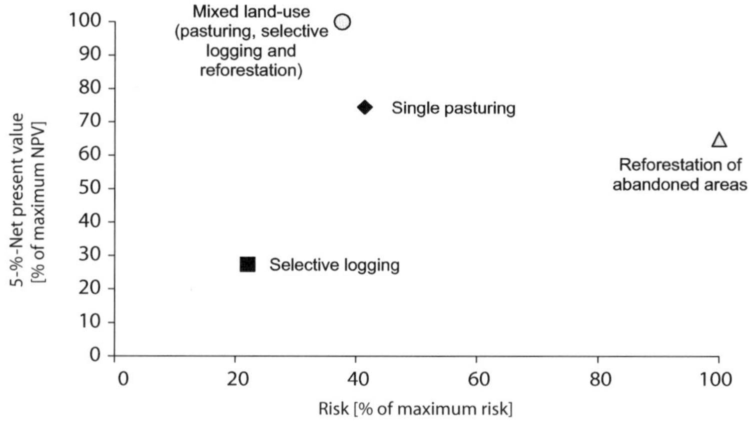

Fig. 17.3 Financial return and risk for various farm-level land-use options in Southern Ecuador (adopted from Knoke and Huth 2011, with alterations); 5 %-net present value is the sum of all net revenues over a 40-year period discounted with an interest of 5 %

be made should depend on the individual level of risk aversion: given that farmers act consistently as risk-averters, the payments should be smaller than those derived under the exclusion of risks (Baumgärtner and Quaas 2010; Knoke et al. 2011). This implies that diversified land-use concepts can reduce the amount of money required for compensation payments.

One example of a diversified land-use strategy was carried out for South Ecuadorian subsistence farmers. The inclusion of both a financial and a sustainability perspective enabled us to develop a concept of Ecological-Economic Farm Diversification (EFFD) that is capable of convincing farmers of the benefits of sustainable practices and halting deforestation (Knoke et al. 2009a, b). The financial success of this new concept compares well with that derived from the available single land-use options (Fig. 17.3). EFFD produced the highest financial returns at a risk lower than that respective of single pasturing and reforestation.

17.4 Outlook and Conclusions

As an outcome of our research, we present diversified land-use concepts including that of intensified agricultural land use. We presented a brief proposal for a long-term, economically favorable mixed land-use approach. However, the problems with all the conceptual approaches hitherto presented include high up-front investment for financing transitioning to diversified land uses. One opportunity would be to include ecosystem services financed by means of compensation payments (Knoke et al. 2009b). Another limitation is that we have up till now considered only subsistence agriculture. There is a strong need to consider intensified agriculture as well. This delivers the bulk of the food production required to feed the nonagricultural population. Only concepts which aim at producing large quantities of food are suitable for keeping food prices low.

Our main conclusion is that we must focus more intensively on agriculture to save considerable areas of tropical forests. Agriculture is a risky business due to various hazards and vulnerabilities stemming from both biophysical conditions and the markets. Farmers thus usually appreciate any alternative to their current situation that reduces their exposure to risk (Knoke et al. 2011). Diversification on the agricultural level is a practice that reduces risks, but is one that is often disregarded. Growing a diverse portfolio of crops is a natural insurance, because pest outbreaks and plagues have stronger impacts in monocultures than in diversified orchards (Torquebiau 2000; van der Werf et al. 2007; Nair et al. 2008). If a land-use portfolio also includes some timber species, the increment in biomass production and soil organic matter content can be significant. Planting additional timber and/or non-timber forest products, e.g., fruit trees, is also a key strategy to maintain food sovereignty and to manage risk on the household level (Bacon 2005). Moreover, a diversified landscape facilitates the flow of species and improves the connectivity among meta-populations (Perfecto and Vandermeer 2010).

Agrobiodiversity provides natural insurance for risk-averse farmers (Baumgärtner and Quaas 2010). To cope with financial uncertainty, farmers establish orchards which contain a wide variety of species. Knoke et al. (2009a) highlight the financial benefits provided by the combination of assets with independent return volatility: one option may generate considerable returns, while another asset might have a poorer performance. A portfolio optimization adapted to land-use problems is a helpful tool for farmers, because it identifies the best shares of each available asset, maximizing the returns as well as keeping the risk at an acceptable level. The application of this technique may thus lead to more sustainable land allocation and utilization.

The diversification concept may also help to mitigate environmental damage resulting from intensification of agricultural land use. Many of the impressive increases in agricultural production in the past may be attributed to intensification (Burney et al. 2010). Some scientists even see land sparing (another word for "intensification") as a superior concept for conserving biodiversity (Phalan et al. 2011). This point of view has also been criticized, however, because it is unclear how intensification may ever be sustainable (Fischer et al. 2011). The concept of diversified land uses could help to combine intensified and sustainable land management, while mitigating erosion by wind and water (Haber 1990). However, it is still open as to how diversified and intensified land-use concepts should be conceptualized in detail and brought together.

Diversification concepts and their economic valuation must be politically acknowledged, extended, and further developed. Their potential for improving the effectiveness of strategies to conserve natural forests under the special consideration of abandoned lands is, for example, demonstrated in Chap. 25.

References

Bacon C (2005) Confronting the coffee crisis: can fair trade, organic, and specialty coffees reduce small-scale farmer vulnerability in Northern Nicaragua? World Dev 33(3):497–511
Baumgärtner S, Quaas M (2010) Managing increasing environmental risks through agro-biodiversity and agro-environmental policies. Agric Econ 41:483–496
Benton TG, Vickery JA, Wilson JD (2003) Farm land biodiversity: is habitat heterogeneity the key? Trends Ecol Evol 18:182–188
Bommert W (2012) Bodenrausch. Bastei Lübbe, Köln
Burney JA, Davis SJ, Lobell DB (2010) Greenhouse gas mitigation by agricultural intensification. Proc Natl Acad Sci USA 107:12052–12057
Daly HE (1992) Allocation, distribution, and scale: towards an economics that is efficient, just, and sustainable. Ecol Econ 6:185–193
Eliasch J (2008) Climate change: financing global forests. Eliasch Review. The Stationary Office, Crown
FAO (2012) FAOSTAT (agriculture and food statistics). http://faostat.fao.org/default.aspx. Accessed 10 Mar 2012
Fearnside PM (2001) Saving forests as a global warming countermeasure: an issue that divides the environmental movement. Ecol Econ 39:167–184
Ferraro PJ, Kiss A (2002) Ecology: direct payments to conserve biodiversity. Science 298:1718–1719
Fischer J, Batary P, Bawa KS et al (2011) Conservation: limits of land sparing. Science 334:593
FRA (2010) Global forest resource assessment 2010. FAO Forestry Paper 163, Rome
Gerique Zipfel A (2010) Biodiversity as a resource: plant use and land use among the Shuar, Saraguros, and Mestizos in tropical rainforest areas of southern Ecuador. PhD Thesis, Universität Erlangen-Nürnberg, Erlangen
Günter S, Gonzalez P, Álvarez G, Aguirre N, Palomeque X, Haubrich F, Weber M (2009) Determinants for successful reforestation of abandoned pastures in the Andes: soil conditions and vegetation cover. For Ecol Manage 258:81–91
Haber W (1990) Basic concepts of landscape ecology and their application in land management. Physiol Ecol Japan 27(special issue):131–146
Hildebrandt P, Knoke T (2011) Investment decisions under uncertainty – a methodological review on forest science studies. For Policy Econ 13:1–15
Hubacek K, van den Bergh JJM (2006) Changing concepts of 'land' in economic theory: from single to multi-disciplinary approaches. Ecol Econ 56:5–27
Knoke T, Huth A (2011) Modelling forest growth and finance: often disregarded tools in tropical land management. Tropical forestry, vol 8, Silviculture in the tropics, Part 3. Springer Berlin, pp 129–142. doi:10.1007/978-3-642-19986-8_11
Knoke T, Stimm B, Weber M (2008) Tropical farmers need productive alternatives. Nature 452:934
Knoke T, Calvas B, Aguirre N, Román-Cuesta R, Günter S, Stimm B, Weber M, Mosandl R (2009a) Can tropical farmers reconcile subsistence needs with forest conservation? Front Ecol Environ 7:548–554
Knoke T, Weber M, Barkmann J, Pohle P, Calvas B, Medina C, Aguirre N, Günter S, Stimm B, Mosandl R, Walter F, Maza B, Gerique A (2009b) Effectiveness and distributional impacts of payments for reduced carbon emissions from deforestation. Erdkunde 63:365–384
Knoke T, Steinbeis O, Bösch M, Roman-Cuesta R, Burkhardt T (2011) Cost-effective compensation to avoid carbon emissions from forest loss: an approach to consider price-quantity effects and risk-aversion. Ecol Econ 70:1139–1153
Koellner T, Schmitz OJ (2006) Biodiversity, ecosystem function, and investment risk. Bioscience 56:977–985
Mosandl R, Günter S, Stimm B, Weber M (2008) Ecuador suffers the highest deforestation rate in South America. In: Beck E, Bendix J, Kottke I, Makeschin F, Mosandl R (eds) Gradients in a

tropical Mountain Ecosystem of Ecuador, vol 198, Ecological studies. Springer, Berlin, pp 37–40

Nair PKR, Gordon AM, Mosquera-Losada MR (2008) Agroforestry. In: Sven EJ, Brian F (eds) Encyclopedia of ecology. Academic, Oxford, pp 101–110

Perfecto I, Vandermeer J (2010) The agroecological matrix as alternative to the landsparing/ agriculture intensification model. Proc Natl Acad Sci USA 107:5786–5791

Phalan B, Onial M, Balmford A, Green RE (2011) Reconciling food production and biodiversity conservation: land sharing and land sparing compared. Science 333:1289–1291

Torquebiau EF (2000) A renewed perspective on agroforestry concepts and classification. Comptes Rendus de l'Académie des Sciences – Series III – Sciences de la Vie 323, 1009–1017

Torras M (2000) The total economic value of Amazonian deforestation, 1978–1993. Ecol Econ 33:283–297

Van der Werf W, Keesman K, Burgess P, Graves A, Pilbeam D, Incoll LD, Metselaar K, Mayus M, Stappers R, van Keulen H, Palma J, Dupraz C (2007) Yield-SAFE: a parameter-sparse, process-based dynamic model for predicting resource capture, growth, and production in agroforestry systems. Ecol Eng 29:419–433

Walter C, Stützel H (2009) A new method for assessing the sustainability of land-use systems (II): evaluating impact indicators. Ecol Econ 68:1288–1300

Wise M, Calvin K, Thomson A, Clarke L, Bond-Lamberty B, Sands R, Smith SJ, Janetos A, Edmonds J (2009) Implications of limiting CO_2 concentrations for land use and energy. Science 324:1183–1186

Wunder S, Engel S, Pagiola S (2008) Taking stock: a comparative analysis of payments for environmental services programs in developed and developing countries. Ecol Econ 65:834–852

Part III
Future Environmental Changes and Their Impacts on Biodiversity and Ecosystem Services

Chapter 18
Climate Change: Effects on Biodiversity and Ecosystem Functioning

Florian A. Werner, Nele Jantz, Valentyna Krashevska, Thorsten Peters, Hermann Behling, Mark Maraun, Stefan Scheu, and Gunnar Brehm

18.1 Introduction

Human activities are profoundly altering the Earth's climate. Temperatures in the tropical Andes have already increased by 0.6 °C over pre-industrial levels. Climate models anticipate temperatures to increase by 2–5 °C during the course of this century and predict that major shifts in precipitation patterns will occur in tropical South America as in many other regions (IPCC 2007). While rainfall is projected to increase in some regions, it is likely to diminish in others, such as our study region in the Andes of southern Ecuador (Chaps. 2 and 19). Moreover, condensation levels are expected to rise, affecting patterns of horizontal precipitation by fog and wind-driven rain in tropical mountains (Marengo et al. 2011). Global climate change also affects precipitation seasonality. Extreme climate events (e.g., torrential rains, droughts) are increasing in frequency and severity in many parts of the world. Diminishing cloud cover may further exacerbate the effects of drought. These

F.A. Werner (✉)
Functional Ecology, Institute of Biology and Environmental Sciences, Carl von Ossietzky University Oldenburg, Carl-von-Ossietzkystraße 9-11, 26111 Oldenburg, Germany
e-mail: florian.werner@uni-oldenburg.de

N. Jantz • H. Behling
Palynology and Climate Dynamics, Albrecht von Haller Institute of Plant Sciences, Georg August University of Göttingen, Untere Karspüle 2, 37073 Göttingen, Germany

V. Krashevska • M. Maraun • S. Scheu
J.F. Blumenbach Institute of Zoology and Anthropology, Georg August University Göttingen, Göttingen, Germany

T. Peters
Institute of Geography, FAU Erlangen-Nuremberg, Kochstraße 4/4, 91054 Erlangen, Germany

G. Brehm
Institute of Systematic Zoology and Evolutionary Biology with Phyletic Museum, Erbertstraße 1, 07743 Jena, Germany

J. Bendix et al. (eds.), *Ecosystem Services, Biodiversity and Environmental Change in a Tropical Mountain Ecosystem of South Ecuador*, Ecological Studies 221,
DOI 10.1007/978-3-642-38137-9_18,

macroclimatic changes are likely to have severe effects on the biodiversity and functioning of tropical Andean ecosystems (Herzog et al. 2011).

Water availability (precipitation) and plant water demand (as a function of the vapor pressure deficit) change rapidly with altitude and exposure, and these gradients are frequently amplified by rain shadows (lee effect). As a result, water stress for primary producers is distributed very heterogeneously across the Andes (Richter et al. 2009) and in the study area (Fries et al. 2012). It is assumed that this remarkable mesoclimatic complexity has strongly facilitated speciation by effectively separating populations (Fjeldså 1994), and today the tropical Andes foster an unsurpassed density of species (Myers et al. 2000; Mutke and Barthlott 2005; Herzog et al. 2011). Many Andean taxa are restricted to narrow ranges of temperature and humidity, resulting in high rates of endemism and species turnover across altitude and surface space (Kessler 2002; Brehm et al. 2003). It is therefore anticipated that climate change will require compensatory species range shifts (Colwell et al. 2008), processes which are obstructed by widespread anthropogenic habitat fragmentation (Chap. 7; Bush 2002; Hole et al. 2011). Understanding the resistance of plant and animal communities to climate change effects is critical for predicting the future of tropical biodiversity. There is growing evidence that a high degree of biodiversity and the presence of functionally "redundant" species are not a luxury, but rather a necessity for providing ecosystem resilience to environmental fluctuations (Isbell et al. 2011; Maestre et al. 2012; Zhang et al. 2012). The resistance of individual species and entire guilds or functional types of microorganisms, animals, and plants to changes in climate may therefore also predict the stability of key ecosystem processes and services such as the provision of soil fertility, water interception, carbon storage, and forest products for human use (Chaps. 4 and 7).

Decomposition of organic matter is one such critical ecosystem process. Effective cycling of nutrients through decomposition is vital for maintaining ecosystem productivity. Soil microfauna play a key role in the decomposition of plant matter and respond rapidly to changes in soil moisture (Vitousek et al. 1994; Epstein et al. 2002). Decomposer communities in acidic tropical forest soils mainly comprise microorganisms and microfauna including protists, especially testate amoebae, whereas macrofauna play a subordinate role (Schönborn 1973; Illig et al. 2005; Krashevska et al. 2007). Testate amoebae play an important role in carbon and nutrient cycling by consuming bacteria, fungi, algae, other protists, and small metazoans (Wilkinson and Mitchell 2010; Krashevska et al. 2010).

The single most pronounced discontinuity along altitudinal gradients is the transition from forest to grassland at the forest line, defining ecosystems of contrasting water, carbon and mineral nutrient dynamics, and species composition. The tree line is determined by the mean growing season temperature in many parts of the world (Körner 2007) and is generally expected to rise with global warming. However, the upper limit of tropical forest growth often appears to be codetermined by additional environmental parameters such as high levels of solar radiation, wind stress, water-logging of soils, aluminum toxicity, and nutrient deficiency (Kessler and Hohnwald 1998; Bader et al. 2007a, b; Peters 2009; Chap. 12). In some places such additional environmental driving forces may entirely replace temperature as determinants of the forest line position (Allen and Walsh 1996; Holtmeier 2003). The Andean depression of northern Peru and southern Ecuador is characterized by an unusually low forest line. Knowing which environmental parameters determine its particular position is a prerequisite for predicting the behavior of this line in the future. Pollen preserved in soils, bogs, and lake sediments comprise an archive of vegetation responses to past climate oscillation. Studying the consequences of past climate change events through palynological techniques therefore offers a unique opportunity for predicting the response of plant communities to future climate change.

A general review of the effects of climate change on tropical Andean biodiversity and ecosystem services is beyond the scope of this chapter and has been addressed thoroughly elsewhere (e.g., Herzog et al. 2011). In this article we present original data from field studies at the Reserva Biológica San Francisco (RBSF) and its environs in southern Ecuador that address climate change effects on biodiversity, vegetation structure, and ecosystem processes. We specifically address the following questions respective of the Andes of southern Ecuador:

(a) How have the structure and composition of vegetation responded to climate fluctuations in the past?
(b) How may continued atmospheric warming affect the altitudinal position of the forest line?
(c) How are the abundance and diversity of soil microfauna (testate amoebae) affected by drought?

These questions are addressed by (1) palynological studies of vegetation responses to climate fluctuations in the past, (2) analysis of tree species composition at the timber line, and (3) simulating the effects of drought on testate amoebae in forest soil.

18.2 Material and Methods

18.2.1 Palynological Studies of Past Vegetation Dynamics

We studied four sediment cores. Peat bog cores were taken with a Russian peat corer, lake cores with a Livingston piston corer. All cores were dated by accelerator mass spectrometry radiocarbon dating. Core 1 (127 cm, sampled at intervals of 2 cm) was taken at a peat bog situated at the pass El Tiro (Niemann and Behling 2008) on the eastern Cordillera on the border of the Podocarpus National Park (2,810 m a.s.l.; 03°59′25.9″ S, 79°08′43.2″ W). This area is covered today by Subpáramo vegetation, with important taxa being *Puya nitida* (Bromeliaceae), *Brachyotum rotundifolium* (Melastomataceae), and *Oritrophium peruvianum* (Asteraceae). Core 2 (114 cm, sampled at intervals of 2 cm) was taken at Tres

Lagunas (Jantz and Behling 2012), a bog in the Quimsacocha volcanic basin (3,780 m; 03°02′50.9″ S, 79°14′29.9″ W) on the eastern slope of the western Cordillera. The prevailing vegetation at this site is at present grass Páramo dominated by tussock grasses (*Stipa*, *Calamagrostis*). Two additional cores were taken at two lakes: Lagunas Natosas Turbera (3,482 m, 04°43′56.6″ S, 79°25′41.2″ W, 208 cm, sampled at intervals of 4 cm, very close to the core Lagunas Natosas Bosque discussed in Chap. 5) and Laguna Rabadilla de Vaca (3,312 m, 04°15′19.7″ S, 79°06′43.7″ W, 490 cm, sampled at intervals of 10 cm, very close to the core Rabadilla de Vaca mire discussed in Chap. 5) (Niemann et al. 2009). These lakes are now surrounded by herbaceous Páramo dominated by *Neurolepis* and *Calamagrostis* (Poaceae). A detailed description of methods for sampling and sample processing is provided by Fægri and Iversen (1989). To determine the nature, severity, and frequency of the vegetation changes evident from our cores, we conducted Principal Component Analysis (PCA) with the programs CANOCO and CanoDraw (ter Braak and Šmilauer 2002). The velocity, or rate, of vegetation change is defined as the amount of change having taken place in a community composition per unit time (Orlóci et al. 2002). The determination of this rate can be helpful in detecting past environmental change; we calculated it using the software Multiv (Pillar 2001).

18.2.2 Tree Species Diversity at the forest line

Tree communities were studied at ten forest line sites in southern Ecuador and northernmost Peru (Fig. 18.1; Table 18.1). Transects were laid out in closed-canopy forest stands near the transition to subpáramo where maximum stem diameter still commonly exceeded 5 cm. At each site the number of species of trees having a stem diameter $>$ 5 cm at 0.3 m above ground level were recorded for eight transects of 50 $\times$ 2 m each following Gentry (1982; see Peters 2009 for details). Data loggers recorded soil (10 cm depth) and air (2 m above ground) temperature every 10 min and stored the hourly mean values. The known altitudinal distribution ranges of tree species were retrieved from the Tropicos database of the Missouri Botanical Garden (2011).

18.2.3 Drought Effects on Testate Amoebae

We reduced the amount of rainfall impacting on investigation plots by installing white plastic roofs of 1.5 $\times$ 1.5 m area at a height of 1.5 m. On the upslope sides of the plots, 50 cm high plastic fences were dug 30 cm deep into the soil to prevent near-surface water from flowing into the plots. All sides of the plots were open to allow free air circulation. Roofs of control plots were covered by 5 $\times$ 5 mm mesh. This allowed free water entry but prevented litter fall, since this was also excluded

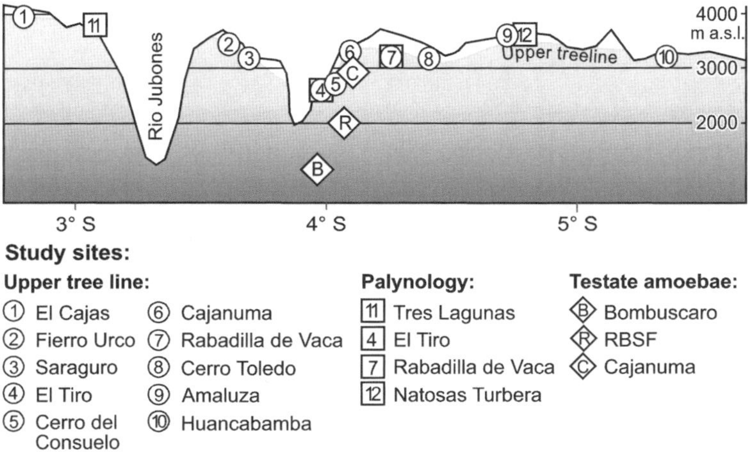

Fig. 18.1 Profile of the Eastern Andean Cordillera across the Andean depression ("Huancabamba Depression") of southern Ecuador and northern Peru, showing the ten study sites at the forest line. Digital elevation model based on United States Geological Survey (SRTM) dataset 2008

from the plots covered by the closed plastic roofs. Four replicates of the meshed and closed roofs were established at each of ~ 1,000 m (Bombuscaro, Podocarpus NP; 04°11′46.0″ S, 078°96′82.9″ W), 2,000 m (RBSF; 03°98′14.4″ S, 079°08′28.2″ W), and 3,000 m altitude (Cajanuma, Podocarpus NP; S 04°10′93.7″ S, 079°17′99.4″ W), resulting in a total of 24 plots. Samples from the litter and fermentation layers were taken 15 months later from each plot to a depth of 5 cm using a soil corer (Ø 5 cm). Testate amoebae were extracted by washing samples over a filter of 500 μm mesh and then back-sieving the filtrate through 20 μm mesh. Microscopic slides were prepared from the final filtrate and testate amoebae were identified and counted (for details see Krashevska et al. 2007, 2012). Density of testate amoebae was analyzed by two-factor randomized complete block analysis of variance (ANOVA) with the fixed factors rain exclusion (with and without) and altitude (1,000, 2,000 and 3,000 m). We used Discriminant Function Analysis (DFA) as implemented in STATISTICA 7.0 to identify drought effects on testate amoeba communities.

18.3 Results and Discussion

18.3.1 Palynological Studies of Past Vegetation Dynamics

Each of the four study sites has been subjected to particular individual climate changes (Fig. 18.2). Even though regional and global climates can be key predictors of vegetation dynamics, specific microclimatic conditions as well as other abiotic

Table 18.1 Characteristics of the upper forest line areas studied in southeastern Ecuador and northern Peru

Study site (time period, altitude)	Geographical position	Transect altitude [m a.s.l.]	T_{mean} [°C]	Number of total tree species (mean species number per transect)
El Cajas	2°46′S; 79°12′W	3,470–4,140		22 (5.3)
Fierro Urco (23.5.07–18.2.08; 3,500 m)	3°43′S; 79°19′W	3,480–3,600	Ta 6.4 Ts 9.1	28 (7.5)
Saraguro (23.5.07–18.2.08; 3,350 m)	3°40′S; 79°14′W	3,210–3,300	Ta 7.4 Ts 11.3	66 (16.4)
El Tiro	3°58′S; 79°8′W	2,790–2,820		39 (11.5)
Cerro del Consuelo (1.1.98–17.9.06; 2,930 m)	3°59′S; 79°3′W	2,760–3,030	Ta 9.6 Ts 11.4	40 (10.4)
Cajanuma (1.1.98–17.9.06; 3,400 m)	4°7′S; 79°9′W	3,170–3,320	Ta 6.8 Ts 10.5	20 (6.6)
Rabadilla de Vaca	4°15′S; 79°7′W	3,150–3,350		44 (9.3)
Cerro Toledo	4°22′S; 79°6′W	3,030–3,280		43 (12.6)
Amaluza (7.10.06–3.3.07; 3,650 m)	4°44′S; 79°25′W	3,480–3,590	Ta 5.6 Ts 8.9	46 (12.6)
Huancabamba	5°20′S; 79°32′W	3,040–3,270		46 (10.0)

Ta = T_{air} (+2 m); Ts = T_{soil} (−10 cm). The time period during which climatic data were recorded and the altitude at which the temperature measurements were made are shown in brackets after the name of the respective study site. The measurement periods at the study sites varied due to the fact that climate stations were rotated among the sites. All sites are located in Ecuador except for Huancabamba (Peru)

and biotic factors exert an influence on sites, which should not be neglected when interpreting paleo-environmental records. Every core thus has its own story and should be interpreted individually.

In the late Pleistocene (20,000–11,500 cal. yr [calibrated years] before present [BP]) herbaceous taxa dominated the pollen assemblages of the El Tiro record (Fig. 18.2). These taxa indicate that the area was covered by a grass Páramo at that time, hence reflecting a lower forest line than that of today, which was related to cooler conditions. Together with *Alnus acuminata*, the only tree species of major importance during this period, these herbaceous taxa may indicate disturbance or primary succession on highly minerogenic soils (Weng et al. 2004; Brunschön and Behling 2009).

A clear directional change towards sub-Páramo and upper montane forest communities is visible at the end of the Pleistocene after ~11,800 cal. yr BP. The increasing importance of woody taxa and ferns indicates not only warming but also the prevalence of moist conditions during this period at El Tiro (Fig. 18.2). However, the presence of tree pollen during the glacial period should be interpreted

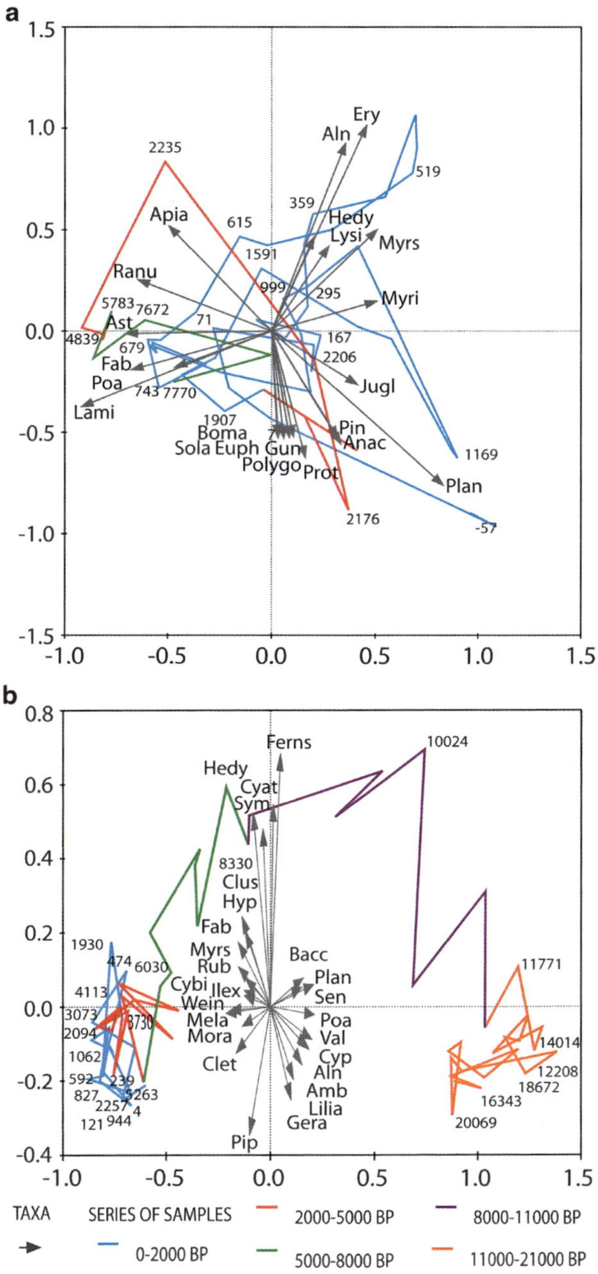

Fig. 18.2 Time trajectory of pollen data from the sediment cores at (**a**) El Tiro and (**b**) Tres Lagunas. PCA ordination (eigenvalues 0.681 and 0.055 for El Tiro, 0.249 and 0.187 for Tres Lagunas) follows Orlóci et al. (2002). Taxa scores are marked by *arrows*. Taxon abbreviations are as follows: *Aln Alnus*[2], *Amb Ambrosia*-type[1], *Anac* Anacardiaceae[2], *Apia* Apiaceae[1],

cautiously, since strong wind fields would likely have caused increased intake of tree pollen from lower, forested altitudes. While turnover velocity at El Tiro did not change markedly during the Pleistocene, the Lagunas Natosas Turbera core (3,600 m a.s.l.) evidenced a high rate of change towards the end of the Pleistocene as compared to the early Holocene (Fig. 18.3). Forest taxa at this site were less abundant during the Holocene than during the Pleistocene, which in turn may be due to a changing wind regime and/or warmer climate. Increasing temperature, here indicated by the steady increase in the number of forest taxa, is a typical climate signal for the period between the late Pleistocene and the early Holocene and can be found in many other records throughout the globe. This directional warming trend in the trajectory at El Tiro continued until approximately 6,000 cal. yr BP as indicated by an upslope shift of montane forest taxa from ~2,100 m to ~2,600 m a.s.l. (Brunschön and Behling 2010). Data from Tres Lagunas indicate that the climate was relatively dry and possibly warmer than at present during the period of from 7,600 to ~2,500 cal. yr BP, since taxa of dry Páramo vegetation (e.g. Fabaceae, Lamiaceae, Poaceae) prevail. The sediment itself also points to a relatively dry environment, since >5,000 years are represented in only 11 cm of sediment (110.5 cm = 7,672 ± 37 cal. yr BP to 98.7 cm = 2,242 ± 66 cal. yr BP). Even though we cannot exclude a hiatus or mixing of the sediment, the notion that sediment accumulation was low during the mid-Holocene is supported by very high pollen and charcoal concentrations, likely caused by rapid peat decomposition in a warmer climate (Coûteaux et al. 2002). Strong decomposition was likewise evident at El Tiro. Drier conditions may have resulted in a high sensitivity of peat composition to temperature and an intense loss of soil organic carbon (Ise et al. 2008). A mid-Holocene dry event has previously been recorded throughout the northern and central Andes region (Ecuador, Peru, and Bolivia) (Hansen et al. 2003; Paduano et al. 2003; Weng et al. 2006; Niemann et al. 2009; Niemann and Behling 2009) and for the Amazon lowlands (Behling and Hooghiemstra 2000; Mayle et al. 2000). However, this dry event is not reflected in pollen records to a similar extent everywhere in the eastern Andes of southern Ecuador, and some cores (e.g., Cerro Toledo, Brunschön and Behling 2009) even indicate rather moist conditions. The velocity curves of all four records show marked peaks at ~7,000 cal. yr BP, corresponding to a warming event at this time (Behling and Pillar 2007).

Fig. 18.2 (continued) Ast Asteraceae[1], *Bacc Baccharis*-type[1], *Boma Bomarea*-type[1], *Clet Clethra*-type[2], *Clus Clusiaceae*, *Cya* Cyatheaceae[2], *Cybi Cybianthus*[2], *Cyp* Cyperaceae[1], *Ery Eryngium*-type[1], *Euph* Euphorbiaceae[2], *Fab* Fabaceae[3], *Fern* Filicatae (excluding Cyatheaceae)[3], *Gera Geranium*[1], *Gun* Gunneraceae[3], *Hedy Hedyosmum*[2], *Hyp Hypericum*[1], *Ilex Ilex*[2], *Jugl Juglans*[2], *Lami* Lamiaceae[1], *Lilia* Liliaceae[1], *Lysi Lysipomia*[1], *Mela* Melastomataceae[3], *Mora* Moraceae/Urticaceae[2], *Myri Myrica*[2], *Myrs Myrsine*[2], *Pin Pinus*[3], *Pip* Piperaceae[2], *Plan Plantago*[1], *Poa* Poaceae[1], *Polygo* Polygonaceae[1], *Prot* Proteaceae[3], *Ranu* Ranunculaceae[1], *Rub* Rubiaceae[3], *Sen Senecio*-type[1], *Sola Solanum*-type[2], *Sym Symplocos*[2], *Wein Weinmannia*-type[2]. [1]Taxa mainly occurring in Páramo, [2]taxa mainly occurring in forest, [3]taxa occurring in both forest and Páramo. Note that *Pinus* is an introduced timber tree

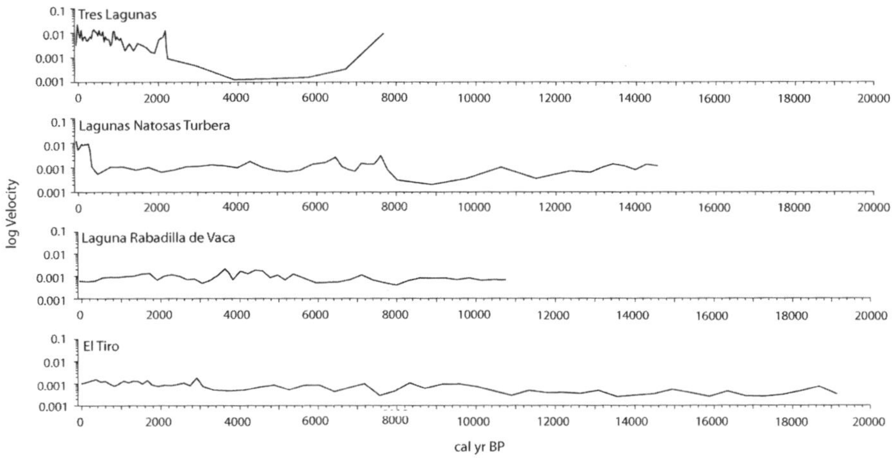

Fig. 18.3 Rate of pollen compositional change (velocity) of four different sediment cores from southeastern Ecuador. Computation is based on chord distances between adjacent sampling units, using percentage data brought to the lowest common taxonomic level. The timescale is based on the linear interpolation of calibrated years BP (cal. yr BP)

During the mid- and late Holocene period (from ~6,000 yr BP until today), vegetation trajectories no longer show a directional pattern. Human impact may have blurred climatic signals, e.g., through the increased incidence of forest fires. A chaotic pattern is also reflected by the turnover of all cores, which exhibit irregular oscillations and poorly pronounced peaks. Both cores show an increasing importance of woody taxa throughout the last approximately 2,000 years. Stronger human impact during the last 100 year is visible particularly at Tres Lagunas in terms of an increase in introduced taxa (*Juglans*, *Pinus*, *Schinus*) and high turnover velocities. At El Tiro and Tres Lagunas, charcoal records indicate high frequency of fire at least since the beginning of the late Holocene (Jantz and Behling, 2012, Niemann and Behling 2009), and tracks of human influence in the region around Loja can be dated back to ~ 4,600 BP (see Chap. 5).

Due to the local microclimatic influences, most observations are not valid for all sites but rather show a regional to local picture of environmental change. Nevertheless, time trajectory analyses and velocity values indicate that three main periods of environmental change altered the composition and structure of vegetation (see also Chap. 5):

1. The climatic transition from the Pleistocene to the Holocene at ~11,500 cal. yr BP which caused an upslope shift of the upper forest line.
2. The early to mid-Holocene dry and warm event from ~8,000 to 5,000 cal. yr BP which led to an increase of drought tolerant taxa and/or to a further upward shift of the upper forest line.
3. The growing human impact during the last ~2,000 cal. yr BP which led to the increase of fire intensity and the spread of taxa that indicate disturbance and/or agricultural activity.

During the late Holocene, the vegetation development indicates a strong human impact at all sites. It seems that since that time anthropogenic influence has a stronger effect on vegetation changes than does climate since.

18.3.2 Tree Species Diversity at the forest line

We found 239 tree species from 39 different families at ten forest line sites. The most species-rich families were Melastomataceae (29 spp.), Asteraceae (19 spp.), and Cunoniaceae (12 spp.), while Melastomataceae and Cunoniaceae were the most abundantly represented, contributing 20 % and 14 % of all individuals, respectively. The highest number of 66 tree species was found at the site Saraguro, where human disturbance has promoted high densities of thin stems. El Cajas stands out in exhibiting the highest forest line (Table 18.1) and is the only site dominated by *Polylepis* spp. (Rosaceae). Together with Cajanuma this site also had the lowest number of species (20 in each case). Unlike in the case of Cajanuma, however, the low species richness at El Cajas was not related to low stem density. Considering that the Andean forest line is typically composed of one to only a few tree species (Baumann 1988; Kessler 1995), the overall tree species richness at our study sites was remarkably high (Table 18.1). Of the 139 species identified, 80 % have previously been registered at higher altitudes in Ecuador and Peru (Fig. 18.4), and only 20 % were found at locations higher than their formerly known upper distribution limits. Given that the mean soil temperatures at our sites were well above 5.5 °C (Table 18.1)—the postulated global threshold value for tree growth within the tropics (Körner 2007)—temperature is unlikely to pose a principal and general limitation to tree competitiveness at the low forest line in the region of our study. Other environmental factors such as extraordinary high incidence of solar irradiation (Emck and Richter 2008; Peters 2009), quasi-permanent and stiff trade winds, and high amounts of annual precipitation may pose influential co-controls over the present position of the forest line. While high global irradiation causes radiation stress for tree seedlings outside closed forest stands (Bader et al. 2007b), strong easterly winds constitute an additional constraint to the establishment of forests, especially near crest lines (Richter et al. 2008). Moreover, precipitation is exceptionally copious (up to 5,000 mm year^{-1}) at high elevations in the center of the Andean Depression (Emck 2007; Rollenbeck and Bendix 2011; Chap. 1). This may lead to excessive nutrient leaching, water logging, and soil oxygen depletion.

Our conclusion that the low forest line in southern Ecuador and northern Peru is not primarily temperature driven is also supported by the altitudinal distribution ranges of the tree species at our sites. Many species have previously been recorded to occur at substantially higher elevations elsewhere in Ecuador and Peru (Fig. 18.4), suggesting that their upper altitudinal distribution limit at our forest line sites is not a function of low temperature. Most of the tree species have also

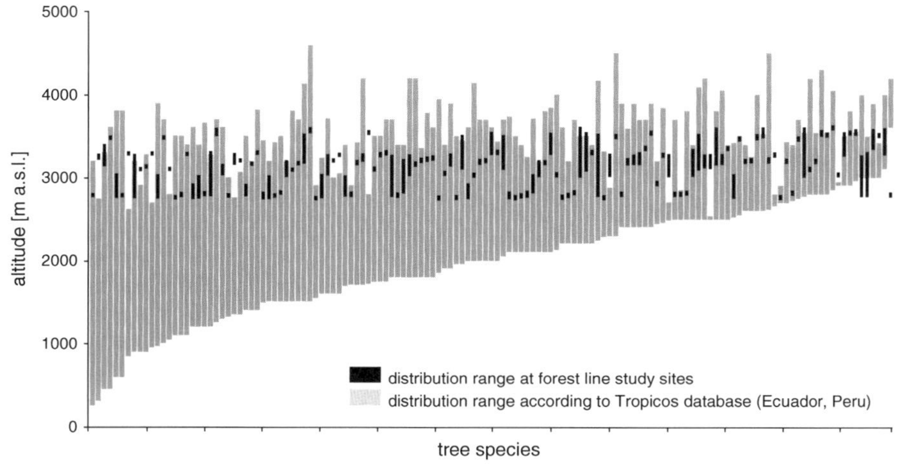

Fig. 18.4 Tree species occurrence at forest line sites in southeastern Ecuador and northern Peru (ranges shown by *black bars*) relative to their distribution ranges in Ecuador and Peru according to the Tropicos database of the Missouri Botanical Garden (*gray bars*). Because El Cajas is located outside of the Huancabamba depression of the Andes, data from this site are excluded. Refer to Peters (2009; Appendix XXXXVI) for details on the respective taxa

been recorded at substantially lower elevations elsewhere (Fig. 18.4), suggesting that warming alone should have but a moderate direct effect on the vertical distribution of many resident tree species, although it may facilitate the establishment of competitor species from lower elevations. However, this conclusion should be viewed with caution. The extraordinarily large altitudinal ranges of some of the species according to the database Tropicos (Fig. 18.4) point to the presence of artifacts stemming from the misidentification of plant specimens or a lumping of poorly differentiated species. Detailed and reliable distribution records are a vital prerequisite for the construction of meaningful and trustworthy ecoclimatic species envelopes and the prediction of biodiversity responses to climate change. The low number of experts for the taxonomy of complex tropical plant taxa has created a bottle-neck for progress in our knowledge of species ranges and the projection of their response to climate change (Gotelli 2004; Brehm et al. 2008).

18.3.3 Drought Effects on Testate Amoebae

A total of 112 taxa of testate amoebae were identified. Only 66 taxa represent live specimens. Communities of testate amoebae found at 3,000 m differed markedly from those at 2,000 and 1,000 m. The sites at 2,000 and 3,000 m a.s.l. shared 50 % of the total number of live taxa, while those at 1,000 and 2,000 m shared 36 % and

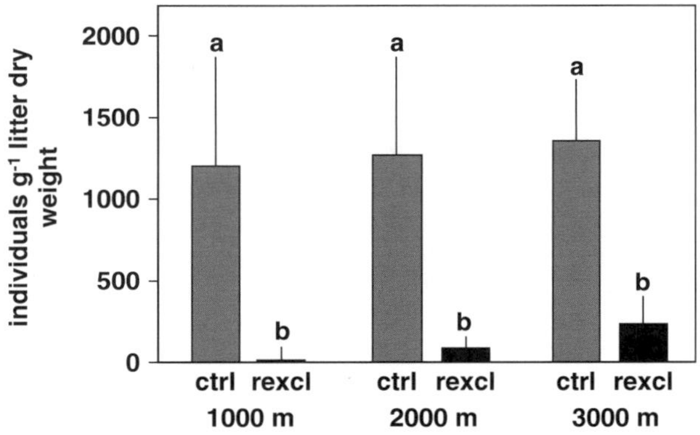

Fig. 18.5 Effect of experimental rain exclusion on the density of live cells of testate amoebae in the litter and fermentation layers (0–5 cm depth) at three altitudes (1,000, 2,000, and 3,000 m a.s.l.). Means and standard deviations ($n = 4$) are shown for control (ctrl) and rain exclusion treatments (rexcl). *Bars* sharing the same letter do not differ significantly (Tukey's HSD test following ANOVA, $p < 0.05$)

those at 1,000 and 3,000 m only 28 %. Overall, rain exclusion did not affect the species number of live testate amoebae significantly; however, the density of live specimens in rain exclusion plots was greatly reduced relative to control plots at all sites (Fig. 18.5). By contrast, the density of testate amoebae cysts was significantly higher in rain exclusion plots than in control plots at 2,000 and 3,000 m (ANOVA followed by Tukey HSD, $p < 0.01$), suggesting that testate amoebae become inactive at low soil moisture and form cysts (see Krashevska et al. 2012 for details).

Overall, 25 taxa of testate amoebae representing 37 % of the total number of live specimens were affected by rain exclusion, i.e., were less commonly or not found in exclusion plots. The species composition of testate amoebae in exclusion plots was markedly different from control plots at 1,000 and 2,000 m but not at 3,000 m (Fig. 18.6). At this high elevation, the vapor pressure deficit is lowest (Emck 2007) and fog is an important source of precipitation (Bendix et al. 2004), which may have buffered the effects of rainfall exclusion.

The results of this study show that rain exclusion strongly affects testate amoebae density irrespectively of altitude. Microorganisms as a potential food resource for testate amoebae also were reduced by rain exclusion (Krashevska et al. 2012). However, the reduction was not as strong as that observed here in testate amoebae, suggesting that reduced moisture adversely affects testate amoebae not only via reduced food supply but also directly, possibly by restricting their movement (Sleigh 1989). The sensitive response of testate amoebae to moisture reduction indicates that trophic levels higher than those of microorganisms need to be included when monitoring environmental changes. Changes in microbial activity and biomass as well as changes in the community composition and density of higher trophic levels which result from changing climatic conditions are likely to alter essential ecosystem processes profoundly.

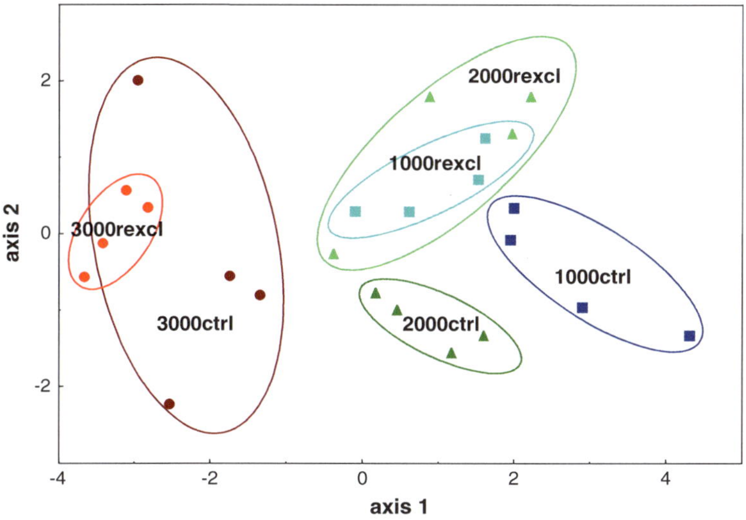

Fig. 18.6 Discriminant function analysis of live cells of soil testate amoeba along an altitudinal transect (axis 1) and between control and rain exclusion treatments (axis 2). Control and rain exclusion treatments at 1,000 (1000ctrl, 1,000rexcl), 2,000 (2000ctrl, 2000rexcl), and 3,000 m altitude (3000ctrl, 3000rexcl) in S-Ecuadorian montane forest; *ellipses* represent confidence ranges at $p = 0.05$

18.4 General Conclusions

Our vegetation studies yielded partly ambiguous results. As expected, pollen records suggest that major warming events of the late Pleistocene and Holocene resulted in a shifting of the southern Ecuador tree line to a higher altitude. In contrast, our analysis of present-day tree assemblages suggests that rising temperature alone is unlikely to cause a pronounced uplift of the region's forest line. However, a combination of warming and drying may have more dramatic effects on the tree line positioning in the near future. Our study region is extraordinarily humid at high altitudes (Emck 2007), which probably helped mitigate the adverse effects of reduced precipitation during the early to mid-Holocene. However, ongoing deforestation of the Amazon lowlands may cause a substantial reduction of precipitation in the eastern Cordillera over the coming decades (Malhi et al. 2008). If the strong anthropogenic pressure of the present day persists, more frequent drought events and high frequency of human-lit fires may soon overstrain the resistance of vegetation to climate change. As a result, large tracts of forests of high carbon density (Chap. 10) could be replaced permanently by more resilient grass- and scrublands.

Simulated drought effects on soil testate amoebae further suggest that top-down forces in microbial food webs rely heavily on precipitation and associated high soil moisture. As testate amoebae are among the dominant grazers of microorganisms, changes in their community composition and density are likely to alter nutrient cycling and thus plant nutrient acquisition. Nutrients in Andean forest ecosystems are in large locked up in organic material that forms thick organic layers (Ließ 2010) resulting in strong nutrient limitations (Chap. 11). Under drier climatic conditions, reduced grazing on soil microorganisms by testate amoebae is likely to aggravate nutrient deficiency and nutrient limitation for primary producers and thereby reduce ecosystem productivity. Drying is also likely to affect numerous other ecosystem processes and services, e.g., strongly diminished abundance, survival, and recruitment of epiphytes on isolated remnant trees in our study region (Nöske et al. 2008; Werner and Gradstein 2008; Werner 2011) suggest that a general drying trend may also diminish epiphyte biomass in intact forest, affecting the interception of water and nutrients by forest canopies (Tobón et al. 2010).

The extent to which climate change will affect future providing ecosystem services in our study region must remain open due to the ambiguity and limited scope of our study results. Our chapter thus confirms that more effort ought to be invested in taxonomy, ecological inventories and experiments to project the consequences of global change for tropical Andean biodiversity and ecosystem functioning into the future.

References

Allen TR, Walsh SJ (1996) Spatial and compositional pattern of alpine treeline, Glacier National Park, Montana. Photogramm Eng Remote Sensing 62:1261–1268

Bader MY, Rietkerk M, Bregt AK (2007a) Vegetation structure and temperature regimes of tropical alpine treelines. Arctic Antarctic Alpine Res 39:353–364

Bader MY, van Geloof I, Rietkerk M (2007b) High solar radiation hinders tree regeneration above the alpine treeline in northern Ecuador. Plant Ecol 191:33–45

Baumann F (1988) Geographische Verbreitung und Ökologie südamerikanischer Hochgebirgspflanzen. Beitrag zur Rekonstruktion der quartären Vegetationsgeschichte der Anden. PhD thesis, Geographisches Institut der Universität Zürich, Zürich, Switzerland

Behling H, Hooghiemstra H (2000) Holocene Amazon rain forest–savanna dynamics and climatic implications: high resolution pollen record Laguna Loma Linda in eastern Colombia. J Quat Sci 15:687–695

Behling H, Pillar VD (2007) Late Quaternary vegetation, biodiversity and fire dynamics on the southern Brazilian highland and their implication for conservation and management of modern Araucaria forest and grassland ecosystems. Phil Trans Roy Soc B 362:243–251

Bendix J, Fabian P, Rollenbeck R (2004) Gradients of fog and rain in a tropical montane cloud forest of southern Ecuador and its chemical composition. In: Proceedings 3rd international conference on Fog, Fog Collection and Dew, Cape Town ZA, 11–15 October 2004

Brehm G, Homeier J, Fiedler K (2003) Beta diversity of geometrid moths (Lepidoptera: Geometridae) in an Andean montane rainforest. Divers Distrib 9:351–366

Brehm G, Homeier J, Fiedler K, Kottke I, Illig J, Nöske NM, Werner FA, Breckle S-W (2008) Are mountain rain forests in southern Ecuador a hotspot of biodiversity? Limited knowledge and diverging patterns. In: Beck E, Bendix J, Kottke I, Makeschin F Mosandl R (eds) Gradients in a tropical mountain ecosystem of Ecuador. Ecological studies, vol 198. Springer, Berlin, pp 15–23

Brunschön C, Behling H (2009) Late Quaternary vegetation, fire and climate history reconstructed from two cores at Cerro Toledo, Podocarpus National Park, southeastern Ecuadorian Andes. Quat Res 72:388–399

Brunschön C, Behling H (2010) Reconstruction and visualization of upper forest line and vegetation changes in the Andean depression region of southeastern Ecuador since the last glacial maximum – a multi-site synthesis. Rev Palaeobot Palynol 163:139–152

Bush MB (2002) Distributional change and conservation on the Andean flank: a paleoecological perspective. Glob Ecol Biogeogr 11:463–473

Colwell RK, Brehm G, Cardelús C, Gilman AC, Longino JT (2008) Global warming, elevational range shifts, and lowland biotic attrition in the wet tropics. Science 322:258–261

Coûteaux MM, Sarmiento L, Bottner P, Acevedo D, Thiéry JM (2002) Decomposition of standard plant material along an altitudinal transect (65–3968 m) in the tropical Andes. Soil Biol Biochem 34:69–78

Emck P (2007) A climatology of south Ecuador. With special focus on the major Andean Ridge as Atlantic-Pacific Climate Divide. PhD thesis, Universität Erlangen-Nürnberg, Erlangen, Germany

Emck P, Richter M (2008) An upper threshold of enhanced global shortwave irradiance in the troposphere derived from field measurements in tropical mountains. J Appl Meteorol Clim 47:2828–2845

Epstein HE, Burke IC, Lauenroth WK (2002) Regional patterns of decomposition and primary production rates in the US Great Plains. Ecology 83:320–327

Fægri K, Iversen J (1989) In: Fægri K, Kaland PE, Krzwywinski K (eds) Textbook of pollen analysis. Wiley, Chichester, 328 p

Fjeldså J (1994) Geographical patterns for relict and young species of birds in Africa and South America and implications for conservation priorities. Biodivers Conserv 3:207–226

Fries A, Rollenbeck R, Nauß T, Peters T, Bendix J (2012) Near surface air humidity in a megadiverse Andean mountain ecosystem of southern Ecuador and its regionalization. Agr Forest Meteorol 152:17–30

Gentry AH (1982) Neotropical floristic diversity: Phytogeographical connections between Central and South America, Pleistocene climatic fluctuations, or an accident of the Andean orogeny? Ann Mo Bot Gard 69:557–593

Gotelli NJ (2004) A taxonomic wish-list for community ecology. Phil Trans Roy Soc Lond B 359:585–597

Hansen B, Rodbell D, Seltzer G, Leon B, Young K, Abbott M (2003) Late-glacial and Holocene vegetational history from two sites in the western Cordillera of southwestern Ecuador. Palaeogeogr Palaeoclimatol Palaeoecol 194:79–108

Herzog SK, Martínez R, Jørgensen PM, Tiessen H (2011) Climate change and biodiversity in the Tropical Andes. Inter-American Institute for Global Change Research (IAI) and Scientific Committee on Problems of the Environment (SCOPE), Paris, 348 p

Hole DG, Young KR, Seimon A, Gomez C, Hoffmann D, Schutze K, Sanchez S, Muchoney D, Grau HR, Ramirez E (2011) Adaptive management for biodiversity conservation under climate change – a tropical Andean perspective. In: Herzog SK, Martínez R, Jørgensen PM, Tiessen H (eds) Climate change effects on the biodiversity of the tropical Andes: an assessment of the status of scientific knowledge. Inter-American Institute of Global Change Research (IAI) and Scientific Committee on Problems of the Environment (SCOPE), Paris, pp 19–46

Holtmeier FK (2003) Mountain timberlines. Ecology, patchiness and dynamics. Advances in Global Research, vol 14. Kluwer, Doordrecht, 369 p

Illig J, Langel R, Norton RA, Scheu S, Maraun M (2005) Where are the decomposers? Uncovering the soil food web of a tropical montane rain forest in southern Ecuador using stable isotopes (^{15}N). J Trop Ecol 21:589–593

IPCC (2007) Climate change 2007: the physical science basis. Contribution of working group I to the fourth assessment report of the intergovernmental panel on climate change. Cambridge University Press, Cambridge

Isbell F, Calcagno V, Hector A, Connolly J, Harpole WS, Reich PB, Scherer-Lorenzen M, Schmid B, Tilman D, van Ruijven J, Weigelt A, Wilsey BJ, Zavaleta ES, Loreau M (2011) High plant diversity is needed to maintain ecosystem services. Nature 477:199–203

Ise T, Dunn AL, Wofsy SC, Moorcroft PR (2008) High sensitivity of peat decomposition to climate change through water-table feedback. Nat Geosci 1:763–766

Jantz N, Behling H (2012) A Holocene environmental record reflecting vegetation, climate, and fire variability at the Páramo of Quimsacocha, southwestern Ecuadorian Andes. Veg Hist Archaeobot 21:169–185

Kessler M (1995) *Polylepis*-Wälder Boliviens: Taxa, Ökologie, Verbreitung und Geschichte. Dissertationes Botanicae 246, 303 p

Kessler M (2002) The elevational gradient of Andean plant endemism: varying influences of taxon-specific traits and topography at different taxonomic levels. J Biogeogr 29:1159–1165

Kessler M, Hohnwald S (1998) Bodentemperaturen innerhalb und außerhalb bewaldeter und unbewaldeter Blockhalden in den bolivianischen Hochanden. Ein Test der Hypothese von Walter & Medina (1969). Erdkunde 52:54–62

Körner C (2007) Climatic treelines: conventions, global patterns, causes. Erdkunde 61:316–324

Krashevska V, Bonkowski M, Maraun M, Scheu S (2007) Testate amoebae (Protista) of an elevational gradient in the tropical mountain rain forest of Ecuador. Pedobiologia 51:319–331

Krashevska V, Maraun M, Ruess L, Scheu S (2010) Micro- and macroscale changes in density and diversity of testate amoebae of tropical montane rain forests of southern Ecuador. Acta Protozoola 49:17–28

Krashevska V, Sandmann D, Maraun M, Scheu S (2012) Consequences of exclusion of precipitation on microorganisms and microbial consumers in montane tropical rain forests. Oecologia 170:1067–1076

Ließ M (2010) Soil-landscape modelling in an Andean mountain forest region in southern Ecuador. PhD thesis, University of Bayreuth, Bayreuth, Germany

Maestre FT, Quero JL, Gotelli NJ, Escudero A, Ochoa V, Delgado-Baquerizo M, García-Gómez M, Bowker MA, Soliveres S, Escolar C, García-Palacios P, Berdugo M, Valencia E, Gozalo B, Gallardo A, Aguilera L, Arredondo T, Blones J, Boeken B, Bran D, Conceição AA, Cabrera O, Chaieb M, Derak M, Eldridge DJ, Espinosa CI, Florentino A, Gaitán J, Gatica MG, Ghiloufi W, Gómez-González S, Gutiérrez JR, Hernández RM, Huang X, Huber-Sannwald E, Jankju M, Miriti M, Monerris J, Mau RL, Morici E, Naseri K, Ospina A, Polo V, Prina A, Pucheta E, Ramírez-Collantes DA, Romão R, Tighe M, Torres-Díaz C, Val J, Veiga JP, Wang D, Zaady E (2012) Plant species richness and ecosystem multifunctionality in global drylands. Science 335:214–218

Malhi Y, Roberts JT, Betts RA, Killeen TJ, Li W, Nobre CA (2008) Climate change, deforestation, and the fate of the Amazon. Science 319:169–172

Marengo JA, Pabón JD, Díaz A, Rosas G, Ávalos G, Montealegre E, Villacis M, Solman S, Rojas M (2011) Climate change: evidence and future scenarios for the Andean region. In: Herzog SK, Martínez R, Jørgensen PM, Tiessen H (eds) Climate change and biodiversity in the Tropical Andes. Inter-American Institute for Global Change Research (IAI) and Scientific Committee on Problems of the Environment (SCOPE), Paris, pp 110–127

Mayle F, Burbridge R, Killeen TJ (2000) Millennial-scale dynamics of southern Amazonian rain forests. Science 290:2291–2294

Missouri Botanical Garden (2011) Tropicos database: http://www.tropicos.org Accessed 11 Aug 2011

Mutke J, Barthlott W (2005) Patterns of vascular plant diversity at continental to global scales. In: Friis I, Balslev H (eds) Plant diversity and complexity patterns – local, regional and global dimensions, vol 55. Biologiske Skrifter, Copenhagen, pp 521–537

Myers N, Mittermeier RA, Mittermeier CG, Da Fonseca GAB, Kent J (2000) Biodiversity hotspots for conservation priorities. Nature 403:853–858

Niemann H, Behling H (2008) Late Quaternary vegetation, climate and fire dynamics inferred from the El Tiro record in the southeastern Ecuadorian Andes. J Quat Sci 23:203–212

Niemann H, Behling H (2009) Late Pleistocene and Holocene environmental change inferred from the Cocha Caranga sediment and soil records in the southeastern Ecuadorian Andes. Palaeogeogr Palaeoclimatol Palaeoecol 276:1–14

Niemann H, Haberzettl T, Behling H (2009) Holocene climate variability and vegetation dynamics inferred from the (11700 cal. yr BP) Laguna Rabadilla de Vaca sediment record, southeastern Ecuadorian Andes. Holocene 19:307–316

Nöske NM, Hilt N, Werner FA, Brehm G, Fiedler K, Sipman HJM, Gradstein SR (2008) Disturbance effects on epiphytes and moths in a montane forest in Ecuador. Basic Appl Ecol 9:4–12

Orlóci L, Pillar VD, Anand M, Behling H (2002) Some interesting characteristics of the vegetation process. Community Ecol 3:125–146

Paduano G, Bush M, Baker P, Fritz S, Seltzer G (2003) A vegetation and fire history of Lake Titicaca since the Last Glacial Maximum. Palaeogeogr Palaeoclimatol Palaeoecol 194:259–279

Peters T (2009) Struktur und ökologische Merkmale der oberen Waldgrenze in der Andinen Depression. PhD thesis, Universität Erlangen-Nürnberg, Erlangen

Pillar, VD (2001) SYNCSA; application program for multivariate analysis of trait-based community data. UFRGS, Departamento de Ecologia, Porto Alegre

Richter M, Diertl KH, Peters T, Bussmann RW (2008) Vegetation structures and ecological features of the upper timberline ecotone. In: Beck E, Bendix J, Kottke I, Makeschin F, Mosandl R (eds) Gradients in a tropical mountain ecosystem of Ecuador. Ecological studies, vol 198. Springer, Heidelberg, pp 123–136

Richter M, Diertl KH, Emck P, Peters T, Beck E (2009) Reasons for an outstanding plant diversity in the tropical Andes of southern Ecuador. Landscape Online 12:1–35

Rollenbeck R, Bendix J (2011) Rainfall distribution in the Andes of southern Ecuador derived from blending weather radar data and meteorological field observations. Atmos Res 99/277–289

Schönborn W (1973) Humusform und Testaceen-Besatz. Pedobiologia 13:353–360

Sleigh MA (1989) Protozoa and other protists. Edward Arnold, London

ter Braak CJF, Šmilauer P (2002) CANOCO Reference manual and CanoDraw for Windows User's Guide: software for canonical community ordination (version 4.5). Microcomputer Power, Ithaka

Tobón C, Köhler L, Frumau KFA, Bruijnzeel LA, Burkhard R, Schmid S (2010) Water dynamics of epiphytic vegetation in a lower montane cloud forest: fog interception, storage, and evaporation. In: Bruijnzeel L, Scatena FN, Hamilton LS (eds) Tropical montane cloud forests: science for conservation and management. Cambridge University Press, Cambridge, pp 261–267

Vitousek PM, Turner DR, Parton WJ, Sanford RL (1994) Litter decomposition on the Mauna Loa matrix: patterns, mechanisms, and models. Ecology 75:418–429

Weng C, Bush MB, Chepstow-Lusty AJ (2004) Holocene changes of Andean alder (*Alnus acuminata*) in highland Ecuador and Peru. J Quat Sci 19:685–691

Weng C, Bush MB, Curtis JH, Kolata AL, Dillehay TD, Binford MW (2006) Deglaciation and holocene climate change in the western Peruvian Andes. Quat Res 66:87–96

Werner FA (2011) Reduced growth and survival of vascular epiphytes on isolated remnant trees in a recent tropical montane forest clear-cut. Basic Appl Ecol 12:172–181

Werner FA, Gradstein SR (2008) Seedling establishment of vascular epiphytes on isolated and enclosed forest trees in an Andean landscape, Ecuador. Biodivers Conserv 17:3195–3207

Wilkinson D, Mitchell EAD (2010) Testate amoebae and nutrient cycling with particular reference to soil. Geomicrobiol J 27:520–533

Zhang Y, Chen HYH, Reich PB (2012) Forest productivity increases with evenness, species richness and trait variation: a global meta-analysis. J Ecol 100:742–749. doi:10.1111/j.1365-2745.2011.01944.x

Chapter 19
Global Climate Change Impacts on Local Climate and Hydrology

Lutz Breuer, Jean-François Exbrayat, Ina Plesca, Wouter Buytaert, Theresa Ehmann, Thorsten Peters, Edison Timbe, Katja Trachte, and David Windhorst

19.1 Introduction

The significant warming trend forecasted by climate models for the tropical Andes is emphasized with increasing altitudes (Urrutia and Vuille 2009). Based on reanalysis data from the National Centers for Environmental Prediction (NCEP)/ National Center for Atmospheric Research (NCAR) these warming trends are already visible for the grid cell of southern Ecuador, showing a warming trend of +0.22 °C per decade since 1948 (Bendix et al. 2010; see also Fig. 2.4). Contradicting this trend, local measurements at the meteorological station at Estación Científica San Francisco (ECSF) within the Rio San Francisco valley show a significant net cooling ($\tau = -0.045$, $P < 0.0001$, $n = 4{,}532$, Seasonal Mann–Kendall test used for daily temperature series) during 1998–2010 (Fig. 19.1). This cooling trend can be ascribed to an increase in daily temperature ranges ($\tau = 0.124$, $P < 0.0001$, $n = 4{,}532$, Seasonal Mann–Kendall test used for daily temperature series) resulting in a decrease in daily minimum temperatures

L. Breuer (✉) • I. Plesca • T. Ehmann • D. Windhorst
Institute for Landscape Ecology and Resources Management, University of Gießen, Gießen, Germany
e-mail: lutz.breuer@umwelt.uni-giessen.de

J.-F. Exbrayat
Climate Change Research Centre, University of New South Wales, Sydney, Australia

W. Buytaert
Department of Civil and Environmental Engineering, Imperial College London, London, UK

T. Peters
Geographic Institute, University of Erlangen, Erlangen, Germany

E. Timbe
Grupo de Ciencias de la Tierra y del Ambiente, Universidad de Cuenca, Cuenca, Ecuador

K. Trachte
Faculty of Geography, University of Marburg, Marburg, Germany

J. Bendix et al. (eds.), *Ecosystem Services, Biodiversity and Environmental Change in a Tropical Mountain Ecosystem of South Ecuador*, Ecological Studies 221,
DOI 10.1007/978-3-642-38137-9_19, © Springer-Verlag Berlin Heidelberg 2013

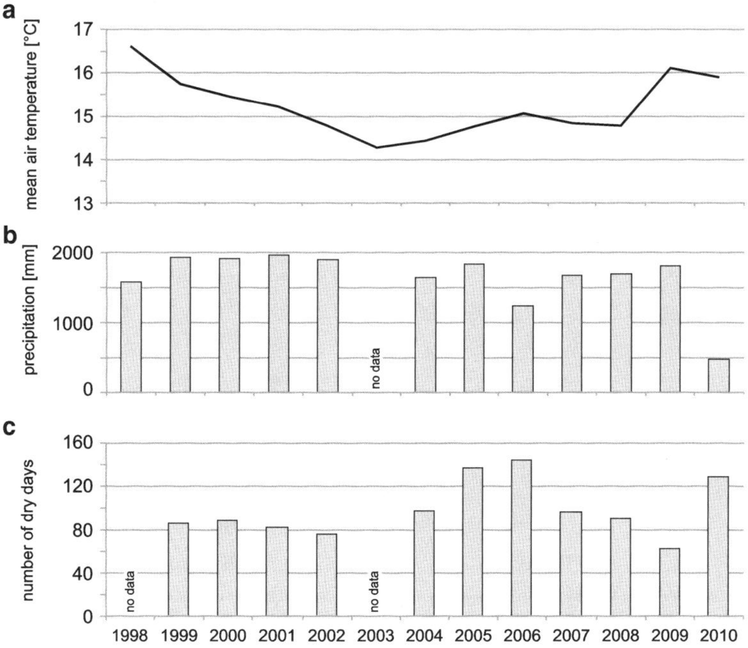

Fig. 19.1 Local climate trends measured during the last 12 years at the ECSF meteorological station (see Fig. 1.2): (**a**) mean annual air temperature, (**b**) precipitation, and (**c**) number of dry days

($\tau = -0.148$, $P < 0.0001$, $n = 4{,}532$) as well as an increase in daily maximum temperatures ($\tau = 0.085$, $P < 0.0001$, $n = 4{,}532$).

In comparison with temperature, projections for changes in precipitation are spatially much less cohesive, with regions of decreased and increased precipitation scattered across the Andean mountain chain (Urrutia and Vuille 2009). Analyses of local precipitation trends show a strong interannual variation (Fig. 19.1) and a clear negative trend during 1998–2010 ($\tau = -0.048$, $P < 0.0001$, $n = 4{,}171$, Seasonal Mann–Kendall test used for daily precipitation series). This can be caused by changes in the intensity and frequency of precipitation events as well as by changes in the number of dry days (see also Fig. 2.4). The latter is probably true for the study region, where a nonsignificant increase in the number of dry days (daily precipitation = 0 mm) is observed ($\tau = 0.08$, $P = 0.235$, $n = 154$, Seasonal Mann–Kendall test used for monthly dry day series, Fig. 19.1). This may have more extensive impacts on environmental and human systems than the registered temperature changes.

The question remains how likely future global climate change will impact local climate and hydrology. To investigate these potential effects for the Reserva Biologica San Francisco (RBSF, see Fig. 1.1) we conducted a sensitivity study using simple, statistically downscaled future projections based on General Circulation model (GCM) scenarios for Ecuador and used this dataset to force a set of hydrological catchment rainfall-runoff models to calculate possible changes of regional discharge.

19.2 Material and Methods

19.2.1 Future Climate Scenarios

We used a simple, but very effective, statistical downscaling method based on the Delta method (Fowler et al. 2007), which has been successfully applied to Ecuadorian landscapes before (Buytaert et al. 2009). The Delta method provides future scenarios for precipitation and temperature. The latter was also used to calculate potential evapotranspiration using the Penman–Monteith approach. For the input variables wind speed and radiation, the values for 2007/2008 were used. The Delta method utilizes differences between the current and future GCM simulations. The anomalies between the average value of each variable of the modeled historical time slice (1960–1990) and the future time slice are then applied to given observations for 2007/2008 by simple scaling. This method has the shortcoming that variability remains static and spatial patterns are not considered. However, it has its advantages in cases where only a limited amount of data is available as it is the case in our study. Additionally, the local climate system of the southern Andes is highly complex, preventing the application of more sophisticated downscaling methods.

Projections for precipitation and maximum and minimum temperature were generated for three time slices (2010–2039, 2040–2069, 2070–2099) and two emission scenarios (A1B and A2) from IPCC Assessment Report 4 from eight GCMs (BCM2, CSMK3.0, CSMK3.5, GFCM20, GFCM2.1, INCM3, MIMR, NCCCSM3). GCM data were obtained from http://www.ipcc-data.org. The generated time series for the three time slices reflect a representative time series during that period; they are not meant to be predictions for a particular year within each time slice.

19.2.2 Hydrological Modeling

With regard to the already observed structural model uncertainty to predict rainfall-runoff reaction in tropical mountainous catchments (Plesca et al. 2012), we applied a set of seven hydrological models to predict the likely impacts of climate change on hydrology in tropical montane rainforests of southern Ecuador. The complexity of the participating models ranges from the most simple, lumped (HBV-light, LASCAM, CHIMP, HEC-HMS, NAM) to more distributed (HBV-N-D) and more process-oriented models (SWAT). The major differences between the models are the spatial representation (regular pixels, semi-lumped subcatchments, or fully lumped concepts representation) and the type of infiltration processes used by the models. For a more detailed model description the reader is referred to Plesca et al. (2012). In addition, we also included the NAM model in our study. NAM is a conceptual lumped rainfall-runoff model, simulating the flow by its water balance components. NAM's performance showed a Nash–Sutcliffe efficiency of 0.59 and 0.52 as well as a log-biased constrained efficiency of 0.50 and 0.44 during calibration and validation, respectively (Plesca et al. 2012).

In order to simulate the climate change impact on the hydrological regimes of the river San Francisco, we ran the models under the A1B and A2 scenarios and the seven GCMs, following the hypothesis that the application of an ensemble of models generally increases the credibility of prediction (Tebaldi et al. 2010). To reduce the heterogeneity of model forcing data we applied the same input data and the same spatial interpolation procedures for precipitation and temperature (the inverse distance weighted method). In this sense we also developed a common potential evapotranspiration dataset used by all models to focus on the different structures in rainfall-runoff generation rather than on any other potentially effective model input data.

Since measured discharge data are available for only 14 months, all models have been calibrated to the first 10 months and were validated for the 4 remaining months (Fig. 19.2). We used a 12 months spin up time for all models by simply copying the available time series as a pre-run time period.

We consider the output of mean monthly average runoff as well as the maximum, minimum, and mean discharge from the seven hydrological models as a combined information for this model ensemble and observations to determine the response of runoff under the A1B and A2 scenarios.

19.3 Results and Discussion

19.3.1 Effects on Local Climate

The results of the downscaling method are displayed in Fig. 19.3. To illustrate the predicted local climate change impact for the three periods 2010–2039, 2040–2069, and 2070–2099 of the IPCC scenarios A1B and A2, simulated data are compared to measured data from the ECSF climate station for a 14-month period in 2007–2008. Results for both emission scenarios A1B and A2 revealed similar dynamics. However, increases in precipitation are substantially higher for A2. For the sake of clarity, the presented results focus on the A1B scenario.

The monthly means of temperature show a clear increase over the three periods. Until 2070–2099, the annual average temperature is expected to rise by about 2.9 °C. It is worth to note that the biggest monthly differences of the mean temperatures occur in June and March, with temperature increases of 3.4 °C and

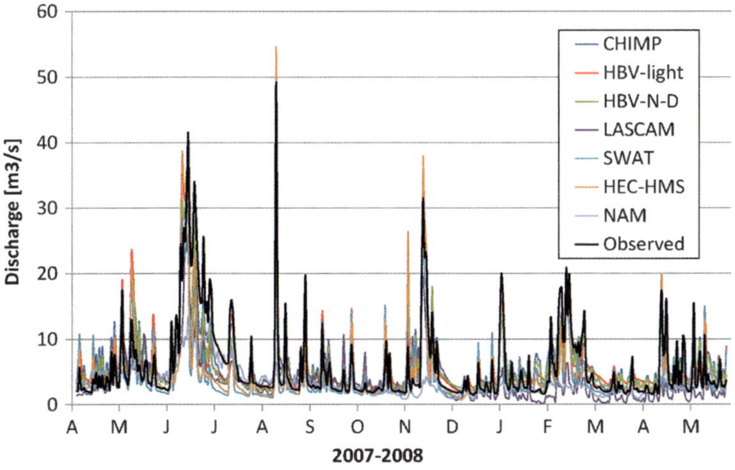

Fig. 19.2 Simulated discharge [m^3 s^{-1}] for the reanalyzed (RA) period: 2000–2009. The *black curve* shows the observed discharge for a period of 14 months. For more information about each model simulation results, see Plesca et al. (2012). See text for further information about the applied models

3.5 °C, respectively. The change of the minimum and maximum temperature is nearly equivalent to the mean temperature and shows a steady upward trend. Annual mean minimum temperature increases from 11.2 °C in the reference period to 14.2 °C in the period 2070–2099, and the annual mean maximum temperature rises from 20.7 °C to 23.4 °C, respectively. The lines of minimum and maximum temperatures indicate the lower and upper bounds of predicted changes from the eight GCMs used in this study.

Boxplots in Fig. 19.3 depict the changes in monthly precipitation. Each boxplot shows the range of the calculated monthly mean precipitation of the applied GCMs. It is noteworthy that boxplots for May and June exhibit a considerably greater dispersion than boxplots for the other months (average standard deviation 25 mm) with 157 mm and 300 mm, respectively.

The annual rainfall from April 2007 until March 2008 amounted to 1.873 mm. For the three simulated time periods, annual rainfall increases up to 2.395 mm. Until the first future period (2010–2039), rainfall raises by about 11 %, approximately by additional 4 % in the second period and in 2070–2099 by further 10 %, compared to the previous time period. Mainly affected are months with extreme precipitation, i.e., May and June for which the amount of rainfall is expected to double in comparison to the reference period. This strong increase goes along with a general decreasing trend in precipitation from September to November. As in the reference period, the increase in precipitation in May and June is accompanied by a strong decrease in the maximum and an increase in the minimum temperature.

In general, the calculated time series for the climate change scenarios follow the typical precipitation regime of the reference period. Of course, this is due to the simple approach of downscaling that does not consider changes in variability or extremes. The peak in May and June becomes much more distinct in the future

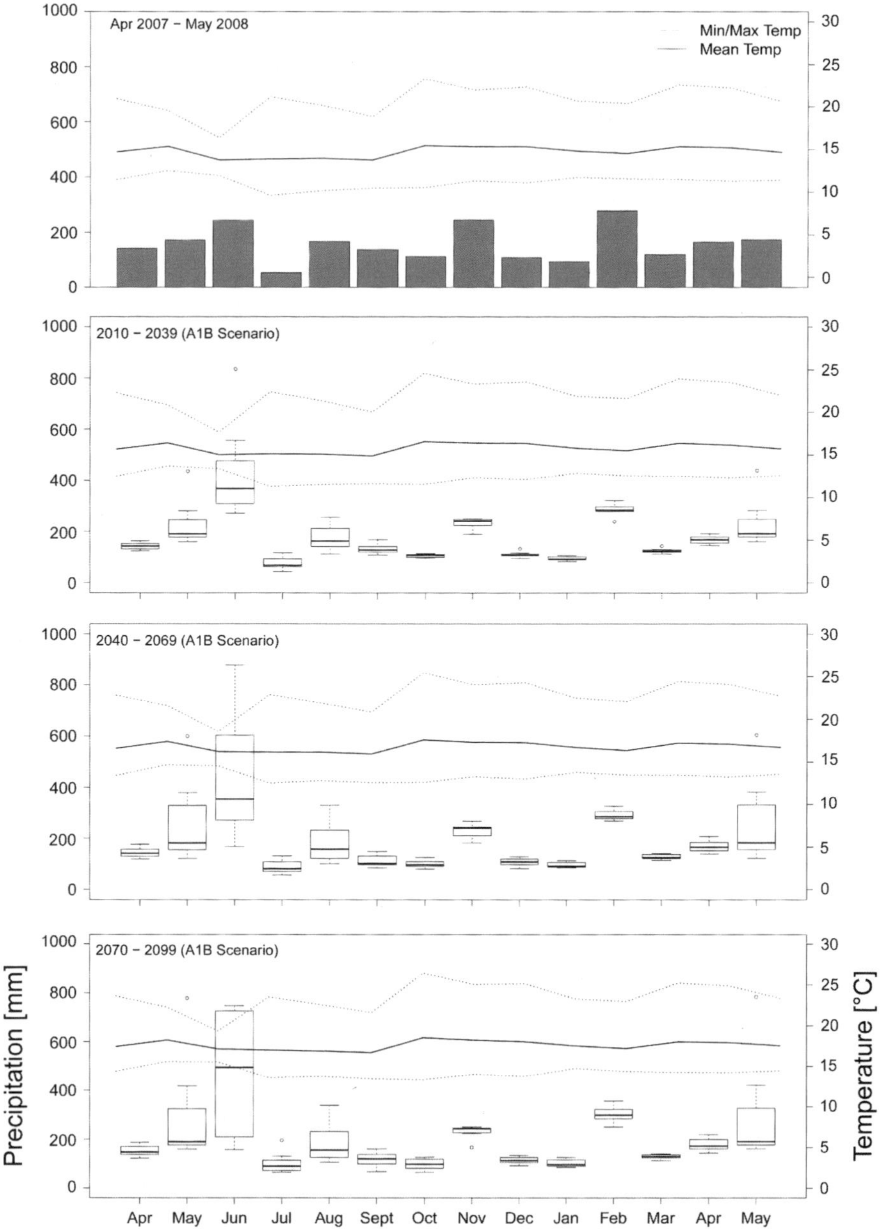

Fig. 19.3 Measured and predicted temperature (*lines*) and precipitation (*bars*) for the climate station ECSF. Reference period is April 2007–May 2008 and predictions are 2010–2039, 2040–2069, and 2070–2099 for the IPCC emission scenario A1B. Predictions represent the same 14-month time series under climate change impact (see details in text for the method applied). *Dotted lines* show minimum and maximum temperatures calculated by the eight down-scaled GCMs. *Boxplots* represent results of predicted precipitation by the same ensemble

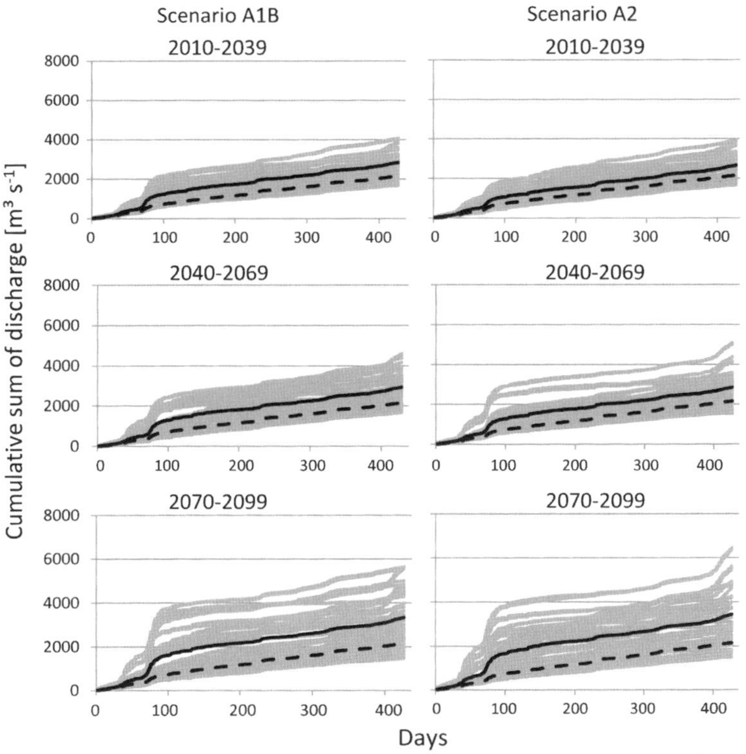

Fig. 19.4 Average discharge (*solid black line*) based on seven hydrological models and eight GCM realization for the A1B (*left*) and A2 (*right*) emission scenarios. Single model forecasts ($n = 56$) are represented by the cohort of *gray lines*. Observed discharge under current conditions is shown as *dashed black line*

simulations, and in the already dryer period (from September to November), rainfall decreases slightly by 10 % on average.

One has to keep in mind that the anomalies of the climate change scenarios were derived by utilizing the time series from 1960 to 1990 as a reference. However, we assume that this period is comparable to the available observed time series 2007–2008 and that basically no changes occurred between the two time slices which is probably untrue. Ideally, the anomalies should be calculated with reference to a matching 30-year period that encompasses the observed period. Unfortunately, such data are not available for the ECSF station.

19.3.2 Effects on Local Hydrology

Both emission scenarios A1B and A2 lead to similar predictions with regard to discharge dynamics. Figure 19.4 shows the cumulative sum of daily discharge for the simulation period of 14 months for all models ($n = 7$) and all downscaled

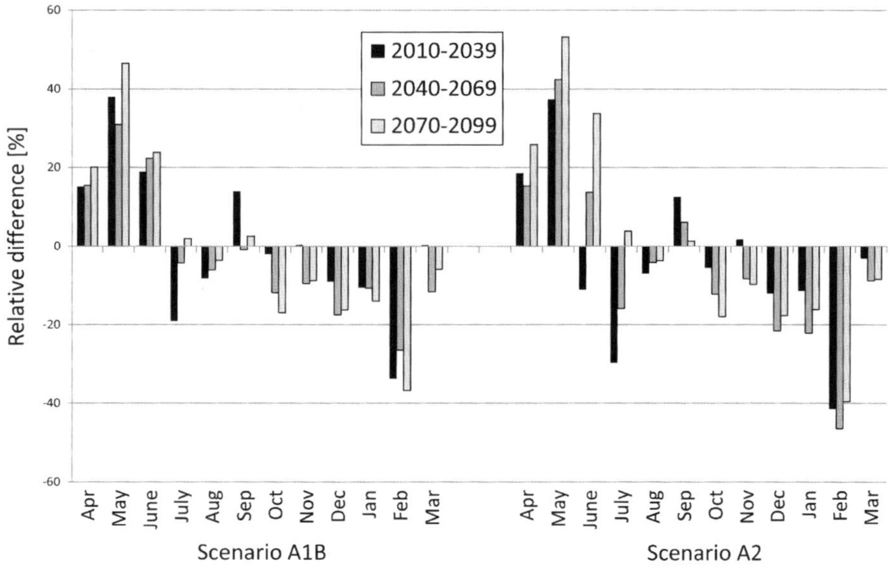

Fig. 19.5 Relative differences [%] between mean monthly observed discharge in 2007–2008 and the projected time slices 2010–2039, 2040–2069, and 2070–2099 for emission scenarios A1B (*left*) and A2 (*right*)

GCMs ($n = 8$). Mean discharge is calculated based on the cohort of all model runs ($n = 56$). For the first time slice 2010–2039 forecasts for both emission scenarios are well in agreement with respect to both the mean of all forecasts and the behavior of single model runs. As can be seen in comparison to the cumulative discharge of the current situation (2007–2008) total discharge is substantially increasing by around 35–40 %. The overall pattern of single model forecasts is somehow changing for the following time slice, where despite similar mean predictions of the model ensemble a larger deviation between single model runs can be depicted. This trend is even further amplified in the last time slice 2070–2099 where more and more models tend to predict larger discharge volumes for the A2 emission scenario, especially towards the end of the simulation period. In comparison to the observed discharge under current conditions, increases are forecasted to around 80 %.

In order to find out which periods of the year have the highest impacts on discharge under current and future climate conditions, we analyzed the relative differences between mean monthly discharge for the observation period 2007–2008 and the predictions of the hydrological models (Fig. 19.5). Similar results were found for both scenarios in the case of relative differences. Monthly discharges for April, May, and June expect an increase in the order of 13–53 %, which is compensated by reduced discharge from October to February (Fig. 19.5). On average, following the A1B scenario, mean discharges are predicted to increase by 0.3 % for 2010–2039, whilst for 2040–2069 and 2070–2099 reduced discharge of −2.55 and −0.63 % are expected, respectively. While the general pattern of

Table 19.1 Expected impact of future climate trends on water-related ecosystem services in the San Francisco River valley (definitions in agreement with the Millennium Ecosystem Assessment 2005; see also Chap. 9)

	Climate change impact	Note
Provisioning service: fresh water	↑	Hydropower generation can expect increasing benefits due to increasing amount of available water
Regulating service: water purification and waste treatment	→	Water quality will remain stable at high-quality levels with sufficient amount of water available for dilution
Regulating service: water regulation	↓	Increasing amounts of discharge and number of extreme precipitation events especially in May and June are likely to increasing flood occurrences and potentially erosion events. However, due to low local population number, the economic losses are expected to be of minor value; this might be different in downstream catchments with larger population

Climate change increases provisioning services, if human consumption of the service is likely to increase. An increase for regulating service is predicted, if climate change improves the service from the human perspective. Supporting services are not directly used and climate change impacts the supply of all other services. If these indirect effects are listed, supporting service would be double counted. ↑ increasing/enhancing, → no change, ↓ decreasing

changes in discharge is similar for the emission scenario A2, absolute changes are somewhat more pronounced. This is, in fact, contrasting with the general increase in precipitation shown in Fig. 19.3. Increasing temperatures and resulting larger losses through evapotranspiration (data not shown) may be a reason for this seaming discrepancy. It has to be noted that other parameters like wind speed, relative humidity, or global radiation that strongly influence evapotranspiration were also not altered during the simulations as no GCM projections were available. Especially global radiation might severely affect the estimation of evapotranspiration as it is usually the most sensitive parameter in the Penman–Monteith equation. Increasing cloud cover in the future might reduce incoming global radiation and thus have a negative feedback on evapotranspiration rates.

19.4 Conclusion: Impact of Climate Change on Water-Related Ecosystem Services in the Future

Based on the results obtained in the hydrological scenarios considering climate change impact, we conclude that most water-related ecosystem services are only affected to a minor degree, as summarized in Table 19.1. The largest change among all is expected for provisioning services. Here increasing water volumes can be directly used to increase hydropower generation. Water quality will remain on a very high level, as nutrients are effectively retained in the ecosystem and subsequent losses are small. But even if nutrients are lost, they are diluted due to the availability of large water volumes. Increasing precipitation might trigger flood events and landslide occurrences, effecting regulating services. From an anthropogenic viewpoint, economic losses will remain low in the research area due to the low density of population and infrastructure.

References

Bendix J, Behling H, Peters T, Richter M, Beck E (2010) Functional biodiversity and climate change along an altitudinal gradient in a tropical mountain rainforest. In: Tscharnke T, Leuschner C, Veldkamp E, Faust H, Guhardja E, Bidin A (eds) Tropical rainforests and agroforests under global change. Springer, Heidelberg, pp 239–268

Buytaert W, Celleri R, Time L (2009) Predicting climate change impacts on water resources in the tropical Andes: effects of GCM uncertainty. Geophys Res Lett 36, L07406. doi:10.1029/2008GL037048

Fowler HJ, Blenkinsop S, Tebaldi C (2007) Linking climate change modelling to impacts studies: recent advances in downscaling techniques for hydrological modelling. Int J Climatol 27:1547–1578

Millennium Ecosystem Assessment (2005) Ecosystems and human well-being: synthesis. Island Press, Washington, DC

Plesca I, Timbe E, Exbrayat J-F, Windhorst D, Kraft P, Crespo P, Vaché KB, Frede H-G, Breuer L (2012) Model intercomparison to explore catchment functioning: results from a remote montane tropical rainforest. Ecol Model 239:3–13. doi:10.1016/j.ecolmodel.2011.05.005

Tebaldi C, Smith RL, Nychka D, Mearns LO (2010) Quantifying uncertainty in projections of regional climate change: a Bayesian approach to the analysis of multimodel ensembles. J Climate 18:1524–1540

Urrutia R, Vuille M (2009) Climate change projections for the tropical Andes using a regional climate model: temperature and precipitation simulations for the end of the 21st century. J Geophys Res 114, D02108. doi:10.1029/2008JD011021

Chapter 20
Impacts of Local Land-Use Change on Climate and Hydrology

David Windhorst, Brenner Silva, Thorsten Peters, Hanna Meyer, Boris Thies, Jörg Bendix, Hans-Georg Frede, and Lutz Breuer

20.1 Introduction

Land-use change and associated collateral damages are one of the major driving forces of alteration of ecosystem services in the tropics (Foley et al. 2007). Ecuador is one of the countries with the highest deforestation rate in South America. Apart from land-use changes for intensive, monoculture-type agricultural production in the lowlands, areas in the highlands of Ecuador are prone to changes due to deforestation for wood harvesting and providing land for extensive pasturalism (see Chap. 2 for more details and a specific analysis of the conditions in the research area of RBSF).

The effects of land-use change on natural resources such as soils (de Koning et al. 2004), water (Bruijnzeel 2004), climate (Fearnside 2000), as well as biodiversity (Barlow et al. 2007) have been on the research agenda for decades. In this study we focus on the implications of land-use change on water and climate relevant fluxes, such as energy transfer between the soil–biosphere–atmosphere systems of tropical montane forests. Despite the general textbook knowledge that deforestation leads to higher discharges and reduced evapotranspiration rates, paired site studies have shown that this general assumption is sometimes too simplistic (Bruijnzeel 2004).

The objectives of this study are to provide evidence for the differences of hydrological and climatic conditions in the two dominating land-use types within

D. Windhorst • H.-G. Frede • L. Breuer (✉)
Institute for Landscape Ecology and Resources Management, University of Gießen, Gießen, Germany
e-mail: Lutz.Breuer@umwelt.uni-giessen.de

B. Silva • H. Meyer • B. Thies • J. Bendix
Faculty of Geography, University of Marburg, Marburg, Germany

T. Peters
Institute for Geography, University of Erlangen, Erlangen, Germany

J. Bendix et al. (eds.), *Ecosystem Services, Biodiversity and Environmental Change in a Tropical Mountain Ecosystem of South Ecuador*, Ecological Studies 221,
DOI 10.1007/978-3-642-38137-9_20,

the wider RBSF area, i.e., grassland and forest. Based on current observations and modeling approaches we further aim to predict these differences and extrapolate our results to potential future land-use distributions.

20.2 Material and Methods

Estimating the effect of land-use change on climate and hydrology requires a sound knowledge of the current state of the system and how it is influenced by different land-use types. To gain this knowledge we first investigate the microclimatic conditions across different land-use types by means of local meteorological observations and compare and complete these findings with modeling results.

For this study we focus on describing the two land-use types in the study area which dominate the current and probably also the future landscape: forest and grassland. During the years 1987–2001, forest and grassland together represent over 80 % of the area's land cover, and—as our simulations indicate—are most likely to dominate the region in the future as well. We will focus our analysis on an area of 3,860 km^2, as outlined in Chap. 1.

20.2.1 Climate Conditions

In order to investigate local climate differences between the dominant land uses of RBSF, two fully equipped automatic climate stations (Thies Clima) were installed at a small open site surrounded by natural forest (ECSF open natural forest, 1,957 m a.s.l.) and an open grassland (ECSF grassland, 1,950 m a.s.l.) site close to the ECSF station. Meteorological data were recorded every 5 and 30 min arithmetic means (for wind direction vectorial) were calculated. At both sites air temperature and humidity were measured at 2 m and precipitation at 1 m above ground level. Wind velocities were measured at 2.5 m above ground. The data reported here cover the period from 01.07.2010 to 30.09.2011. Data shown cover those parameters that we consider to be land use dependent, i.e., wind speed, daily maximum and minimum temperature, and daily relative humidity.

As an indicator for exchange between land and atmosphere, the ratio of the sensible to latent heat turbulent fluxes (Bowen ratio) is used as a quantitative approach to assess the regulation functioning of a given land cover. The Bowen ratio is a good indicator for the climate variability and susceptibility to extreme conditions, since for instance with an increasing Bowen ratio, the land cover tends towards more arid conditions, which increases the probability of fires and soil degradation.

20.2.2 Prediction of Land-Use Change

Two Landsat scenes from 1987 to 2001 were preprocessed and classified to derive land-use classification for both time steps. Changes that happened during this 14-year period were derived from a post-classification change detection analysis (see Chap. 2). Based on the detected land-use changes in this training period a scenario was developed to project future changes in the study area until 2070.

The land-use change projection method used is based on a stochastic-empirical approach using remote sensing data originally developed for the IDRISI Andes *Land Use Change Modeller* extension. For such a projection it is necessary to distinguish between projections of the (1) quantity and (2) allocation of change.

(1) The quantity of change is determined by demands for land-based products. It was projected through Markov Chain Analysis (e.g., Turner et al. 1995). A "business-as-usual" scenario was chosen assuming ongoing land-use change as observed for the training period. The protected area of the Podocarpus National Park was excluded from the projection.
(2) The allocation of change requires identification of the natural and cultural landscape attributes and their spatial relationship as allocation factors that spatially determine land-use change. Artificial Neural Networks (ANN) offer the advantage of dealing well with nonlinear behavior as well as with complex relationships (Mas et al. 2004). Further, they can also be applied in cases of multicollinearity between input variables (Mas et al. 2004) which is advantageous in modeling land-use change where the variables are supposed to be connected to each other. In this study a Multi-Layer Perceptron (MLP) ANN is used. Eight variables are included as allocation factors: the Euclidean distance to roads, rivers, the provincial capital Loja, other villages, and forest edges, respectively, as well as elevation, slope, and aspect. Maps of roads and villages date back to 1992. The distance from forest edges is directly retrieved from the Landsat scene classification for 2001. The other variables were calculated from a digital elevation model (DEM) and respective GIS layers. Topographic variables are derived from the SRTM (Shuttle Radar Topography Mission) 90 m elevation data. In addition, a layer of the protected area of the Podocarpus National Park was considered. The selected allocation factors are used in combination with the land-use change detected between 1987 and 2001 to train the MLP ANN and to reveal spatial regularities of land-use change. The output of the MLP learning process was used in connection with the respective allocation factors to calculate the transition potential maps for each land use.

To achieve a spatial allocation of land-use change from 2001 to 2010, the transition potential maps for 2001 computed by the MLP ANN were combined with the Markov Chain. The Markov Chain Analysis matrix contains the expected amount of change for the desired time step for each of the projected transitions. The amount of change for each transition projected with the Markov Chain was allocated according to its spatial potential using the multi-objective land allocation procedure.

20.2.3 Coupled Hydro-Meteorological Modeling

At first, we explored the dependency of the regulating services provided by these two land-use types on local climate and hydrology. Building on the developed land-use scenarios, we then used this knowledge to quantify the state of those regulating services in the year 2050 and compared it to the current state.

To estimate the local hydro-meteorological fluxes affecting the regulating services of forests and grasslands we deployed a Soil–Vegetation–Atmosphere Transfer Model (SVAT) called *Southern Bracken Competition Model* (SoBraCoMo, Silva et al. 2012) and coupled it to the hydrological model *Catchment Modeling Framework* (CMF, Kraft et al. 2011).

The SoBraCoMo is based on the Community Land Model (CLM) and operates with atmospheric forcing data. The canopy module (energy and water balance) requires site- and species-specific parameterization. Site-specific radiation partitioning is described in Bendix et al. (2010). Species-specific physiological parameters (quantum and carboxylation efficiency) for the grass *Setaria sphacelata* are taken from field measurements (Silva et al. 2012). Functional types for broad-leaf evergreen plants obtained from CLM are used to describe the physiological parameters of the tree species *Alnus acuminata*. In the comparative and coupled SoBrCoMo/CMF simulation runs, terrain and soil conditions are the same for forest and grass patches since both vegetation types occur under the same topographic conditions within the San Francisco valley. Species-specific physiological model parameters of the considered plants were changed according to field measurements, while site-specific parameterization (e.g., atmospheric forcing data) remained constant throughout all simulations for the coupled SoBraCoMo - CMF model framework.

The Catchment Modeling Framework (CMF) is a programming library comprising a wide range of software objects, which can be used to set up a range of Finite Volume Method based hydrologic models to predict water flows. In the scope of this study, CMF was used to simulate the hydrological fluxes within a one-dimensional soil column. In the case of infiltration excess or seepage, the water leaving the soil column is directly routed to the groundwater or stream using a Dirichlet boundary condition with a constant negative pressure (i.e., no groundwater capillary rise). The movement of soil water within each layer is controlled by the Richards equation.

To simulate the two prevailing land-use types we defined two base scenarios: In the first base scenario, the grassland is represented by an extensive managed *Setaria sphacelata* meadow (see also Chaps. 15, 22 and 26), which will most likely prevail in the future. *Alnus acuminata* is used to represent the forested areas in a second scenario, since it is a native species with potential for sustainable reforestation (Weber et al. 2008) and thus tends to cover a larger area in the future. It is important to note that only the climate regulation function of forest areas is considered in the present simulation, which supports the use of *Alnus acuminata* as a representing species for forest areas. Results in either case reflect a fully developed stand to represent a well-established forest or grassland side, respectively. The simulations run with the climate data of 2008 (annual precipitation input amounts to 1979 mm) as a base line. Atmospheric forcing data (i.e., horizontal wind vectors, air pressure, temperature and humidity, precipitation, incident shortwave and longwave radiation, leaf temperature, soil temperature, and moisture) used were recorded by an automatic weather station at 2,100 m a.s.l. in the vicinity to established *Setaria sphacelata* field trials (Silva et al. 2012).

Subsequently, the plot scale model results are extrapolated to describe the current (2010) and future (2050) land-use allocation in the study area.

20.3 Results and Discussion

20.3.1 Climatic Differences Between Land Uses

Local wind speeds and air temperatures vary markedly between grassland and forested sites (Fig. 20.1). Average air temperatures are about 1.2 °C higher on the natural open forest site than on the grassland site. Differences between day and night air temperatures are approximately 10 °C, while the average annual temperature is around 15.5 °C. Especially daily maximum temperatures differ clearly between manmade and natural ecosystems, whereas minimum temperatures are almost identical. The main reasons are differences in the energy balance (see Table 20.2) because more latent heat is released over the forest than over the pasture which generates more sensible heat and thus higher temperatures. Average wind speeds are about 0.8 m s^{-1} higher on grassland leading to an increased air mass exchange.

Relative air humidity is highly variable on both sites ranging between 23 and 99.9 % during the day, with an average value of 82 %. Different patterns are shown for daily mean values which are also related to different wind speeds at our measuring sites. In most cases, when wind speed is similar at open grassland and forest sites, relative air humidity is higher at the grassland site. Grassland sites are more prone to higher wind speeds than forest, which decreases the relative air humidity at grassland to values lower than that at the forest site. This means that the relative vapor saturation deficit over the pasture is higher and might pose problems for organisms during periods of extreme droughts. By comparing the climate inside the forest under a canopy shelter with the climate in the open pasture, it could be shown that lowest minimum, average, and maximum relative humidity values occur on the pasture site and the humidity regulation function of the open site is clearly reduced in comparison to the forest interior (Fries et al. 2012).

Average annual rainfall adds up to 2,150 mm (data not shown) in the forested site with an additional input of approx. 120 mm water intake by fog (Bendix et al. 2008). This is around 17 % higher as compared to the grassland site. The main reason is the distance from the Cordillera del Consuelo in upwind direction which

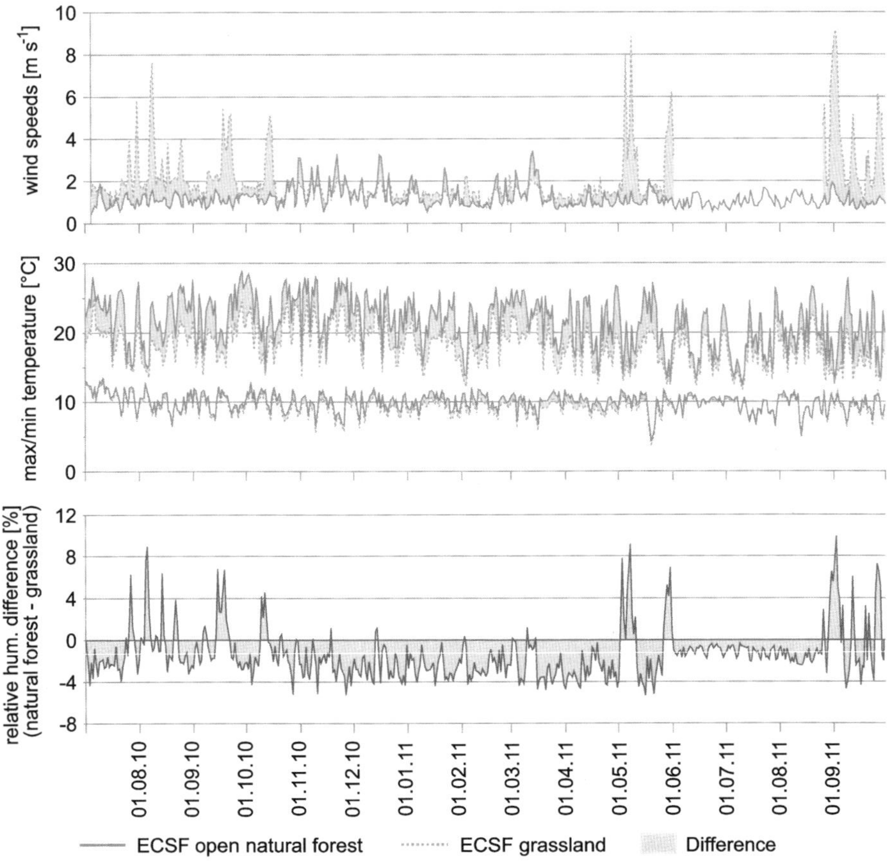

Fig. 20.1 Climate of open natural and anthropogenic sites within the Rio San Francisco valley based on daily data. Max and min refer to the maximum and minimum temperature of 1 day

shelters the study sites from the surging easterlies. While the forest station, which is closer to the Cordillera, still benefits from rainfall spillover, the more distant pasture site already receives less rainfall due to rain shadow effects.

20.3.2 Future Land-Use Change in the San Francisco Catchment

The results of the land-use change prediction show the expected land-use distributions for the years 2010–2070 (Fig. 20.2). According to the business as usual scenario, the forest cover steadily decreases while at the same time grassland and urban areas are expanding. From 2030 on, grassland is expected to be the dominant land-use class in the study area.

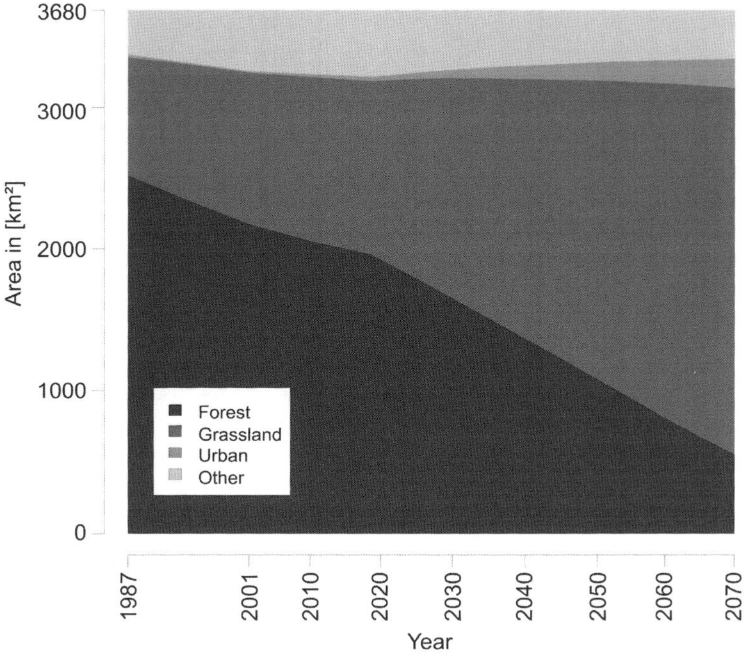

Fig. 20.2 Detected land-use changes between 1987 and 2001 and projected changes of land use to 2070

The approach presented here allows to project feasible land-use changes with a focus on forest development. It shows how land would develop under processes of deforestation, reforestation, and urbanization as observed over the last decades. Thus, the presented projection follows a "business as usual" storyline. Consequently, this Land Use Cover Change (LUCC) projection allows for the identification of hotspots of special interest in the study area, where changes potentially affect the natural resource base through impacts on soil and water (de Koning et al. 1999). Following this projection, all forest resources will be cleared until 2070 except those of the Podocarpus National Park (552.1 km^2). Apart from shrinking forest cover, the urban area is expanding around Loja and along the eastern river dales (209.2 km^2 in 2070). From 2030 on, grassland is expected to be the dominant LUC class in the study area (1,702 km^2 in 2030; 2,583 km^2 in 2070).

20.3.3 Effects on Local Hydro-Meteorological Fluxes

Different vegetation types will directly alter hydro-meteorological fluxes and the ecosystem services associated with them due to, e.g., different transpiration rates, rooting depths, and leaf area. To summarize the effects of the two dominant land-use types in the study area, we present the conclusions drawn from the coupled model runs of SoBraCoMo and CMF and compare the local hydro-meteorological fluxes of the years 2010 and 2050 based on the expected land use.

An extensive review of the ecosystem services related to the hydrological cycle of the research area is presented in Chap. 9. Within this section we concentrate on the ecosystem services relevant for the study area, and how they are affected by land-use change.

We found the following water-related services to be influenced by land-use change and thus suitable for description by the hydrological fluxes simulated in this study: Water supply for hydroelectric power and drinking water (provisioning service) described by the potential discharge, water supply for plants (provisioning service) described by evapotranspiration (ET), and the prevention of erosion (regulating service) as a function of reduced surface water flow. In this context, the potential discharge (equivalent to fresh water supply as defined in Chap. 9) is the sum of simulated groundwater recharge and surface water flow.

The annual hydrological fluxes for the base scenario of the tree plantation (*Alnus acuminata*) and the grassland (*Setaria sphacelata*) are shown in Table 20.1. In comparison, the *Alnus acuminata* stand is characterized by a higher water demand (higher ET) and a slightly higher surface water flow. Due to the increased water demand of *Alnus acuminata*, the groundwater recharge, and as a direct consequence the potential discharge, is lower than in the grassland scenario. Both scenarios generate only a relatively small amount of surface flow. In nature, the steep slope of the valley will increase the vulnerability to erosion and the generation of surface water flow, but this process can only partly be captured by the 1D model setup.

The hydrological and climate indicators presented in Table 20.1 are in good agreement with data published by Oke (1987), mentioning typical values of ET of 1,752 mm a^{-1} for a tropical rainforest and 876 mm a^{-1} for grasslands, respectively. Typical ranges of Bowen ratio are 0.1–0.3 for tropical rainforests and 0.4–0.8 for grasslands (Oke 1987). The differences in energy fluxes between forest and grassland are primarily a result of the presented ET values. The higher ET values over the forest clearly cause an enhanced exchange of latent heat, supported by a higher generation of turbulence (momentum flux) due to the greater roughness of the trees in comparison to grass bunches. The latter is also responsible for the generally lower energy exchange between the surface and the atmosphere over grass. The higher Bowen ration over grass points out that more absorbed solar radiation energy is left for the conversion to sensible heat flux due to less energy invested for ET. All together, this leads to a reduced ability for thermal regulation over grass. Thus, converting forest to grass means a reduction in the thermal regulation function of the ecosystem which is also observed in the field by comparing 2 m measurements above the pasture and inside (stem room) the forest (refer to Sect. 20.3.1 and Fries et al. 2009).

Knowing the potential services which different ecosystem types can provide in the future is of importance for civil planning and conservation considerations. Next to the envisaged climate change presented in Chap. 19, the expected land-use change until 2050 (Fig. 20.2) is likely to exert a strong effect on these services. To illustrate the potential changes in ecosystem services we extrapolated the results

Table 20.1 Simulated annual hydrological fluxes and climate indicators relevant for ecosystem services of the two base scenarios for forests and grassland

	Hydrological fluxes				Climate indicators		
	ET [mm a^{-1}]	Groundwater recharge [mm a^{-1}]	Surface water flow [mm a^{-1}]	Potential discharge [mm a^{-1}]	Sensible heat flux [W m^{-2}]	Latent heat flux [W m^{-2}]	Bowen ratio
Forest	1,433	215	87	302	8.9	113.3	0.08
Grassland	945	783	52	835	26.1	68.0	0.38

Table 20.2 Change in the hydrologic and climatic indicators for changes in ecosystem services of the study area as a response to land-use change

Year	Land cover	Surface water flow [m^3/h^{-1}]	Potential discharge [m^3/h^{-1}]		Sensible heat flux [$\times 10^{10}$ W]	Latent heat flux [$\times 10^{10}$ W]	Bowen ratio
2010	Forest	5.7	19.9		1.85	23.5	
	Grassland	1.9	30.1		2.97	7.73	
	Sum	7.6	50.0	Average	2.41	15.6	0.15
2050	Forest	3.0	10.3		0.95	12.1	
	Grassland	3.5	56.7		5.59	14.6	
	Sum	6.5	67.0	Average	3.27	13.3	0.25

of the forest and grassland base scenario to the wider research area of the RBSF (see Chap. 1 for details). From 2010 to 2050, forest is likely to decline by almost 50 % from 2,075 to 1,070 km^2. At the same time, grassland is projected to almost double its cover to around 2,140 km^2.

If grasslands become the dominant land-use type in the study area, this might have far-reaching consequences for hydrological fluxes (Table 20.2). The fresh water supply per hour generated in the study area, which could be used to generate hydropower or serve as drinking water, will increase by over 33 % to 67 $m^3 h^{-1}$. The vulnerability to water erosion as expressed by the amount of surface water flow will remain relatively small with 2 % (52 mm a^{-1}) of the incident rainfall for grassland and 4 % (87 mm a^{-1}) for forested areas, respectively. The observed decrease by 15 % can be attributed to the relatively higher water uptake by *Setaria sphacelata* in the upper soil layers which leads to a higher saturation deficit in the top soil. According to Bruijnzeel (2004), who reviewed a series of studies concerning the hydrological impacts brought about by the conversion of tropical forests to other land-use types, in nature this effect will be overruled by a decreasing infiltration capacity of the grassland soil.

Even though the simulated increase in fresh water supply following the conversion from forest to grassland (533 mm a^{-1}) is above the values presented by Bruijnzeel (2004) (150–300 mm per year), it is still in good agreement with the general assumption that short vegetation leads to a higher water yield due to its reduced capacity to intercept and evaporate rainfall (van Dijk and Bruijnzeel 2001) and to extract water from deeper soil layers (Hodnett et al. 1995).

Regarding the climatic indicators for changes in ecosystem services, Table 20.2 shows the results in sensible and latent heat fluxes, which were used to calculate the Bowen ratio. Regarding the scenario for 2050, the mean Bowen ratio will increase by 0.1 in relation to 2010, which means lower exchange of moisture between surface and atmosphere. As a consequence, the susceptibility to droughts increases, as well as the potential occurrence of fires (natural and man-triggered) and soil degradation. Although it is not possible to quantify the consequences of changes in climate regulation due to land cover changes, the results presented here point to a deterioration of related ecosystem services in the future.

Table 20.3 Expected impact of future land-use change trends on water- and climate-related ecosystem services in the Rio San Francisco valley (definitions in agreement with Millennium Ecosystem Assessment 2005; see also Chap. 9)

	Land-use change impact	Note
Provisioning service: Freshwater	↑	Hydropower generation can expect increasing benefits due to increasing amounts of available water
Regulating service: Water purification and waste treatment	→	Water quality will remain stable and not be impacted by local population
Regulating service: Water regulation	→	Amount of surface water as an indicator for erosion slightly decreases but on a very low level
Regulating service: Climate regulation	↓	Deforestation will increase susceptibility to droughts

Land-use change increases provisioning services, if human consumption of the service is likely to increase. An increase for regulating service is predicted, if land-use change improves the service from the human perspective. Supporting services are not directly used and land-use change impacts the supply of all other services. If these indirect effects are listed, supporting service would be double counted. ↑ Increasing, → No change, ↓ Decreasing

20.3.4 Outlook and Conclusion: Impact of Land-Use Change on Water-Related Ecosystem Services

Based on the results obtained in this study, we conclude that hydrologic and climatic related ecosystem services are likely to be severely impacted by land-use change—at least under the projected very intense land-use changes outlined in the future scenarios of this study. One may criticize that the assumptions we made are simplifying the very complex and intrinsic interaction of vegetation, soils, atmosphere, and hydrosphere, as we only considered one dominant land cover type per land use and disregarded lateral fluxes and interactions on the landscape scale. Hence, we see our results more in the sense of a virtual sensitivity study rather than as a quantitative forecasting experiment. Table 20.3 summarizes the conclusions we draw with regard to ecosystem services.

References

Barlow J, Gardner TA, Araujo IS, Ávila-Pires TC, Bonaldo AB, Costa JE, Esposito MC, Ferreira LV, Hawes J, Hernandez MIM, Hoogmoed MS, Leite RN, Lo-Man-Hung NF, Malcolm JR, Martins MB, Mestre LAM, Miranda-Santos R, Nunes-Gutjahr AL, Overal WL, Parry L, Peters SL, Ribeiro-Junior MA, da Silva MNF, da Silva MC, Peres CA (2007) Quantifying the biodiversity value of tropical primary, secondary, and plantation forests. Proc Natl Acad Sci USA 104:18555–18560

Bendix J, Rollenbeck R, Fabian P, Emck P, Richter M, Beck E (2008) Climate variability. In: Beck E, Bendix J, Kottke I, Makeschin F, Mosandl R (eds) Gradients in a tropical mountain ecosystem of Ecuador, vol 198, Ecological studies . Springer, Berlin, pp 280–291
Bendix J, Silva B, Roos K, Göttlicher DO, Rollenbeck R, Nauss T, Beck E (2010) Model parameterization to simulate and compare the PAR absorption potential of two competing plant species. Int J Biometeorol 54:283–295
Bruijnzeel LA (2004) Hydrological functions of tropical forests: not seeing the soil for the trees? Agric Ecosyst Environ 104:185–228
de Koning GHJ, Verburg PH, Veldkamp A, Fresco LO (1999) Multi-scale modelling of land use change dynamics in Ecuador. Agric Syst 61:77–93
de Koning GHJ, Veldkamp E, López-Ulloa M (2004) Quantification of carbon sequestration in soils following pasture to forest conversion in northwestern Ecuador. Global Biogeochem Cycles 17:1098–1110
Fearnside PM (2000) Global warming and tropical land-use change: greenhouse gas emission from biomass burning, decomposition and soils in forest conversioin, shifting cultivation and secondary vegetation. Clim Change 46:115–158
Foley JA, Asner GP, Costa MH, Coe MT, DeFries R, Gibbs HK, Howard EA, Olson S, Patz J, Ramankutty N, Snyder P (2007) Amazonia revealed: forest degradation and loss of ecosystem goods and services in the Amazon Basin. Front Ecol Environ 5:25–32
Fries A, Rollenbeck R, Göttlicher D, Nauß T, Homeier J, Peters T, Bendix J (2009) Thermal structure of a megadiverse mountain ecosystem in southern Ecuador, and its regionalization. Erdkunde 63:321–335
Fries A, Rollenbeck R, Nauß T, Peters T, Bendix J (2012) Near surface air humidity in a megadiverse Andean mountain ecosystem of southern Ecuador and its regionalization. Agric For Meteorol 152:17–30
Hodnett MG, da Silva LP, da Rocha HR, Cruz Senna R (1995) Seasonal soil water storage changes beneath central Amazonian rainforest and pasture. J Hydrol 170:233–254
Kraft P, Vaché KB, Frede H-G, Breuer L (2011) CMF: a hydrological programming language extension for integrated catchment models. Environ Model Softw 26:828–830
Mas J, Puig H, Palacio J, Sosa-Lopez A (2004) Modelling deforestation using GIS and artificial neural networks. Environ Model Softw 19:461–471
Millennium Ecosystem Assessment (2005) Ecosystems and human well-being. Synthesis Island Press, Washington, DC
Oke TR (1987) Boundary layer climates, 2nd edn. Methuen, London
Silva B, Roos K, Voss I, König N, Rollenbeck R, Scheibe R, Beck E, Bendix J (2012) Simulating canopy photosynthesis for two competing species of an anthropogenic grassland community in the Andes of southern Ecuador. Ecol Modell 239:14–26
Turner BL, Skole D, Sanderson S, Fischer G, Fresco L, Leemans R (1995) Land-use and land-cover change – science/research plan. IGBP report No. 35, HDP Report No. 07, Stockholm and Geneva
van Dijk AIJM, Bruijnzeel LA (2001) Modelling rainfall interception by vegetation of variable density using an adapted analytical model. Part 2. Model validation for a tropical upland mixed cropping system. J Hydrol 247:239–262
Weber M, Günter S, Aguirre N, Stimm B, Mosandl R (2008) Reforestation of abandoned pastures: silvicultural means to accelerate forest recovery and biodiversity. In: Beck E, Bendix J, Kottke I, Makeschin F, Mosandl R (eds) Gradients in a tropical mountain ecosystem of Ecuador, vol 198, Ecological studies. Springer, Berlin, pp 431–441

Chapter 21
Current and Future Variations of Nutrient Depositions and Influences on Tree Growth

Rütger Rollenbeck, Insa Otte, Peter Fabian, Wolfgang Wilcke, Darwin Pucha Cofrep, Achim Bräuning, and Jörg Bendix

21.1 Introduction

Future trends and variations of nutrient inputs into tropical mountain forests change the local nutrient cycle (Boy et al. 2008) and likely also affect biodiversity (Harpole and Tilman 2007), ecosystem integrity, and ecosystem services.

An important contribution to these nutrient inputs can be attributed to forest fires in the Amazon basin which cause enhanced release of plant nutrients, especially N (Allen and Miguel 1995; Artaxo et al. 2002; Da Rocha et al. 2005). After release, these nutrients are transported with the tropical easterlies to the study forest in south Ecuador (Boy et al. 2008). The remote fertilization affects the local nutrient cycling (Chap. 11) and might have effects on biomass productivity (Wullaert et al. 2010, Chap. 12). Altered productivity resulting from eutrophication of this nutrient-scarce ecosystem may also have detrimental effects on local biodiversity (Sala et al. 2000; Phoenix et al. 2006; Harpole and Tilman 2007). While the influence of remote fertilization by long-distance atmospheric transport has been demonstrated for the Amazon lowland rain forest (Artaxo et al. 1990, 1998; Graham et al. 2003), its

R. Rollenbeck (✉) • I. Otte • J. Bendix
Faculty of Geography, University of Marburg, Deutschhausstr. 10, 35032 Marburg, Germany
e-mail: rollenbeck@lcrs.de

P. Fabian
Life Science Center Weihenstephan, TU München, Am Hochanger 13, 85354 Freising, Germany

W. Wilcke
Geographic Institute of the University of Bern (GIUB), University of Bern, Hallerstrasse 12, 3012 Bern, Switzerland

D.P. Cofrep • A. Bräuning
Institute of Geography, University of Erlangen-Nuremberg, Kochstraße 4/4, 91054 Erlangen, Germany

J. Bendix et al. (eds.), *Ecosystem Services, Biodiversity and Environmental Change in a Tropical Mountain Ecosystem of South Ecuador*, Ecological Studies 221,
DOI 10.1007/978-3-642-38137-9_21, © Springer-Verlag Berlin Heidelberg 2013

effect on tropical mountain ecosystems has just recently gained attention (Fabian et al. 2005, 2009; Boy et al. 2008).

Long-term observations of nutrient deposition in tropical mountain regions are generally scarce (Veneklaas 1990; Clark et al. 1998) and the existing studies do not link observed nutrient inputs to atmospheric transport processes. Therefore, the deposition record (>10 year) from the Reserva Biologica San Francisco (RBSF) reported here enables—in combination with modeling and remote sensing approaches—a comprehensive analysis of the controls and effects of remote fertilizing in the Amazonia-exposed northern Andes. To assess aerosol transport in the atmosphere, trajectory modeling has been widely used (Andreae et al. 2001). This tool can be applied to follow the impact of individual emission events like forest fires (Damoah et al. 2004) or volcanic eruptions. Here, we use trajectory modeling to link global environmental changes to local impacts with a long-term perspective. The special report on emission scenarios (SRES) of the IPCC (IPCC 2000; Nakicenovic and Swart 2000) has provided a basis to assess trends and developments of future emissions. Our objective was to use these data as input for trajectory-based modeling of long-range nutrient transport in the atmosphere and to calibrate the model with observations of the past decade, in order to develop scenarios of future trends in remote fertilization.

The history of local fertilization is analyzed by means of nutrient concentrations in tree rings. This is possible because trees are fixing specific nutrient elements in the wood, consequently forming annual and thus datable tree rings which are suitable indicators to trace back the history of remote fertilization (Donnelly et al. 1990).

21.2 Methods

21.2.1 Sampling of Atmospheric Nutrients and Identification of Potential Sources

The analysis of current and future nutrient inputs to the ecosystem of the RBSF (Reserva Biologica San Francisco) is based on a combination of local observations, geostatistical modeling, and time series evaluation. The local observations are provided by rain and fog samples which have been taken weekly while precipitation quantities are registered each hour since March 2002. Three sites cover an altitudinal gradient in the research area RBSF (cf. Fig. 1.2).

All samples have been analyzed for pH, electrical conductivity, and the concentrations of K^+, $Na,^+$, Ca^{2+}, and Mg (Inductively coupled Plasma-Optical Emission Spectrometry) and NH_4^+, Cl^-, SO_4^{2-}, NO_3^-, and PO_4^{3-} (Ion Chromatography). The data of all three stations were composed to an area average, by assigning each station a fraction of the whole area, according to its altitude level. This fraction of the total area of the RBSF is used to derive a weighted mean of the analyzed matter concentration. Fog and rain measurements were averaged according to their contribution to the precipitation total.

The geostatistical tool *Emisstrajekt* (Rollenbeck et al. 2006 and Rollenbeck 2010) was used to calculate potential transport of nutrients as provided by different emission inventories like RETRO ("REanalysis of the TROpospheric chemical composition over the past 40 years") and EDGAR ("Emission Database for Global Atmospheric Research") (Schultz 2007). These inventories provide daily emissions of NO_x by vegetation fires and industrial sources. The output of Emisstrajekt is a time series of potential matter transport (= transmission concentration) on a daily basis, which can be related to variations of local matter concentration in the precipitation samples.

Since nitrogen is one of the most important nutrients for linking human activities and plant growth, the analyses presented here are focused on a comparison of observed NH_4^+ and NO_3^- concentrations with emissions of NO_x from industry, urban areas and agriculture, and biomass burning (Sect. 2.4). The latter will probably show the largest anthropogenically caused variations in the future. A smaller NO_3^- fraction can be linked to lightning, but this contribution is assumed to be nearly constant over time. To consider the influence of changes in atmospheric circulation, the observed concentrations of NO_3^- and NH_4^+ in precipitation are additionally related to a time series of the dominant wind direction.

21.2.2 Projections of Future Deposition

To assess the future emission trends for nitrogen and other nutrients, a four-step procedure is applied.

1. Time series of monthly transmission concentration for the pixels of the field observations (here: meteorological stations at the ECSF, TS1, Cerro del Consuelo; for locations refer to Chap. 1) is calculated by means of the trajectory analysis as described above.
2. This data is used to derive a linear regression model where the predictor variable is the transmission concentration resulting from the trajectory analysis and the response variable is the measured nitrogen concentration at site. The models show fair to high correlations for the single stations.
3. The future transmission concentration for the pixels of the stations is calculated using the emission data provided by the IPCC SRES A1B and B1 story lines (IPCC 2000; Nakicenovic and Swart 2000). A1B is the most likely emission scenario presuming a fast growing world population and a balanced use of fossil and nonfossil fuels. The more optimistic B1 story line describes a transformation to a less industrialized world of a more service-oriented society, while assuming the same population growth (cf. Sect. 2.3). These scenarios also address the potential development of agriculture and hence the potential emissions from biomass burning.

4. The regression model of step two is applied to the future transmission concentrations in order to derive the future development of depositions at the specific sites.

21.2.3 Element Concentration History in Plants

To determine the local impact of the variations in nutrient transport into the ecosystem of the Andean mountain forest on decadal time scales, samples of two increment cores from each of six *Cedrela montana* Moritz ex Turcz (Meliaceae) trees from RBSF were collected at 2,000 m a.s.l. (see Chap. 5; Bräuning et al. 2009). After proper cross-dating and visual and statistical age alignment, annual tree rings were separated with a scalpel for chemical analyses which was carried out for the elements Ca, Mg, and K with an Atomic Absorption Spectrometer and for P with a Continuous Flow Analyzer after acid digestion with concentrated HNO_3 in a microwave oven (Pucha 2007). Due to translocation processes in radial direction, the uptake time of elements concentrations in a specific ring does not necessarily exactly correspond to the age of the dated ring (Donnelly et al. 1990; Watmough 1999), and their higher concentrations in the last rings could be an effect of more cellular activity in the rings near to the phloem. However, the lateral mobility of different elements varies considerably: Whereas large portions of Ca are largely immobile in the cell walls, P, K, and Mg may be more easily translocated along the radial ray parenchyma.

21.3 Results and Discussion

21.3.1 Past and Current Remote Fertilization

Deposition rates of nutrients are mainly driven by rainfall because falling droplets are effective in scavenging atmospheric aerosols and the higher mass flux of rainfall is greater than in other pathways (Fig. 21.1).

Deposition rates therefore mainly reflect the precipitation regime while concentrations in precipitation better reflect variations of immission of nutrients from the atmosphere. Besides the seasonal variations in N concentrations of rainfall related to the burning season in the Amazon basin (Boy et al. 2008, Fig. 11.1a), peak N concentrations in rainfall are observed at the end of the (little pronounced) rainy season when the easterlies are weakened and westerly wind direction occurs more frequently (Fig. 21.2). This suggests a stronger influence of anthropogenic emissions in the inner Andean valley which is more densely populated than the east facing slopes of the eastern cordillera and the adjacent Amazon lowland area.

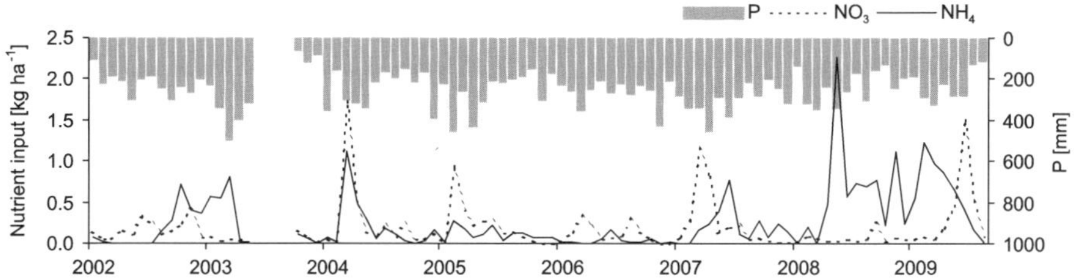

Fig. 21.1 Time series of nitrate and ammonium inputs (area-weighted mean deposition in fog- and rainwater) in the RBSF area between 2002 and 2009

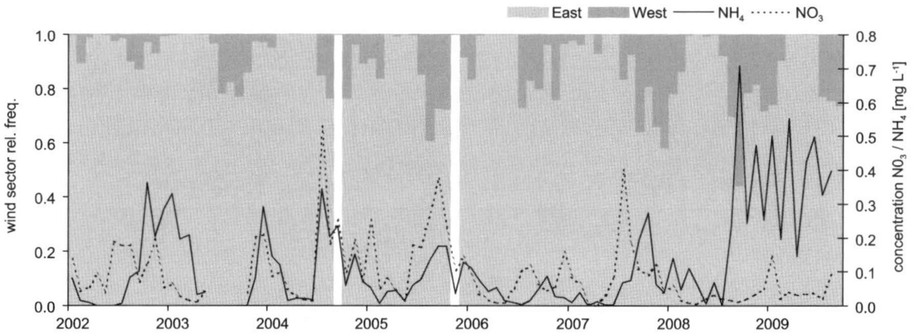

Fig. 21.2 Time series of area-weighted mean concentrations of NO_3^- and NH_4^+ and dominant wind directions (west = *dark gray bars*, east = *light gray bars*) in the RBSF between 2002 and 2009

The concentrations of nitrogen compounds apparently show some influence of varying atmospheric circulation, because peak values seem to prevail during changes of the dominant wind direction. Usually, the highest concentrations are observed at the end of the rain season, when the dominance of the easterlies ceases.

21.3.2 Future Scenarios of Remote Fertilization

The results for the simulated future development of nitrogen (NO_3–N) deposition are presented in Fig. 21.3. It is obvious that the most likely scenario A1B with its higher underlying consumption of fossil fuels and more intensive land-use changes causes quantitatively higher future nitrogen depositions. The more optimistic B1 scenario with an accelerated change to an energy-saving service society consequently generates lower depositions, particularly after 2050. Both scenarios are reaching peak depositions between 2040 and 2050. While the deposition remains stable on a high level for the A1B scenario, the societal reorganization underlying the B1 story line leads to a decline of nitrogen depositions by 2100 towards a just slightly higher deposition level than today.

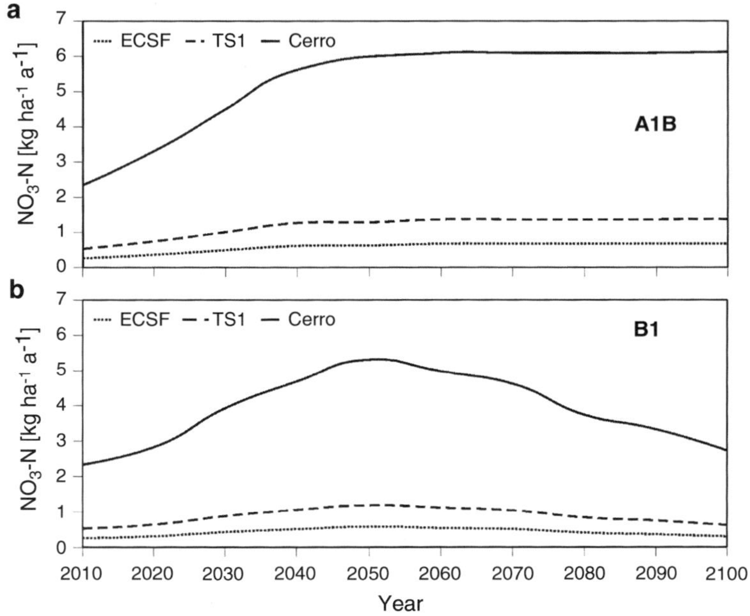

Fig. 21.3 Local deposition scenario of NO_3–N according to IPCC special report emission scenarios (SRES) A1B (**a**) and B1 (**b**) for three different altitudes in the San Francisco valley (ECSF 1,960 m a.s.l.; TS1 2,660 m a.s.l.; Cerro del Consuelo: 3,180 m a.s.l.)

Interestingly, there is a clear altitudinal gradient of future deposition trends in the San Francisco Valley (Chap. 11). For both scenarios, the lowest intensity of future deposition is simulated close to the valley bottom (ECSF met station at 1,960 m a.s.l.). Up to 2,660 m a.s.l. (met station TS1), the magnitude of future deposition increases slightly. A big leap in the order of magnitude can be observed for the crest line area around the Cerro fog gauge station at 3,180 m a.s.l. where the sixfold of the annual deposition at the valley bottom is projected to be reached by the end of the century for the most likely A1B scenario. The soaring deposition in the upper parts is due to the saltational change of climate conditions. Directly above the TS1 station, the synoptic circulation impinges the crest line nearly unsheltered, causing high flow and condensation rates and, consequently, high deposition by rain and occult cloud fog precipitation (e.g., Bendix et al. 2008).

21.3.3 Impact of Variations in Long-Term Remote Fertilization on Tree Growth

The average annual increment of the studied *Cedrela* trees shows an increasing trend during the past 60 years. The reason for this growth increase is probably a

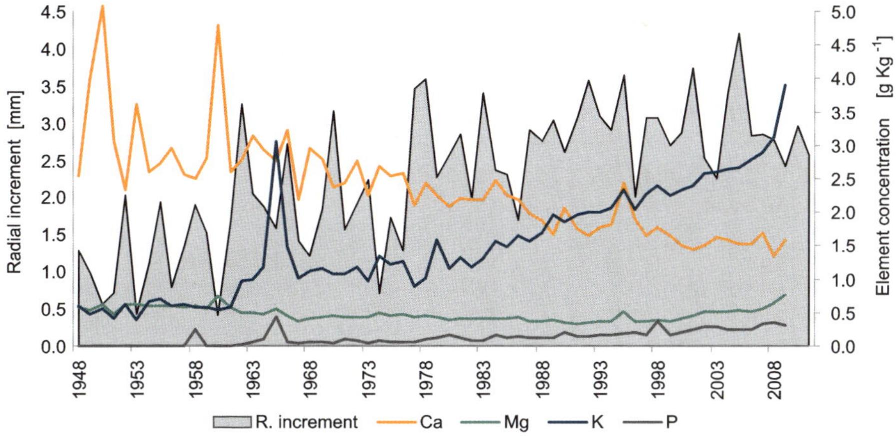

Fig. 21.4 Trends of radial (R.) increment and element concentrations of Ca, Mg, K, and P in tree rings of *Cedrela montana*. Measurements are mean values of six trees

combination of remote fertilization growth of the trees into the upper canopy with better light availability and increased productivity. In parallel, mean concentrations of K, P, and Mg show higher concentrations in the outer parts of the stems close to the cambium (Fig. 21.4). Higher concentrations of K in the younger rings are probably at least partly associated with their proximity to the cambium which is physiologically highly active. In contrast, Ca concentrations strongly decreased after the 1960s, although values before 1960 should be interpreted with caution due to low sample replication. We did not find any significant correlations between element distribution and climate parameters, but it is noteworthy that all elements show randomly (i.e., noncyclic) distributed concentration peaks. For example, all elements except Ca show high concentrations following the La Niña event of 1964. After removing the long-term trends by dividing the ring width and the Ca series by 5-year moving standard deviations, the two series were negatively correlated ($r = -0.29$; $p < 0.05$), indicating that wider rings tend to have lower Ca concentrations. Peak concentrations of K and Mg are often associated with narrower rings that indicate drier conditions. However, care should be taken concerning the mobility of Mg, K, and P. Due to translocation processes in radial direction, the uptake time of elements concentrations in a specific ring does not necessarily exactly correspond to the age of the dated ring (Donnelly et al. 1990; Watmough 1999).

For the last decade, the trends in element concentrations of tree rings coincide with those of nutrient depositions of the corresponding elements. While Ca concentrations in precipitation show decreasing trends between 2002 and 2009 (linear trend: -0.35 kg ha^{-1} year^{-1}; Fig. 21.5), K concentrations show an increasing trend (+0.18 kg ha^{-1} year^{-1}). The concentrations of P and Mg do not have clear trends, but some oscillations with two peak periods from 2002 to 2004 and 2007 to 2009 can be seen. Furthermore, the higher concentrations of PO_4 in rain and fog in

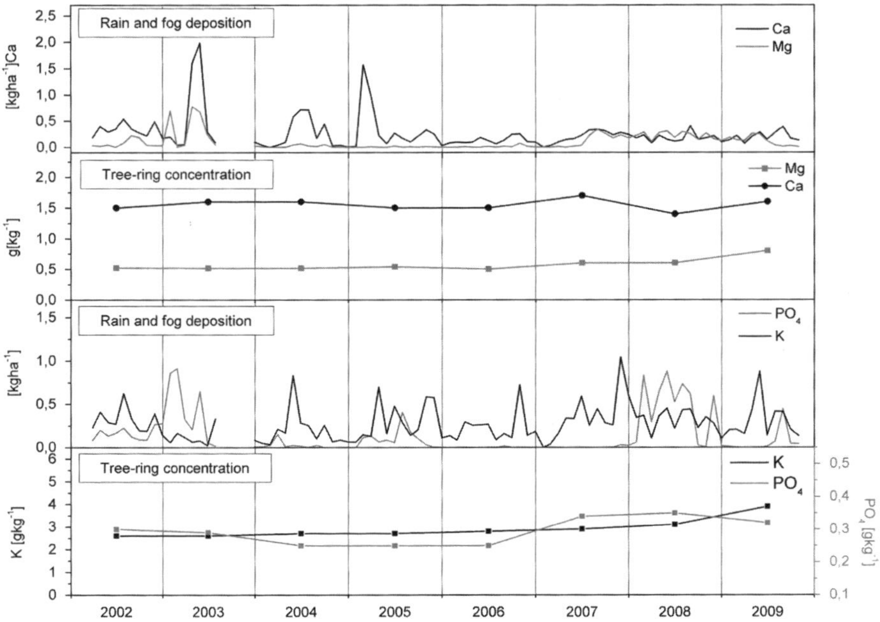

Fig. 21.5 Monthly time series of area-weighted mean concentrations in bulk precipitation (rain and fog) for Mg, Ca, K, and PO_4 and in annual values of concentrations of these elements in tree rings of *Cedrela montana* (mean of six trees)

2002–2003 and 2007–2008 coincide with the increments of P in the tree rings, despite that P has a high mobility in the plants (Hochmuth et al. 1991).

21.4 Conclusion

The study area is affected by nutrient deposition via atmospheric transport. We observed deposition of N released during the fire season in the Amazon basin transported with the tropical easterlies to our study site. Furthermore, there were N inputs from the inter-Andean valley during the short periods of westerly winds in November/December.

The influence of regional circulation is reflected by the fact that nutrient input varies with the dominant wind direction. This substantiates that significant contributions of local depositions are caused by long-range transport and not by local phenomena, which would certainly be less dependent on the regional atmospheric flow.

Considering the IPCC SRES emission scenarios, N deposition will probably increase in the near future and, depending on the underlying socioeconomic story line, either remain stable at an elevated level until 2100 (A1B scenario) or peak in the 2050s and decline thereafter (B1 scenario). As today, the highest parts of the area exposed to the synoptic circulation reveal highest deposition rates also in the future projections.

Nutrient concentrations in tree rings show variations, which were related to measured deposition rates during the last decade. This illustrates a relationship between remote fertilization and local tree growth. Overall a growth-stimulating effect of increasing nutrient input seems likely. The expected eutrophication of the nutrient-scarce ecosystem also bears the potential risk of losses in biodiversity. This loss in biodiversity may adversely affect buffering capabilities and possibly cause a loss of regulating services (Chap. 4), especially with respect to pest outbreaks and internal ecosystem stability.

Acknowledgments We like to thank Marcelo Alonso (INPE) for supplying the Emission inventories of EDGAR/RETRO and Sandro Makowski for adapting them to the *Emisstrajekt*-Program.

References

Allen AG, Miguel AH (1995) Biomass burning in the Amazon – characterization of the ionic component of aerosols generated tom flaming and smoldering rain-forest and savanna. Environ Sci Technol 29:486–493

Andreae MO, Artaxo P, Fischer H, Freitag SR, Gregoire J-M, Hansel A, Hoor P, Kormann R, Krejci R, Lange L, Lelieveld J, Lindinger W, Longo K, Peters W, Dereus M, Scheeren B, Silvia Dias MAF, Ström J, Van Velthoven PFJ, Williams J (2001) Transport of biomass burning smoke to the upper troposphere by deep convection in the equatorial region. Geophys Res Lett 28:951–954

Artaxo P, Maenhaut W, Storms H, Van Grieken R (1990) Aerosol characteristics and sources for the Amazon basin during the wet season. J Geophys Res 95:16971–16985

Artaxo P, Fernandes E, Martins J, Yamayoe M, Hobbs P, Maenhaut W, Longo K, Castanho A (1998) Large-scale aerosol source apportionment in Amazonia. J Geophys Res 103:31837–31847

Artaxo P, Martins J, Yamasoe M, Procópio A, Pauliquevis T, Andreae M, Guyon P, Gatti L, Leal A (2002) Physical and chemical properties of aerosols in the wet and dry seasons in Rondónia, Amazonia. J Geophys Res 107(D20):8081–8085

Bendix J, Rollenbeck R, Göttlicher D, Nauß T, Fabian P (2008) Seasonality and diurnal pattern of very low clouds in a deeply incised valley of the eastern tropical Andes (South Ecuador) as observed by a cost effective WebCam system. Meteorol Appl 15:281–291

Boy J, Rollenbeck R, Valarezo C, Wilcke W (2008) Amazonian biomass burning-derived acid and nutrient deposition in the north Andean montane forest of Ecuador. Glob Biogeochem Cycle 22, GB4011. doi:10.1029/2007GB003158

Bräuning A, Volland-Voigt F, Burchardt I, Ganzhi O, Nauss T, Peters T (2009) Climatic control of radial growth of *Cedrela montana* in a humid mountain rain forest in southern Ecuador. Erdkunde 63(4):337–345

Clark KL, Nadkarni NM, Schaefer DR, Gholz WL (1998) Cloud water and precipitation chemistry in a tropical mountain forest, Montaverde, Costa Rica. Atmos Environ 32:1595–1603

Da Rocha GO, Allen AG, Cardoso AA (2005) Influence of agricultural biomass burning on aerosol size distribution and dry deposition in southeastern Brazil. Environ Sci Technol 39:5293–5301

Damoah R, Spichtinger N, Forster C, James P, Mattis I, Wandinger U, Berle S, Wagner T, Stohl A (2004) Around the world in 17 days – hemispheric scale transport of forest fire smoke from Russia in May 2003. Atmos Chem Phys 4:1311–1321

Donnelly J, Shane J, Schaberg P (1990) Lead mobility within the xylem of red spruce seedlings: implications for the development of pollution histories. J Environ Qual 19:268–271

Fabian P, Kohlpaintner M, Rollenbeck R (2005) Biomass burning in the Amazon – fertilizer for the mountaineous rain forest in Educador. Environ Sci Pollut Res 12:290–296

Fabian P, Rollenbeck R, Spichtinger N, Brothers L, Dominguez G, Thiemens M (2009) Sahara dust, ocean spray, volcanoes, biomass burning: pathways of nutrients into Andean rainforests. Adv Geosci 22:85–94

Graham B, Guyon P, Maenhaut W, Taylor PE, Ebert M, Matthias-Maser S, Mayol-Bracaro OL, Godoi R, Artaxo P, Van Grieken R, Glovsky MM, Flagan RC, Andreae MO (2003) Composition and diurnal variability of the natural Amazonian aerosol. J Geophys Res 108 (D24):4765–4783

Harpole WS, Tilman D (2007) Grassland species loss due to reduced niche dimension. Nature 446:791–793

Hochmuth G, Maynard D, Vavrina C, Hanlon E, Simonne E (1991) Plant tissue analysis and interpretation for vegetable crops in Florida. University of Florida Cooperative Extension Service Special Series HS964

IPCC-SRES (2000) Intergovernmental panel on climate change – special report on emission scenarios. Geneva

Nakicenovic N, Swart R (eds) (2000) Special report on emissions scenarios: a special report of working group III of the intergovernmental panel on climate change, Cambridge University Press, Cambridge, 599 pp

Phoenix GK et al (2006) Atmospheric nitrogen deposition in world biodiversity hotspots: the need for a greater global perspective in assessing N deposition impacts. Glob Change Biol 12:470–476

Pucha D (2007) Variación a largo plazo de la deposición del calcio en el bosque lluvioso montano bajo de la “Estación Científica San Francisco”. Diploma Thesis, University National of Loja

Rollenbeck R (2010) Global sources – local impacts: natural and anthropogenic matter deposition in the Andes of Ecuador. Geoöko 1–2(2010):1–25

Rollenbeck R, Fabian P, Bendix J (2006) Precipitation dynamics and chemical properties in tropical mountain forests of Ecuador. Adv Geosci 6:73–76

Sala OE et al (2000) Global biodiversity scenarios for the year 2100. Science 87:1770–1774

Schultz M (ed) (2007) REanalysis of the TROpospheric chemical composition over the past 40 years (RETRO). A long-term global modeling study of tropospheric chemistry funded under the 5th EU Framework Programme. Final Report, Jülich/Hamburg

Veneklaas EJ (1990) Nutrient flux in bulk precipitation and throughfall in two montane tropical rain forests, Colombia. J Ecol 78:974–992

Watmough S (1999) Monitoring historical changes in soil and atmospheric trace metal levels by dendrochemical analysis. Environ Pollut 106:391–403

Wullaert H, Homeier J, Valarezo C, Wilcke W (2010) Response of the N and P cycles of an old-growth montane forest in Ecuador to experimental low-level N and P amendments. For Ecol Manage 260:1435–1445

Chapter 22
Nutrient Additions Affecting Matter Turnover in Forest and Pasture Ecosystems

Ute Hamer, Karin Potthast, Wolfgang Wilcke, Hans Wullaert, Carlos Valarezo, Dorothee Sandmann, Mark Maraun, Stefan Scheu, and Jürgen Homeier

22.1 Introduction

Globally, the input of nutrients into terrestrial ecosystems is increasing and this is projected to accelerate in the next decades (Lamarque et al. 2005; Phoenix et al. 2006). Especially in tropical regions, N is expected to significantly affect biogeochemical cycles (Matson et al. 1999; Galloway et al. 2004). Hietz et al. (2011) recently have presented indications that the availability of N for plants in tropical forests increased during the last decades because of anthropogenic N deposition. As shown in Sects 2.4, 11.3.1 and 21.3 in particular, increasing frequency of forest fires in Amazonia because of slash-and-burn agriculture is expected to increase nutrient

U. Hamer (✉)
Institute of Landscape Ecology, WWU-University of Muenster, Muenster, Germany
e-mail: ute.hamer@uni-muenster.de

K. Potthast
Institute of Soil Science and Site Ecology, Dresden University of Technology, Dresden, Germany

W. Wilcke
Geographic Institute of the University of Bern (GIUB), University of Bern, Bern, Switzerland

H. Wullaert
Earth System Science Research Center, Geographic Institute, Johannes Gutenberg University Mainz, Mainz, Germany

C. Valarezo
Dirección General de Investigaciones, Universidad Nacional de Loja, Loja, Ecuador

D. Sandmann • M. Maraun • S. Scheu
J.F. Blumenbach Institute of Zoology and Anthropology, Georg August University Göttingen, Göttingen, Germany

J. Homeier
Albrecht von Haller Institute of Plant Sciences, Georg August University Göttingen, Göttingen, Germany

J. Bendix et al. (eds.), *Ecosystem Services, Biodiversity and Environmental Change in a Tropical Mountain Ecosystem of South Ecuador*, Ecological Studies 221,
DOI 10.1007/978-3-642-38137-9_22,

deposition including N and P in the tropical mountain rainforest region of south Ecuador (Boy et al. 2008; Wullaert et al. 2010). Occasional dust deposition from the Sahara has been identified as source of Ca and Mg input (Boy and Wilcke 2008).

The response of regulating ecosystem services, such as the cycling of nutrients in biogeochemical cycles (Fig. 4.1), to nutrient additions is complex due to interactions between several co-occurring above- and belowground matter transformation processes. The major drivers of nutrient cycling processes are water fluxes, plants, animals, and microorganisms. Depending on the element, throughfall (+stemflow) or litterfall are the largest element fluxes to soil (Likens and Bormann 1995). Nutrient deposition from the atmosphere will first reach the canopy where it can be retained on the surface or taken up by plants and phyllosphere organisms (Ulrich 1983). Superficially retained nutrients will finally be transferred to soil via litterfall. Herbivores contribute to the nutrient cycle by causing the fall of plant fragments, via their feces, as animal cadavers, and by enhancing the leaching of nutrients from damaged plant tissue, all influencing the quality of throughfall, stemflow, and litterfall (Hunter 2001).

Microbial communities affect processes occurring in the forest canopy and in soil. Schwarz et al. (2011) detected nitrification in the canopy converting deposited NH_4^+ to NO_3^-. Reduced litter decomposition in a northern hardwood ecosystem after chronic N deposition was related to changes in the saprotrophic fungal community of the forest organic layer and to a downregulation of the expression of the ligninolytic gene lcc (Edwards et al. 2011). Contrasting effects of fertilization have also been reported for grassland soils, mainly in temperate regions, ranging from increases in fungal abundance (Bardgett et al. 1999; Rousk and Baath 2007) to decreases (Grayston et al. 2004; de Vries et al. 2006) or no change (Denef et al. 2009). Important parameters determining the ecosystem response to nutrient additions presumably include the a priori nutrient availability as well as the structure of the soil microbial community within different land-use systems.

Nutrient additions might provoke different responses of plants and soil organisms because of differences in nutrient demands and acquisition strategies and/or varying capacities to adapt by changing their community composition (Reed et al. 2011). The thickness of the organic layer in tropical rainforests of Latin America was determined by both P concentration and total mass of litterfall input (Kaspari and Yanoviak 2008). Tree reproduction increased within 6 years of N fertilization in lowland tropical forests in Panama (Kaspari et al. 2008). In contrast, decomposition of cellulose and leaf litter did not respond to N fertilization but increased after fertilization with P and K (cellulose) or with P and micronutrients (leaf litter). This suggests co-limitation by multiple nutrients (Kaspari et al. 2008). In their recent review (Harpole et al. 2011) highlighted that nutrient limitation at the community or ecosystem level is complex, since metabolic pathways might be limited by different elements either simultaneously or consecutively. Strict co-limitation of N and P occurred in 28 % of the 641 studies conducted in terrestrial, freshwater, and marine ecosystems followed by serial limitation. During the last two decades of ecosystem research the concept of co-limitation by multiple nutrients challenged the paradigm of Liebig's law of the minimum at least for species-rich (tropical) ecosystems (Harpole et al. 2011).

In our research area, much of the tropical montane rainforest has been converted to pastures (cf. Sect. 2.4, Figs. 15.3, and 20.2) which presumably results in different nutrient limitations between forest and pasture ecosystems. As presented in more detail in Sect. 11.3.3 and Table 15.1 the availability of nutrients for plant and microbial growth is significantly higher in mineral soils under pasture compared to those under natural forest (Makeschin et al. 2008; Potthast et al. 2011). The increase in soil pH and exchangeable Ca—because of the input of alkaline ashes from forest clearing by burning—has been identified as main driver for the shift in soil microbial community structure after forest to pasture conversion (Potthast et al. 2011). Within forest soil profiles, the organic layer provides highest amounts of available nutrients (Wilcke et al. 2002; Potthast et al. 2011). Nutrient availability was most important for explaining the differences in microbial community composition between organic layer and forest mineral soil (Potthast et al. 2011).

To explore nutrient limitations in pastures and forests of the study region we experimentally manipulated the input of N, P, and Ca, elements assumed to be of prime importance for plant production, litter decomposition, and microbial community structure and functioning. The effects of fertilization with N, P, and Ca on the biogeochemical cycles of these elements in both ecosystems (forest and pasture) were investigated by determining (1) nutrient concentrations in tissues of dominant plant species and (2) soil chemical and (3) soil microbial properties. Effects of fertilization on C dynamics are discussed in Chaps. 23 (forest) and 26 (pasture). We expected biogeochemical cycles, plants, and microorganisms of the pasture to be less responsive to nutrient additions than those of the forest because of a higher nutrient input into pastures and differences in soil microbial community composition.

22.2 Material and Methods

Fertilization experiments have been established in 2008 in the natural forest (nutrient manipulation experiment, NUMEX) and on active pasture sites (pasture fertilization experiment, FERPAST) at about 2,000 m a.s.l. (the sites of both experiments are given in Fig. 1.2). Both ecosystems received 50 kg N ha^{-1} a^{-1} (urea) and/or 10 kg P ha^{-1} a^{-1} (forest: NaH_2PO_4, pasture: rock phosphate). In the forest, Ca was applied in addition (10 kg Ca ha^{-1} a^{-1} as $CaCl_2$). At the pasture site, Ca is an important constituent of rock phosphate (27.3 kg ha^{-1} a^{-1}). At both sites, Cambisols were classified as the predominant soil type (WRB, 2006) developed from Palaeozoic phyllites, quartzites, and metasandstones (Makeschin et al. 2008).

22.2.1 NUMEX (Site and Methods)

A nutrient manipulation experiment (NUMEX) was conducted on 20 plots of 400 m^2 consisting of five treatments (+N, +P, +N+P, +Ca, control) in four replicates which were randomly assigned in natural montane forest stands at 2,020–2,120 m a.s.l. (see Chap. 23 for more details). The soils are covered by a thick organic layer with a mean C/N ratio of 21, C/S ratio of 356, and C/P ratio of 1,120. Soils are strongly acid, partly anaerobic (subsoil), and nutrient-poor with a low effective cation-exchange capacity (16–54 $mmol_c\ kg^{-1}$) and very low base metal concentrations in the organic layer and base saturation in the mineral soil (2.3–8.8 %) (Wullaert et al. 2010) (Chap. 23).

Zero-tension lysimeters ($n = 3$) were used to collect leachate from the organic layer termed "litter leachate" (Wullaert et al. 2010). Litter leachates were sampled fortnightly and bulked to a mixed sample per plot and collection date. After field collection, litter leachate pH (Sentix HWS, WTW GmbH, Weilheim, Germany) was immediately measured in an aliquot of each sample. Water samples and digests of the litterfall were analyzed for concentrations of Ca with flame atomic absorption (Varian AA240FS, Mulgrave, Australia), total dissolved N (water samples only), and total dissolved P using continuous flow analysis after acid digestion (CFA, Bran+Luebbe GmbH, Norderstedt, Germany) (Wullaert et al. 2010). Litterfall was collected monthly from six litterfall traps (0.60 m $\times$ 0.60 m) per plot and bulked to a composite sample per plot.

Furthermore, leaf samples from sun-exposed branches from four tree species (*Alchornea lojaensis* Secco, *Graffenrieda emarginata* (Ruiz & Pav.) Triana, *Hieronyma fendleri* Brig., and *Myrcia* sp. nov.) were collected to quantify changes in foliar nutrient concentrations 2 years after the onset of the NUMEX experiment. These tree species are the most representative for determination of nutrient uptake, since they frequently occur in the research area. Therefore sufficient replicate analyses were possible. For tree species selection in Chap. 23 other criteria had to be fulfilled enabling the measurement of increments. Leaves were dried at 60 °C to constant mass. The concentrations of total C and N of leaf and litterfall samples were determined with a CNS analyzer (vario EL III, Elementar-Analysensysteme, Hanau, Germany). The concentrations of total P were analyzed using an Inductively coupled Plasma-Optical Emission Spectrometer (Optima 5300DV ICP-OES, Perkin Elmer) after digesting the samples with concentrated HNO_3.

One year after establishment of the experiment, in spring 2009, litter (L/F material) of the upper 5 cm of the organic layers of the experimental plots was sampled. Before measurements, roots >2 mm were removed and the remaining material was chopped to pieces of <25 mm^2, homogenized, and pre-incubated at 20 °C for 5 days. Microbial biomass C was measured by substrate-induced respiration (MBC_{SIR}) (Anderson and Domsch 1978; Joergensen and Scheu 1999). Measurements were done using an automated O_2-microcompensation system (Scheu 1992). The mean of the three lowest hourly measurements within the first 10 h was taken as the maximum initial respiratory response (MIRR).

MBC_{SIR}(mg g^{-1}) was calculated as 38 × MIRR (μl O_2 g^{-1} soil dry weight h^{-1}) according to Beck et al. (1997). Lipid extraction for phospholipid fatty acid (PLFA) analysis followed Frostegard et al. (1991); for details see Krashevska et al. (2010). The fatty acids i15:0, a15:0, 15:0, i16:0, 16:1ω7, 17:0, i17:0, cy17:0, 18:1ω7, and cy19:0 were assumed to represent bacteria (Zelles 1999); 16:1ω5c was used as an indicator for arbuscular mycorrhizal fungi (AMF) (Olsson et al. 1995). The concentration of the fungal specific fatty acid 18:2ω6,9 was used as a fungal marker (Frostegard et al. 1996). The sum of 16:1ω7, cy17:0, and cy19:0 was used as marker for Gram-negative bacteria and the sum of i15:0, a15:0, i16:0, and i17:0 as marker for Gram-positive bacteria (Zelles 1999).

22.2.2 FERPAST (Site and Methods)

The pasture fertilization experiment (FERPAST) was conducted on a 17-year-old active pasture site. The site is dominated by the tropical grass *Setaria sphacelata* (Schum.) Stapf & C.E. Hubb. (98 %) and has been extensively grazed by dairy cattle (<1 cow per ha) (Potthast et al. 2012). The soil of the pasture site (0–20 cm) is moderately acid (pH 5.2) and characterized by a base saturation of 46 % and a low C/N and C/P ratio of 13.0 and 63.3, respectively. Due to these favorable soil conditions, the amount of microbial biomass is among the highest in tropical pasture soils (Potthast et al. 2011). In a fenced area (0.5 ha) six replicate plots (25 m^2 each) of control (X) and of each fertilizer treatment (+N, +P; +N+P) were randomly arranged as described for the NUMEX plots. Every 3 months grass biomass samples were collected from each plot. Afterwards, total plots were harvested simulating the system of rotational grazing used by the farmers (person. com.). The yield of total dry matter (40 °C) was determined. Grass biomass samples were analyzed for total N concentration by a CNS Analyzer (vario EL III/ elementar) and for total element concentrations by acid digestion and subsequent ICP-OES (CIROS, Spectro).

Soil samples were taken from mineral topsoil (0–5, 5–10, 10–20 cm) at each plot 1.5 years after starting the fertilization. Total C and N (TN) concentration (CNS Analyzer, vario EL III/elementar) as well as easily available ammonium (NH_4^+–N) and nitrate (NO_3^-–N) concentrations were determined (Skalar Analytik GmbH, Germany) (Potthast et al. 2012). Furthermore, a ^{15}N-isotope pool dilution method (Barraclough 1995; Mulvaney et al. 1997) was used to determine gross N mineralization and NH_4 consumption rates (Hamer et al. 2009). Concentrations of soil microbial biomass C (MBC) and N (MBN) were determined by 0.5 M K_2SO_4 fumigation–extraction (CFE) (1:5 soil:solution ratio) (Vance et al. 1987) applying conversion factors of $k_{EC} = 0.43$ and $k_{EN} = 0.45$. Soil microbial biomass P (MBP) was determined by NH_4F (0.03 M) and HCl (0.025 M) fumigation–extraction (1:5 soil:solution ratio, $k_{EP} = 0.4$) (Brookes et al. 1982; Chen and He 2004). To determine the structure of the microbial community, total phospholipid fatty acids ($PLFA_{tot}$) were extracted and analyzed according to Zelles (1995). A detailed description of the applied method is published in Hamer et al. (2007). To compare specific PLFA biomarkers of the forest organic layer (upper 5 cm) with those of the pasture topsoil (0–5 cm), the same assignment as for forest organic layer, which is described above, was used.

22.3 Results and Discussion

22.3.1 Nutritional Status of Selected Plant Species

The tree species in the NUMEX plots had foliar nutrient concentrations in the lower range of values known for tropical trees (Soethe et al. 2008). Foliar N/P ratios of the unfertilized trees ranged from 28 to 40, thus pointing to a limitation of tree growth mainly by P (Townsend et al. 2007). After 2 years of nutrient addition (in total 100 kg N ha^{-1} and/or 20 kg P ha^{-1}), three tree species (an exception was *Myrcia* sp. nov. after addition of P) showed slightly higher foliar concentrations of N and P after addition of the respective nutrient (Table 22.1). However, concentrations were only significantly increased by P addition in the two Euphorbiaceae *A. lojaensis* and *H. fendleri*. Furthermore, the latter species showed significantly higher foliar concentrations of N (after NP addition) and of Mg (after P addition). Fertilization with N alone increased N/P ratios in all species, but after addition of P or NP the N/P ratios decreased in three species and were lower than those in the control trees (again with the exception of *Myrcia* sp. nov. which did not respond to P addition). Calcium addition had no effect on the foliar nutrient concentrations of the studied species.

At the pasture site fertilization with rock phosphate significantly increased the P concentration in the pasture grass *S. sphacelata* within 2 years after starting the fertilization (Table 22.1). This increase was higher in the plots receiving P only than in plots receiving NP. The Ca concentration in the pasture grass showed the same trend as that of P; however, this increase was not statistically significant. There was no response of the N, Mg, and K concentrations in *S. sphacelata* to fertilization. The total grass biomass significantly increased in NP-fertilized plots (Chap. 26). Hence, *S. sphacelata* is co-limited by N and P. Thus, the data indicate that the response of plant species to nutrient addition is species specific. However, most species responded to P addition with an increase in P concentration of plant tissue.

22.3.2 Soil Chemical Properties

Until 1.5 year after the start of fertilizer application, the mean biweekly water, N, P, and Ca fluxes from the organic layer into the mineral soil were not significantly different among all treatments of NUMEX (Fig. 22.1). Thus, the applied nutrients

Table 22.1 Development of element concentrations (N, P, Ca, Mg, K) in tree (forest) and pasture grass leaves (*S. sphacelata*) after fertilization with urea (+N), NaH_2PO_4 (forest)/rock phosphate (pasture) (+P), or the combination of both +N+P compared to the unfertilized control (mean values, with SE in parenthesis

	Control		+N		+P		+N+P		+Ca	
N [mg g^{-1}]										
Alchornea lojaensis	11.92	(0.67)	13.07	(0.81)	13.37	(0.78)	12.47	(0.84)	12.94	(0.79)
Graffenrieda emarginata	10.08	(0.53)	11.37	(0.90)	9.82	(0.99)	10.54	(0.75)	9.92	(0.61)
Hieronyma fendleri	13.34	(0.19)[a]	14.53	(0.53)[ab]	13.53	(0.41)[a]	15.88	(0.48)[b]	13.48	(0.26)[a]
Myrcia sp. nov.	10.47	(0.60)	11.34	(0.27)	9.91	(0.41)	11.37	(0.27)	10.21	(0.25)
Setaria sphacelata	13.39	(0.41)	14.57	(0.69)	13.72	(0.32)	12.68	(0.43)	–	
P [mg g^{-1}]										
Alchornea lojaensis	0.30	(0.02)[a]	0.29	(0.02)[a]	0.39	(0.04)[ab]	0.52	(0.07)[b]	0.39	(0.03)[ab]
Graffenrieda emarginata	0.25	(0.03)	0.30	(0.02)	0.33	(0.05)	0.38	(0.07)	0.26	(0.03)
Hieronyma fendleri	0.48	(0.03)[a]	0.44	(0.02)[a]	0.68	(0.04)[b]	0.62	(0.03)[b]	0.46	(0.03)[a]
Myrcia sp. nov.	0.33	(0.03)	0.31	(0.02)	0.28	(0.02)	0.34	(0.02)	0.30	(0.02)
Setaria sphacelata	1.79	(0.10)[ab]	1.49	(0.06) [a]	2.51	(0.09)[c]	2.10	(0.07)[b]	–	
Ca [mg g^{-1}]										
Alchornea lojaensis	3.25	(0.47)	3.62	(0.38)	4.34	(0.45)	4.18	(0.39)	4.62	(0.48)
Graffenrieda emarginata	1.42	(0.09)	1.09	(0.16)	1.30	(0.15)	1.04	(0.09)	1.45	(0.18)
Hieronyma fendleri	1.68	(0.25)	1.35	(0.15)	2.15	(0.14)	1.75	(0.27)	1.87	(0.22)
Myrcia sp. nov.	0.82	(0.13)	0.77	(0.07)	0.76	(0.08)	0.70	(0.10)	1.00	(0.11)
Setaria sphacelata	4.32	(0.21)	4.17	(0.24)	4.91	(0.19)	4.61	(0.33)	–	
Mg [mg g^{-1}]										
Alchornea lojaensis	1.65	(0.10)	2.21	(0.33)	2.30	(0.06)	2.18	(0.17)	2.47	(0.11)
Graffenrieda emarginata	2.75	(0.19)	2.07	(0.23)	2.69	(0.26)	2.06	(0.27)	2.20	(0.17)
Hieronyma fendleri	1.60	(0.14)[ab]	1.46	(0.10)[a]	1.90	(0.11)[b]	1.75	(0.08)[ab]	1.60	(0.04)[ab]
Myrcia sp. nov.	0.88	(0.14)	1.09	(0.21)	1.03	(0.08)	1.06	(0.11)	1.12	(0.10)
Setaria sphacelata	2.45	(0.23)	2.63	(0.19)	2.45	(0.17)	2.51	(0.26)	–	
K [mg g^{-1}]										
Alchornea lojaensis	2.69	(0.09)	2.53	(0.15)	2.48	(0.16)	2.59	(0.31)	2.30	(0.15)

(continued)

Table 22.1 (continued)

	Control		+N		+P		+N+P		+Ca	
Graffenrieda emarginata	5.48	(0.49)	4.28	(0.44)	4.76	(0.59)	4.42	(0.54)	4.14	(0.54)
Hieronyma fendleri	4.79	(0.95)	4.74	(0.43)	4.07	(0.56)	4.12	(0.20)	3.59	(0.41)
Myrcia sp. nov.	2.64	(0.49)	2.78	(0.33)	2.64	(0.21)	2.71	(0.28)	2.88	(0.40)
Setaria sphacelata	31.97	(1.23)	28.09	(1.60)	32.07	(0.95)	31.03	(1.50)	–	

Different letters in each of the rows indicate significant differences at $p < 0.05$, Tukey Test). Nutrient concentrations of tree leaves after 2 years of fertilization collected in January 2010, $n = 4–6$. Nutrient concentrations of the pasture grass leaves are weighted means of the values measured between January 2009 and February 2010 and take the duration of each particular growing period into account, $n = 36$

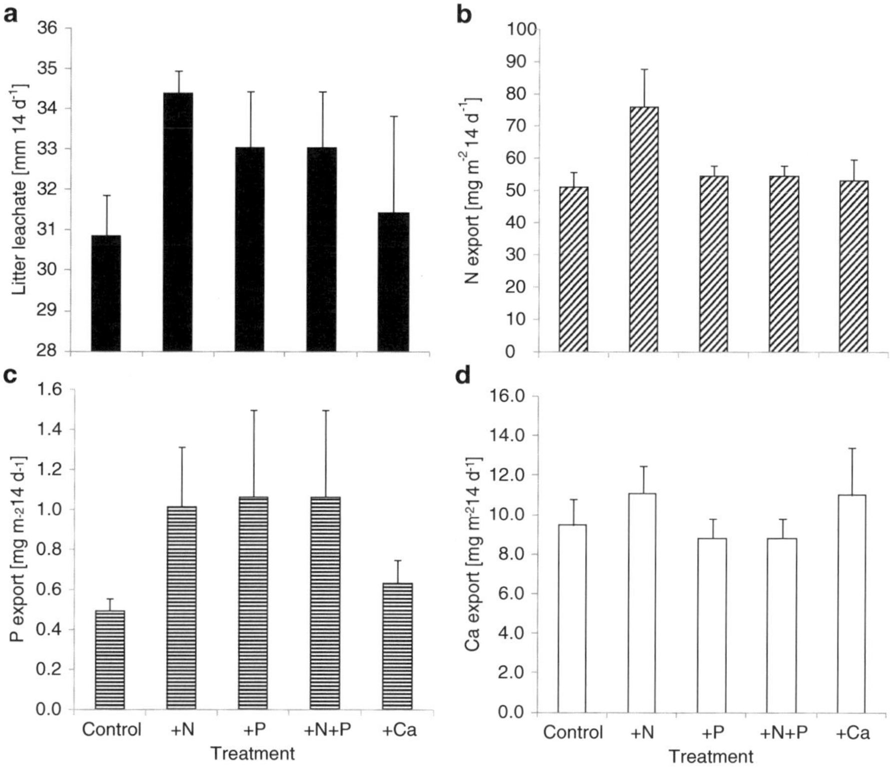

Fig. 22.1 Mean biweekly (**a**) water fluxes, (**b**) N export, (**c**) P export, and (**d**) Ca export from the organic layer into the mineral soil of the four treatments and the control of NUMEX. *Error bars* represent standard errors. Means have been tested for significant differences with Tukey's HSD test and there were no significant differences

did not increase nutrient export from the organic layer which harbors almost all roots (Soethe 2006; own observation at the NUMEX site). This is particularly surprising for Ca because of the strongly acid soil reaction which usually favors Ca leaching. Our findings concerning N are in contrast to reports from other tropical montane forest sites where N fertilizer usually resulted rapidly in increased N concentrations in soil solutions such as in Hawaii (Lohse and Matson 2005), Panama (Corre et al. 2010), and Puerto Rico (Cusack et al. 2011). Cumulative losses calculated as difference between the nutrient addition treatments and the control accounted for indicated that <10 % (N), <1 % (P), and <4 % of the added nutrients left the organic layer with litter leachate. Furthermore, N_2O–N losses were small (on average <1.5 kg N_2O–N ha^{-1} $year^{-1}$ until 2009 and only responded positively to N addition in the second year after the start of fertilizer application (Martinson 2011). This suggests that the added nutrients were (1) taken up by the vegetation, (2) immobilized by soil microorganisms, e.g., because of the wide C to nutrient ratios, and (3) in the case of P potentially also retained by precipitation as

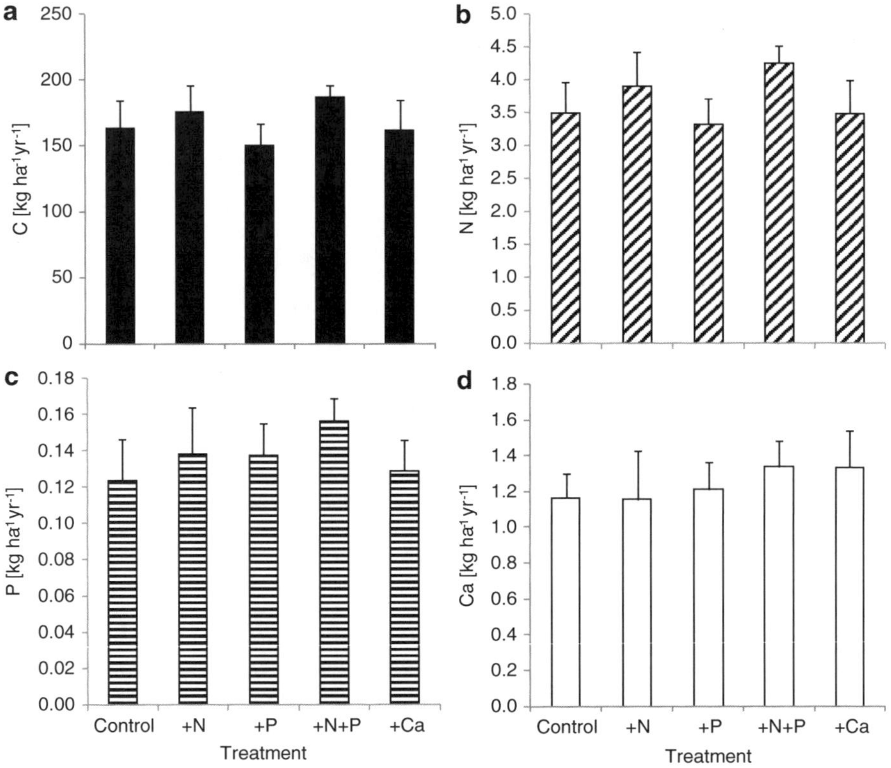

Fig. 22.2 Mean monthly (**a**) C, (**b**) N, (**c**) P, and (**d**) Ca fluxes with litterfall to the soil of the four treatments and the control of NUMEX. *Error bars* represent standard errors. Means have been tested for significant differences with Tukey's HSD test and there were no significant differences

Al phosphate or by sorption to Fe oxides, although the mineral soil contribution to the organic layer was on average <15 mass %.

The assumption that part of the nutrient retention in the vegetation is explained by plant uptake is supported by the fact that nutrient fluxes with litterfall of N increased in the +N and +N+P treatments, those of P in the +N+P treatment, and those of Ca in the +N+P and +Ca treatments, although none of the differences were significant (Fig. 22.2). In the +N and +N+P treatments, litterfall returned 11 $\pm$ 13 (SD) and 15 $\pm$ 10 % of the applied N to the soil, respectively, in the +P and +N+P 1.3 $\pm$ 2.4 and 3.2 $\pm$ 2.2 % of the applied P, respectively, and in the +Ca treatment 20–27 % of the applied Ca (Wullaert 2010). Interestingly, the application of N and P together increased the Ca flux with litterfall similarly as the application of Ca, suggesting that the increased uptake of N and P also increased the requirement of other nutrients. In a secondary Mexican forest ecosystem, Campo et al. (2007) also observed a fast appearance of fertilized N in litterfall while higher P concentrations in litterfall were also reported from a tropical rain forest in Borneo

(Mirmanto et al. 1999) and from a P-limited forest in Hawaii (Harrington et al. 2001) where the addition of P and N+P also resulted in increased P fluxes.

At the pasture site, there is indirect evidence that fertilizer N and P also remain to a large extent in the above- and belowground plant–soil system as shown for NUMEX in the forest. According to the difference method by Syers et al. (2008) a maximum of 43 % of yearly added N and 69 % of yearly added P was stored in the aboveground biomass of *S. sphacelata* (Potthast et al. 2012). Nutrient losses by leaching were assumed to be small since, generally, streams of catchments dominated by extensive pastures in the study area had lower nitrate concentrations and annual nitrate export rates than the forested streams (Bücker et al. 2011). Assuming comparable gaseous N-losses of about 14 % (N_2O–N + NO–N) as in a pasture fertilization experiment with urea in the humid tropics of Costa Rica (Veldkamp et al. 1998), a notable proportion of N must be stored in the fine-root biomass as well as adsorbed by soil organic matter (SOM) or clay minerals. Under pasture management conditions with regular cattle grazing between 75 and 90 % of ingested nutrients are excreted in form of urine and feces, and are returned to the soil system (Dao and Schwartz 2011). However, it has to be taken into account that these animal excreta are not distributed homogeneously (Dias-Filho et al. 2001) leading to a small-scale spatial heterogeneity of soil nutrient availability at the pasture site.

22.3.3 Soil Microbial Properties

The assumption that part of the added nutrients to the forest is immobilized by soil microorganisms is supported only in the +N+P treatments. Here, microbial biomass C as measured by SIR reached a maximum. However, microbial biomass was reduced by the addition of N (Fig. 22.3a). This decrease is also supported by PLFA analysis. The biomass of each fungal, Gram(−), and Gram(+) bacterial PLFAs were at a minimum in the +N treatment (Fig. 22.3b). Notably, this decline was negated in the +N+P treatment, suggesting that P counteracted the detrimental effect of N. Notably, the reduction was more pronounced in bacteria than in fungi, resulting in a relative increase in the fungal-to-bacterial ratio in the N addition treatment. PCA of the response pattern of PLFAs confirmed that fungi were associated with the N addition treatment, but it also reflected that Gram-negative and Gram-positive bacteria were mainly associated with P and NP addition (Fig. 22.4).

Pasture fertilization (+N; +P; +N+P) neither stimulated microbial growth (Fig. 22.5a) nor induced an additional uptake of nutrients by soil microorganisms (Fig. 22.5b, c). Thus, soil microorganisms of the active pasture soil were neither limited by N nor P (Potthast et al. 2012). Soil pH tended to slightly decrease in the +N-treatment compared with the other treatments (Fig. 22.5d). However, these differences did not influence the microbial biomass or activity, as also shown in

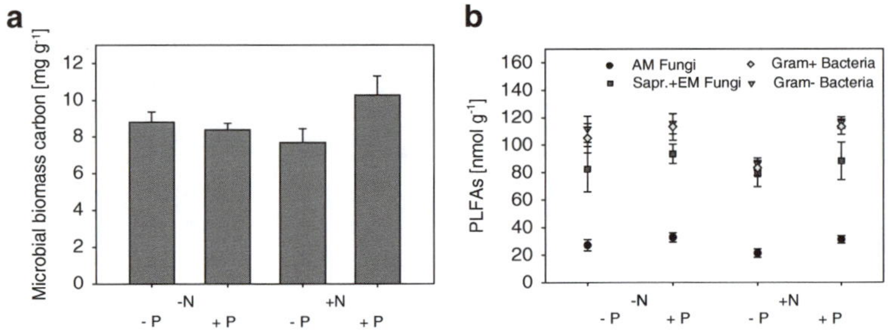

Fig. 22.3 Impact of fertilization on the amount of soil microbial biomass carbon (MBC_{SIR}) (**a**) and on specific PLFA biomarkers (Gram (+), Gram (−), saprotrophic + ectomycorrhizal (EM) fungi, and arbuscular mycorrhizal (AM) fungi) (**b**) of the forest organic layer (upper 0–5 cm). Means with SE

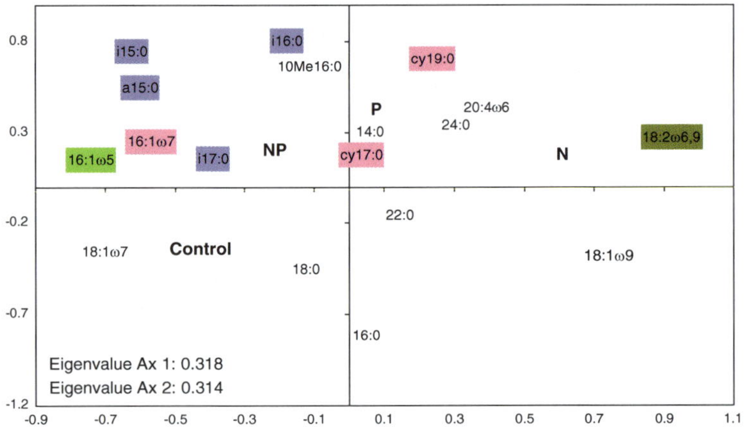

Fig. 22.4 Principal Components Analysis of phospholipid fatty acid (PLFA) profiles [mol%] in litter material of the forest (NUMEX) including the control and fertilizer treatments (+N, +P, +N+P). Treatments were included as passive variables. Fungal marker PLFA 18:2ω6,9 is highlighted in *dark green*, marker PLFA for arbuscular mycorrhiza in light *green*, marker PLFA for Gram-positive bacteria in *lilac*, and for Gram-negative bacteria in *pink*

Fig. 22.5e, f by similar gross N mineralization and NH_4-consumption rates. In contrast, the relative and absolute abundance of the saprotrophic fungal PLFA biomarker 18:2ω6,9 increased in the order: control < +N < +N+P < +P and primarily responded to P addition. In old-growth tropical forests in China, which were N saturated because of long-term N deposition, the soil microbial community markedly responded to P fertilization. Total microbial biomass and especially fungal PLFA marker as well as soil respiration increased (Liu et al. 2012), primarily suggesting a P-limitation for soil microbes. In contrast, P fertilization of pine forests and mixed forests, both N limited, did not affect soil microorganisms (pine) or only increased the fungal/bacterial ratio and soil respiration (mixed) (Liu et al. 2012).

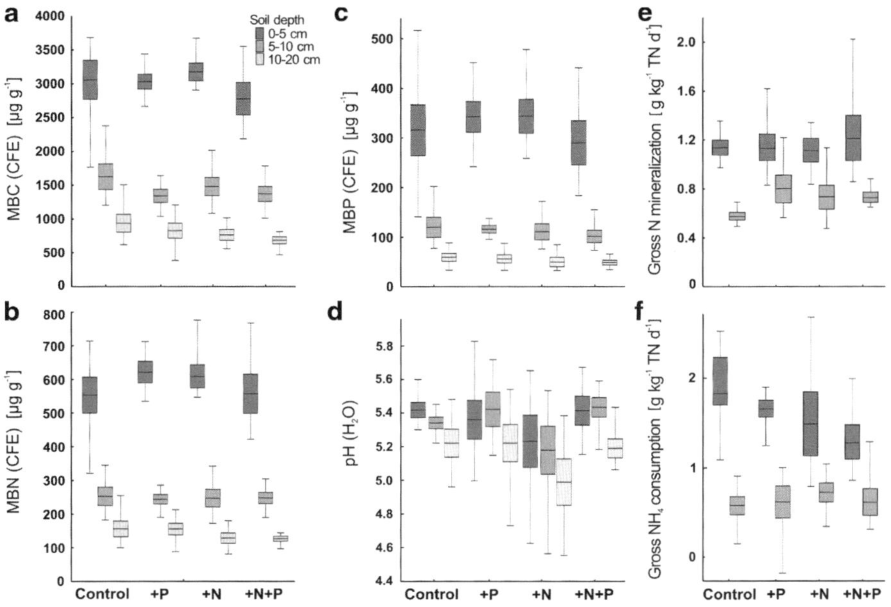

Fig. 22.5 Comparison of biochemical soil properties between control and different fertilizer treatments (+N, +P, +N+P) depending on mineral soil depth (0–5, 5–10, 10–20 cm), 1.5 years after the start of the FERPAST experiment (mean, box = SE, whisker = min/max, $n = 6$), data from (Potthast et al. 2012). Microbial biomass C (**a**), microbial biomass N (**b**), and microbial biomass P (**c**) were determined by chloroform–fumigation extraction (CFE), soil pH (**d**) at a soil: water ratio of 1:2.5, and gross N mineralization (**e**) and gross NH_4 consumption (**f**) using the ^{15}N isotope dilution method

In the study area, microbes of the forest litter layer were co-limited by N and P in contrast to the pasture topsoil. The microbial community structure was affected on both sites by fertilization. At the forest site N-addition only decreased the absolute abundance of Gram(+) and Gram(−) bacteria and thus increased the relative abundance of fungi whereas at the pasture site the abundance of fungi was significantly influenced by P addition (Potthast et al. 2012).

22.4 Conclusions

There were fast but statistically not significant responses of the forest to the nutrient addition 1.5 years after the start of nutrient application. We attribute the current lack of statistical significance to the high heterogeneity of soil conditions in the study area (Wilcke et al. 2002). Nevertheless, our measurements of the nutrient cycling in response to +N, +P, +N+P, and +Ca additions to the old-growth tropical montane forest revealed that most of the added nutrients remained in the system after a cumulative application of 75 kg ha^{-1} N, 15 kg ha^{-1} P, and 15 kg ha^{-1} Ca.

Soon after application the fertilized elements partly showed up in leaves and litterfall, indicating that the plants took up the added nutrients. There were only for the NP treatment indications for microbial immobilization of the added nutrients. At the pasture site a significant uptake of N and P by the grass *S. sphacelata* and a concomitant increase of grass biomass after +N+P addition revealed a co-limitation. Since pasture fertilization neither induced an additional uptake of nutrients by soil microorganisms nor stimulated microbial growth or microbial mediated N mineralization rates, soil microbes were not N and/or P limited. Distinct effects after fertilization on soil microbial community structure were found in both ecosystems with, e.g., N input negatively affecting microorganisms in forest organic layer, in particular bacteria. Increasing remote fertilization in the future (cf. Chap. 21, Fig. 21.3) is expected to eutrophicate the forest with detrimental effects on biodiversity above- and belowground while it might be beneficial for pasture productivity. Thus, future ecosystem services in forest and pasture ecosystems will vary greatly. The fact that all added mineral nutrients were retained simultaneously in the ecosystem supports the view of Kaspari et al. (2008) that multiple and temporally variable nutrient limitations can coexist in tropical ecosystems. Results will become clearer in the near future, since resamples will be taken at later stages of the experiments.

References

Anderson JPE, Domsch KH (1978) A physiological method for the quantitative measurement of microbial biomass in soils. Soil Biol Biochem 10:215–221

Bardgett RD, Mawdsley JL, Edwards S, Hobbs PJ, Rodwell JS, Davies WJ (1999) Plant species and nitrogen effects on soil biological properties of temperate upland grasslands. Funct Ecol 13:650–660

Barraclough D (1995) ^{15}N isotope dilution techniques to study soil nitrogen transformations and plant uptake. Fert Res 42:185–192

Beck T, Joergensen RG, Kandeler E, Makeschin F, Nuss E, Oberholzer HR, Scheu S (1997) An inter-laboratory comparison of ten different ways of measuring soil microbial biomass C. Soil Biol Biochem 29:1023–1032

Boy J, Wilcke W (2008) Tropical Andean forest derives calcium and magnesium from Saharan dust. Global Biogeochem Cycles 22(1), GB1027

Boy J, Rollenbeck R, Valarezo C, Wilcke W (2008) Amazonian biomass burning-derived acid and nutrient deposition in the north Andean montane forest of Ecuador. Global Biogeochem Cycles 22, GB4011

Brookes PC, Powlson DS, Jenkinson DS (1982) Measurement of microbial biomass phosphorus in soil. Soil Biol Biochem 14:319–329

Bücker A, Crespo P, Frede H-G, Breuer L (2011) Solute behaviour and export rates in neotropical montane catchments under different land-uses. J Trop Ecol 27:305–317

Campo J, SolÃ-s E, Valencia MG (2007) Litter N and P dynamics in two secondary tropical dry forests after relaxation of nutrient availability constraints. For Ecol Manage 252:33–40

Chen GC, He ZL (2004) Determination of soil microbial biomass phosphorus in acid red soils from southern China. Biol Fertil Soils 39:446–451

Corre MD, Veldkamp E, Arnold J, Joseph Wright S (2010) Impact of elevated N input on soil N cycling and losses in old-growth lowland and montane forests in Panama. Ecology 91:1715–1729

Cusack DF, Silver WL, Torn MS, McDowell WH (2011) Effects of nitrogen additions on above- and belowground carbon dynamics in two tropical forests. Biogeochemistry 104:203–225

Dao TH, Schwartz RC (2011) Effects of manure management on phosphorus biotransformations and losses during animal production. In: Bünemann E, Oberson A, Frossard E (eds) Phosphorus in action: biological processes in soil phosphorus cycling. Springer, Berlin, pp 407–429

de Vries FT, Hoffland E, van Eekeren N, Brussaard L, Bloem J (2006) Fungal/bacterial ratios in grasslands with contrasting nitrogen management. Soil Biol Biochem 38:2092–2103

Denef K, Roobroeck D, Manimel Wadu MCW, Lootens P, Boeckx P (2009) Microbial community composition and rhizodeposit-carbon assimilation in differently managed temperate grassland soils. Soil Biol Biochem 41:144–153

Dias-Filho MB, Davidson EA, de Carvalho CJR (2001) Linking biogeochemical cycles to cattle pasture management and sustainability in the Amazon Basin. In: McClain ME, Victoria RL, Richey JE (eds) The biogeochemistry of the Amazon basin. Oxford University Press, New York, pp 84–105

Edwards IP, Zak DR, Kellner H, Eisenlord SD, Pregitzer KS (2011) Simulated atmospheric N deposition alters fungal community composition and suppresses ligninolytic gene expression in a northern hardwood forest. PLoS One 6(6):e20421

Frostegard A, Tunlid A, Bååth E (1991) Microbial biomass measured as total lipid phosphate in soils of different organic content. J Microbiol Methods 14:151–163

Frostegard A, Tunlid A, Bååth E (1996) Changes in microbial community structure during long-term incubation in two soils experimentally contaminated with metals. Soil Biol Biochem 28:55–63

Galloway JN, Dentener FJ, Capone DG, Boyer EW, Howarth RW, Seitzinger SP, Asner GP, Cleveland CC, Green PA, Holland EA, Karl DM, Michaels AF, Porter JH, Townsend AR, Vorosmarty CJ (2004) Nitrogen cycles: past, present, and future. Biogeochemistry 70:153–226

Grayston SJ, Campbell CD, Bardgett RD, Mawdsley JL, Clegg CD, Ritz K, Griffiths BS, Rodwell JS, Edwards SJ, Davies WJ, Elston DJ, Millard P (2004) Assessing shifts in microbial community structure across a range of grasslands of differing management intensity using CLPP, PLFA and community DNA techniques. Appl Soil Ecol 25:63–84

Hamer U, Unger M, Makeschin F (2007) Impact of air-drying and rewetting on PLFA profiles of soil microbial communities. J Plant Nutr Soil Sci 170:259–264

Hamer U, Makeschin F, An S, Zheng F (2009) Microbial activity and community structure in degraded soils on the Loess Plateau of China. J Plant Nutr Soil Sci 172:118–126

Harpole WS, Ngai JT, Cleland EE, Seabloom EW, Borer ET, Bracken MES, Elser JJ, Gruner DS, Hillebrand H, Shurin JB, Smith JE (2011) Nutrient co-limitation of primary producer communities. Ecol Lett 14:852–862

Harrington RA, Fownes JH, Vitousek PM (2001) Production and resource use efficiencies in N- and P-limited tropical forests: A comparison of responses to long-term fertilization. Ecosystems 4:646–657

Hietz P, Turner BL, Wanek W, Richter A, Nock CA, Wright SJ (2011) Long-term change in the nitrogen cycle of tropical forests. Science 334:664–666

Hunter MD (2001) Multiple approaches to estimating the relative importance of top-down and bottom-up forces on insect populations: Experiments, life tables, and time-series analysis. Basic Appl Ecol 2:295–309

Joergensen RG, Scheu S (1999) Response of soil microorganisms to the addition of carbon, nitrogen and phosphorus in a forest Rendzina. Soil Biol Biochem 31:859–866

Kaspari M, Yanoviak SP (2008) Biogeography of litter depth in tropical forests: evaluating the phosphorus growth rate hypothesis. Funct Ecol 22:919–923

Kaspari M, Garcia MN, Harms KE, Santana M, Wright SJ, Yavitt JB (2008) Multiple nutrients limit litterfall and decomposition in a tropical forest. Ecol Lett 11:35–43

Krashevska V, Maraun M, Ruess L, Scheu S (2010) Carbon and nutrient limitation of soil microorganisms and microbial grazers in a tropical montane rain forest. Oikos 119:1020–1028
Lamarque JF, Kiehl JT, Brasseur GP, Butler T, Cameron-Smith P, Collins WD, Collins WJ, Granier C, Hauglustaine D, Hess PG, Holland EA, Horowitz L, Lawrence MG, McKenna D, Merilees P, Prather MJ, Rasch PJ, Rotman D, Shindell D, Thornton P (2005) Assessing future nitrogen deposition and carbon cycle feedback using a multimodel approach: analysis of nitrogen deposition. J Geophys Res Atmos 110, D19303
Likens GE, Bormann FH (1995) Biogeochemistry of a forested ecosystem. Springer, New York
Liu L, Gundersen P, Zhang T, Mo J (2012) Effects of phosphorus addition on soil microbial biomass and community composition in three forest types in tropical China. Soil Biol Biochem 44:31–38
Lohse KA, Matson P (2005) Consequences of nitrogen additions for soil losses from wet tropical forests. Ecol Appl 15:1629–1648
Makeschin F, Haubrich F, Abiy M, Burneo JI, Klinger T (2008) Pasture management and natural soil regeneration. In: Beck E, Bendix J, Kottke I, Makeschin F, Mosandl R (eds) Gradients in a tropical mountain ecosystem of Ecuador. Springer, Berlin, pp 397–408
Martinson GO (2011) Trace gas fluxes from tropical montane forests of southern Ecuador. Doctoral thesis, Georg-August University Göttingen, Göttingen
Matson PA, McDowell WH, Townsend AR, Vitousek PM (1999) The globalization of N deposition: ecosystem consequences in tropical environments. Biogeochemistry 46:67–83
Mirmanto E, Proctor J, Green J, Nagy L, Suriantata (1999) Effects of nitrogen and phosphorus fertilization in a lowland evergreen rainforest. Philos Trans R Soc Lond B Biol Sci 354:1825–1829
Mulvaney RL, Khan SA, Stevens WB, Mulvaney CS (1997) Improved diffusion methods for determination of inorganic nitrogen in soil extracts and water. Biol Fertil Soils 24:413–420
Olsson PA, Baath E, Jakobsen I, Soderstrom B (1995) The use of phospholipid and neutral lipid fatty acids to estimate biomass of arbuscular mycorrhizal fungi in soil. Mycol Res 99:623–629
Phoenix GK, Hicks WK, Cinderby S, Kuylenstierna JCI, Stock WD, Dentener FJ, Giller KE, Austin AT, Lefroy RDB, Gimeno BS, Ashmore MR, Ineson P (2006) Atmospheric nitrogen deposition in world biodiversity hotspots: the need for a greater global perspective in assessing N deposition impacts. Global Change Biol 12:470–476
Potthast K, Hamer U, Makeschin F (2011) Land-use change in a tropical mountain rainforest region of southern Ecuador affects soil microorganisms and nutrient cycling. Biogeochemistry 111(1-3):151–167. doi:10.1007/s10533-011-9626-7
Potthast K, Hamer U, Makeschin F (2012) In an Ecuadorian pasture soil the growth of *Setaria sphacelata*, but not of soil microorganisms, is co-limited by N and P. Appl Soil Ecol 62:103–114. doi:10.1016/j.apsoil.2012.08.003
Reed SC, Vitousek PM, Cleveland CC (2011) Are patterns in nutrient limitation belowground consistent with those aboveground: results from a 4 million year chronosequence. Biogeochemistry 106:323–336
Rousk J, Baath E (2007) Fungal and bacterial growth in soil with plant materials of different C/N ratios. FEMS Microbiol Ecol 62:258–267
Scheu S (1992) Automated measurement of the respiratory response of soil microcompartments: active microbial biomass in earthworm faeces. Soil Biol Biochem 24:1113–1118
Schwarz MT, Oelmann Y, Wilcke W (2011) Stable N isotope composition of nitrate reflects N transformations during the passage of water through a montane rain forest in Ecuador. Biogeochemistry 102:195–208
Soethe N (2006) Structure and function of root systems at different altitudes of a south Ecuadorian montane forest. Faculty of Agriculture and Horticulture. Humboldt-Universität zu Berlin, Berlin pp 146
Soethe N, Lehmann J, Engels C (2008) Nutrient availability at different altitudes in a tropical montane forest in Ecuador. J Trop Ecol 24:397–406

Syers JK, Johnston AE, Curtin D (2008) Efficiency of soil and fertilizer phosphorus use. Food and Agriculture Organization of the United Nations, Rome

Townsend AR, Cleveland CC, Asner GP, Bustamante MMC (2007) Controls over foliar N:P ratios in tropical rain forests. Ecology 88:107–118

Ulrich B (1983) Interaction of forest canopies with atmospheric constituents: SO_2, alkali and earth alkali cations and chloride. Effects of accumulation of air pollutants in forest ecosystems. Reidel, Dordrecht, Göttingen, pp 33–45

Vance ED, Brookes PC, Jenkinson DS (1987) An extraction method for measuring soil microbial biomass C. Soil Biol Biochem 19:703–707

Veldkamp E, Keller M, Nunez M (1998) Effects of pasture management on N2O and NO emissions from soils in the humid tropics of Costa Rica. Global Biogeochem Cycles 12:71–79

Wilcke W, Yasin S, Abramowski U, Valarezo C, Zech W (2002) Nutrient storage and turnover in organic layers under tropical montane rain forest in Ecuador. Eur J Soil Sci 53:15–27

WRB IWG (2006) World reference base for soil resources 2006—a framework for international classification, correlation and communication. In: World soil resources reports. Food and Agriculture Organization of the United Nations, Rome, p 128

Wullaert H (2010) Response of nutrient cycles of an old-growth montane forest in Ecuador to experimental low-level nutrient amendments. Doctoral thesis, Johannes Gutenberg University Mainz

Wullaert H, Homeier J, Valarezo C, Wilcke W (2010) Response of the N and P cycles of an old-growth montane forest in Ecuador to experimental low-level N and P amendments. For Ecol Manage 260:1434–1445

Zelles L (1995) Fatty acid patterns of microbial phospholipids and lipopolysaccharides. In: Schinner F, Öhlinger R, Kandeler E, Margesin R (eds) Methods in soil biology. Springer, Berlin, pp 80–93

Zelles L (1999) Fatty acid patterns of phospholipids and lipopolysaccharides in the characterisation of microbial communities in soil: a review. Biol Fertil Soils 29:111–129

Chapter 23
Effects of Nutrient Addition on the Productivity of Montane Forests and Implications for the Carbon Cycle

Jürgen Homeier, Christoph Leuschner, Achim Bräuning, Nixon L. Cumbicus, Dietrich Hertel, Guntars O. Martinson, Susanne Spannl, and Edzo Veldkamp

23.1 Introduction

Nitrogen (N) and phosphorous (P) are key nutrient elements for plant growth. The ecological cycles of these elements that were previously free of anthropogenic disturbances have recently been severely affected by human activities. As a consequence, P and N in particular are often more readily available for plant growth than in "undisturbed" ecosystems, and a better understanding of how net primary production is affected by higher nutrient availability is urgently needed (Elser et al. 2007; Gruber and Galloway 2008; Xia and Wan 2008). The results of recent meta-analyses show that N is limiting for plant productivity in most terrestrial ecosystems of the world, but limitation by P can be equally important. Moreover, the addition of both N and P to ecosystems results in strongly positive synergistic reactions (Elser et al. 2007; LeBauer and Treseder 2008; Xia and Wan 2008).

J. Homeier (✉) • C. Leuschner • D. Hertel
Albrecht von Haller Institute of Plant Sciences, University of Göttingen, Untere Karspüle 2, 37073 Göttingen, Germany
e-mail: jhomeie@gwdg.de

A. Bräuning • S. Spannl
Institute of Geography, University of Erlangen, Kochstraße 4/4, 91054 Erlangen, Germany

N.L. Cumbicus
Institute of Ecology, Universidad Tecnica Particular de Loja, San Cayetano Alto, Loja, Ecuador

G.O. Martinson
Max Planck Institute for Terrestrial Microbiology, Karl-von-Frisch-Straße 10, 35043 Marburg, Germany

E. Veldkamp
Soil Science of Tropical and Subtropical Ecosystems, Büsgen Institute, University of Göttingen, Büsgenweg 2, 37077 Göttingen, Germany

J. Bendix et al. (eds.), *Ecosystem Services, Biodiversity and Environmental Change in a Tropical Mountain Ecosystem of South Ecuador*, Ecological Studies 221,
DOI 10.1007/978-3-642-38137-9_23,

Nutrient manipulation experiments (NUMEXs) carried out in tropical forests have in most cases shown that traits related to aboveground productivity such as litter production or stem growth are positively influenced by N and P addition (Tanner et al. 1998). In contrast, no clear picture has emerged with respect to the response of belowground processes to fertilization. Some studies found a slowing down of belowground carbon (C) turnover after N and P addition (e.g., Giardina et al. 2004), while others noted enhanced soil C sequestration (Cusack et al. 2010; Li et al. 2006). Further studies revealed large soil CO_2 losses upon P addition (Cleveland and Townsend 2006), a slowing down of belowground C turnover upon N and P addition (Giardina et al. 2004) and an increase in soil carbon stocks upon N and P addition (Gamboa et al. 2010). Increased deposition of NH_4^+ may also increase soil acidification due to nitrification and nitrate leaching, which possibly could reduce forest productivity (Lewis et al. 2004).

These manifold effects point to complex interactions between soil nutrient availability and C turnover in tropical forests and highlight the incompleteness of our understanding of how the tropical C cycle will respond to changes in nutrient availability (Cleveland and Townsend 2006; Cleveland et al. 2006; Wright 2005). One scenario resulting from increasing N (and P) deposition in tropical forests is that the system is driven toward N saturation, as has been observed in a number of temperate forest ecosystems (e.g., Aber et al. 1998). Comprehensive multidiscipline studies on physiological and biogeochemical responses to continuous high N input have been conducted in several temperate and boreal forests (e.g., Högberg et al. 2006; Magill et al. 2004), but are virtually lacking for tropical forests. A recent study in a Panamanian lowland forest showed that N addition may increase soil N losses through both enhanced nitrate leaching and N-oxide emission where the soils are relatively fertile (Corre et al. 2010). However, the nutrients were added in high amounts in most of these experiments ($\geq$100 kg N ha^{-1} and/ or 50 kg P ha^{-1}). The reported responses accordingly allow no reliable prediction as to possible long-term effects of the increases in atmospheric N and P deposition expected for tropical forests, which would represent a more moderate input distributed over a longer time period.

This chapter presents initial results of the ongoing NUMEX experiment that is being conducted in southern Ecuador to improve our understanding of the effects of continuous moderate N and P addition to tropical montane forest ecosystems. Both carbon storage and sequestration are major ecosystem services provided by forests (regulating services; see Chap. 4), and the NUMEX study aims to identify the underlying mechanisms for the variation of these services as affected by future changes in nutrient availability.

23.2 Material and Methods

23.2.1 The Ecuadorian Nutrient Manipulation Experiment

A full-factorial nutrient addition experiment was conducted in old-growth montane forest stands in southern Ecuador to examine the effects of moderate addition of N and P on forest productivity and biogeochemical cycles. NUMEX was set up at three different elevations on the eastern slope of the Andes. Two of the sites are in the Podocarpus National Park (1,000 m site: 990–1,100 m a.s.l., S 4°7′ W 78°58′ and 3,000 m site: 2,900–3,050 m a.s.l., S 4°7′ W 79°11′) and one is in the San Francisco Reserve (2,000 m site: 2,020–2,120 m a.s.l., S 3°58′ W 79°04′) close to Loja (Fig. 23.1). All three study sites are located in protected forest areas.

The research area has a tropical humid climate, and the mean annual precipitation increases along the altitudinal gradient (see Chap. 1). Rainfall shows little seasonality, and there are no regularly occurring dry periods (Emck 2007).

The annual bulk N deposition from precipitation ranged from 9.5 to 10 kg N ha^{-1} during the period of 1998–2003 (Boy et al. 2008), but recent data based on the monitoring of bulk and dry deposition between 1998 and 2010 (Wilcke et al. unpublished data) indicate higher annual depositions of 14–45 kg N ha^{-1}, as well as 0.4–4.9 kg P ha^{-1}.

Paleozoic metamorphic schist and sandstone with some quartz veins form the parent material for soil development at the 2,000 m and 3,000 m sites, whereas the soil parent material at the 1,000 m site consists of deeply weathered granodioritic rock of the Jurassic Zamora granitoide formation (Wolf et al. 2011). Thick organic soil layers (Oi + Oe + Oa) are found at the 2,000 m and 3,000 m sites, whereas the mineral soil at the 1,000 m site is only covered by a thin organic litter layer (Oi). A description of the soil characteristics and forest structures is given in Table 23.1.

The three study sites harbor three different forest types (Homeier et al. 2008). At 1,000 m, in the transition zone between the tropical lowland and the lower montane forest, an evergreen premontane forest reaches up to 40 m in height. Common tree families of this forest type are Fabaceae, Melastomataceae, Moraceae, Myristicaceae, Rubiaceae, and Sapotaceae. The evergreen lower montane forest at 2,000 m reaches a canopy height of 18–22 m. Characteristic tree families are Euphorbiaceae, Lauraceae, Melastomataceae, and Rubiaceae. The upper montane forest at 3,000 m reaches up to the tree line and canopy height is rarely higher than 8–10 m. Dominant tree families are Aquifoliaceae, Clusiaceae, Cunoniaceae, Lauraceae, and Melastomataceae. There is a complete tree species turnover from 1,000 m to 3,000 m, and less than five species are shared between the study plots at 2,000 m and either of the two other sites.

At each elevation level we established 16 plots of 400 m^2 each (20 m × 20 m). These accommodated four treatments (N, P, N + P, control) in four replicates each. One replicate of each treatment was arranged randomly in one of four blocks (Fig. 23.1). The plots were installed in old-growth, closed-canopy stands that

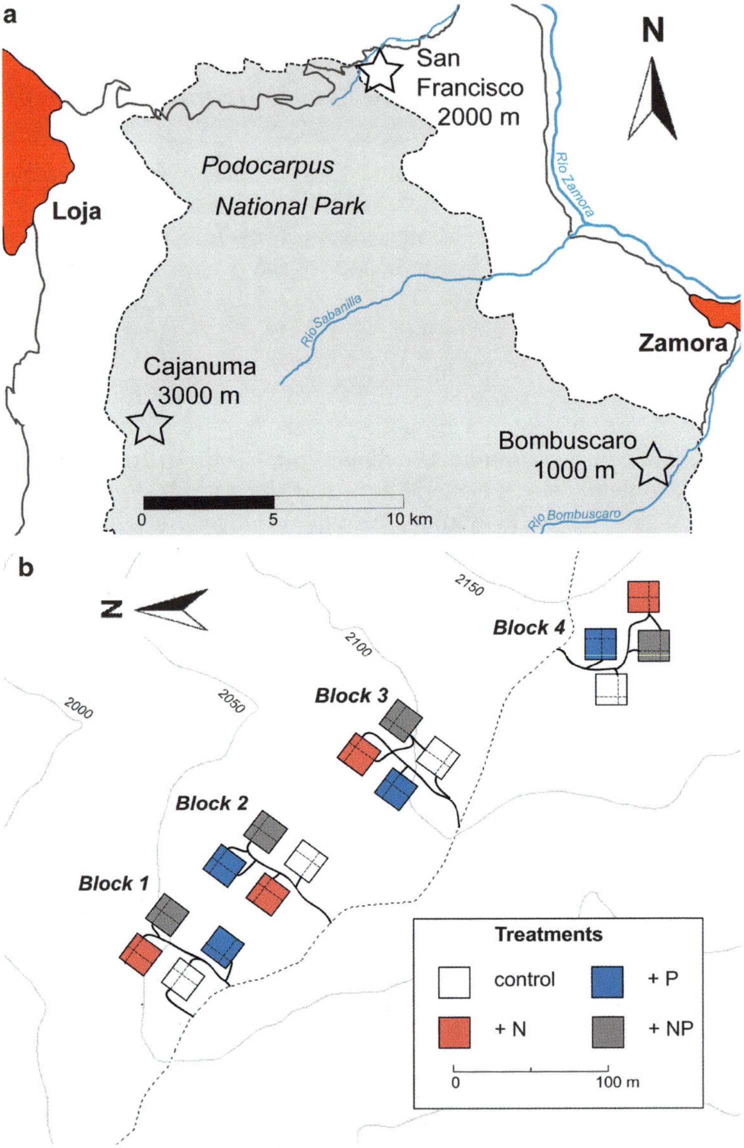

Fig. 23.1 (**a**) The location of the study sites of the Ecuadorian nutrient manipulation experiment (NUMEX). The position of each of the three sites named along with their altitude is marked with a *star*. (**b**) Scheme of the study design and plot distribution at the 2,000 m a.s.l. site in the San Francisco Reserve (redrawn after Homeier et al. 2012)

Table 23.1 Characteristics of the soil and the forest stands (including only trees of ≥10 cm dbh) at the three NUMEX sites. The figures quoted are means (±SE). Soil characteristics were determined in November 2007 prior to the first nutrient application

	Sites		
Parameters	Bombuscaro (1,000 m)	San Francisco (2,000 m)	Cajanuma (3,000 m)
Soil type	Dystric Cambisol	Stagnic Cambisol	Stagnic Histosol
Org. layer depth [cm]	<1	10–30	10–40
Mineral topsoil (0–5 cm)			
pH (H_2O)	4.6 (0.06)	3.8 (0.03)	4.7 (0.03)
Base saturation [%]	16.5 (1.1)	4.0 (1.5)	12.0 (1.1)
Total C [mg g^{-1}]	6.0 (1.2)	10.6 (1.1)	9.6 (1.4)
Total N [mg g^{-1}]	0.6 (0.1)	0.7 (0.1)	0.7 (0.1)
C/N ratio	9.3 (0.4)	17.2 (2.2)	14.2 (0.5)
Total P [mg g^{-1}]	0.14 (0.00)	0.04 (0.01)	0.51 (0.04)
Stand characteristics			
Canopy height [m]	20–25	10–14	6–8
Tree density [trees ha^{-1}]	748	1,143	1,305
Tree basal area [m^2 ha^{-1}]	33.4	22.8	25.5

evidenced no visible signs of human or natural disturbance and were separated by a distance from each other of at least 10 m.

Most of the research was conducted within six subplots of 4 m^2 each that were randomly placed along two perpendicular transects selected at random inside each of the 400 m^2 plots.

N and P were added to the plots at an annual rate of 50 kg N ha^{-1} (as urea, CH_4N_2O) and 10 kg P ha^{-1} (as NaH_2PO_4). The nutrients were dispersed homogeneously within the plots on two application dates per year. NUMEX has been started in February 2008.

23.2.1.1 Soil Respiration

Soil respiration was measured using static vented chambers. Four permanent chamber bases with a base area of 0.04 m^2 and a base height of 0.25 m each were randomly placed in four of the six subplots of each plot (16 per block; 48 per elevation). Gas samples were taken once every month from January 2008 until December 2008. Gas sampling is described in detail in Wolf et al. (2011). Gas fluxes were calculated from the linear increase in gas concentration multiplied by the density of the air and the volume of the chamber headspace. The value for the air density was adjusted for soil temperature and air pressure measured at the time of sampling.

23.2.1.2 Fine Root Biomass

For quantifying fine root biomass (roots < 2 mm in diameter), we took six root samples per plot to a depth of 20 cm using a soil corer of 3.5 cm diameter in January 2009. The soil samples were transferred to plastic bags and transported to the laboratory, where they were stored at 4 °C and processed within 6 weeks. The samples were soaked in water and freed from soil residues using a sieve with a mesh size of 0.25 mm. Only the fine roots of trees were considered for analysis. Live fine roots (biomass) were separated from dead rootlets (necromass) under the stereomicroscope. The fine root biomass was dried at 70 °C for 48 h, weighed, and expressed as g m^{-2} ground area.

23.2.1.3 Fine Litter Production

Six litter traps (surface area 0.36 m^2 each, positioned 1 m aboveground) were installed per plot, one in each subplot. The fallen litter was collected every 2 (at 1,000 m) to 4 weeks (at 2,000 and 3,000 m). The samples were oven-dried at 60 °C and the dry weights were determined.

23.2.1.4 Leaf Area Index

The leaf area index (LAI) of the stands was estimated with the LAI-2000 Plant Canopy Analyzer (LI-COR Inc., Lincoln, NE, USA). The measurements were conducted with two instruments in the remote mode, i.e., by synchronous readings below the canopy at 2 m height above the forest floor and in a nearby open area ("above-canopy" reading). Six measurements were made directly above the litter traps in each plot in January 2008 (before the first fertilization with N and/or P: see Sect. 23.2.1) and again in January 2009 (1 year after the first fertilization).

23.2.1.5 Tree Diameter Growth and Plot Basal Area Increment

The stem diameter growth of all trees with a diameter at breast height (dbh) $\geq$ 10 cm present in a plot was monitored every 6 weeks in each of the 48 plots using permanently installed girth-increment tapes (D1 dendrometers, UMS, Munich). The annual increase in the cumulative basal area of a plot was calculated by adding the basal area increment of all of the trees of a plot during the period from February 2008 (after the first fertilization) to January 2009.

23.2.1.6 Drought Sensitivity of Tree Growth After Nutrient Addition

To investigate the short- and medium-term growth response of important tree species to drought, we conducted a preliminary study with electronic high-resolution dendrometers (Type DR, Ecomatic, Munich) on N-fertilized and non-fertilized trees of similar size of the species *Podocarpus oleifolius* D. Don ex Lamb., *Graffenrieda emarginata* (Ruiz & Pav.) Triana, *Alchornea lojaensis* Secco, and *Prunus* sp. These tree species were selected because of their appropriate wood anatomy.

Radial stem diameter variation was measured at 30-min intervals for each of one fertilized and one non-fertilized tree per species. Cumulative growth and daily stem radius change (dA) were calculated from these measurements (Deslauriers et al. 2007; Bräuning et al. 2009; Volland-Voigt et al. 2011). We defined drought spells as periods of at least four consecutive days without precipitation as registered at a nearby meteorological station. From July 2010, when the dendrometers were installed, to December 2010, nine dry spells occurred, the longest one of which showed no rainfall for 11 consecutive days (21.10.2010 to 31.10.2010). Altogether, tree data from 48 sampling days (sum of all days in drought spells) were subjected to analysis.

23.3 Results and Discussion: Rapid Effects of N and/or P Addition on Forest Productivity

23.3.1 Soil Respiration

Soil CO_2 emission rates in the control forest plots differed among the three studied elevations in 2008 and were at the lower limit of other soil CO_2 emission rates reported for tropical montane forests (Koehler et al. 2009; McGroddy and Silver 2000) (Fig. 23.2a). The decrease of soil CO_2 emission rates with increasing elevation suggests that soil respiration was controlled mainly by soil temperature. Lower soil temperature at higher elevation sites may hamper soil microbial activity and metabolic CO_2 release and lead to accumulation of thick organic layers.

Soil CO_2 emissions did not differ between the control and nutrient-addition plots at the 1,000 m site. In contrast, N and N + P addition led to increased CO_2 efflux at the 2,000 m site ($p < 0.02$; Tukey HSD test) as did N addition at the 3,000 m site ($p < 0.01$). The enhanced soil respiration with N addition at 3,000 m may be due to an increase in microbial biomass, as has been found by Corre et al. (2010) in thick soil organic layers of Panamanian tropical montane forests following N addition. But in our study at 2,000 m only after N + P addition an increase of microbial biomass C was recorded whereas N addition even reduced the microbial biomass (see Sect. 22.3). We were not able to distinguish between autotrophic and

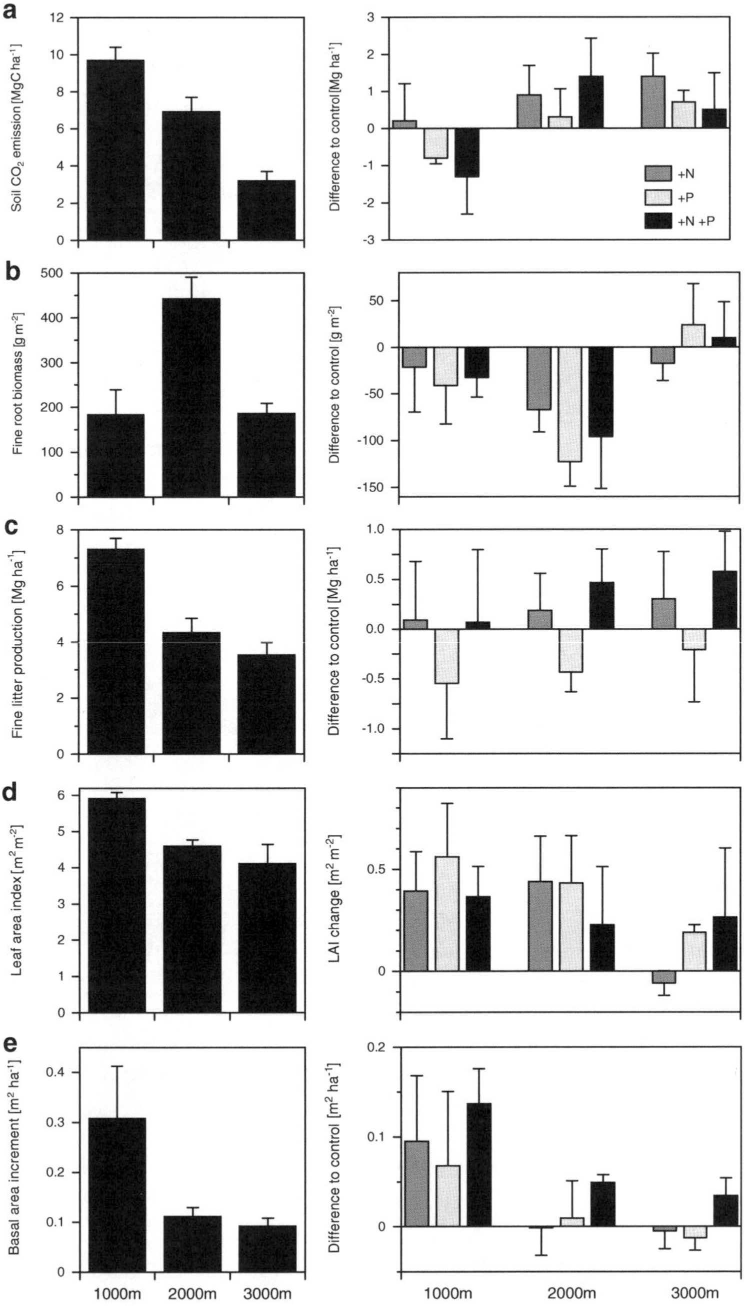

Fig. 23.2 Cumulative effects of 1 year of nutrient addition on selected stand parameters related to the ecosystem carbon cycle. (**a**) Soil CO_2 emissions, (**b**) fine root biomass, (**c**) annual fine litter production, (**d**) leaf area index (LAI), and (**e**) plot basal area increment (trees with dbh $\geq$ 10 cm).

heterotrophic respiration and their responses to N addition, since we measured only gross soil CO_2 emission rates.

23.3.2 Fine Root Biomass

This study found the highest fine root biomasses at the 2,000 m elevation. This was unexpected, since previous studies had recorded a continuous increase in fine root biomass along the transition from 1,000 m to 3,000 m (Röderstein et al. 2005; Moser et al. 2011). The present results show that the topographic position has a strong impact on root biomass in the study area. The high values at 2,000 m can be explained in terms of the location of the study plots on an upper slope, in contrast to the mid-slope position of the plots at the two other study sites. All three fertilization treatments at 1,000 m and 2,000 m resulted in a marked reduction in standing fine root biomass, with the effect being strongest after P addition (Fig. 23.2b). A decline of fine root biomass and a concurrent increase of dead root mass after nutrient addition have also been found in other montane forests when N was added (Cusack et al. 2011) or upon the addition of either N, P, or N + P (Gower and Vitousek 1989).

23.3.3 Fine Litter Production

A decrease in fine litter production with increasing elevation in tropical mountains is well documented (e.g., Moser et al. 2007) and may be explained by declines in tree leaf area and leaf production with increasing altitude coupled with higher mean leaf longevities. In our study the total amount of fine litter decreased after the first year of P addition at all elevations, whereas N and N + P addition had positive effects on litter production (Fig. 23.2c). An increase in litter production after N addition was also reported for other tropical montane forests by Adamek et al. (2009, Panama) and Tanner et al. (1992, Venezuela). Mirmanto et al. (1999) found that the addition of N, P, and the combination of N and P had positive effects on fine litter production in a lowland forest in Borneo.

Fig. 23.2 (continued) The figures on the *left-hand side* illustrate the average values (±SE) of the control plots for each of the three elevation levels. The figures on the *right-hand side* show the mean effects (±SE) of each of the three nutrient treatments

23.3.4 Leaf Area Index

The LAI of the control plots decreased by 1.8 units upon an increase in altitude from 1,000 m to 3,000 m (Fig. 23.2d). This is well in accordance with the LAI decrease of about 1.1 units per km of elevation increase (from 500 to 2,000 m) reported by Unger et al. (2013) from northern Ecuador and with the average decrease of 1 unit per km of elevation increase reported in a pan-tropical review by Moser et al. (2007). We found LAI increases of 0.2–0.4 units after N, P, or N + P addition (except for the N-treatment at 3,000 m). In combination with the slightly higher foliar N concentrations after N addition (see Sect. 22.3) this should affect leaf photosynthetic capacity and ecosystem C sequestration positively (see Chap. 10, and Hyvönen et al. 2007). Since P addition increased the LAI but reduced fine litter production at all three elevations, mean leaf longevity must have been increased by the addition of 10 kg P ha^{-1} $year^{-1}$.

23.3.5 Tree Growth

The annual stem diameter increment of montane forest trees is generally lower than that of tropical lowland trees. We accordingly found the highest mean radial growth rates and also the largest increases in the cumulative basal area of the plots at 1,000 m. At this altitude N addition as well as N + P addition resulted in a 20–30 % boost of the plot basal area increment in the first year after the start of the experiment (Fig. 23.2e). At 2,000 m and 3,000 m the addition of N or P led to only small positive or negative effects on basal area increment relative to the control. In contrast, the addition of both N and P had a marked positive effect, though less pronounced than that observed at 1,000 m.

A rapid increase in stem diameter growth of mature trees as a response to N addition has also been observed in other fertilization experiments, such as those in Jamaican montane forests (Tanner et al. 1990), in Hawaiian *Metrosideros* forests (Vitousek and Farrington 1997), and in Panamanian premontane forests (Adamek et al. 2009). However, Cavelier et al. (2000) found increased stem diameter growth in a Colombian elfin forest only when N and P were added together, and not when N or P was added alone.

23.3.6 Drought Sensitivity of Tree Growth After Nutrient Addition

The study of short-term variation in radial growth on eight trees of four species using high-resolution dendrometers indicates that the response to N addition differs largely among the species. All four tree species showed strong synchronicity in

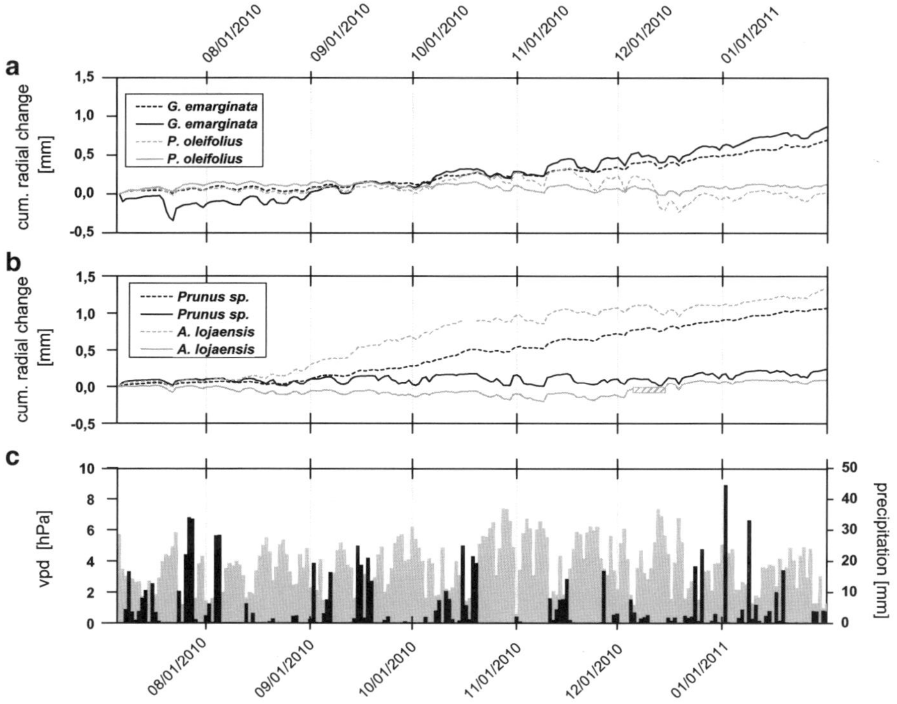

Fig. 23.3 Stem radial growth of four tree species and the course of climatic variables at 2,000 m a. s.l. from July 2010 until February 2011. Cumulative daily radial change of (**a**) *Graffenrieda emarginata* and *Podocarpus oleifiolius* and (**b**) *Prunus* sp. and *Alchornea lojaensis*. Tree individuals growing on N-manipulated plots are depicted with *dashed lines*, and individuals on non-manipulated plots with *continuous lines* (one tree per species and treatment). A data gap is indicated by a *gray hatched box*. (**c**) Daily totals of precipitation (*black columns*) and vapor pressure deficit (*gray columns*)

stem diameter variations with respect to diameter shrinkage during dry spells (Fig. 23.3). With the exception of *G. emarginata* (the dominant tree species at 2,000 m), the fertilized trees grew more vigorously than did the control trees (Table 23.2).

23.4 Conclusions: Implications for the Carbon Cycle in the Future

The rapid responses of the studied Andean montane forests to N and P addition observed at this early stage of the experiment illustrate the vulnerability of the forests to higher nutrient deposition. They also emphasize the urgent need for

Table 23.2 Stem diameter characteristics and variation of N-fertilized (+N) and control (con) trees during dry spells over the period of July to December 2010 ($n = 1$ tree per species and treatment)

	Dbh 2011 [cm]		Cum. growth [mm]		Mean dA [mm]		dA/VPD[a]		dA/VPD[b]	
	+N	con	+N	con	+N	con	+N	con	+N	con
Graf. em.	15.30	19.00	1.076	1.782	0.039***	0.075	0.37**	0.68	0.32*	0.68
Prun. sp.	11.05	14.02	1.088	0.290	0.069*	0.079	0.76**	0.85	0.78**	0.79
Alch. lo.	17.58	10.03	1.530	0.130	0.070	0.064	0.56**	0.87	0.58**	0.88
Podo. ol.	10.69	10.06	0.546	0.241	0.169***	0.128	0.86**	0.80	0.91**	0.73

cum. growth cumulative growth, *dA* daily amplitude of radial stem diameter variation, *VPD* vapor pressure deficit

Graf. em = *Graffenrieda emarginata*, *Prun.* sp. = *Prunus spec.*, *Alch. lo.* = *Alchornea lojaensis*, *Podo. ol.* = *Podocarpus oleifolius*

Differences between N-fertilized and control trees are significant at $^{*}p < 0.05$, $^{**}p < 0.01$, or $^{***}p < 0.001$

[a]Pearson correlation coefficient

[b]Spearman correlation coefficient

experimental studies on the effects of likely future alterations in the cycles of key elements in these forest ecosystems.

Since the forest canopy will become denser (increased LAI) upon fertilization and the trees will be relieved of growth constraints due to limited N and/or P availability, competition for light will increase among the trees with increasing input of nutrients. These changes may reduce the competitive ability of the seedlings and saplings of the currently abundant tree species in the understory and will in the long-term probably result in their eventual replacement by species adapted to more fertile soils. Changes in tree species composition (from slow-growing species adapted to nutrient-poor soils to faster growing species adapted to more fertile soils) will probably accelerate the projected shifts in the C cycle by increasing the biomass turnover rate.

The different tree species may respond in contrasting ways to elevated nutrient inputs. It is known for temperate forests that species which have a lower demand for N and are adapted to poorer soils are impacted more strongly by increases in N availability (Pardo et al. 2007). Drought sensitivity is another important tree trait that will shape future tree communities and is most likely to be affected by changes in nutrient availability.

Given the large stocks of carbon bound in the organic soil layers and also in the mineral soil (see Chap. 10), these forest ecosystems could represent significant sources of CO_2 for the atmosphere in the future, since lower litter C/N ratios resulting from fertilization typically lead to faster decomposition rates. The large amounts of soil organic matter stored in these high-elevation forests make it likely that any change in the conditions controlling decomposition will result in large changes in the ecosystem C pool.

The results presented here summarize the short-term effects of nutrient addition evident at the end of the experiment's first year. Given that these initial trends will persist, continued addition of substantial amounts of N and P will probably result in taller forests with a higher aboveground biomass but smaller belowground biomass. However, the belowground response of the system to nutrient addition is only poorly understood.

References

Aber J, McDowell W, Nadelhoffer K, Magill A, Berntson G, Kamakea M, McNulty S, Currie W, Rustad L, Fernandez I (1998) Nitrogen saturation in temperate forest ecosystems – hypotheses revisited. Bioscience 48(11):921–934

Adamek M, Corre MD, Hölscher D (2009) Early effect of elevated nitrogen input on above-ground net primary production of a lower montane rain forest, Panama. J Trop Ecol 25:637–647

Boy J, Rollenbeck R, Valarezo C, Wilcke W (2008) Amazonian biomass burning-derived acid and nutrient deposition in the north Andean montane forest of Ecuador. Glob Biogeochem Cycles 22, GB4011

Bräuning A, Volland-Voigt F, Burchardt I, Ganzhi O, Nauss T, Peters T (2009) Climatic control of radial growth of Cedrela montana in a humid mountain rain forest in southern Ecuador. Erdkunde 63:337–345

Cavelier J, Tanner E, Santamaria J (2000) Effect of water, temperature and fertilizers on soil nitrogen net transformations and tree growth in an elfin cloud forest of Colombia. J Trop Ecol 16:83–99

Cleveland CC, Townsend AR (2006) Nutrient additions to a tropical rain forest drive substantial soil carbon dioxide losses to the atmosphere. Proc Natl Acad Sci USA 103:10316–10321

Cleveland CC, Reed SC, Townsend AR (2006) Nutrient regulation of organic matter decomposition in a tropical rain forest. Ecology 87:492–503

Corre MD, Veldkamp E, Arnold J, Wright SJ (2010) Impact of elevated N input on soil N cycling and losses in old-growth lowland and montane forests in Panama. Ecology 91:1715–1729

Cusack D, Torn MS, McDowell WH, Silver WL (2010) The response of heterotrophic activity and carbon cycling to nitrogen additions and warming in two tropical soils. Glob Chang Biol 16:2555–2572

Cusack DF, Silver WL, Torn MS, McDowell WH (2011) Effects of nitrogen additions on above- and belowground carbon dynamics in two tropical forests. Biogeochemistry 104:203–225

Deslauriers A, Rossi S, Anfodillo T (2007) Dendrometer and intra annual tree growth: what kind of information can be inferred? Dendrochronologia 25:113–124

Elser JJ, Bracken MES, Cleland EE, Gruner DS, Harpole WS, Hillebrand H, Ngai JT, Seabloom EW, Shurin JB, Smith JE (2007) Global analysis of nitrogen and phosphorus limitation of primary producers in freshwater, marine and terrestrial ecosystems. Ecol Lett 10:1135–1142

Emck P (2007) A climatology of South Ecuador. With special focus on the major Andean ridge as Atlantic-Pacific climate divide. Dissertation, Universität Erlangen-Nürnberg, 275 pp

Gamboa AM, Hidalgo C, de Leon F, Etchevers JD, Gallardo JF, Campo J (2010) Nutrient addition differentially affects soil carbon sequestration in secondary tropical dry forests: early- versus late-succession stages. Restor Ecol 18(2):252–260

Giardina CP, Binkley D, Ryan MG, Fownes JH, Senock RS (2004) Belowground carbon cycling in a humid tropical forest decreases with fertilization. Oecologia 139:545–550

Gower ST, Vitousek PM (1989) Effects of nutrient amendments on fine root biomass in a primary successional forest in Hawaii. Oecologia 81:566–568

Gruber N, Galloway N (2008) An earth-system perspective of the global nitrogen cycle. Nature 451:293–296

Högberg P, Fan H, Quist M, Binkley D, Tamm CO (2006) Tree growth and soil acidification in response to 30 years of experimental nitrogen loading on boreal forest. Glob Chang Biol 12:489–499

Homeier J, Werner FA, Gradstein SR, Breckle S-W, Richter M (2008) Potential vegetation and floristic composition of Andean forests in South Ecuador, with a focus on the RBSF. In: Beck E, Bendix J, Kottke I, Makeschin F, Mosandl R (eds) Gradients in a tropical mountain ecosystem of Ecuador. Ecological studies, vol 198. Springer, Berlin, pp 87–100

Homeier J, Hertel D, Camenzind T, Cumbicus NL, Maraun M, Martinsin GO, Poma LN, Rillig MC, Sandmann D, Scheu S, Veldkamp E, Wilcke W, Wullaert H, Leuschner C (2012) Tropical Andean forests are highly susceptible to nutrient inputs - rapid effects of experimental N and P addition to an Ecuadorian montane forest. PLoS One 7:e47128

Hyvönen R, Agren GI, Linder S, Persson T, Cotrufo MF, Ekblad A, Freeman M, Grelle A, Janssens IA, Jarvis PG, Kellomäki S, Lindroth A, Loustau D, Lundmark T, Norby RJ, Oren R, Pilegaard K, Ryan MG, Sigurdsson BD, Strömgren M, van Oijen M, Wallin G (2007) The likely impact of elevated CO_2, nitrogen deposition, increased temperature and management on carbon sequestration in temperate and boreal forest ecosystems: a literature review. New Phytol 173:463–480

Koehler B, Corre MD, Veldkamp E, Sueta JP (2009) Chronic nitrogen addition causes a reduction in soil carbon dioxide efflux during the high stem-growth period in a tropical montane forest but no response from a tropical lowland forest on a decadal time scale. Biogeosciences 6:2973–2983

LeBauer DS, Treseder KK (2008) Nitrogen limitation of net primary productivity in terrestrial ecosystems is globally distributed. Ecology 89:371–379

Lewis SL, Malhi Y, Phillips OL (2004) Fingerprinting the impacts of global change on tropical forests. Phil Trans R Soc Lond B 359:437–462

Li Y, Xu M, Zou X (2006) Effects of nutrient additions on ecosystem carbon cycle in a Puerto Rican tropical wet forest. Glob Chang Biol 12:284–293

Magill AH, Aber JD, Currie WS, Nadelhoffer KJ, Martin ME, McDowell WH, Melillo JM, Steudler P (2004) Ecosystem response to 15 years of chronic nitrogen additions at the Harvard Forest LTER, Massachusetts, USA. For Ecol Manage 196:7–28

McGroddy M, Silver WL (2000) Variations in belowground carbon storage and soil CO_2 flux rates along a wet tropical climate gradient. Biotropica 32:614–624

Mirmanto E, Proctor J, Green J, Nagy L, Suriantata (1999) Effects of nitrogen and phosphorus fertilization in a lowland evergreen rainforest. Phil Trans R Soc Lond 354:1825–1829

Moser G, Hertel D, Leuschner C (2007) Altitudinal change in LAI and stand leaf biomass in tropical montane forests: a transect study in Ecuador and a pan-tropical meta-analysis. Ecosystems 10:924–935

Moser G, Leuschner C, Hertel D, Graefe S, Soethe N, Iost S (2011) Elevation effects on the carbon budget of tropical mountain forests (S Ecuador): the role of the belowground compartment. Glob Chang Biol 17:2211–2226

Pardo LH, McNulty SG, Boggs JL, Duke S (2007) Regional patterns in foliar 15N across a gradient of nitrogen deposition in the northeastern US. Environ Pollut 149:293–302

Röderstein M, Hertel D, Leuschner C (2005) Above- and below-ground litter production in three tropical montane forests in southern Ecuador. J Trop Ecol 21:483–492

Tanner EVJ, Kapos V, Freskos S, Healey JR, Theobald AM (1990) Nitrogen and phosphorus fertilization of Jamaican montane forest trees. J Trop Ecol 6:231–238

Tanner EVJ, Kapos V, Franco W (1992) Nitrogen and phosphorus fertilization effects on Venezuelan montane forest trunk growth and litterfall. Ecology 73:78–86

Tanner EVJ, Vitousek PM, Cuevas E (1998) Experimental investigation of nutrient limitation of forest growth on wet tropical mountains. Ecology 79:10–22

Unger M, Homeier J, Leuschner C (2013) Relationships among leaf area index, below-canopy light availability and tree diversity along a transect from tropical lowland to montane forests in NE Ecuador. Trop Ecol 54(1):33–45

Vitousek PM, Farrington H (1997) Nutrient limitation and soil development: experimental test of a biogeochemical theory. Biogeochemistry 37:63–75

Volland-Voigt F, Bräuning A, Ganzhi O, Peters T, Maza H (2011) Radial stem variations of Tabebuia chrysantha (Bignoniaceae) in different tropical forest ecosystems of southern Ecuador. Trees 25:39–48

Wolf K, Veldkamp E, Homeier J, Martinson GO (2011) Nitrogen availability links forest productivity, soil nitrous oxide and nitric oxide fluxes of a tropical montane forest in southern Ecuador. Glob Biogeochem Cycles 25, GB4009

Wright SJ (2005) Tropical forests in a changing environment. Trends Ecol Evol 20:553–560

Xia J, Wan S (2008) Global response patterns of terrestrial plant species to nitrogen addition. New Phytol 179:428–439

Chapter 24
Climate Change and Its Impact on Current and Future Vegetation Dynamics and Carbon Cycling

Brenner Silva, Claudia Dislich, Ingo Voss, Kristin Roos, Renate Scheibe, Peter Vorpahl, Boris Schröder, Andreas Huth, Erwin Beck, and Jörg Bendix

24.1 Introduction

The future development of vegetation dynamics and components of the carbon cycle sensitive to environmental change should be considered separately for both parts of the ecosystems in the San Francisco Valley, i.e. the natural forest and the pastures as a widespread anthropogenic replacement system. Useful tools to predict an impact of environmental change on vegetation are numerical growth models (Bonan et al. 2003). Process-oriented photosynthesis models allow for an estimation of changes in photosynthesis, biomass, and thus, carbon allocation, which is applied to the pasture system. To analyse the dynamics of forest stands under environmental change impacts, process-based forest-gap models can be used. In addition, soil mechanical and hydrological properties as well as tree morphology should be considered, because of the intensity of landslides in the mountain forest. Both model types need extensive parameterization to adapt the models to the local situation. The parameter values were derived from field surveys for single species

B. Silva (✉) • J. Bendix
Faculty of Geography, University of Marburg, Marburg, Germany
e-mail: silvab@geo.uni-marburg.de

C. Dislich • A. Huth
Department of Ecological Modelling, Helmholtz Centre for Environmental Research UFZ Leipzig, Leipzig, Germany

I. Voss • R. Scheibe
Department of Plant Physiology, University of Osnabrück, Osnabrück, Germany

K. Roos • E. Beck
Department of Plant Physiology and Bayreuth Centre of Ecology and Environmental Research, University of Bayreuth, Bayreuth, Germany

P. Vorpahl • B. Schröder
Landscape Ecology, Technische Universität München, Munich, Germany

University of Potsdam, Potsdam, Germany

J. Bendix et al. (eds.), *Ecosystem Services, Biodiversity and Environmental Change in a Tropical Mountain Ecosystem of South Ecuador*, Ecological Studies 221,
DOI 10.1007/978-3-642-38137-9_24,

or representative plant functional types. In this chapter we present results on modelling the carbon cycle of the two most representative land use options for the study area (cf. Sect. 2.2): pasture and forest.

Climate change might affect the forest and the pasture in the *Reserva Biológica San Francisco* (RBSF, see Fig. 1.1) in different ways:

- On pastures, an increase of the temperature might equally stimulate the growth of the C4-type pasture grass *Setaria sphacelata* and the C3-type bracken (*Pteridium arachnoideum*) and thus might not alter the competitive interaction of both species. This was investigated by means of scenarios-driven runs of the Southern Bracken Competition Model (*SoBraCoMo*, see Sect. 24.2.1).
- In the natural forest, increase of precipitation may lead to a higher landslide activity, and thus to an enhanced forest turnover with impacts on carbon sequestration. The effect of the natural disturbance factor 'landslides' on the dynamics of the natural forest was investigated with the forest-gap model FORMIND (see Sect. 24.2.2).

24.2 Methods

24.2.1 The SoBraCoMo *Model*

The *SoBraCoMo* is a canopy-photosynthesis model based on the two-big-leaf approach (Dai et al. 2004). As described by Silva et al. (2012), the general components of the *SoBraCoMo* are (1) a radiation module, including canopy radiation and radiation partitioning, (2) a canopy-photosynthesis module, (3) a respiration module for leaves and subterraneous plant parts, and (4) a module for conversion of net assimilated CO_2 into biomass (dry matter). All modules are parameterized for two species, the main pasture grass *Setaria* and the Southern Bracken. For the radiation scheme, species-specific optical and morphological plant traits were derived from field measurements (Bendix et al. 2010). The parameterization of the photosynthesis model was done by determining species-specific values of carboxylation rate and quantum yield. These parameters were derived from leaf porometry of the photosynthetic response to CO_2 and to photosynthetic active radiation of each species. To upscale from leaf to canopy a "two-big-leaf" model was considered, which required parameterization of sunlit and shade leaves, partitioned radiation input (diffuse and direct radiation), and field measurements of leaf area index (Silva et al. 2012). To assess the thermal influence on biomass production and the competitive strength at leaf and canopy level, leaf measurements were made under controlled conditions and used to parameterize the *SoBraCoMo*, which is run at canopy level using meteorological forcing data of the year 2008. Further parameterization concerned respiratory carbon losses and the conversion of net assimilation to dry matter. Environmental forcing of the model is conducted with data of global radiation, air temperature, relative humidity, atmospheric pressure, leaf temperature, wind speed, precipitation, soil water content, and soil temperature, which were continuously recorded in the field. In the following, the model is applied to calculate scenarios at different ambient temperatures. To force the model, downscaled meteorological data sets of air temperature (for more information on climate scenarios refer to Sect. 19.2.1) are used on a daily basis. All variables which are directly dependent on air temperature (soil and leaf temperatures) are projected to the IPCC-SRES (Special Report on Emissions Scenarios) A1B scenario periods (Nakicenovic and Swart 2000) by regression analysis. In the regression analysis, the residuals were tested for normal distribution (Shapiro-Wilk test >0.99) and a linear model was fitted by least squares between the temperatures of the soil, or the leaf and the average air temperature (reference period).

24.2.2 The FORMIND Model

FORMIND belongs to the class of process-based forest-gap models. It is designed to analyse the dynamics of uneven-aged, species-rich forest stands with a focus on the impact of natural or anthropogenic disturbances on forest structure and composition (Köhler and Huth 2004; Huth and Ditzer 2001). It has been successfully applied to various forests throughout the tropics (e.g. Rüger et al. 2008; Köhler and Huth 2010; Pütz et al. 2011). The main processes of the model are the competition for light and space which influences tree growth, tree mortality, and the establishment of young trees. To handle the high number of tree species in the San Francisco valley (many of them are extremely rare), tree species within the model are grouped into seven plant functional types (PFTs) according to selected attributes like maximum attainable diameter or maximum height of tree species. In this chapter we analysed one forest type, the ridge forest (1,900–2,100 m a.s.l.) harbouring ~70 tree species. A parameterization of the FORMIND model for the ridge forest within the RBSF tropical mountain forest was developed and the predicted forest succession dynamics were examined (Dislich et al. 2009). Data from repeated field inventories (2004, 2005, and 2007) of 4.88 ha, in which the diameter at breast height (1.3 m) of all trees with diameters >20 cm was measured, and from 0.12 ha, in which all trees above 5 cm were measured, were used for model calibration (Homeier 2004).

Landslides are an important natural disturbance factor in the RBSF forest and a strong driver of spatio-temporal vegetation turnover (Restrepo et al. 2003; Wilcke et al. 2003; Bussmann et al. 2008). Apart from different soil mechanical, hydrological, and vegetation-related factors, precipitation is an important external trigger of landslides (Sidle 1992; Stoyan 2000; Muenchow et al. 2012, see also Sect. 12.3). The current rainfall in our study area lays around 1,800 mm a^{-1} (Rollenbeck et al. 2007). Since the downscaled climate scenarios suggest an increase in rainfall by ~+515 mm a^{-1} (IPCC-SRES scenario A1B, see also Sect. 19.2.1), future changes in frequency and/or magnitude of extreme rain events will also affect landslide dynamics. Consequently, we assume shifting landslide regimes under climate change.

In the FORMIND model, landslides are implemented as a spatially explicit form of disturbance (Fig. 24.3a). The local forest succession after landslide disturbance might follow different trajectories (Velázquez and Gómez-Sal 2008; Dislich and Huth 2012). In the current chapter it is assumed that reduced tree growth on landslide sites occurs due to nutrient limitation (Wilcke et al. 2003). Since it is not clear, to which extent increasing rainfall will affect landslide frequencies, we applied the current landslide frequency as well as doubled and a fourfold landslide frequencies to analyse the effect of changing landslide activity on aboveground biomass and tree species composition.

24.3 Results and Discussion

24.3.1 Biomass Production and Competition on the Pastures upon Local Warming

For the comparison of photosynthetic productivity of the two species in the pasture system, temperature is an essential environmental parameter. Photosynthetic performance strongly depends on temperature, especially when comparing a high-temperature C4-type grass (*Setaria*) with a moderate-temperature C3-type species (bracken). Due to a higher photosynthetic efficiency C4-plants often prevail in habitats with warm and humid or dry conditions and high or moderate irradiation, whereas under cooler conditions C3-plants have the edge on photosynthetic biomass production and growth (Sage et al. 1999). In our days, the daily average air temperature is 14.8 °C on the pastures (Silva et al. 2012) which is suboptimal for *Setaria*, and thus, should favour the competitive strength of bracken.

The results of the laboratory measurements at leaf level (Fig. 24.1a–c) show that the photosynthetic performance of bracken is better than that of *Setaria* at temperatures below 20 °C when both plants were cultivated at 20 °C (Fig. 24.1a). For measurements with leaf temperatures above 20 °C, the photosynthetic production of *Setaria* increases and outperforms that of the bracken fern. Even more pronounced trends of photosynthetic temperature dependence were observed for plants which were grown and examined at 10, 20, and 30 °C, respectively (Fig. 24.1b). In that case the amount of bracken-RubisCO appears to correlate inversely with the temperature, being higher at 10 °C growth temperature than at 20 and 30 °C (Fig. 24.1c). Such an effect of the growth temperature has also been described by Hurry et al. (1995). In contrast, no temperature-related change in the RubisCO amount is apparent in the fodder grass.

In contrast to single leaf measurements of photosynthetic activity, the model describes net photosynthetic CO_2 uptake at the canopy level (Fig. 24.1d, e) which generally confirms the contrasting thermal dependency of CO_2 assimilation for bracken and *Setaria* (Fig. 24.1a, b). For biomass production, the temperature dependent mitochondrial respiration (leaf and root tissues) was calculated as

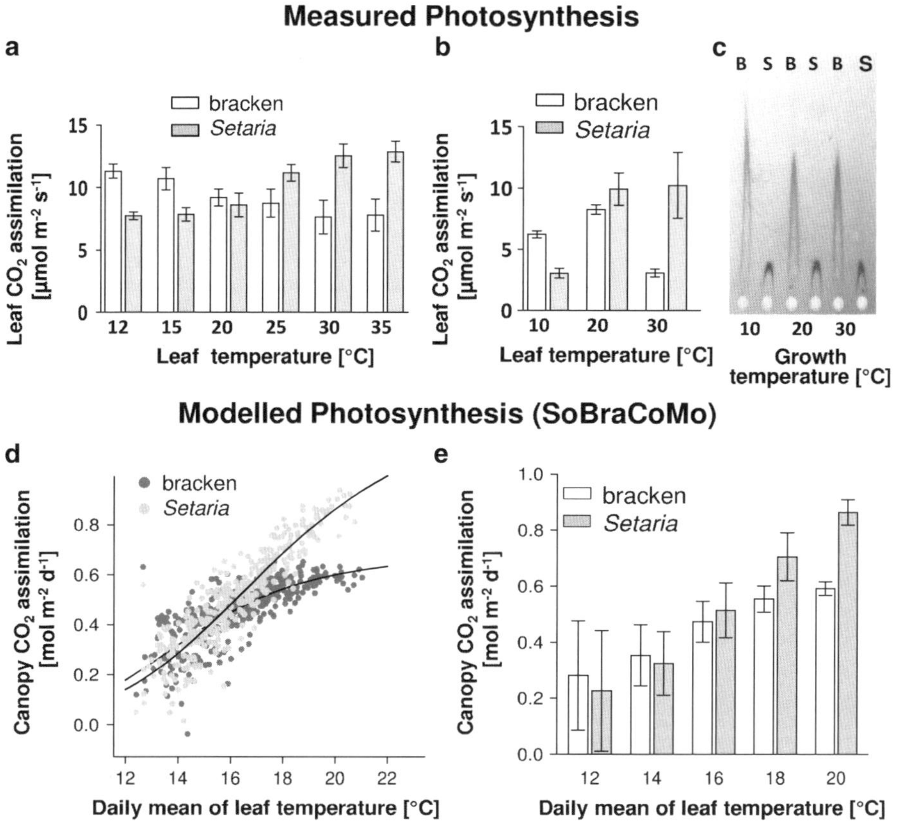

Fig. 24.1 Temperature dependency of photosynthesis in Southern Bracken (*Pteridium arachnoideum*, C3-plant) and *Setaria sphacelata* (C4-plant). *Measurements* (**a**) Southern Bracken (*white bars*) and *Setaria* (*grey bars*) were cultivated at 20 °C, and net CO_2 assimilation was determined at different leaf temperatures at 1,200 µmol quanta m^{-2} s^{-1}. (**b**) Southern Bracken (*white bars*) and *Setaria* (*grey bars*) were cultivated at 10, 20, and 30 °C, respectively, at 500 µmol quanta m^{-2} s^{-1}, and CO_2 assimilation was measured at the growth temperatures. (**c**) Amount of RubisCO protein in leaves of bracken and *Setaria* which were grown under 10, 20, and 30 °C, respectively. The relative amounts of the protein were determined by rocket immuno-electrophoresis of leaf extracts. *Simulations* (**d**) Scatterplot of canopy net CO_2 assimilation with daily means of leaf temperature in the year 2008. (**e**) Canopy net CO_2 assimilation depending on leaf temperature

0.23 and 0.16 µmol CO_2 m^{-2} s^{-1} for bracken and *Setaria* leaves, respectively, and as 0.9 and 2.0 µmol CO_2 kg^{-1} s^{-1} for rhizomes and roots of the two species at 20 °C (Silva et al. 2012). At canopy level, i.e. considering an average observed leaf area index of 2.6 m^2 leaf m^{-2} ground for both species, the turning point towards a growth advantage of the C4-plant was at a daily mean of the leaf temperatures between 14 and 16 °C. Of course mean daily leaf temperatures are characteristic of the special experimental sites in the San Francisco valley.

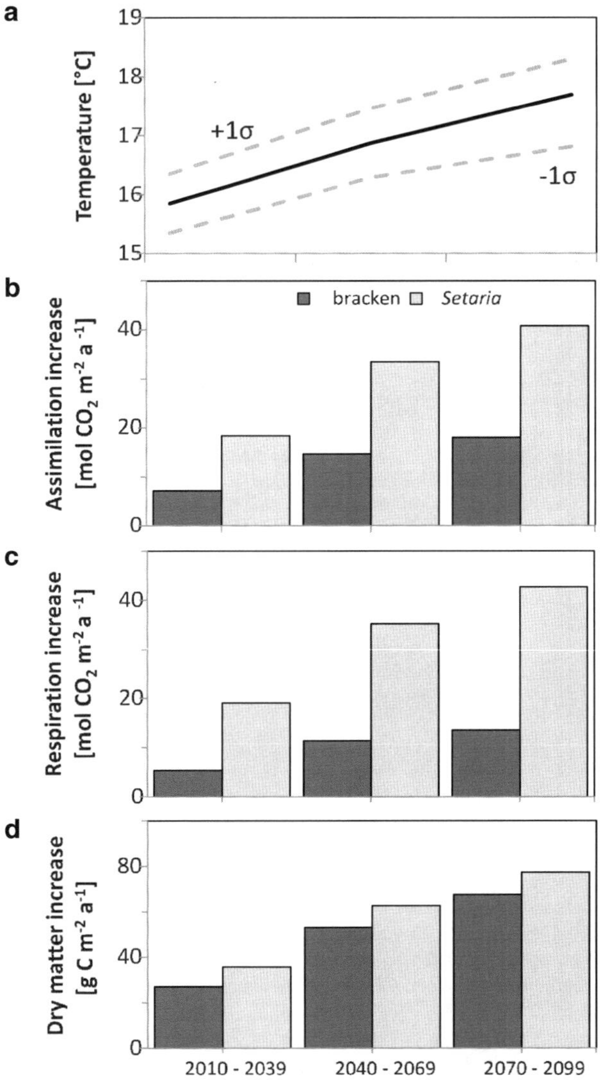

Fig. 24.2 Modelled influence of increasing temperatures on *Setaria* and bracken for future years. (**a**) Air temperature of the research site (at 2 m height) according to the IPCC-SRES A1B scenario (average and standard deviation σ of 10 models) and (**b–d**) resulting changes (=difference between scenario – present-day values) of gross photosynthesis (**b**), respiration (sum of maintenance and growth respiration) (**c**), as well as dry matter production for Southern Bracken and the pasture grass *Setaria* (**d**)

Following the most likely IPCC-SRES A1B emission scenario—which presumes a fast-growing world population and a balanced use of fossil and non-fossil fuels—and the corresponding ensemble calculations, ambient air temperature at the altitudinal level of the pasture (RBSF meteorological station) will increase by 1.1 °C ($\sigma = 0.5$) by 2039, 2.1 °C ($\sigma = 0.6$) by 2069, and 2.9 °C ($\sigma = 0.9$) by 2099 (Fig. 24.2a). Simultaneously, the atmospheric CO_2 in the San Francisco valley will increase from 362 ppm by 2010 to 662 ppm by 2099. According to field measurements a factor of 0.91 was used to consider the altitude (1,900 m a.s.l.) in the projected concentrations of the IPCC-SRES A1B (BERN 2011).

As a result, global warming might increase the growth potential of both bracken and *Setaria*. However, there are two different reasons for this common increase. For *Setaria* the high loss due to mitochondrial respiration will be compensated by the exponential increase in carboxylation with temperature. For bracken, there will be no significant increase in carboxylation with temperature, but in the case of elevated atmospheric CO_2, photorespiration loss will be reduced. Figure 24.2b–d reveals that particularly *Setaria* can respond to an increase of the temperature by 2.9 °C (year 2099) with an increase of gross CO_2 assimilation from 301 to 342 mol CO_2 uptake per m^2 and year (+14 %) whereas increase of bracken photosynthesis will be from 212 to 230 mol CO_2 m^{-2} a^{-1} (+9 %). At the same time, the increase in respiration losses will be different in both species, being higher in *Setaria* (+19 %) than in bracken (+11 %). Consequently, biomass production of both *Setaria* and bracken will almost equally increase until 2099. Due to a steeper rate of temperature increase between 2010 and 2069, the increase in biomass production of *Setaria* and bracken will be higher than in the course of the following decades.

24.3.2 Forest Dynamics After Natural Disturbance

The simulated impact of landslides is presented in Fig. 24.3. On the entire surface of the landslide, the complete tree biomass is removed and remains lower than the original one during the first 50 years after the event (Fig. 24.3c). The subsequent accumulation of new tree biomass initially originates from the fast-growing pioneer species (PFT 1 and 2) with an increasing share from mid-successional species (PFT 3 and 4) and eventually also from slow-growing species (PFT 5, 6, and 7). After 75 years, the overall biomass attains a close to pre-disturbance level, but even after 150 years the composition of the different plant functional types is biased towards fast-growing tree species. It takes around 350 years, until all species groups reach their mature biomass. Interestingly, even in the mature ridge forest, the fast-growing species (PFT 1 and 2) account for a major share of tree biomass which might be attributed to a generally high frequency of disturbances and poor edaphic conditions that prevent a dominance of late successional species.

On the landscape level, landslides increase the heterogeneity of biomass distribution and therefore increase landscape heterogeneity by creating a mosaic landscape structure consisting of forest patches with different successional stages (Yamamoto et al. 1995; Geertsema and Pojar 2007). The simulation experiment demonstrates the effect of landslides on aboveground carbon stocks in the RBSF forest ecosystem. Applying a realistic current landslide frequency (0.02 landslides ha^{-1} $year^{-1}$) and landslide size distribution which were derived from aerial photographs, the average aboveground tree biomass is predicted as 94 tons ha^{-1} (Fig. 24.3d). An increase of the current landslide frequency following an increase of precipitation reduces the aboveground carbon stocks considerably—doubling or even quadruplicating the current landslide frequency would result in 12 % or 37 % less tree biomass, respectively. In the opposite case of diminishing

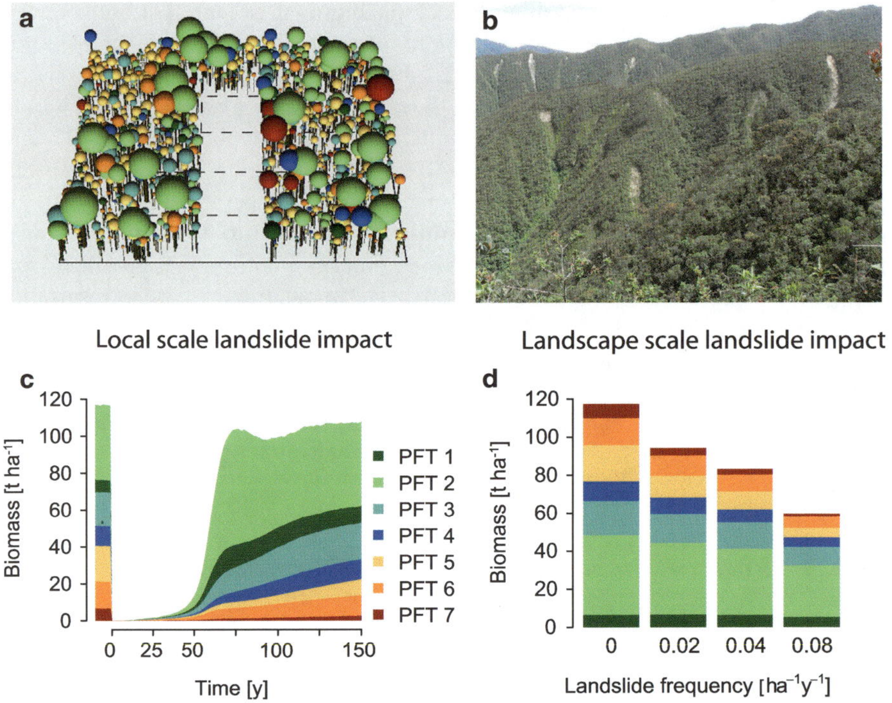

Fig. 24.3 Simulated impact of landslides on the forest. (**a**) Visualisation of the forest model (area 1 ha) with recent landslide disturbance (cf. Dislich and Huth 2012, Fig. 24.1). (**b**) Photograph of the RBSF forest with several visible traces of landslides. (**c**) Simulated accumulation of aboveground tree biomass after landslide disturbance from different plant functional types (PFT). (**d**) Aboveground tree biomass under different landslide frequencies (current landslide frequency is 0.02 ha^{-1} $year^{-1}$). All simulation runs apply a reduced growth rate of trees on landslide-disturbed areas due to nutrient limitation. The functional types of the tree species are classified as fast-growing pioneer species (PFT 1 and 2, *green*), mid-successional species (PFT 3 and 4, *blue*), and slow-growing species (PFT 5, 6, and 7, *orange* and *red*)

landslide frequencies, the upper limit of tree biomass was calculated as 117 tons ha^{-1}, representing a 25 % increase compared to the current biomass. Depending on the landslide frequency the model predicts changes in the community composition. With increasing disturbance frequency, the relative abundance of fast-growing pioneer tree species (PFT 1 and 2) slightly increases at the cost of slow-growing species (PFT 5, 6, and 7). The species composition of each PFT group is summarized in Chap. 8.

The influence of landslide disturbances on forest productivity is ambiguous—while the area with undisturbed forest productivity decreases due to landslides, the disturbed area provides new space for forest succession, but under unfavourable environmental conditions, like low nutrient availability, instable substrates, and strong exposition to wind (Walker 1994; Walker and del Moral 2003). The investigation of the effect of landslides on forest productivity with the model showed moderate reductions in forest productivity due to this type of disturbance (Dislich and Huth 2012).

24.4 Conclusion

The performance of the C4-pasture grass *Setaria* under current day thermal conditions of the research area shows that the grass is not optimally adapted for the competition with the C3-bracken fern. Currently, *Setaria* is striving close to the thermal turning point towards a better competitive strength in comparison to bracken which is doing very well. With the expected increase in temperature and atmospheric CO_2 without significant decrease in humidity (see Sect. 19.3.2) the growth potential of both the C4-grass and the C3-bracken will increase in the same proportion. Planting a high productive C3-grass instead of *Setaria* would possibly improve pasture yields, as long as the increase of temperature and atmospheric CO_2 does not affect other climate variables. Regarding to supporting and regulating services, this means that global warming will foster the growth of the pasture grass, leading to an enhanced carbon sequestration by the pastures. Concurrently, the Southern Bracken will profit from higher levels of atmospheric CO_2 and thus also increase in productivity. Hence, no significant change is expected in the growth of *Setaria* pasture in relation to the growth of Southern Bracken. Furthermore, the higher productivity of both *Setaria* pasture and Southern Bracken should be considered in pasture management, particularly bracken control.

Future changes in precipitation amounts and intensity might affect landslide regimes in the RBSF area. Increased landslide frequencies reduce the aboveground carbon stocks, e.g. a fourfold increase would result in a 37 % reduction of aboveground tree biomass. This reduction would be accompanied by changes in the community composition towards a higher fraction of early successional species and species that can tolerate the harsh conditions on landslide sites (e.g. *Clethra revoluta*, *Tibouchina lepidota*). Naturally, landslide occurrence does not only depend on the rainfall regime. Various other biotic and abiotic triggering factors of landslides, for example soil characteristics and wind, will also have to be taken into account to predict changes in future landslide frequencies (Sect. 12.2, Vorpahl et al. 2013). To analyse the effect of landslides on the entire ecosystem carbon cycle, the influence of this particular disturbance type on soil carbon dynamics is an additional important aspect that needs to be considered. Landslides induce increased soil erosion and contribute to landscape evolution over long timescales (Walker and Shiels 2008; Muenchow et al. 2012).

The investigated changes of landslide regimes are just one facet of change that might affect forest carbon cycles in the future: increasing temperatures might lead to an upward shift of species, and changes in precipitation as well as remote fertilization (cf. Sect. 21.3) might change species composition. Predicting such changes is much more difficult for the highly diverse forest ecosystem compared to the pasture system due to the extremely complex interactions (across and within different trophic levels). In this sense we are still at an early stage of understanding future regulations of important ecosystem services like carbon sequestration and carbon storage capacity in tropical mountain forests.

Acknowledgments The first author B. Silva would like to thank the Brazilian Council of Technological and Scientific Development (CNPq) for research grants (GDE 290033/2007-1). We thank Rütger Rollenbeck (Univ. Marburg) and Nicolas König (Univ. of Osnabrück) for their help in the field experiments on the pastures.

References

Bendix J, Silva B, Roos K, Göttlicher D, Rollenbeck R, Nauß T, Beck E (2010) Model parameterization to simulate and compare the PAR absorption potential of two competing plant species. Int J Biometeorol 54:283–295

Bern Carbon Cycle Model (2011) Intergovernmental panel on climate change, Geneva, CH. http://www.ipcc-data.org/ddc_co2.html. Accessed 23 Jan 2013

Bonan GB, Levis S, Sitch S, Vertenstein M, Oleson KW (2003) A dynamic global vegetation model for use with climate models: concepts and description of simulated vegetation dynamics. Glob Change Biol 9:1543–1566

Bussmann RW, Wilcke W, Richter M (2008) Landslides as important disturbance regimes: causes and regeneration. In: Beck E, Bendix J, Kottke I, Makeschin F, Mosandl R (eds) Gradients in a tropical mountain ecosystem of Ecuador. Ecolological studies, vol 198. Springer, Berlin, pp 319–330

Dai Y, Dickinson RE, Wang Y-P (2004) A two-big-leaf model for canopy temperature, photosynthesis, and stomatal conductance. J Climate 17:2281–2299

Dislich C, Huth A (2012) Modelling the impact of shallow landslides on forest structure in tropical montane forests. Ecol Model 239:40–53

Dislich C, Günter S, Homeier J, Schröder B, Huth A (2009) Simulating forest dynamics of a tropical montane forest in South Ecuador. Erdkunde 63:347–364

Geertsema M, Pojar JJ (2007) Influence of landslides on biophysical diversity – a perspective from British Columbia. Geomorphology 89:55–69

Homeier J (2004) Baumdiversität, Waldstruktur und Wachstumsdynamik zweier tropischer Bergregenwälder in Ecuador und Costa Rica. PhD Thesis, University of Bielefeld

Hurry V, Keerberg O, Pärnik T, Gardeström P, Öquist G (1995) Cold-hardening results in increased activity of enzymes involved in carbon metabolism in leaves of winter rye (Secale cereale L.). Planta 195:554–562

Huth A, Ditzer T (2001) Long-term impacts of logging in a tropical rain forest – a simulation study. For Ecol Manage 142:33–51

Köhler P, Huth A (2004) Simulating growth dynamics in a South-east Asian rainforest threatened by recruitment shortage and tree harvesting. Clim Change 67:95–117

Köhler P, Huth A (2010) Towards ground-truthing of spaceborne estimates of above-ground life biomass and leaf area index in tropical rain forests. Biogeosciences 7:2531–2543

Muenchow J, Brenning A, Richter M (2012) Geomorphic process rates of landslides along a humidity gradient in the tropical Andes. Geomorphology 139:271–284

Nakicenovic N, Swart R (eds) (2000) Special report on emissions scenarios: a special report of working group III of the intergovernmental panel on climate change. Cambridge University Press, Cambridge, 599 pp

Pütz S, Groeneveld J, Alves LF, Metzger JP, Huth A (2011) Fragmentation drives tropical forest fragments to early successional states: a modelling study for Brazilian Atlantic forests. Ecol Model 222:1986–1997

Restrepo C, Vitousek P, Neville P (2003) Landslides significantly alter land cover and the distribution of biomass: an example from the Ninole ridges of Hawaii. Plant Ecol 166:131–143

Rollenbeck R, Bendix J, Fabian P, Boy J, Wilcke W, Dalitz H, Oesker M, Emck P (2007) Comparison of different techniques for the measurement of precipitation in tropical montane rain forest regions. J Atmos Oceanic Technol 24:156–168

Rüger N, Williams-Lineara G, Kissling WD, Huth A (2008) Long-term impacts of fuelwood extraction on a tropical montane cloud forest. Ecosystems 11:868–881

Sage RF, Wedin DA, Li M (1999) The biogeography of C4 photosynthesis: patterns and controlling factors. In: Sage RF, Monson RK (eds) C4 plant biology. Academic, Toronto, pp 313–373

Sidle RC (1992) A theoretical model of the effects of timber harvesting on slope stability. Water Resour Res 28:1897–1910

Silva BSG, Roos K, Voss I, König N, Rollenbeck R, Scheibe R, Beck E, Bendix J (2012) Simulating canopy photosynthesis for two competing species of an anthropogenic grassland community in the Andes of southern Ecuador. Ecol Mod 239:14–26. doi:10.1016/j.ecolmodel.2012.01.016

Stoyan R (2000) Aktivität, Ursachen und Klassifikation der Rutschungen in San Francisco/Südecuador. Diploma Thesis, University of Erlangen-Nürnberg

Velázquez E, Gómez-Sal A (2008) Landslide early succession in a neotropical dry forest. Plant Ecol 199:295–308

Vorpahl P, Dislich C, Elsenbeer H, Märker M, Schröder B (2013) Biotic controls on shallow translational landslides. Earth Surf Proc Land 38:198–212

Walker LR (1994) Effects of fern thickets on woodland development on landslides in Puerto Rico. J Veg Sci 5:525–532

Walker LR, del Moral R (2003) Primary succession and ecosystem rehabilitation. Cambridge University Press, UK

Walker LR, Shiels AB (2008) Post-disturbance erosion impacts carbon fluxes and plant succession on recent tropical landslides. Plant Soil 313:205–216

Wilcke W, Valladarez H, Stoyan R, Yasin S, Valarezo C, Zech W (2003) Soil properties on a chronosequence of landslides in montane rain forest, Ecuador. Catena 53:79–95

Yamamoto S, Nishimura N, Matsui K (1995) Natural disturbance and tree species coexistence in an old-growth beech – dwarf bamboo forest, southwestern Japan. J Veg Sci 6:875–886

Chapter 25
Sustainable Use of Tropical Forests: A Plea for a Landscape View

Thomas Knoke, Baltazar Calvas, Patrick Hildebrandt, Michael Weber, Bernd Stimm, Sven Günter, Nikolai Aguirre, and Reinhard Mosandl

25.1 Introduction: Forests as Part of the Landscape

The Marrakesh Accords (UNFCCC 2002) define sustainable forest management as "practices for stewardship and use of forest land aimed at fulfilling relevant ecological (including biological diversity), economic and social functions of the forest in a sustainable manner". Knoke and Weber (2006) showed how this guideline can be implemented in forest management planning. However, a limitation of the guideline to forests is not sufficient. Demanding sustainability for forest management practices implies that no other ecosystem goods or ecosystem services (see Chap. 4) may be compromised by the use of forests. This can only be assured by integrating forests and their management into a comprehensive landscape approach in which all relevant land-use options are considered simultaneously to account for possible interactions.

Landscape approaches include pragmatic perspectives that would also accept mixed land-use strategies (Gardner et al. 2009). However, such landscape approaches are only rarely used to analyse forestry land-use options. The lack of

T. Knoke (✉) • B. Calvas
Institute of Forest Management, Center of Life and Food Sciences Weihenstephan, Technische Universitaät München, Hans-Carl-von-Carlowitz-Platz 2, 85354 Freising, Germany
e-mail: knoke@forst.wzw.tum.de

P. Hildebrandt • M. Weber • B. Stimm • R. Mosandl
Institute of Silviculture, Center of Life and Food Sciences Weihenstephan, Technische Universität München, Hans-Carl-von-Carlowitz-Platz 2, 85354 Freising, Germany

S. Günter
Tropical Agricultural Research and Higher Education Center, Latin-American Chair for Protected Areas and Biological Corridors "Kenton Miller", CATIE 7170, Turrialba 30501, Costa Rica

N. Aguirre
Center for Studies and Development of Amazonia (CEDAMAZ), National University of Loja, Loja, Ecuador

J. Bendix et al. (eds.), *Ecosystem Services, Biodiversity and Environmental Change in a Tropical Mountain Ecosystem of South Ecuador*, Ecological Studies 221,
DOI 10.1007/978-3-642-38137-9_25,

comprehensive approaches may result in ineffective or even counterproductive recommendations for policies to control land use for achieving more sustainable practices. If important trade-offs, but also potential synergies, are disregarded by mutually exclusive analyses of land-use options, one cannot expect really feasible, acceptable and also sustainable land-use strategies to develop. A prominent example is the obvious conflict between climate-friendly land-use strategies that imply halting further deforestation (Godfray et al. 2011) and the meeting of the resource demands of growing populations, a need that requires more rather than less agricultural area.

A focus on the sustainable future use of tropical forest ecosystems requires other land-use options being considered as well. Approaches that enable us to analyse multiple land-use options and to simultaneously come up with optimal land-use combinations are thus imperative for sustainable and really feasible forestry practices. Examples of these are the low-impact management of natural forests and the protection of natural forests, as well as the reforestation of degraded lands. Such sustainable forestry strategies include the slowing or even halting of further conversion of natural forests into agricultural lands, while simultaneously addressing the production of enough food and other resources to satisfy increasing human demands. Moreover, a sustainable land-use strategy must also consider the livelihood needs of the local population. In this regard, payments to ensure the protection of the natural forests and their ecosystem services are essential to counter the effect of market signals (e.g. increasing prices when agricultural products become scarce) that could stimulate the tendency to convert the forests into agricultural land. These market signals may, however, have to be modelled and considered on a national rather than on a local or regional scale. In this sense, the following chapter aims at demonstrating a national approach to land-use modelling (Optimised Land-Use Diversification) that is based mainly on land-use options investigated in the San Francisco Valley (see Chap. 1) and in the wider study area in southern Ecuador. In our attempt to develop a national perspective on the interplay between product prices and product supply, we must also consider cropland practices (maize cultivation, in our case) that are not very common in the San Francisco Valley, but are in other regions of Ecuador. The approach has been conceptualized to determine how land-use practices should be spatially distributed to enable sustainable forest (and landscape) use. The model approach may thus help to investigate the following hypothesis:

> "A comprehensive land-use approach that simultaneously considers various agricultural and forestry options will enable the development of sustainable forest strategies, which would not compromise future agricultural production."

Table 25.1 Sources used to model the consideration of land-use options

Land-use option (land cover)	Comment	Sources
Natural forests (managed)	Low-impact forest management	Günter et al. (2008), Oesker et al. (2008), Knoke et al. (2009b), Chap. 13
Natural forests (unmanaged)	Natural forests left untouched	Pohle et al. (2010)
Forest plantations (on potential croplands)	Andean alder (*Alnus acuminata*) considered as a native tree species	Dunn et al. (1990), Grau and Veblen (2000), Fehse et al. (2002), Olschewski and Benitez (2005), Weber et al. (2008), Knoke et al. (2011), Chap. 13
Croplands	Represented by maize fields	Castro (2010), Knoke et al. (2011)
Pastures (active)	Represented by pastures with milk as main product	Knoke et al. (2009a, b, 2011)
Repasturization	Options to be considered on abandoned lands	Roos et al. (2010), Chap. 15
Reforestation		Dunn et al. (1990), Grau and Veblen (2000), Fehse et al. (2002), Olschewski and Benitez (2005), Weber et al. (2008), Knoke et al. (2009a, b, 2011), Chap. 13

25.2 Material and Methods

Plans for improving future land use often ignore the large land expanses occupied by already cleared but then subsequently abandoned areas (e.g. Phalan et al. 2011). One focus of our Research Unit (RU) has thus been to investigate possible future use options for abandoned (and often degraded) lands, such as reforestation (see Chap. 13) or repasturization (see Chaps. 15 and 26). Both options will be included in our comprehensive land-use strategy called "Optimised Land-Use Diversification". These options complement the already existing approach described by Knoke et al. (2011). All calculations are based on the research being carried out by several working groups of the RU (Table 25.1).

The sustainable and low-impact management of still existing natural forests (also treated in Chap. 13) is a further alternative to be integrated into future land-use planning. Leaving the natural forest untouched (total protection) is also an option. Of course, the actually dominating land-use form of pasturing is also included in our considerations, as well as the presence of croplands to enable the necessary national perspective. The studies we used to derive productivities, prices and costs are summarized in Table 25.1.

Our "Optimised Land-Use Diversification" is described in detail by Knoke et al. (2011). As already mentioned in brief, we have now modified this approach by integrating the activities "repasturization" and "reforestation" as options for use on abandoned lands. We prefer to use the perspective of risk-avoiding tropical landowners for viewing land-use problems instead of the eyes of a person interested in maximizing profits (Knoke et al. 2009b). Our risk-avoiding farmers all have the same land-use options (Table 25.1) and the same expectations. Moreover, in contrast to Knoke et al. (2011) we now constrain the cropland area to 10 % of the total land cover even under forest conservation scenarios. This constraint is important for satisfying future food demand. In addition to taking payments for protecting still existing natural forests into consideration, we analyse the influence of subsidies for the repasturization of abandoned lands on the effectiveness of forest conservation policies.

Possible future land-use distribution was estimated by means of non-linear programming under consideration of economic factors, an approach which closely follows the concept of "Optimised Land-Use Diversification". To hedge the adverse effects of uncertainties such as price fluctuations, risk-avoiding farmers will choose a land-use distribution in which financial land yield (Y_L)—minus the yield of a riskless benchmark investment (Y_R) per unit of risk—is at a maximum (this follows from classical financial theory). To pay due to this perspective of a risk-averting farmer, a land-use distribution is selected that maximizes the *reward-to-variability ratio* (R_L) introduced by Sharpe (1966, 1994) (Eq. 1). The financial yield which is thereby achieved in excess of the riskless yield is maximized per unit of risk (S_L: standard deviation of financial yield).

$$\max R_L = \frac{Y_L - Y_R}{S_L} \tag{1}$$

The model makes a simplification in assuming that all farmers consider the same land-use options and have the same future expectations. This would result in every farmer theoretically exercising the same structural portfolio of land uses, independently of his/her individual degree of risk aversion. From this simplification and from Tobin's theorem of separation (Tobin 1958) it follows that the individual degree of risk aversion comes into play only on the decision as to how much to invest in risky land management and how much in the riskless asset. Farmers may sell or purchase land, either to distribute their financial funds between riskless (*financial*) and risky (*natural*) assets (this is the case when selling land and investing in a safe asset), or to invest borrowed money to enlarge their risky natural assets (this is the case when purchasing land). In the case that land is sold, the obtained land price (US$ 1,000 per ha) is invested in a safe asset so that the riskless yield is determined by the land price multiplied by a riskless interest factor (5 %). The resulting Y_R is thus US$ 50 per ha per year in our case. Note that 5 % is a very high riskless interest. For comparative purposes, Wöhe and Döring (2010, p. 663) describe the riskless interest as ranging between 2 and 3 %. However, we use the figure of 5 % to express the usually high time preference of farmers.

Financial yield and price uncertainties as inputs for Eq. 1 follow from Eqs. 2 and 3 (vectors printed in bold).

$$
\begin{aligned}
Y_{\mathrm{L}} &= \mathbf{y}^{\mathrm{T}}\mathbf{a} = \sum_{i \in L} y_i a_i \\
&\text{s.t.} \\
y_i &= f(q_i, n_i, c_i) \\
\mathbf{1}^{\mathrm{T}}\mathbf{a} &= \sum_{i \in L} a_i = 1 \\
a_i &\geq 0
\end{aligned} \tag{2}
$$

In Eq. 2, $\mathbf{y}^{\mathrm{T}}$ is the transposed vector of the estimated annual financial yield per hectare for all considered land uses, where the yield of a given land use i (y_i) is an inverse demand function, $f(q_i,n_i,c_i)$, that declines with increasing country-wide production (q_i). The function further depends on productivity (n_i) as well as production costs (c_i). Finally, $\mathbf{a}$ is the vector of shares (i.e. percentual proportions: a_i) of the land-use options, the sum of which must be 1 (we consider the shares of all active land uses), and L is the set of all available land-use options.

$$
\begin{aligned}
S_{\mathrm{L}} &= \sqrt{\mathbf{a}^{\mathrm{T}}\Sigma\mathbf{a}} = \sqrt{\sum_{i \in L} \sum_{j \in L} a_i a_j \mathrm{cov}_{i,j}} \\
&\text{s.t.} \\
\mathbf{1}^{\mathrm{T}}\mathbf{a} &= 1 \\
\mathrm{cov}_{i,i} &:= \mathrm{var}_i \\
\mathrm{cov}_{i,j} &= k_{i,j} s_i s_j \\
a_{i,j} &\geq 0
\end{aligned} \tag{3}
$$

In Eq. 3, S_{L} is the standard deviation of the uncertain portfolio yield, Σ, is the covariance matrix that consists of the variances (var_i) and covariances ($\mathrm{cov}_{i,j}$) of uncertain financial yields caused by the price volatility for each land-use type i, where covariances between every possible land-use combination must be considered. Moreover, $k_{i,j}$ is the correlation coefficient between the yield of land-use types i and j, and s_i is the standard deviation of the yield for land-use i (assumed to be normally distributed).

For details on the estimation of the inverse demand curves and other technical issues see Knoke et al. (2011). Upfront subsidies to cover the initial investment for repasturization (US$ 1,500 per ha) are to be financed by credits (5 % interest) issued by development banks. The landowner would thus use the credit money to regenerate abandoned lands for pasture use (see Roos et al. 2010 for feasibility studies) and then face a yearly cost to pay back the credit inclusive interest. In our scenarios the money to pay back the credit is subsidized to various extents that analyse the influence of stimulating repasturization on the effectiveness of conservation strategies.

Note that since cropland food production (represented by maize cultivation) is necessary for Ecuador as a whole but is not a really representative asset in the San Francisco Valley, we want to adopt a more comprehensive perspective on this land-use form. Our approach is thus "top-down" and conceptualized for a perspective on the national level of a South American country like Ecuador.

25.3 Results and Discussion

25.3.1 Reforestation on Abandoned Lands Protects Natural Forests

Driven only by economic forces and without taking any conservation payments or subsidies into consideration, our generalizing model predicts that natural forests will shrink to an extent of only ~1 % of the total land area. This is due to the fact that there is no (unmanaged) or only a small (managed) flow of net revenues from the natural forests to farmer households (Table 25.2, Fig. 25.1 and Knoke et al. 2011).

Both croplands and pastures will increase significantly in area (the middle scenario in Fig. 25.1). The increase in pasture area will be larger, even though cropland is the more profitable alternative. This fact illustrates the risk-avoiding attitude of the farmers, because pastures bear a far lower risk than do croplands (standard deviation of net revenues ±18.3 versus ±46.1). The high proportion of forest plantations on potential agriculture land also reflects the risk-avoiding attitude of the farmers, who recognize the poor correlation between forest plantation and agricultural yields and thus use forests to diversify their income (Mills and Hoover 1982). Although a reduction of natural forests to only ~1 % of the total land area does not appear to be realistic, it accurately describes (at least in principle) the direction in which economic forces will drive future use of the land. It may thus represent a worst-case future scenario against which sustainable land-use strategies may be tested, and provides a suitable reference scenario for the commencement of our analysis.

It does not, however, appear to be a wise choice to use potential agriculture lands for forest plantations when a growing population is demanding more food and timber resources. Increasing food prices, timber demand and land scarcity will direct interest to the recultivation of abandoned areas. If either repasturization or reforestation is being considered for abandoned pastures, reforestation is the better choice. Forest plantations will thus be shifted from potentially agricultural lands to more marginal areas (right scenario in Fig. 25.1). This will also result in more natural forest persisting in the future: about 6 % of the present forest area should remain intact in place of the about 1 % predicted above. In our simulation, the reforestation of abandoned lands would thus have the secondary effect of conserving ~1.4 million ha of natural forests in Ecuador.

Table 25.2 Average net revenues for land-use options at their given spatial distribution, adopted from Knoke et al. (2011) with some alterations (e.g. for repasturization and reforestation)

Land-use option, i	Average financial yield (y_i) at initial distribution [US$ per ha per year]	Standard deviation (s_i) of financial yield at initial distribution
Unmanaged natural forest	–	–
Managed natu ral forest	31.8	±14.5
Forest plantation	105.6	±30.4
Cropland	208.4	±46.1
Pasture	130.5	±18.3
Abandoned land	–	–
Reforestation	105.6	±30.4
Repasturization	13.6	±18.3

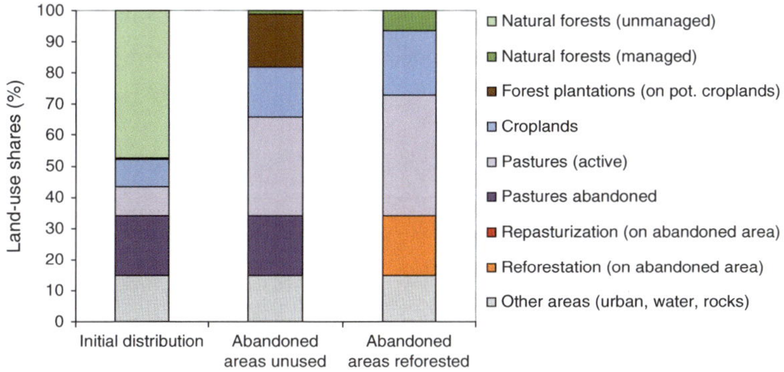

Fig. 25.1 Present (*left*) and future land-use shares—meaning the spatial distribution of land-use practices—excluding (*middle*) and including (*right*) the use of abandoned lands for reforestation

25.3.2 The Effect of Conservation Payments

All the future land-use scenarios described above are certainly not overly climate-friendly because they involve considerable amounts of carbon emissions. Knoke et al. (2011) computed a total emission of 5.2 billion tons CO_2 for the above-mentioned scenario of future land-use without reforestation on abandoned lands (Fig. 25.1, middle column). Compensation payments for forest protection should thus be considered as a counter measure to prevent emissions resulting from forest loss.

If cropland area is limited to at least 10 % of Ecuador's land cover and farmers are compensated with US$ 76 per ha and year for protecting their still existing natural forests, a constant area of natural forests is predicted by our model.

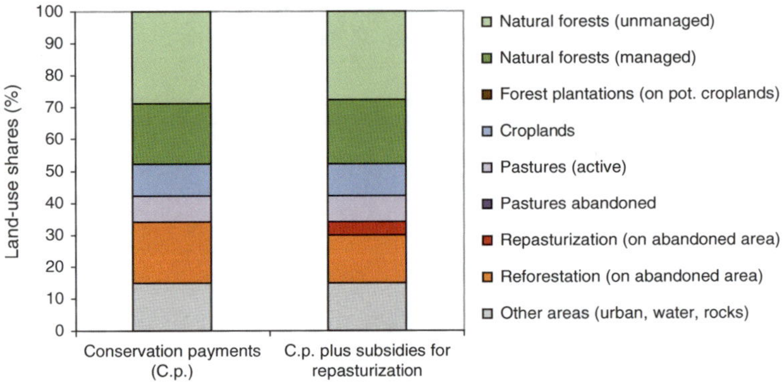

Fig. 25.2 Future land-use shares resulting from conservation payments (C.p.) as a reward for the conservation of natural forests without (*left*) and with additional subsidies for repasturization (*right*)

Approximately one-third of the natural forest area would be under low impact management in this scenario (Fig. 25.2), while the other two-thirds would remain untouched.

25.3.3 Subsidizing Repasturization Will Relax the Pressure on Natural Forests and Save Costs

Repasturization is an interesting option for forest conservation policies, although it is not very profitable for farmers without subsidies (Table 25.2). This option might mitigate the scarcity of agricultural lands and also lead to decreased prices for agricultural products. The pressure on natural forests could in turn also be lower than without any repasturization of abandoned lands.

The implementation of repasturization by subsidies will obviously have an impact on the overall land-use portfolio. If farmers would be provided with subsidies of US$ 1,000 per ha (68 % of total upfront investments), the area of repasturized parcels would increase to a proportion of 4.2 % (Fig. 25.2). The inclusion of repastured areas further diversifies the land-use portfolio. The increased area of repastured lands also has the above-assumed indirect, but important impact on the effectiveness of forest conservation. In fact, it relaxes the pressure on natural forests to the extent that only ~US$ 60 per ha and year is required as a conservation payment to keep the area of natural forests constant. This is instead of the US$ 76 per ha and year which would be necessary when repasturization is not subsidized.

If both compensation payments to protect the forest (US$ 60 per ha per year) and subsidies for repasturization (US$ 1,000 per ha) are integrated into our model, total conservation costs would be up to 12 % smaller than in the case of no subsidies for

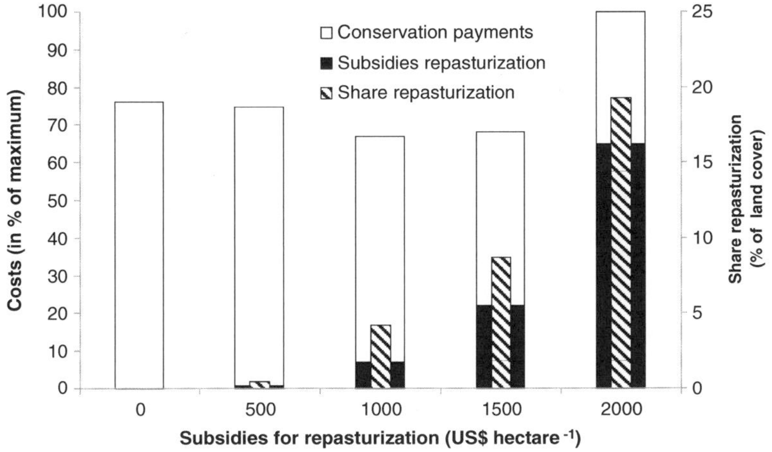

Fig. 25.3 Effect of subsidies for repasturization on the total costs for conservation policies

repasturization. Assuming an increase in the subsidy to US$ 1,500 per ha, the total costs would still be 11 % lower than without repasturization (Fig. 25.3). In this case, the area of repasturization would increase to 8.7 % of the total land-use area. The repasturization area could even be boosted to 19.3 % (which is the total area of the currently abandoned lands) by providing landowners with a subsidy of US$ 2,000 per ha (Fig. 25.3). However, this would increase the total conservation costs to 131 % of the costs incurred without considering the repasturization option. In conclusion, a moderate degree of repasturization stimulated by subsidies would lead to only modest expenditures for conservation while at the same time providing more food for the people.

25.4 Conclusions

According to Wise et al. (2009) we may state: "Improved land-use management and improved agricultural practices could reduce upward pressure on crop prices and the costs of emissions mitigation. However, the allocation of scarce land resources to competing ends will remain a major challenge for the 21st century." This statement justified the testing of our initial hypothesis:

> "A comprehensive land-use approach that simultaneously considers various agricultural and forestry options will enable the development of sustainable forest strategies, which would not compromise future agricultural production."

In support of the above hypothesis we have shown how a combination of conservation payments to protect the existing natural forest and subsidies to re-establish pastures on abandoned lands, as well as forestry options (reforestation on abandoned lands and sustainable low-impact management in part of the natural forest area), may lead to the sustainable use and conservation of tropical forests and their ecosystem services, such as carbon storage. This finding was provided by integrative and risk-sensitive land-use modelling, in which all available land-use options are considered simultaneously to make use of synergies. This was done in a way such that indirect effects, for example between subsidies for repasturization and conservation payments, could be mirrored. This approach shows great potential. Our still general and aggregated model must be supplemented by spatially explicit approaches in the future. Moreover, the still static approach must be further developed to consider the transition periods required for one land-use distribution to change into another (dynamic approach). These improved models could then be used for landscape zoning to designate priority areas for cropland production, pasturing, reforestation and repasturization activities.

References

Castro LM (2010) Conservation payments for agroforestry systems in southwest Ecuador. Master's Thesis, Center of Life and Food Sciences Weihenstephan, Technische Universität München

Dunn WW, Lynch AM, Morgan P (1990) Benefit-cost analysis of fuelwood management using native alder in Ecuador. Agrofor Syst 11:125–139

Fehse J, Hofstede R, Aguirre N, Paladines S, Kooijman A, Svenik J (2002) High altitude tropical secondary forests: a competitive carbon sink? For Ecol Manage 163:9–25

Gardner TA, Barlow J, Chazdon R, Ewers RM, Harvey CA, Peres CA, Sodhi NS (2009) Prospects for tropical forest biodiversity in a human-modified world. Ecol Lett 12:561–582

Godfray HCJ, Pretty J, Thomas SM, Warham EJ, Beddington JR (2011) Linking policy on climate and food. Science 331:1013–1014

Grau HR, Veblen TT (2000) Rainfall variability, fire and vegetation dynamics in neotropical montane ecosystems in north-western Argentina. J Biogeogr 27:1107–1121

Günter S, Cabrera O, Weber M, Stimm B, Zimmermann M, Fiedler K, Knuth J, Boy J, Wilcke W, Iost S, Makeschin F, Werner F, Gradstein R, Mosandl R (2008) Natural forest management in neotropical mountain rain forests – an ecological experiment. In: Beck E, Bendix J, Kottke I, Makeschin F, Mosandl R (eds) Gradients in a tropical mountain ecosystem of Ecuador. Ecological studies, vol 198. Springer, Berlin, pp 347–359

Knoke T, Weber M (2006) Expanding carbon stocks in existing forests – a methodological approach for cost appraisal on the enterprise level. Mitig Adapt Strat Glob Chang 11:579–605

Knoke T, Weber M, Barkmann J, Pohle P, Calvas B, Medina C, Aguirre N, Günter S, Stimm B, Mosandl R, Walter F, Maza B, Gerique A (2009a) Effectiveness and distributional impacts of payments for reduced carbon emissions from deforestation. Erdkunde 63:365–384

Knoke T, Calvas B, Aguirre N, Román-Cuesta RM, Günter S, Stimm B, Weber M, Mosandl R (2009b) Can tropical farmers reconcile subsistence demands with forest conservation? Front Ecol Environ 7:548–554

Knoke T, Steinbeis OE, Bösch M, Román-Cuesta RM, Burkhardt T (2011) Cost-effective compensation to avoid carbon emissions from forest loss: an approach to consider price–quantity effects and risk-aversion. Ecol Econ 70:1139–1153

Mills WL, Hoover WL (1982) Investment in forest land: aspects of risk and diversification. Land Econ 58:33–51

Oesker M, Dalitz H, Günter S, Homeier J, Matezki S (2008) Spatial heterogeneity patterns – a comparison between gorges and ridges in the upper part of an evergreen lower montane forest.

In: Beck E, Bendix J, Kottke I, Makeschin F, Mosandl R (eds) Gradients in a tropical mountain ecosystem of Ecuador. Ecological studies, vol 198. Springer, Berlin, pp 267–274

Olschewski R, Benitez PC (2005) Secondary forests as temporary carbon sinks? The economic impact of accounting methods on reforestation projects in the tropics. Ecol Econ 55:380–394

Phalan B, Onial M, Balmford A, Green RE (2011) Reconciling food production and biodiversity conservation: land sharing and land sparing compared. Science 333:1289–1291

Pohle P, Gerique A, Park M, López Sandoval MF (2010) Human ecological dimensions in sustainable utilization and conservation of tropical mountain forests under global change in southern Ecuador. In: Tscharntke T, Leuschner C, Veldkamp E, Faust H, Guhardja E, Bidin A (eds) Tropical rainforests and agroforests under global change, Environmental science and engineering. Springer, Berlin, pp 477–509

Roos K, Rödel HG, Beck E (2010) Short- and long-term effects of weed control on pastures infested with *Pteridium arachnoideum* and an attempt to regenerate abandoned pastures in South Ecuador. Weed Res 51:165–176

Sharpe WF (1966) Mutual fund performance. J Bus 39:119–138

Tobin J (1958) Liquidity preferences as a behavior towards risk. Rev Econ Stud 25:65–86

UNFCCC (2002) The Marrakesh accords and the Marrakesh declaration. In: Report of the conference of the parties on its seventh session, Marrakesh, 29 Oct-10 Nov 2001

Weber M, Günter S, Aguirre N, Stimm B, Mosandl R (2008) Reforestation of abandoned pastures: silvicultural means to accelerate forest recovery and biodiversity. In: Beck E, Bendix J, Kottke I, Makeschin F, Mosandl R (eds) Gradients in a tropical mountain ecosystem of Ecuador. Ecological studies, vol 198. Springer, Berlin, pp 431–441

Wise M, Calvin K, Thomson A, Clarke L, Bond-Lamberty B, Sands R, Smith SJ, Janetos A, Edmonds J (2009) Implications of limiting CO_2 concentrations for land use and energy. Science 324:1183–1186

Wöhe G, Döring U (2010) Einführung in die allgemeine Betriebswirtschaftslehre, 24th edn. Vahlen, Munich

Chapter 26
Future Provisioning Services: Repasturisation of Abandoned Pastures, Problems, and Pasture Management

Erwin Beck, Jörg Bendix, Brenner Silva, Rütger Rollenbeck, Lukas Lehnert, Ute Hamer, Karin Potthast, Alexander Tischer, and Kristin Roos

26.1 Introduction

In the Rio San Francisco valley, three types of pastures were differentiated, the *pastos azules*, the *Melinis* pastures, and the *Setaria* pastures (*pastos mieles*, Chap. 15). Here, we concentrate on the latter which is the dominating type in the research area and beyond. *Pastos mieles* are highly susceptible to invasion by weeds, mainly by the tropical bracken fern. *Setaria sphacelata* is an exotic pasture grass from tropical Africa which is planted manually in horizontal rows at a density of around 32,400 tufts per hectare. Although amply flowering, it does hardly produce viable seeds. Instead, it rapidly propagates by tillers and most of the *Setaria* pastures represent clones (Chap. 15). Performing C4-type photosynthesis, it can produce high biomass harvests of up to 37 Mg ha^{-1} a^{-1} (Mizumachi et al. 2011). In the research area, however, biomass production is commonly lower (see Sects. 26.3.1.1 and 26.3.5). *Setaria* pastures are established mainly on sloping and well-drained ground where light can penetrate through the canopy down to the ground. This allows weeds to germinate and establish. To keep control of the weeds and at the same time stimulate regrowth of the grass, farmers once in a while set fire on the

E. Beck (✉) • K. Roos
Department of Plant Physiology and Bayreuth Centre of Ecology and Environmental Research, University of Bayreuth, 95440 Bayreuth, Germany
e-mail: erwin.beck@uni-bayreuth.de

J. Bendix • B. Silva • R. Rollenbeck • L. Lehnert
Laboratory of Climatology and Remote Sensing, University of Marburg, Deutschhausstr. 10, 35032 Marburg, Germany

U. Hamer
Institute of Landscape Ecology, University of Muenster, Robert-Koch-Str. 2848149 Muenster, Germany

K. Potthast • A. Tischer
Institute of Soil Science and Site Ecology, Dresden University of Technology, 01737 Tharandt, Germany

J. Bendix et al. (eds.), *Ecosystem Services, Biodiversity and Environmental Change in a Tropical Mountain Ecosystem of South Ecuador*, Ecological Studies 221,
DOI 10.1007/978-3-642-38137-9_26, © Springer-Verlag Berlin Heidelberg 2013

pastures. Burning, however, stimulates not only growth of the grass, but, unfortunately also of special weeds which after repeated burning increasingly suppress the grass (Hartig and Beck 2003; Beck et al. 2008). Finally a type of shrubland results, termed succession fallow (Makeschin et al. 2008) where a few remaining grass tufts indicate the former pasture. Although not attaining a real climax, this type of secondary vegetation is fairly stable showing resilience against episodic burning. This is due to a high proportion of weed species which in addition to an enormous seed production can readily re-sprout from the root stocks after a fire. Bracken has the potential of ample sexual propagation by spores, too (Page 1986). Its dense rhizome system, growing at various soil depths, additionally furnishes its resistance to many exogenous impacts. Large former pasture areas have thus developed into shrubland (Chap. 15). Some of these areas are said by farmers to be fallow on intention where the pasture shall be regenerated after burning. However, once the weeds have established, sustainable repasturisation, simply by burning of the vegetation, is not possible. Nevertheless, the vast areas of abandoned pastures in the San Francisco valley may represent a resource for potential pastoral use, if an effective method for repasturisation can be elaborated. To that end, understanding the eco-physiological and ecological causes and processes of pasture degradation is indispensable.

Addressing the future improvement of the ecosystem and its services, this chapter reports an extended rehabilitation experiment for abandoned pastures. The long-term experiment was accompanied by an investigation of the competition between the pasture grass and the primary invader bracken. Growth of both species was modelled as depending on the weather conditions, and followed over 1 year. In addition the effect of burning on the regrowth of both species was documented, as well as the reaction of the competitors to fertilisation and simulated grazing. Finally, a fertilisation experiment (FERPAST) of an active pasture is presented, addressing the prevention of negative nutrient balances of active pastures by moderate fertilisation with an emphasis on soil quality, especially of nitrogen and phosphorus availability, grass yield, and quality.

26.2 Materials and Methods

26.2.1 *Competition Between* Setaria *and Bracken: The SoBraCoMo and Competition for Nutrients*

Growth competition between the C3-species Southern Bracken and the C4-species *Setaria sphacelata* was simulated with a canopy-photosynthesis model (*SoBraCoMo*, see Sect. 24.2.1). In short, the *SoBraCoMo* consists of four modules which were parameterised with data from field and laboratory measurements for both species (Bendix et al. 2010; Silva et al. 2012). Environmental forcing was conducted with concomitant measurement of solar radiation, precipitation, soil water content, and leaf temperature which were continuously recorded with a temporal resolution of 10 min in the field.

Nutrient analysis of leaf samples of both species was realised by a CNS Analyser (vario EL III/elementar; total N concentration) as well as by ICP-OES (CIROS, Spectro; total element concentrations after acid digestion). For details refer to Chap. 22.

26.2.2 Regrowth of Bracken and *Setaria* After Burning: Monitoring by Automated Vegetation Analysis

Farmers set repeatedly fire on their pastures to stimulate growth of the grass and destroy the weeds. Such treatment was simulated in November 2009, by burning an abandoned pasture area of 50 × 20 m on a 40° steep slope where *Setaria* and bracken grew together. To lit the fire, dry plant material harvested from a fire protection girdle around the experimental area was heaped up on the area and imbued with ethanol (instead of the diesel fuel which is commonly used by the farmers). For complete burning, remaining patches were manually treated with gas burners.

Before and several times after burning, the vegetation was automatically recorded by two cameras (normal and near-infrared) mounted in series on a balloon kept at a flight height of 35 m aboveground. Individual aerial photos were merged into a mosaic image of the whole site, applying geometric correction and stitching techniques. For geometric rectification selected ground control points were established which allowed transformation to a planar projection. Identification of bracken and *Setaria* was by spectral signatures which had been verified by ground truthing (Bendix et al. 2010).

26.2.3 Pasture Rehabilitation Experiment

A three-step pasture rehabilitation experiment (Fig. 26.4) was performed on a slope (25°–40° steep, at ca. 2,000 m a.s.l.) that was heavily infested by bracken. We started with bracken control applying mechanical (cutting), chemical (asulam; glyphosate; picloram + metsulfuron methyl = PM; pinolene, a transpiration blocker), and physical treatments (covering with transparent or black plastic sheeting) and combinations thereof. In total 13 methods were examined. Treatments were administered over a time period of 23 months in a series of six repetitions in intervals of 4–6 months, depending on the weather conditions. After a short fallow, *Setaria* was re-planted in a second step. Bunches of the grass were obtained from local farmers, shortened to a length of 15–20 cm and cut into small rooted portions of approx. 3 cm diameter. In each experimental plot of 25 m^2, 81 portions were planted in the usual distances of 50 cm (32,400 plants ha^{-1}). Bracken fronds were cut after 9 months to keep it under control.

Growth of bracken (and later of *Setaria*) was monitored monthly. In these records using these records the performance of bracken and grass was followed using a proxy (cover abundance (percent of the area) times frond height (cm)) for the above-ground biomass[1] (for details see Roos et al. 2011). Real above-ground bracken biomass was determined once at the end of the bracken control experiment.

26.2.4 Pasture Management Study

For elaborating management protocols for the recovered pastures the plots were anew randomly arranged in four parallels each (step 3 in Fig. 26.4). Management variables were grazing and fertilisation, both with two intensities and also in combination (Fig. 26.5). Grazing was simulated by cutting the leaves down to a height of 20 cm while also trampling on the bracken fronds. Intensities of "grazing" were 2 and 5 cuts per year. Fertilisation was carried out with a customary pasture fertiliser containing N (21 %), P (12 %), K (15 %), Mg (3 %), and S (4 %). Fertilising intensities were (a) low (50 kg N, 29 kg P, and 36 kg K per ha and year) and (b) intense (150 kg N, 86 kg P, and 107 kg K) in three equal portions over the year. Growth was recorded monthly and the same above-ground biomass proxy as above was applied.

26.2.5 Fertilisation Experiment

The pasture fertilisation experiment (FERPAST) was established at an extensively used *Setaria sphacelata* pasture of 0.5 ha. Moderate amounts of urea (50 kg N ha^{-1} a^{-1}), rock phosphate (10 kg P ha^{-1} a^{-1}), and a combination thereof were applied on six replicate plots (25 m^2 each) which were block-wise arranged along with a control plot of the same size. Rotational grazing was simulated quarterly by cutting the leaf tufts down to approx. 3 cm sward height. Total dry matter yield of the treatments was determined and the nutrient contents of the leaf samples were analysed by the methods described above (Sect. 26.2.1) and also presented in Chap. 22.

[1] A monthly harvest of the above-ground biomass was not possible due to the limited experimental area.

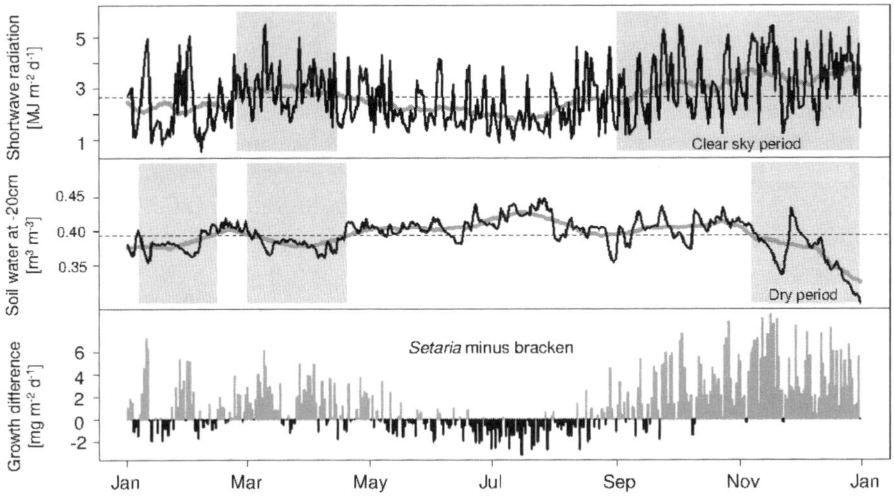

Fig. 26.1 Daily accumulated solar radiation, soil water content (at 20 cm depth), and difference between the potential daily dry matter production of *Setaria* (*grey*) and bracken (*black*) for the year 2008 in the Rio San Francisco valley. The *dashed lines* indicate the annual mean values of solar radiation and soil water content. The *grey lines* in the *upper two panels* represent moving window means of 30 days which were used to define periods of specific weather situations. The *lower panel* indicates when and to which extent bracken or *Setaria* produces more biomass

26.3 Results and Discussion

26.3.1 Competition Between Setaria *and Bracken*

26.3.1.1 Modelling the Response of Growth to Weather Conditions Over the Year: The Southern Bracken Competition Model (SoBraCoMo)

The results of the SoBraCoMo show a potential dry matter production (sum of above- and below-ground) of 55.5 Mg ha^{-1} a^{-1} for southern bracken, which is slightly lower than that of the grass (58.8 Mg ha^{-1} a^{-1}). Comparable potentials of dry matter production were also published by Portela et al. (2009) for bracken and Mizumachi et al. (2011) for *Setaria* (above ground).

As expected, biomass production follows the course of the weather conditions over the year of which solar shortwave radiation and soil moisture are shown exemplarily in the upper part of Fig. 26.1. The lower panel indicates when and to which extent bracken (black) or *Setaria* (grey) produced more biomass. Expectedly, productivity of the C4-plant *Setaria* particularly decreased during colder and rainy periods (May to August and January/February). On the other hand, the grass can produce more biomass than bracken on dry and sunny days (shaded areas in Fig. 26.1). The model shows the potential of *Setaria* to outstrip the fern although bracken is more efficient in absorbing radiation on bright days (Bendix et al. 2010). On the other hand, both species are likewise light demanding, and therefore shading of the grass by bracken fronds would handicap the pasture grass.

Temperature also plays a crucial role, in particular for the C4-grass *Setaria* whose thermal growth response is improving with increasing air temperature up to 25–30 °C. In the study area, the average daily temperature oscillates between 14 and 22 °C. Regarding thermal effects on photosynthesis, C3 (bracken) is favoured over C4 (*Setaria*) below 15–16 °C and *vice versa* above this thermal threshold (Chap. 24). Thus, several 100 m further down or up the valley, which means higher or lower air temperatures compared to the study site, will strengthen or weaken the competitiveness of *Setaria* against bracken.

26.3.1.2 Competition for Nutrients as Modulated by a Pathogenic Fungus

In the course of the research further details of the competition between *Setaria* and bracken turned out. Chemical analyses of the plant material revealed a higher capacity of bracken to extract growth-limiting macronutrients from the soil, namely P and N. *Setaria*, on the other hand, accumulated more K and Ca (Table 26.1). These results are confirmed by visible symptoms of P-, N-, and S-deficiency on the grass leaves (red margins and pale intercostal fields), whereas bracken fronds were vividly green.

In that respect, bracken appears superior to *Setaria*; the more so as the latter is grazed by the cattle, while bracken, as a potentially poisonous plant (Hannam 1986; Fenwick 1989), is spurned and at most trampled by the animals. On the other hand, Southern Bracken appears highly susceptible to infection by a phytopathogenic ascomycete (*Cryptomycella pteridis* (Kalchbr.) Hohn[2] = *Ascochyta pteridis* Bres.) which kills the fronds irrespective of their age. Infection by *Cryptomycella* occurs patchwise (Roos et al. 2011), but the path of infection is still unknown.

All in all, the *Setaria* pastures can achieve a frail balance between the pasture grass and the weed. This balance strongly depends on the intensity of grazing which is a very delicate issue: Heavy grazing shifts the equilibrium in favour of bracken, whereas low grazing, in particular very long intervals between the grazing periods, will induce flowering of the grass.

[2] The pathogen was tentatively identified by R. Berndt, Swiss Federal Institute of Technology, Zurich, Switzerland.

Table 26.1 Nutrient contents (N, P, Ca) and nutrient ratios (C:N, C:P, Ca:P) of *Setaria sphacelata* and bracken leaves from the active (FERPAST, control plots) and abandoned pasture sites (regeneration experiment) as well as of bracken leaves of different age in the Rio San Francisco valley (mean, SE, $n = 6$)

Leave	Site		N [%]		P[%]		Ca[%]		C:N		C:P		Ca:P	
Setaria	Active pasture		1.34	(0.02)	0.18	(0.003)	0.43	(0.02)	33.3	(2.7)	247	(33)	2.4	(0.3)
	Abandoned pasture		0.45	(0.02)	0.053	(0.002)	0.39	(0.04)	100	(4.5)	847	(33)	7.3	(0.6)
Bracken	Active pasture	young	1.5	(0.02)	0.19	(0.002)	0.13	(0.01)	33.0	(2.1)	215	(30)	0.6	(0.03)
		old	1.15	(0.02)	0.14	(0.002)	0.32	(0.02)	42.0	(3.8)	349	(41)	2.3	(0.2)
	Abandoned pasture		0.91	(0.1)	0.12	(0.028)	0.23	(0.02)	55	(4.7)	420	(59)	1.9	(0.2)

For description of analyses, see Chap. 22

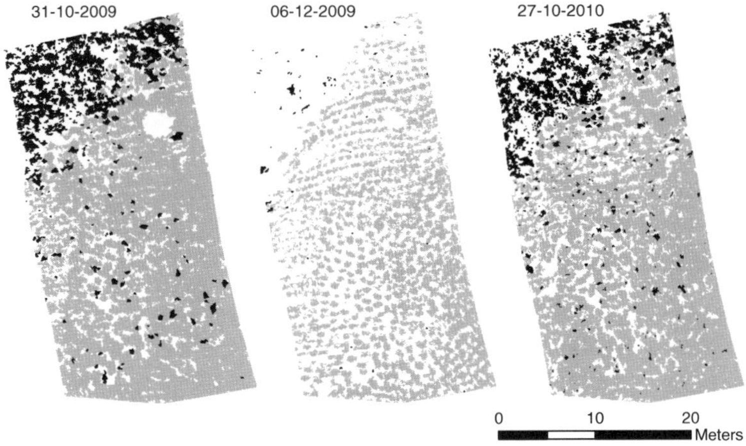

Fig. 26.2 Distribution of *Setaria* (*grey*) and bracken (*black*) on the experimental plot 4 days before (*left*), 1 month (*middle*), and about 1 year (*right*) after controlled burning of the area. *White areas* are either bare soil or covered by other plant species than grass and fern

26.3.2 Regrowth of the Pastures After Burning (Fire Experiment)

If the pastures are well managed by frequent grazing and clearing of weeds, burning can be suspended for years; however due to own observations these conditions rarely apply and burning is conducted once *Setaria* has started to flower or weeds had grown high. Occasionally *Setaria* is planted anew after burning.

We observed vegetation regrowth after experimental burning. Figure 26.2 shows rectified camera shots 4 days before (left), 1 month (middle), and nearly 1 year after the fire experiment (right). While *Setaria* had already developed small tufts 1 month after burning, cover abundance by bracken was still very low but many (more than 50) small dots representing emerging bracken fronds could be spotted. After 1 year cover abundances of *Setaria* and of bracken reached almost the pre-burning state (95 % for both species), but many more bracken leaves had come up (12 % more), and also on sites where bracken fronds were not present before burning. This observation confirms our earlier results (Roos et al. 2010) on the triggering effect of a heat wave on leaf-bud break on the fern rhizomes. The result also demonstrates that bracken rhizomes were everywhere in the ground, but due to competition by the grass could not produce fronds all over the area. In summary, stronger promotion of the fern than of *Setaria* by burning of an infested pasture is quite obvious.

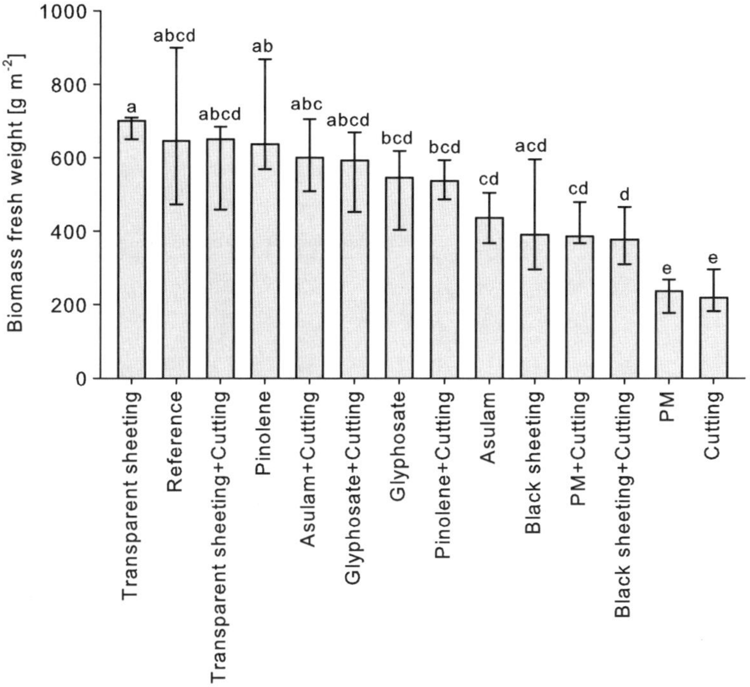

Fig. 26.3 Effects on frond biomass of bracken of six consecutive bracken control treatments ($\chi^2 = 38.54$, df = 13, $P < 0.001$) (PM = application of the herbicide combination picloram and metsulfuron methyl). Medians with 75th and 25th percentiles of four plots per treatment are shown. Different letters indicate statistically significant differences between the effects of treatments (Kruskall–Wallis H with post hoc Mann–Whitney *U* Test $\alpha = 0.05$). From Roos et al. (2011)

26.3.3 The Pasture Rehabilitation Experiment ("Repasturisation")

Thirteen methods of bracken control were applied for 2 years to a former *Setaria* pasture, which was completely (up to 80 % coverage) overgrown by bracken and a few shrubs (Roos et al. 2011). Using bracken above-ground biomass as the most reliable measure, only the treatments with the herbicide mixture PM, and cutting, respectively, resulted in a statistically significant *reduction* of about 65 % (Fig. 26.3). The treatments with black foil, PM + cutting, and black foil + cutting were also very effective; but the effects were statistically not significant.

Using a proxy for above-ground biomass production which could be determined monthly, pronounced differences between the effects of the various treatments on *regrowth* of bracken were observed (not shown here, for details see Roos et al. 2011). Regular cutting with a machete and the herbicide PM were by far the most effective treatments (Fig. 26.4). Spraying with PM showed initially only a slight reduction but the second treatment resulted in a total breakdown from which the

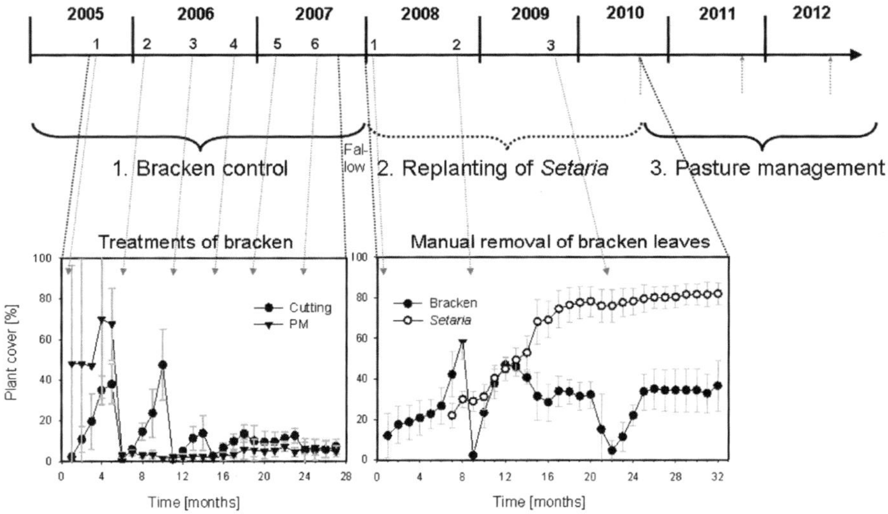

Fig. 26.4 Set-up and timetable of the pasture regeneration experiment showing the three steps: (1) bracken control, (2) replanting of *Setaria*, and (3) pasture management (PM = application of the herbicide combination picloram and metsulfuron methyl). For a detailed description of the area and steps 1 and 2, see Roos et al. (2011). The results of step 3 are presented in Fig. 26.5

weed did not recover during the subsequent 1½ years. The maximum effect was achieved after three consecutive treatments. The combination: cutting—PM—cutting likewise resulted in maximal bracken reduction.

Considering the morphology of bracken with its rhizomatous shoots buried in the soil, complete eradication by above-ground control treatments (even with systemic herbicides) has proven impossible (Marrs et al. 1998). Therefore a realistic aim is to keep bracken at a low level at which it does not seriously affect growth of the pasture grasses. In that respect and in view of practicability and costs the number of repetitive treatments is important, which are sufficient to achieve the maximum effect. On the basis of the results three consecutive treatments with the herbicide PM or four consecutive cuts are recommended which resulted in a reduction of bracken by more than 80 % (Fig. 26.4) (Roos et al. 2011).

After 2 months of fallow *Setaria* was re-planted. Eighteen months later, the grass clearly dominated on all plots, with an average cover abundance of 73 %, while the average cover of bracken was 36 % (Fig. 26.4). With an average tuft height of 69.4 ± 1.3 cm (mean ± SE), *Setaria* overtopped the bracken fronds (64.7 ± 1.3 cm, mean ± SE). This is important, as the fern could not shade the grass.

The repasturisation experiment clearly showed the superiority of the C4-plant *Setaria* in biomass production and growth, once Southern Bracken has been substantially weakened. But bracken showed a tendency to recover (Fig. 26.4) and thus in a few years could again suppress *Setaria*. Therefore the recommendation was, to cut the bracken fronds once per year.

One point of concern were potentially remaining residues of the herbicides in the soil. In spite of small differences in the initial growth rates of *Setaria* on some plots, statistically significant impacts could not be determined. In the soil, 2 years after herbicide application soil microbial activity and biomass were the same as in the control plots (M. Kirsten, unpublished).

26.3.4 Management of Recovered Pastures

After re-establishment of the pasture several management protocols were set up and examined for proposing a regime of sustainable pasture utilisation. The experiment started with removal of re-grown bracken fronds in all plots which is reflected by the initial decline of bracken in October 2010 (Fig. 26.5). In the control plots *Setaria* kept its level slightly below and above the biomass proxy value (see Sect. 26.2.4) of 6,000. The fern recovered after cutting to a value of 2,000 but continued to increase slowly in the course of the year. The simulation (mowing) of low grazing without fertilisation resulted in a lower value of *Setaria* and a higher of bracken which equaled after 1 year. This contrasts with the plots on which *Setaria* was mowed five times: Although after the fourth and fifth cut the grass attained a lower value than after two cuts, bracken was even more affected by the recurrent trampling. Low fertilisation did not improve growth of the competing species in contrast to intense fertilisation which almost doubled growth of *Setaria*. Ample growth of the grass in turn strongly suppressed bracken. Intense fertilisation combined with low "grazing" resulted in a biomass proxy value of 6,000 for *Setaria* and only 1,000 of bracken. Three more cuts reduced the value of the grass to 3,500 but almost abolished the fern. Thus intense fertilisation will allow two or three cuts or grazing periods in the course of a year without the originally proposed annual cut of the bracken fronds. The results of the first year of pasture management experiments were encouraging. Obviously on the original nutrient-depleted soil, *Setaria* could not fully develop its growth potential as C4 grass. It is known that yield increase of *Setaria* requires sufficient nitrogen supply at a developmental stage when the tufts are still small (Hacker and Jones 1969). What remains to be determined are the optimal time points for fertilisation and grazing.

26.3.5 Fertilisation of Active Pastures (FERPAST Experiment)

Fertilisation beyond the "natural fertilisation" by the cowpats is uncommon in the study area. Therefore shortage of some essential nutrients like N and P is not surprising even in the soil of the active pastures and the more so of abandoned sites. Nutrient losses and nutrient immobilisation are triggered especially by burning

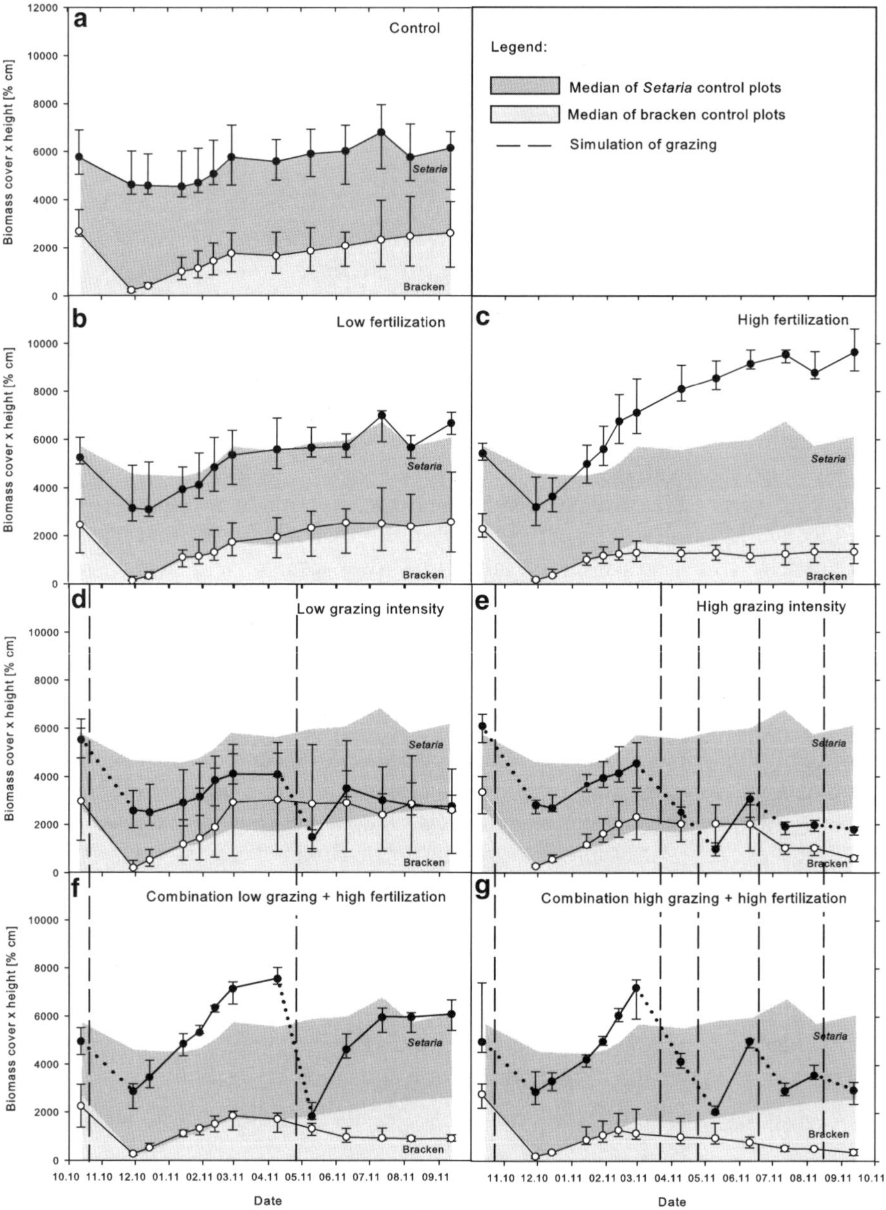

Fig. 26.5 Results from step 3 (see Fig. 26.4) of the pasture regeneration experiment: Optimisation of management of the restored pastures. Development of *Setaria* and bracken without any treatment (Control, **a**) and under six different regimes of management (**b**–**g**). Standing crop was estimated with the product of cover abundance [% cover] and height of the plants [cm] as a proxy for above-ground biomass

(Cleveland et al. 2003; Feigl et al. 2006; Numata et al. 2007). To avoid pasture degradation of active pastures in the San Francisco valley and as a guideline for the future operation of rehabilitated areas the FERPAST experiment was started in 2008.

Deficiency of soil nutrients was very well reflected in the element contents of *Setaria* leaves (Table 26.1). Phosphate contents of *Setaria* growing on abandoned and active pastures were below the critical level of 2.1 mg g^{-1} for normal plant growth (Skerman and Riveros 1990). Following moderate fertilisation with rock phosphate, the P content of *Setaria* leaves increased reaching the range of sufficient supply. At the FERPAST site a significant biomass response of *Setaria* to fertilisation was only observed after combined fertilisation with P and N. The increase in annual above-ground biomass production by one-fourth to 11.2 Mg ha^{-1} a^{-1} suggests that both macronutrients were limiting in the FERPAST soil (Potthast et al. 2012a). The contents of N, P, and Ca as well as the nutrient ratios of the leaves reflected the overall nutrient availability at the particular sites (Table 26.2). Thus, it was hypothesised that the lower nutrient concentrations of *Setaria* at the abandoned pasture compared to the active pasture site are indicative of a multiple nutrient limitation of plant growth.

As described in Sect. 26.3.1 bracken is a strong and successful competitor in the uptake of nutrients. This competitive advantage was observed not only on the abandoned but also on the active pastures. The N and P concentrations of bracken fronds from abandoned areas were approximately 30 % lower than those from active pasture plots. However, N and P concentrations of *Setaria* leaves were 50 % lower compared to *Setaria* at the active pasture. This pattern again reflects a higher nutrient use efficiency and resorption potential (Killingbeck 1996) of the fern of N and P were significantly higher in young than in old fronds (Table 26.1). Thus, if cutting is used as a bracken control method it should be targeted to the young leaves. According to a study by Potthast et al. (2010) litter of young fronds may be easily decomposable by microbes and would lead to higher rates of nutrient mineralisation.

26.4 Conclusion

The widely used pasture grass *Setaria sphacelata* emerged as a powerful competitor of Southern Bracken if managed properly. Rehabilitation of highly bracken-infested abandoned pastures is possible, using the elaborated strategy of combatting bracken to a state in which it can be controlled by the subsequently planted *Setaria*. Crucial for the sustainability of rehabilitated pastures is the surrender of the widely used rejuvenating of pastures by burning, an adjusted fertilisation to promote grass growth, and an optimised grazing rotation to keep the grass vital while weakening the fern by trampling.

A stock density of 0.7–1.5 cattle heads per ha and year has been described for the research area (Chap. 15). Those pastures are only fertilised by cow dung during grazing and therefore net increase of nutrients cannot be expected. If the grass

Table 26.2 Properties of soils (0–5 cm depth, $n = 6$) from active (FERPAST, control plots) and abandoned pasture sites (regeneration experiment)

Site	SOC [mg g^{-1}]	TN [mg g^{-1}]	NH_4^+–N [mg kg^{-1}]	NO_3^-–N [mg kg^{-1}]	PO_4–P [mg kg^{-1}]	Exchangeable Ca [mmolCa kg^{-1}]	pH (H_2O)
Active pasture	106 (21)	7.8 (1.5)	4.47 (0.81)	0.12 (0.30)	4.5 (2.5)	21.9 (9.3)	5.3 (0.2)
Abandoned pasture	96 (0.8)	4.6 (0.5)	1.25 (0.61)	0.27 (0.26)	0.5 (0.4)	8.5 (3.8)	4.4 (0.2)

The content of soil organic carbon (SOC) and total nitrogen (TN) was determined by a CNS analyser; NH_4–N and NO_3–N were extracted by KCl solution and PO_4–P by Bray I solution (NH_4F + HCl); exchangeable Ca in mmol charge per kg was extracted by NH_4Cl (Potthast et al. 2012b). Means and (SE) are shown biomass production can be doubled by fertilisation, at least the upper value of stock density can be attained. In addition weeding would be no longer necessary and the maintenance costs estimated by the farmers (Chap. 15) could be retrenched. In this way pasture farming may become profitable even for small holders.

References

Beck E, Hartig K, Roos K (2008) Forest clearing by slash and burn. In: Beck E, Bendix J, Kottke I, Makeschin F, Mosandl R (eds) Gradients in a tropical mountain ecosystem of Ecuador, vol 198, Ecological studies. Springer, Berlin, pp 371–374

Bendix J, Silva B, Roos K, Göttlicher DO, Rollenbeck R, Nauß T, Beck E (2010) Model parameterization to simulate and compare the PAR absorbtion potential of two competing plant species. Int J Biometeorol 54:283–295

Cleveland CC, Townsend AR, Schmidt SK, Constance BC (2003) Soil microbial dynamics and biogeochemistry in tropical forests and pastures, southwestern Costa Rica. Ecol Appl 13:314–326

Feigl B, Cerri C, Piccolo M, Noronha N, Augusti K, Melillo J, Eschenbrenner V, Melo L (2006) Biological survey of a low-productivity pasture in Rondonia state, Brazil. Outlook Agric 35:199–208

Fenwick GR (1989) Bracken (*Pteridium aquilinum*) – toxic effects and toxic constituents. J Sci Food Agric 46:147–173

Hacker JB, Jones RJ (1969) The *Setaria sphacelata* complex – a review. Trop Grasslands 3:13–34

Hannam DAR (1986) Bracken poisoning in farm animals with special reference to the North York Moors. In: Smith RT, Taylor JA (eds) Bracken: ecology, land use and control technology. The Parthenon Publishing Group, Leeds Lancs, pp 133–138

Hartig K, Beck E (2003) The bracken fern (*Pteridium arachnoideum* (Kaulf.) Maxon) dilemma in the Andes of Southern Ecuador. Ecotropica 9:3–13

Killingbeck KT (1996) Nutrients in senesced leaves: key to the search for potential resorption and resorption proficiency. Ecology 77:1716–1727

Makeschin F, Haubrich F, Abiy M, Burneo JI, Klinger T (2008) Pasture management and natural soil regeneration. In: Beck E, Bendix J, Kottke I, Makeschin F, Mosandl R (eds) Gradients in a tropical mountain ecosystem of Ecuador, vol 198, Ecological studies. Springer, Berlin, pp 397–408

Marrs R, Johnson S, Le Duc M (1998) Control of bracken and restoration of heathland. VI. The response of bracken fronds to 18 years of continued bracken control or 6 years of control followed by recovery. J Appl Ecol 35:479–490

Mizumachi S, Shinjo T, Ishigaki K, Imura S, Sugimoto Y, Kawamoto Y (2011) The effect of nitrogen application and harvesting interval on basal coverage of three tropical grasses. Proceedings of the 3rd international conference on sustainable animal agriculture for developing countries, Nakhon Ratchasima Thailand, pp 3175

Numata I, Chadwick OA, Roberts DA, Schimel JP, Sampaio FF, Leonidas FC, Soares JV (2007) Temporal nutrient variation in soil and vegetation of post-forest pastures as a function of soil order, pasture age, and management, Rondonia, Brazil. Agr Ecosyst Environ 118:159–172

Page C (1986) The strategies of bracken as a permantent ecological opportunist. In: Smith RT, Tayler JA (eds) Bracken: ecology. Land Use and control technology. The Parthenon Publishing Group, Leeds Lancs, pp 173–181

Portela RCQ, Matos DMS, Siqueira LP, Braz MIG, Silva-Lima L, Marrs RH (2009) Variation in aboveground biomass and necromass of two invasive species in the Atlantic rainforest, Southeast Brazil. Acta Bot Bras 23:571–577

Potthast K, Hamer U, Makeschin F (2010) Impact of litter quality on mineralization processes in managed and abandoned pasture soils in Southern Ecuador. Soil Biol Biochem 42:56–64

Potthast K, Hamer U, Makeschin F (2012a) In an Ecuadorian pasture soil the growth of *Setaria sphacelata*, but not of soil microorganisms, is co-limited by N and P. Appl Soil Ecol 62:103–114

Potthast K, Hamer U, Makeschin F (2012b) Land-use change in a tropical mountain rainforest region of southern Ecuador affects soil microorganisms and nutrient cycling. Biogeochemistry 111:151–167

Roos K, Rollenbeck R, Peters T, Bendix J, Beck E (2010) Growth of tropical bracken (*Pteridium arachnoideum*): response to weather variations and burning. Invasive Plant Sci Manag 3:402–411

Roos K, Rödel HG, Beck E (2011) Short- and long-term effects of weed control on pastures infested with *Pteridium arachnoideum* and an attempt to regenerate abandoned pastures in South Ecuador. Weed Res 51:165–176

Silva B, Roos K, Voss I, König N, Rollenbeck R, Scheibe R, Beck E, Bendix J (2012) Simulating canopy photosynthesis for two competing species of an anthropogenic grassland community in the Andes of southern Ecuador. Ecol Model 239:14

Skerman PJ, Riveros F (1990) Tropical grasses. Food and Agriculture Organization of the United Nations, Rome

Part IV
Synopsis

Chapter 27
Synopsis: Towards a Sustainable Land Use Portfolio

Erwin Beck and Jörg Bendix

27.1 Introduction

27.1.1 Core and Matrix: An Ideal Setting for a Comprehensive Ecosystem Study

From Sala et al. (2000) it can be concluded that in tropical mountain ecosystems like the Andes of Ecuador, biodiversity and ecosystem services are threatened by four main factors: climate change (Chap. 18), local land use change following intense road construction (Chap. 2), atmospheric fertilisation with pollutants from human activities in adjacent or even remote areas (Chap. 11) and invasive species encroaching mainly cultivated land (Chap. 26). The main aim of the current book is to investigate biodiversity and all kinds of ecosystem services under the aspect of a howsoever changing environment. The study area, part of which belongs to the Podocarpus National Park in South Ecuador has been identified as one of the outstanding hotspots of biological diversity. Situated in the centre of the Podocarpus-El Condor UNESCO MAB Biosphere Reserve, it comprises both, part of the protected core area as well as part of the surrounding matrix area, the so-called buffer and transition zones (Jha 2006). Here, we can compare the ecosystem services of the undisturbed ecosystem "tropical mountain forest" with those of its anthropogenic replacement systems in the matrix, mostly not only pastures and abandoned pastures, respectively, but also of forests, which have been altered in an ecologically compatible way by purposeful forest management measures

E. Beck (✉)
Department of Plant Physiology and Bayreuth Centre of Ecology and Environmental Research, University of Bayreuth, 95440 Bayreuth, Germany
e-mail: erwin.beck@uni-bayreuth.de

J. Bendix
Laboratory of Climatology and Remote Sensing, Faculty of Geography, University of Marburg, Deutschhausstr. 10, 35032 Marburg, Germany

J. Bendix et al. (eds.), *Ecosystem Services, Biodiversity and Environmental Change in a Tropical Mountain Ecosystem of South Ecuador*, Ecological Studies 221,
DOI 10.1007/978-3-642-38137-9_27,

(Chap. 13) or, on the other hand, of tree plantations. A major part of the study area where the natural forest has been cleared for pastures is not farmed in a sustainable way. Unsustainability results mainly from invasion by a non-eradicative weed, the (tropical) Southern Bracken (Chap. 26).

The basic evil that triggers bracken invasion is the traditional use of fire for forest clearing and pasture management (Hartig and Beck 2003; Beck et al. 2008; Roos et al. 2010). Unsustainable pasture farming also means that pastures on which the grass has been crowded out by the weed are usually abandoned after a few attempts to expel the fern by fire. The obvious consequence is the clearing of another part of the natural forest mostly in areas which are higher up in the mountain and far away from the farm.

On a first glance, it may be surprising that unsustainability of pastures results from an aggressive weed, but apparently other commonly detrimental factors do not play a substantial role. In spite of the high precipitation, the dense vegetation cover and the only extensive cattle grazing prevent the area from severe soil erosion, and landslides, which are frequent in the natural forest but have only been observed as part of the road construction activity in the pasture area. Annual erosion rates on the pastures are even lower than in the natural forest (Chaps. 1 and 9). Only after a fire followed by strong precipitation, erosion rates might rise for a short while (Chap. 11). Also, negative effects on hydrology and nutrient cycles have not been detected (Chap. 19). Nevertheless, proper fertilisation could increase the productivity and stability of the pastures (Chap. 22).

27.1.2 Resilience of Ecosystems and Ecosystem Services Under Global Change

As stated in the beginning the study area which is representative of the humid eastern part of the Ecuadorian Andes is also threatened by a change of the climate and the atmospheric deposition of pollutants, mostly plant nutrients. These changes affect both the natural and the anthropogenic ecosystems. Comparing the impacts of these processes on both types of ecosystems allows conclusions about ecosystem stability, flexibility and capacity. Summarising these aspects will elucidate the importance of the effects of changes, which is one of the endeavours of this book. Regarding actions towards sustainability of land use, the ecosystem approach of the CBD and the Aichi Biodiversity Targets ("living in harmony with nature", http://www.cbd.int/sp/targets/) provide some guidelines for measures (Chap. 3).

When research started 15 years ago, not much was known about the biodiversity and ecology of the study area and time series of relevant observations or monitoring data were completely lacking. Understanding of the landscape history rests therefore on four pillars (1) long-term vegetation and paleoclimate analysis since the late Pleistocene, i.e. since the retreat of the glaciers 12,000 years ago (Chap. 5), (2) regional climate data (INAMHI and NCEP/NCAR re-analysis data) and general geographic observations (Chaps. 2 and 12), (3) archival data about colonisation and civilising processes (Chap. 16) and (4) own records over the past decade and results from ecological experiments (e.g. Chaps. 15 and 26). In a uniquely comprehensive approach, numerous features of these tropical Andean ecosystems have since been recognised and many aspects thereof have been analysed. For a true understanding of the ecosystem, functional aspects are of utmost importance. Ongoing environmental changes, however, are blurring functional relations but on the other hand can be considered an ecological experiment on the resilience of certain traits and services of the ecosystem.

In this chapter, we try to answer the main two questions relevant to the investigated ecosystem:

1. Which are the impacts of environmental change on biodiversity and ecosystem services in the study area?
2. How must sustainable land use systems be shaped in order to preserve central ecosystem services also under external environmental change?

Regarding the complex ecosystem services, we follow the guideline of the Millennium Ecosystem Assessment (2005), differentiating between supporting, provisioning, regulating and cultural services. As discussed in Chap. 4, the latter three services are related to human use of the ecosystem, while supporting services are those which establish an ecosystem and determine its stability and resilience against impacts. Ecological traits of an ecosystem can serve several objects and purposes. Therefore, we cannot strictly structure this chapter by ecosystem services alone. We rather discuss ecological important parameters comparing the natural and the man-made ecosystems and dip into possible future developments.

A wealth of knowledge has accumulated since 1997 when research into the ecosystems of the South-Ecuadorian Andes started. It has been compiled in more than 400 publications in peer-reviewed journals and in 2 books. This knowledge, though necessarily still incomplete, represents a cultural asset for the region but at the same time contributes to ecological sciences and ecosystem theory in general and can thus be classified as *cultural service* of the ecosystem. Our studies which are primarily focused on exploring environmental and biological "archives" for scientific discovery could and still will be extended to contributing to the actual discussion of a sustainable land use portfolio (for more information on the programme refer to http://www.TropicalMountainForest.org).

27.2 Methods

This synopsis discusses the results achieved by the individual projects of the research group. The methods used are project specific and have been described in detail in the individual chapters of the book. In principle, three types of approaches were carried out which are presented below, referring to the respective chapters:

1. Field surveys along altitudinal (=climate) and disturbance (=land use) gradients. This space for time approach can be used to investigate how service will be affected with a changing climate or an altered land use intensity (refer to Chaps. 5, 6, 7, 8, 9, 10, 11, 14, 16 and 18).
2. Ecological experiments, which alter specific conditions (as e.g. nutrient or water availability) to analyse the reaction of the ecosystem services of interest (refer to Chaps. 13, 15, 22, 23 and 26).
3. Numerical model runs, which allow to play through different scenarios of environmental change impacts on ecosystem services and to assess future projections of drivers for changes in ecosystem services (refer to Chaps. 12, 17, 19, 20, 21, 24 and 26).

27.3 Results and Discussion

27.3.1 Climate and Landscape History and Perspectives

Understanding of the actual state of the ecosystem on the basis of our knowledge of its responses to past environmental changes as well as basic traits of functionality is a prerequisite for a reliable assessment of its resilience. Here, we make three points: (1) for a humid tropical region, an altitudinal range from 1,000 to 3,700 m does not reach into extreme stressful and unlivable habitats; (2) the strong and almost steady easterly winds in the top regions of the ranges represent a serious environmental challenge for plants, animals and humans, but nevertheless pastures can be found even there and (3) at present, the upper tree line is interpreted as a result of these strong winds, which carry an immense load of moisture, soaking the soil.

Studies on the vegetation history during the Holocene suggest that precipitation might have played the determinant role for the altitudinal zonation of the forest, the scrub and the grass páramo (Chap. 18), and the upwards or downwards shifts of their mutual borderlines. For the neotropical Andes, a treeline around 3,000 m a.s.l. is unusually low but on the other hand rich in species which is a peculiarity of the Huancabamba depression separating the northern and the central Andes (Richter et al. 2009). In those parts of the Andes, the treeline is more than 1,000 m higher and appears very monotonous as formed only by the Rosaceae *Polylepis* (sometimes accompanied by *Gynoxys* spec.). In the research area, *Polylepis* was present in the early Pleistocene but disappeared probably due to increasing man-triggered fire activity in the Late Holocene (Chap. 5).

With respect to the current warming , the question to what extent warming phases during the Holocene have pushed up borderlines of the vegetation is of particular interest. Given a stable average lapse rate of $-0.61\ °C\ 100\ m^{-1}$ (Bendix et al. 2008), the increase of temperature must result in an uplift of ecothermal belts (Bendix et al. 2010). As a consequence of an assumed temperature increase by 2.7 °C in the course of the twenty-first century, the area with an average air temperature of the ECSF meteorological station (1,860 m a.s.l.; 15.5 °C) would increase by a factor of 4.4 and the current thermal conditions of the ECSF would shift from 1,860 m to 2,300 m a.s.l. (Chap. 2). However, positions of vegetation borderlines such as the treeline "must be assumed to reflect an environmental integral over several hundreds of years" (Koerner 1999) and therefore rapid changes may not be expected, even if warming would be accompanied by a reduction of the annual precipitation. Assuming an increase of the annual average temperature by 2.7 °C until the end of the century (Chap. 19), mid-term structural changes of the woody vegetation formations are not very likely, if not other environmental factors, such as long-lasting drought or atmospheric pollution will foster the effects of warming. But drought- and heat-tolerant species could gain terrain and also invasive species from anthropogenic habitats may be favoured. They readily propagate on pastures, abandoned agricultural land, roadsides and exotic tree plantations wherever there is an open space for establishment. A more serious change could take place if, as calculated from the Intergovernmental Panel on Climate Change (IPCC) scenario A1B, precipitation would increase significantly (Chap. 19).

27.3.2 Biodiversity: Origin, Distribution and Persistence

An interesting question which is associated with climate changes during the past concerns their potential importance for the speciation process, a question that is of special interest for understanding a hotspot of species diversity. Phylogenetic studies (Chap. 6) of an extraordinarily species-rich insect group (moths, mainly not only of the geometrid genus *Eois*, but also other groups) revealed that long-term extreme climate oscillations as the change from the warm Miocene to the cold Pleistocene did not foster massive speciation and radiation, but the Andean uplift concomitant with the development of a mountain vegetation, in particular of mountain forests did so. The concomitant radiation of host plant and herbivore species suggests that biotic interactions were (are) the driving forces of adaptive speciation. The consequence of these findings for a sustainable land use portfolio is that the conservation of sufficiently large areas of natural forest is essential to preserve the high species richness in the Andes of southern Ecuador.

Mutual organismic interactions are usually multidimensional, forming networks. One of the burning questions with considerable outreach into ecosystem stability is the determinateness of these interactions or the degree of specialisation. Instructive examples are plant–mycorrhiza relations: Chap. 14 shows that the diversity of plants is intimately linked with the diversity of mycorrhizal fungi where rare species are interacting with frequent partners of the respective complementary sides. Specific patterns of specialist subsets (few interacting partners) with generalist subsets (many interacting species) suggest a high degree of nestedness in the study area. That high degree of nestedness of biotic interactions, as an integrative indicator for the state of the ecosystem, attests considerable stability and robustness to the ecosystems of the research area. This conclusion is corroborated by the fact that arbuscular fungal root colonisation revealed no obvious divergence between forests and pastures, regarding both, the abundance of arbuscular mycorrhizal (AM) fungal structures and morphological traits. Most of the genomically characterised frequent species occurred in both systems indicating resilience of a high fraction of AM fungi even against complete changes of the ecosystem (Chap. 7). Irrespective of the stability of such mutual biotic interactions, the resilience of the two ecosystems, the tropical mountain rain forest and the pastures, is fundamentally different: While the forest appears as a stable ecosystem, the pastures are highly instable, even in the absence of further external environmental conditions. For improving the establishment of nursery-grown tree seedlings, inoculation with the dominating generalist Glomeromycota already in the nursery is recommended. Carrying on the above consideration, an important point of ecosystem theory is the question whether the degree of organismic diversity applies to all organismic groups, which are connected in a food chain or web. Since such networks of interactions are commonly locally amassed, the question can be generalised to the relation of above- and belowground biological diversity. Contrary to the expectations of a close correspondence between the outstanding plant diversity (Chap. 8) providing a rich litter menu for the decomposers, the belowground invertebrate richness does not match that of the above-ground invertebrates. Also changes of the environment, as portrayed in an elevation gradient or by the already-mentioned conversion of forest to pasture affected the individual functional components of the decomposing systems in different and not predictable ways. Summing up, the analysis of the soil organisms revealed the complexity and functional compartmentalisation of the trophic systems but no direct correlation with the above-ground complexity. Thus, generalisation of functionality of biotic interactions must confine itself to principles like nestedness and stability, species turnover and environmental changes, or drivers for phylogenetic radiation (Chaps. 7 and 14).

27.3.3 Carbon Relations: Forests and Pastures

Quantitative data on carbon turnover and the carbon pools of tropical lowland forests have been presented for paleo- and neotropical lowland forests using biomass inventories, Eddie covariance measurements and models (Grace et al. 1995; Malhi et al. 1999, 2011; Ribeiro et al. 2011; Chave et al. 2005), whereas reports on carbon relations of tropical mountain forests are scarce (Girardin et al. 2010). Apart from methodological problems (e.g. Eddie covariance technology cannot be used in steep mountains), the altitudinal gradient creates a number of difficulties for scale-related metadata analysis and presentation. The gradual temperature decrease and the non-linear moisture gradient affect structure and growth of the trees, photosynthesis, auto- and heterotrophic respiration, and nutrient availability as a result of the temperature-dependent rates of mineralisation and nutrient turnover. Chapter 10 summarises a hitherto unmatched wealth of carbon-related data, collected in an altitudinal transect from 1,000 to 3,000 m a.s.l. of the pristine tropical pre-montane to upper montane rain forest of the southern Ecuadorian Andes. In an attempt to measure all major C pools and C fluxes in a statistically significant approach, it could be shown that these forest ecosystems store at least as much carbon as neotropical lowland rain forest systems and that the decrease in carbon sequestration over an altitudinal range of 2,000 m and the corresponding temperature gradient from 19 to 9 °C (average annual temperature) was less than 15 %. Nevertheless, net primary production (NPP) and net ecosystem biomass production decreased with a rate of 0.23 Mg C ha^{-1} $year^{-1}$ per °C mean annual temperature, which disproves earlier assumptions that net primary production (NPP) of tropical and subtropical forests is temperature insensitive down to an annual average temperature of 10 °C. With respect to the already-demonstrated annual temperature increase, this is an important finding for both *ecosystem supporting services* (biomass production) and for *regulating services* (carbon sequestration). With increasing elevation not only the diversity of tree species declines but also the above-ground biomass while the below-ground matter increases. This altered biomass distribution does not result from a physiological trend of carbon allocation in favour of the root system but from a decrease of metabolic activity (CO_2 release) of roots in favour of (fine)root growth . Based on own measurements and proxies from the literature Chap. 10 presents for the first time an entire and detailed carbon balance of a neotropical mountain rain forest, which allows a comparison with respective data from the anthropogenic ecosystems, namely uncared, intensely managed and abandoned pastures, and plantations of indigenous alder (*Alnus acuminate*) or exotic pine (*Pinus patula*) (Table 27.1).

Productivity of the pastures was at least twice as much as that of the forest and 3–4 times higher than 7-year-old plantations with an indigenous or an exotic tree species. On the other hand, carbon sequestration by the biomass of the anthropogenic ecosystem was by one or even two orders of magnitude lower than that of the natural forest. Carbon stocks in the reforestation were calculated for a 20-year-old managed plantation. The bulk of the carbon stocks, however, is deposited in the soil. The data impressively demonstrate the carbon depletion following the conversion of the natural forest into agricultural areas. They further show that rebuilding of the soil carbon stocks takes much more time than growth of a tree plantation. The carbon pools in the above-ground plant biomass decrease with altitude, while those in the soil (soil organic carbon + carbon in the roots) increase; thus, carbon sequestration as an *ecosystem regulating service* does not exhibit a strong altitudinal reaction as long as the natural forest is considered.

Table 27.1 Productivity (supporting ecosystem service) and carbon sequestration (regulating ecosystem service) of the natural mountain rain forest and several anthropogenic ecosystem variants of it

Ecosystem	Mountain rain forest	Pasture			Reforestation with	
		Managed	Unmanaged	Abandoned	Alder	Pine
Net biomass production [Mg dm ha^{-1} $year^{-1}$]	12.4	28.4	24.1	31.8	7.4	8.3
Carbon in biomass [Mg C ha^{-1}]	128.8	9.6	5.2	33.0	24.6	29.3
Soil organic carbon (+litter) [Mg C ha^{-1}]	191.0	96.3	91.8	87.3	91.7	93.5
Total carbon [Mg ha^{-1}]	318.8	105.9	97.0	120.3	116.3	122.7

All data refer to areas at around 2,000 m a.s.l. For the carbon stocks of the pastures, average values of standing crop with 5 and 3 mowing/grazing events per year were determined. Biomass that was removed upon mowing/grazing was not calculated as carbon stock

Data are from Chaps. 10, 13, 15, 17, 18, 25 and 26

dm dry matter

27.3.4 Carbon Relations: Soils and Food Webs

This picture changes completely when carbon turnover is considered which is directly depending on the temperature: gross photosynthetic production as well as ecosystem respiration decrease by two-thirds over the stretch from 1,000 m to 3,000 m a.s.l. Attributing the decrease of respiratory CO_2 release alone to the temperature effect would, however, be too simple as demonstrated by investigations of the soil micro-arthropods and amoebae as essential elements of the soil food web .

Whereas species diversity of testate amoebae and oribatid mites did not or only slightly decline with altitude, the abundance of the mites decreased almost by two-thirds from 1,000 to 3,000 m elevation (Chap. 7). In a nutrient addition experiment, an increased soil respiration was observed which was interpreted as resulting from an increased microbial biomass (Chap. 23). Thus several factors contribute to the decrease of soil respiration with increasing elevation: temperature, abundance of decomposer guilds and a low nutrient content of the litter. The described effects are significant when comparing data over an annual temperature gradient of 10 °C.

27.3.5 Carbon Relations Under Climate Change

The question is whether the warming trend projected for the tropical Andes (Urrutia and Vuille 2009) will continue with an increase of 0.22 °C per decade (Bendix et al. 2010) or, as shown in Chap. 19 will be interrupted by cooling periods as between 1998 and 2003 (Chap. 19). Using the IPCC scenarios A1B and A2, an increase of

2.9 °C of the mean annual temperature is predicted until the end of the twenty-first century for the research area concomitant with an increase in precipitation by up to 25 % (Chap. 19). To what extent such climate change could affect NPP and the carbon stocks in the area and thus the *carbon-related supporting* and *regulating services* of the respective ecosystems is difficult to predict in a general sense. Individual plant species may profit from a rise of the temperature more than others, e.g. the C4-grass *Setaria sphacelata*, which as the dominating pasture grass is growing far below its temperature optimum and thus suffers from the competitive strength of the C3-type weed bracken. On the other hand, a substantial increase in precipitation would strengthen the fern (Roos et al. 2010) while much less benefiting the grass (Chap. 24). Especially mutualistic interactions, such as competition for nutrients and space as in the mentioned example, will react very sensitive to environmental changes and predictions on a change of the carbon relations must be considered at best as tentative. Under certain constellations, other factors, which are not primarily contributing to the carbon metabolism, may nevertheless affect carbon relations of an ecosystem: in the natural forest, a potential increase in rainfall might foster landslide frequency and thus substantially alter the aboveground carbon stocks. Atmospheric fertilisation as already observed in the study area may accelerate organismic activities above- and belowground and by forcing the carbon cycle alter the *carbon-related regulation services* (Chap. 23).

27.3.6 Hydrology

As pointed out in Sect. 27.1, *supporting, regulating and provisioning services* of an ecosystem are manifold interlaced or merely different aspects of the same trait. Hydrological models parameterised with analytical data collected over several years show that neither water quality nor quantity appears significantly affected by the conversion of natural forest to pasture in the Rio San Francisco valley (Chap. 9). This result is surprising as textbook knowledge attributes an important contribution to the hydrology of an area to its vegetation. Forests are efficient collectors of precipitation especially from clouds and mist, which significantly contribute to the moisture regime of the study area (Bendix et al. 2008, Chap. 2). On the other hand, forests consume more water by evapotranspiration than pastures, which decreases the difference of the hydrological balance of the two ecosystem types. To this adds a topographic effect, as the pastures are mainly localised on the lower parts of the valley slopes which receive considerably less precipitation than the upper parts (Chap. 2) where less of the forest has been cleared. On a first glance, the ecosystem *water provisioning service* (hydropower and potable water generation) appears not to be endangered as long as pollution by human activity and water consumption would not dramatically increase. In contrast, regulation of water discharge by the pastures in the Rio San Francisco catchment is definitely less effective than that of the forest. If the predictions of the IPCC scenarios A1B or A2 are correct, the lower capacity of grassland for discharge regulation will affect the flow also of the rivers bigger than the Rio San Francisco. Thus, an improvement of the *water provisioning service* by an increase of rainfall may be counteracted by the adverse effects of strongly varying water levels and flood peaks during spells of heavy rains (Chap. 19).

27.3.7 Nutrient Cycles

The regulation of nutrient cycles is an outcome of the interactions of the geo-, bio and atmosphere but can severely be affected by human activities. In the natural ecosystem, nutrients are released locally by weathering of the geologically young parent rock, by mineralisation of litter and other organic matter and by atmospheric deposition via rain and cloud water after long-distance transport from remote sources in the Amazon or even from the Sahara. Dissolved nutrients are also transported downhill by surface and near-surface water flow. The shares of these processes to the nutrient supply differ from site to site and especially so with the altitude. Litter and rock decomposition rates decrease with altitude, whereas atmospheric nutrient deposition following the precipitation regime increases with altitude. On the other hand, litter leachate in the water percolating through the organic layer increases with the altitudinal increase in precipitation. Particularly N, K, Ca and Mg is lost during peak flow events following strong rains because of their accumulation in the litter and organic layers. The depth of the organic stratum increases with the altitude from 1 cm at 1,000 m to 10–30 cm at 2,000 m and 10–40 cm at 3,000 m (Chaps. 10 and 11). Decomposition and mineralisation of the debris depends strongly on the moisture regime of the soil. Water-logged soil, dominating at and above the upper treeline are at least temporarily hypoxic, which in addition to all other mentioned factors hampers degradation processes and releases nitrogen predominantly as ammonium, which in turn acidifies the soil solution. Spells of rainless days, on the other hand, reduce soil moisture by evaporation and thus stimulate mineralisation temporarily. As rainfall substantially decreases with decreasing altitude, the soils at 1,000 m are relatively dry and the decomposition and mineralisation rates are high, resulting in the very shallow litter layer mentioned above. At 2,000 m a.s.l. with intermittent soaking and drying of the soil, the annual leaf litter was completely turned over within less than 1.5 years and the whole organic soil layer in a period of 2–3 decades (Wilcke et al. 2008). Even at the relatively low mineralisation rates, there is still a high release of N (ca. 600–800 kg ha^{-1} $year^{-1}$) and P (ca. 30–70 kg ha^{-1} $year^{-1}$), which is roughly 3–6 times the return of N and P with litterfall (Chap. 11).

27.3.8 Atmospheric Deposition

Atmospheric deposition of nutrients showed an increasing tendency during the past decade with up to 45 kg N ha^{-1} $year^{-1}$ and 5 kg P ha^{-1} $year^{-1}$ (Chap. 11). They originate from biomass burning in the Amazon and dust transport out of the Sahara. Future projections point to a further increase of such depositions until approximately 2040 (Chap. 21). As both forest fires and the El Niño Southern Oscillation (ENSO) cycle are prone to changes because of climate change, it can be expected that there will be a changed relationship between acid and base deposition. Compared with the nutrient release from litter decomposition and mineralisation, the annual atmospheric input is not much. However, it takes place in short episodes during which the actual nutrient (and proton) concentrations can be high and can transitorily affect the growth of the trees. The nutrient content in tree rings, developing into line with the atmospheric deposition, shows the clear interaction between nutrient intake and biotic activity (Chap. 21). Generally, the nutrient cycle in the mountain rainforest is rather tight, which due to the atmospheric input results in a positive nutrient budget, at least for anionic nutrients (Wilcke et al. 2008). The budget of cations which in addition to rock weathering and litter decomposition derive mostly from the Sahara dust oscillates between loss and gain, depending on the frequency and strength of the ENSO events.

27.3.9 Fertilising the Forest: An Ecological Experiment

Anticipating an increasing atmospheric input, a fertilisation experiment "NUMEX" (nutrient manipulation experiment, Chaps. 22 and 23) was established in which slightly higher amounts of N, P, NP and Ca than the average atmospheric deposition were administered in two portions per year. In spite of the relatively high nutrient stocks in the soil, the additional nutrients were readily taken up by the trees and short-term effects were noticed with respect to a decline of fine root biomass, an increase of foliar density of the crowns, of fine litter production, of nutrient content and lifespan of the leaves, and of stem growth, which all corroborate the above-mentioned findings of rapid nutrient incorporation in tree rings from atmospheric deposition. Of course, the extents of these responses were species specific (Chap. 22). Although additional fertilisation enhanced litter decomposition and mineralisation, N and P were not immobilised as microbial biomass in the soil before becoming available to the plants. Only fertilisation with N + P resulted in an increase of microbial biomass, whereas both elements alone reduced it. According to the Millennium Ecosystem Assessment (2005), the regulation of the nutrient cycles is seen as a *supporting ecosystem service* of the study forest. Tightness of the cycle indicates a certain degree of nutrient shortage, which might contribute to the maintenance of the outstanding plant diversity of this ecosystem but could be temporarily relieved by atmospheric nutrient input. Promotion of individual tree species might in the long run change the structure of the forest favouring those species, which can make better use of the resources. However, most of the short-term responses were statistically not yet significant and thus a long-term effect is not predictable at the present state. Tightness of the nutrient cycle and retention of the nutrients in the local nutrient cycle is also important for the provision of clean, unpolluted water for human consumption. In that respect the natural ecosystem tropical mountain forest plays a crucial role.

27.3.10 Nutrient Cycle and Land Use Change

Land use change from the mountain forest into pastures comprises several aspects with regard to nutrient cycling:

- Clearing of the forest and conversion into pasture
- The nutrient relations of the active pastures
- The nutrient relations after abandoning of the pasture

Locally, the conversion of forest to pastures produces an enormous amount of ashes, which could substantially fertilise the soil if not washed away by the rain. The cation-rich ashes result in a marked alkalisation of the soil by 1 pH unit or more (Makeschin et al. 2008; Hamer et al. 2012), which together with the nutrients of the ash, favours growth of the planted grass. Instead of the burned litter layer, there is an increase of the carbon, nitrogen, phosphate and sulphate stocks in the upper layer of the mineral soil, which is maintained by the nutrient-rich litter of the leaves and fine roots of the grass. These changes of the physical structure and chemical composition of the soil are accompanied or even effected by a shift in the microbial community structure in favour of the relative abundance of fungi and Gram-negative bacteria (Potthast et al. 2011). Altogether microbial growth was enhanced by an improved availability of nutrients, a narrower C:N ratio, and the increased soil pH. Accumulated microbial biomass containing up to 39 % of the total P in the mineral soil constituted an easily available P reservoir (Hamer et al. 2012). Nutrient cycling on the active pastures is about 2.5 times faster as in the forest due to a higher susceptibility of the grass organic matter to microbial decomposition.

For maintenance of the favourable soil conditions for pasture farming, fertilisation is necessary, at least with nitrogen and phosphate. Fertilisation with nitrogen has a priming effect because due to a higher biomass production also more NH_4 and NO_3 will be released from the soil organic matter pool, which keeps the nutrient turnover high. Without care, however, pastures degrade and after abandoning organic soil carbon, microbial biomass, available nitrogen and phosphate decrease to levels, which are lower than of the forest (Table 27.2).

Deterioration of the soil was observed in reforestation experiments with the indigenous alder and particularly with the exotic *Pinus patula*. Improvement of the nitrogen relations could be expected upon reforestation with *Alnus*; however, the time period was apparently too short, if the symbiosis with actinomycetes will

Table 27.2 Comparison of selected soil properties of the original ecosystem tropical mountain forest and its anthropogenic derivatives: pastures and tree plantations

Parameter	Soil depth [cm]	Natural forest	Pastures		Plantations	
			Fertilised	Abandoned	Alder	Pine
pH [H_2O]	0–5	3.9	5.3	5.6	4.2	3.5
	5–10	4.0	5.3	5.6	4.3	3.7
SOC [%]	0–5	10.9	11.6	7.8	8.9	8.8
	5–10	8.8	5.5	6.3	6.9	4.8
MBC [mg kg^{-1}]	0–5	1,094	3,445	382	1,241	722
	5–10	851	1,631	333	889	429
N-mineralis.	0–5	5.1	12.1		2.7	1.9
[mg N kg^{-1} day^{-1}]	5–10	3.1	3.8		2.4	1.6
PO_4-P [mg kg^{-1}]	0–5	1.8	4.5	1.3	1.4	6.9*
	5–10	1.4	3.0	0.4	1.1	4.6*

Data from Chaps. 11, 15, 26, Hamer et al. (2009, 2012) and Potthast et al. (2011)
Data are average values; for sake of clarity, SE or SD is not shown and may be obtained from the cited chapters or references. An *asterisk* (*) indicates particularly high SD, which is due to the patchiness of that soil
SOC soil organic carbon, *MBC* carbon in microbial biomass

take place at all under the environmental conditions of the study area. The worse soil conditions under *Pinus* might be primarily due to the very acid pH of the soil, which inhibits microbial activity in the soil.

As long as fertilisation prompts an increase in the biomass, the soils are not saturated with nutrients. For the pastures nitrogen, phosphate, sulphur and cations like K, Ca, and Mg may be limiting or co-limiting plant growth (Chap. 26). Therefore, it is not surprising that like in the forest the nutrient cycle even in fertilised pastures is closed and nutrients are not leaching into groundwater or surface flow. The *supporting ecosystem service* of nutrient cycling holds thus also for the pastures. Persistence of this service may be expected as long as plants and soil organisms can make use of the atmospheric nutrient import. However, the effects of an improved nutrient supply on carbon acquisition and plant growth must not be underestimated. With regard to pastures and tree plantations, an increase in biomass production is also an increase of *provisioning services* of the ecosystem.

27.3.11 Landslides

Chapter 12, discussing landslides in the natural forest, shows that an increase in foliage production and litter fall as a consequence of a better nutrient status of the trees can enhance the risk for landslides. Biomass (weight) and the thickness of the organic layer are biotic factors that add to the triggering of a landslide, if abiotic factors like steepness of the terrain, stagnic soils, strong precipitation, earthquakes and the dynamic wind pressure coincide. Landslides, on the other hand, considerably increase the dynamics of the forest and thus, according to the intermediate disturbance hypothesis, contribute to the conservation of its biodiversity (Chaps. 1 and 12). In that respect, landslides are a positive element of the landscape. Under climate change, the nutrient balance of the natural forest would be less resilient against a potential increase in rain intensity. Triggered by heavy rains landslides frequently block the roads on steep slopes where the vegetation and the soil layer have been removed and erosion progressively continues. Enhanced road construction in the humid parts of the Andes will considerably destabilise the slopes and further advance erosion and mass transport into the rivers.

27.3.12 Provisioning Ecosystem Services: Economy, Ecology and Sustainable Use

Provisioning Services in principle provide items for which an economic revenue can be gained. In the study area, these are not only water services (as discussed above) but also products from forestry and agriculture. A monetary assessment of these services is, however, not always possible, because many of the items, especially the non-timber forest products are commonly collected in the forest for captive use, e.g. medicinal or ornamental plants or seeds for planting trees (Chaps. 13 and 16). Also, *provisioning services* may require upfront costs, capital or man power to achieve sustainability or generate mid- or long-term earnings. Protection of the natural forest, e.g. in the Podocarpus National Park causes running costs to be borne by the state, the Province, an NGO or a real owner. For protection of the forest as provider of the ecosystem service "clean water" multi-stakeholder initiatives have been implemented which link the consumers and the protected areas through trust funds, e.g. a water protection funds to remunerate the forest owners for conservation of the forest and its water resources, or watershed rehabilitation by reforestation. Ecotourism has started as another source of income in protected areas, which requires also considerable upfront investment for roads and huts. Current revenues of up to 10 million US$ have been estimated for South Ecuador (Chap. 13), which is in the same order of magnitude as estimated for the annual value of non-timber forest products. In a survey, the Government of Ecuador estimated that about 75 % of Ecuador's forest area is used directly or indirectly by indigenous communities to contribute to their livelihoods (Blaser et al. 2011, Chap. 13). Until 1987, when the Podocarpus National Park was gazetted, timber and non-timber products were extracted from convenient areas and although tree felling has been banned in the National Park illegal poaching of valuable timber trees, e.g. *Podocarpus oleifolius* and *Prumnopitys montana* could not be completely stopped. From a natural and intact forest of 13 ha, harbouring a tree diversity of 130 species with a diameter at breast height (dbh) of more than 40 cm, 1.5 trees could be extracted per year without changing the actual forest structure, i.e. under a conservative management. The annual revenue of approximately 31 US$ is not very attractive and raises the question of a compensation payment for not clear-cutting the forest.

In a forest management experiment, improvement thinning was examined as a method to foster growth of valuable timber species (potential crop trees, PTC) by removing less valuable trees as competitors. The reactions of the PTCs were species specific and not as fast as expected. Another important provisioning service of the natural forest is its resource of seeds for reforestation purposes (Chap. 13). At present, Ecuador's annual reforestation amounts to 2,200 ha, which is roughly 1 % of the annual deforestation rate. In 2006, the Ecuadorian government approved a National forest and reforestation programme for 1 million ha including 100,000 ha of protective plantations until 2026. Chapter 13 shows the demands for seedlings (56 million) and the production capacity, which is needed for that programme (560 nurseries and 4,500 permanent jobs). Even if only a small percentage of indigenous tree species will be used for reforestation, collection of useful seed material might turn out as major problem, resulting from the high tree diversity and the concomitant extremely low abundance of fruiting individuals. To this adds the requirement of a broad genetic amplitude of the seed material to maintain the currently existing genetic variability of the selected species.

In the lower part of the study area, the natural forest has been reduced to less than 50 % of its original occurrence. Most of the cleared area has been converted to agricultural land, mainly pastures (Göttlicher et al. 2009). However, as mentioned in the introduction, pasture farming by the newcomers in the region, the *Mestizos*, is not sustainable and therefore a major part of the cultivated land is no longer used. Notwithstanding the ecological degradation, these areas represent a significant reserve of land, the rehabilitation of which appears as a promising challenge. One option is reforestation with undemanding tree species like *Eucalyptus* (mainly *globulus*) and in particular *Pinus patula*. Eighty percent of the afforestations in Ecuador are with these two exotic tree species, which grow fast and seeds and seedlings of which are everywhere available. Apart from their fast initial growth, the straight boles are benefits of these species, but on the other hand especially *Pinus* plantations often show abnormal growth if older than ca. 20 years. Inappropriate provenance of the seed material may coincide with the negative effect of that tree species on the soil (see Table 27.2). Nevertheless, the exotic trees are useful as shelter trees for plantation of indigenous species like *Tabebuia chrysantha* or *Cedrela montana*, which initially grow extremely slow and therefore are much more endangered by weeds and herbivores than the fast-growing exotics. An exception is the indigenous alder (*Alnus acuminata*), which grows almost as fast as the exotic species (Chap. 13) but does not always produce a straight bole.

The other alternative for getting abandoned agricultural areas back in the production process is repasturisation Roos et al. 2011. Chapter 26 shows that repasturisation even of areas, which are completely overgrown by a stable secondary vegetation of bracken and shrubs can be re-cultivated after extensive control of the weeds, which might require a whole year or even more. Mechanical as well as simple chemical control is affordable for most of the farmers. Refraining from any use of fire as a traditional agricultural tool is, however, imperative. A method for restoring abandoned pastures is presented in Chap. 26, which also shows that harvest can be progressively enhanced by periodic fertilisation. Because bracken cannot be eradicated without deep ploughing, which is not possible on the steep slopes, regrowth of that weed is a matter of concern. Mechanical cutting of the fronds or recurrent trampling are effective methods to keeping the fern at a low level, which does not interfere with the pasture grass. Recurrent trampling by periodic grazing will be the best choice, which however, requires fertilisation, because the potential of bracken for nutrient acquisition at a low nutrient supply is higher than that of the common pasture grass *Setaria sphacelata*.

Chapter 25 examines the economic situation of the rural population of the study area and of the region, referring to the lifestyle and livelihood of the mentioned Mestizos in comparison with those of their neighbouring ethnic groups (Chap. 16). These are the Shuar, Amazonian Indians, typical forest dwellers who inhabit the lower areas (<1,400 m a.s.l.) of the tropical mountain rainforest and the Saraguros, upland Indians, who live as agropastoralists in the mid-altitudes (1,700–2,800 m a.s.l.). The Mestizos as a heterogeneous group of mixed Spanish and indigenous descents, which represent the dominant populace of southern Ecuador, immigrated and still come to the area since 1960s. They lack a long-standing tradition as a rural community and live as cattle farmers and agriculturists in scattered farms or smaller villages (Chaps. 13 and 16). Deforestation of their land and abandoning of such areas should not be considered as resulting only from shifting cultivation. Before 1994, clearing of at least 25 % of the forest was required for getting a land title (Chap. 16). While lifestyle and livelihood of the Shuar and Saraguros are more or less regulated by the respective communities, the Mestizos use their resources more individually and frequently not as their major source of income. Since the 1990s, cattle ranching became the main economic activity, while tree felling and timber extraction declined because profitable timber species became over-exploited. Chapter 25 compares the income from various types of sustainable agriculture and forestry including reforestation and repasturisation of abandoned areas. Cropland, the most labour-intensive mode achieves the highest revenues but with respect to market fluctuations is also risky. Forest plantations and pasture farming are equally profitable if performed in a sustainable way. In the future, a growing population will demand more food and other resources as provisioning services of the ecosystems and increasing food prices, timber demand and land scarcity will create a pressure to re-cultivate abandoned areas. Both afforestation and repasturisation require considerable upfront investments during the first 2 years. Pasture farming in addition needs periodic fertilisation, while forest plantations must be weeded during the initial years. Profits from the tree plantation cannot be expected before 7 years when the first thinning campaign takes place. The main profit is after 20 years when the trees are felled. Long-term investments are risky even with trees.

Abandoned areas, which are in reach of the farm or the village, should rather be subject to repasturisation. This is (1) why cattle ranching is required for the local milk production of the area and cannot be completely replaced by forestry, (2) the local people prefer short distances between their pastures and the villages as well as to the main road and (3) these are the topographically and edaphically most suitable areas compared to more remote pastures on commonly steeper and moister areas uphill.

In Chap. 25, an economic model is presented which from scratch starts with the clearing of all natural forest in the study area and sets up a theoretically possible proposal for an optimised rural economy. The major strategic guideline of the model is minimisation of risks .

Except for buying financial insurance which is beyond the financial capacity of most farmers, diversification of the income portfolio is the only promising way to hedge their economic risk. Growing a reasonable diversity of crops (including medicinal and ornamental species) is a natural insurance against pests and at the same time against market fluctuations. This option is already practiced by the mentioned indigenous ethnic groups. Another option is the careful use of the natural forest for timber and non-timber forest products as described above. Tree plantations and pasture farming are further components to a risk-minimising portfolio of economic farming, which can be supported by recovery of abandoned areas. Finally, conservation of sufficiently large areas of natural forest, especially in water catchment areas, for which compensation payments could be applied for, can roundup the general portfolio for a sustainable and ecologically responsible farming.

A concept for a diversified land-use strategy termed Ecological–Economic Farm Diversification (EEFD) is presented in Chap. 25, which could convince farmers as it combines high financial returns with an economic and ecological risk that is lower than single pasturing and tree plantations.

27.4 Conclusion

The book shows that ecosystem functions and services at all levels are affected by environmental changes by different mechanisms and to different extents.

Climate change. This will most likely result in an ongoing warming of the region by 0.22 °C per decade. But the highly uncertain change in precipitation can exert different major impacts on ecosystem services. If droughts would increase, supporting services might be negatively affected, as e.g. the abundance of soil organisms could decrease, affecting in turn nutrient cycling. On the other hand, spells of dry weather could reduce water logging of the soils in the upper forest and the Páramo zone leading to an enhanced turnover of the amassed organic matter and a release of nutrients. Increase of precipitation, which is predicted by IPCC scenarios but not yet demonstrated, could impair other regulating and supporting services by, e.g. increasing water logging of the soil and enhancing the risk of landslides. On a short-term scale, a tendency for more precipitation in the dry inner Andean valleys and less rain in the very humid study area became recently apparent which in principle would benefit both regions. The change towards more extreme events, as e.g. an increase of heavy rains would negatively affect the nutrient balance of the study area. Assessment of the effects of climate change on ecosystem services shows improvements and deteriorations the balance of which depends on the extent of the respective changes, the ecosystem and the specific sites.

Atmospheric nutrient deposition. This might affect ecosystem services in different ways. Eutrophication events are sporadic and the composition of the atmospheric deposition depends on the origin of the pollutant. Regarding *supporting services* of the natural forest nutrient input accelerates the carbon turnover rate, stimulates tree growth and changes the carbon balance of the ecosystems. On the other hand, increased biomass production and litter fall can increase the risk and frequency of landslides, which at least temporarily decrease carbon sequestration and nutrient turnover, but on the other hand contribute to the maintenance of biodiversity. Several of the depositions come with protons and thus stress the buffering capacity of the soils. Shortage of cations results then in a dramatic acidification as in the *Pinus* plantation. With regard to agricultural provisioning services, atmospheric fertilisation is beneficial but not sufficient to replenish nutrient deficiency of the impoverished soils. Like in the forest, atmospheric nutrient deposition fosters carbon turnover also in the pastures. However, even negative ecological effects must be expected resulting from a change of the community structure of soil organisms.

Land use change. This takes mainly place by the conversion of the natural forest to pasture land. Most of the biodiversity is lost and the soil structure, chemistry and biology are basically altered. Exceptions are AM fungi and special groups of soil organisms, e.g. testate amoebae, populations of which survive slash and burn or re-colonise the new habitat. Microbial biomass even increased and a pronounced shift in the composition was observed triggered by an increase in the nutrient supply. With these changes, stability of the ecosystem and its services is lost. Only by permanent agricultural interference, like fertilisation, cattle grazing and weeding can those ecosystems be maintained over a reasonable period of some 10 years and deliver their *provisioning services*. These ecosystems cannot support themselves and without the above-mentioned measures nutrient cycling rates decrease, nutrients are lost and the artificial ecosystems degrade and are finally abandoned. This sequence of deteriorating reactions is boosted by recurrent burning due to a rain-fed leaching of nutrients.

Invasive species. One reason for the instability of the pastures is that the dominating pasture grasses are not indigenous but introduced, mainly from Africa and that they are not completely adapted to the environmental conditions of the area. In an ecological view, *Setaria sphacelata*, *Melinis minutiflora* and *Pennisetum clandestinum* are invasive species on purpose which on the pastures compete with another species (bracken) which although indigenous to the neotropics is an invasive species for the region following slash and burn of the natural forest. Especially in nutrient-depleted soils, bracken can outcompete the grasses, thus negatively affecting the *provisioning services* of the pastures.

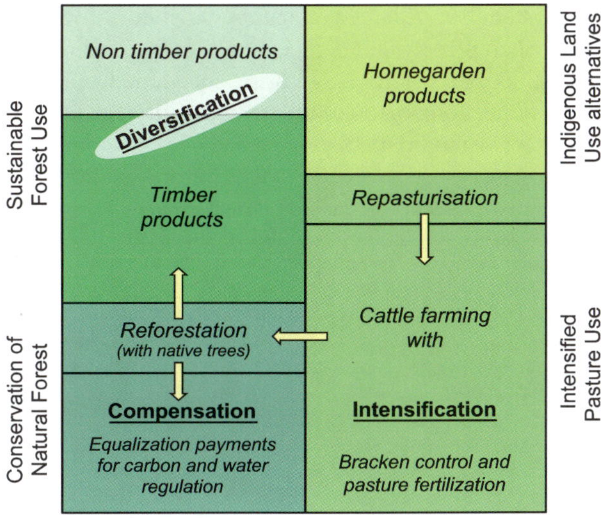

Fig. 27.1 A sustainable land use portfolio for the mountain forest of southern Ecuador based on agricultural and forestry provisioning services

Even if in such a complex system as the study area, the gained knowledge is far from providing a full understanding of the multitude of interactions and feedbacks between the different service levels, the results prove that the current *agricultural provisioning services* are not at all sustainable because they are not accordant with the ecology of the area. They depend on the continuous support by the farmers. In contrast to the natural forest which proves as a remarkably stable ecosystem with a high resilience against environmental changes, the agricultural ecosystems are under risk of environmental change, thus inducing further change by threatening all ecosystem services.

As a result (see also Chap. 17), three main cornerstones are crucial for converting the currently non-sustainable and non-risk resilient land use portfolio into a sustainable one, which harmonises with the mountain rain forest of Ecuador:

- A sustainable agricultural intensification, paired with
- land use diversification as a risk reduction strategy, complemented by
- (external) compensation payments for forest conservation

Figure 27.1 shows the most suitable sustainable land use portfolio as the major outcome of this book for the mountain rain forest of southern Ecuador and its anthropogenic replacement systems. It combines sustainable intensification by warranting a continuous and higher pasture yield by means of fertilisation and bracken control, which halts the pressure to clear new natural forest. Intensification means also carefully using parts of the natural forest by extracting timber and collecting non-timber products as well as including other sustainable land use alternatives in more universal or specialised sustainably managed indigenous home gardens and their products. However, this is not enough. Conservation of sufficiently large parts of the natural forest is a major issue to reach sustainable ecosystem services: equalisation payments for carbon regulation services and watershed protection are incentives for the land owners to take care of their natural forests as ever possible. Restoring abandoned and run wild former agricultural areas are another component of the sustainability portfolio, which will release at least part of the pressure on the native forest.

Acknowledgements The authors wish to cordially thank Prof. Dr. Reinhard Mosandl, Technical University of Munich, for thoughtful comments and useful tips for finalising the synopsis. They also thank MAE for providing the research permission.

References

Beck E, Hartig K, Roos K (2008) Forest clearing by slash and burn. In: Beck E, Bendix J, Kottke I, Makeschin F, Mosandl R (eds) Gradients in a tropical mountain ecosystem of Ecuador. Ecological studies, vol 198. Springer-Verlag, Heidelberg, pp 371–374

Bendix J, Rollenbeck R, Fabian P, Emck P, Richter M, Beck E (2008) Climate variability. In: Beck E, Bendix J, Kottke I, Makeschin F, Mosandl R (eds) Gradients in a tropical mountain ecosystem of Ecuador. Ecological studies, vol 198. Springer-Verlag, Heidelberg, pp 280–291

Bendix J, Behling H, Peters T, Richter M, Beck E (2010) Functional biodiversity and climate change along an altitudinal gradient in a tropical mountain rainforest. In: Tscharntke T, Leuschner C, Veldkamp E, Faust H, Guhardja E, Bidin A (eds) Tropical rainforests and agroforests under global change. Environmental science and engineering, Subseries environmental science. pp 239–268

Blaser J, Sarre A, Poore D, Johnson S (2011) Status of tropical forest management 2011. ITTO Technical Series No 38. International Tropical Timber Organization, Yokohama

Chave J, Andalo C, Brown S, Cairns MA, Chambers JQ, Eamus D, Folster H, Fromard F, Higuchi N, Kira T, Lescure JP, Nelson BW, Ogawa H, Puig H, Riera B, Yamakura T (2005) Tree allometry and improved estimation of carbon stocks and balance in tropical forests. Oecologia 145:87–99

Girardin CAJ, Malhi Y, Aragão LEOC, Mamani M, Huaraca W, Durand L, Feeley KJ, Rapp J, Silva JE, Silman M et al (2010) Net primary productivity allocation and cycling of carbon along a tropical forest elevational transect in the Peruvian Andes. Glob Change Biol 16:3176–3192

Göttlicher D, Obregón A, Homeier J, Rollenbeck R, Nauss T, Bendix J (2009) Land-cover classification in the Andes of southern Ecuador using Landsat ETM+ data as a basis for SVAT modelling. Int J Remote Sens 30(8):1867–1886

Grace J, Lloyd J, McIntyre J, Miranda A, Meir P, Miranda H, Moncrieff J, Massheder J, Wright I, Gash J (1995) Fluxes of carbon dioxide and water vapour over an undisturbed tropical forest in south-west Amazonia. Glob Change Biol 1:1–12

Hamer U, Potthast K, Makeschin F (2009) Urea fertilisation affected soil organic matter dynamics and microbial community structure in pasture soils of Southern Ecuador. Appl Soil Ecol 43:226–233

Hamer U, Potthast K, Burneo JI (2012) Nutrient stocks and phosphorus fractions in mountain soils of Southern Ecuador after conversion of forest to pasture. Biogeochemistry. doi:10.1007/s10533-012-9742-2

Hartig K, Beck E (2003) The Bracken Fern (Pteridium arachnoideum (Kaulf.) Maxon) dilemma in the Andes of Southern Ecuador. Ecotropica 9:3–13

Jha SG (2006) An overview of the Biosphere Reserve concept and its application to Nepal. Banko Janakari 16:21–24

Koerner C (1999) Alpine plant life. Springer, Berlin, 338 p

Makeschin F, Haubrich F, Abiy M, Burneo JI, Klinger T (2008) Pasture management and natural soil regeneration. In: Beck E, Bendix J, Kottke I, Makeschin F, Mosandl R (eds) Gradients in a tropical mountain ecosystem of Ecuador. Ecological studies, vol 198. Springer-Verlag, Heidelberg, pp 397–408

Malhi Y, Baldocchi DD, Jarvis PG (1999) The carbon balance of tropical, temperate and boreal forests. Plant Cell Environ 22:715–740

Malhi Y, Doughty C, Galbraith D (2011) The allocation of ecosystem net primary productivity in tropical forests. Phil Trans Roy Soc B 366(1582):3225–3245

Millennium Ecosystem Assessment (2005) Ecosystems and human well-being: biodiversity synthesis. World Resources Institute, Washington, DC

Potthast K, Hamer U, Makeschin F (2011) Land-use change in a tropical mountain rainforest region of southern Ecuador affects soil microorganisms and nutrient cycling. Biogeochemistry 111:151–167

Ribeiro SC, Fehrmann L, Soares CPB, Jacovine LAG, Kleinn C, de Oliveira Gaspar R (2011) Above- and belowground biomass in a Brazilian Cerrado. For Ecol Manage 262:491–499

Richter M, Diertl KH, Emck P, Peters T, Beck E (2009) Reasons for an outstanding plant diversity in the tropical Andes of Southern Ecuador. Landscape Online 12:1–35. doi:10.3097/LO.200912

Roos K, Rollenbeck R, Peters T, Bendix J, Beck E (2010) Growth of tropical bracken (*Pteridium arachnoideum*): response to weather variations and burning. Inv Plant Sci Manage 3:402–411

Roos K, Rödel HG, Beck E (2011) Short- and long-term effects of weed control on pastures infested with *Pteridium arachnoideum* and an attempt to regenerate abandoned pastures in South Ecuador. Weed Res 51:165–176

Sala OE, Chapin FS III, Armesto JJ, Berlow R, Bloomfield J, Dirzo R, Huber-Sanwald E, Huenneke LF, Jackson RB, Kinzig A, Leemans R, Lodge D, Mooney HA, Oesterheld M, Poff NL, Sykes MT, Walker BH, Walker M, Wall DH (2000) Global biodiversity scenarios for the year 2100. Science 287:1770–1774

Urrutia R, Vuille M (2009) Climate change projections for the tropical Andes using a regional climate model: temperature and precipitation simulations for the end of the 21st century. J Geophys Res 114. doi:10.1029/2008JD011021

Wilcke W, Yasin S, Schmitt A, Valarezo C, Zech W (2008) Nutrient status and fluxes at the field and catchment scale. In: Beck E, Bendix J, Kottke I, Makeschin F, Mosandl, R (eds) Gradients in a tropical mountain ecosystem of Ecuador. Ecological studies, vol 198. Springer-Verlag, Heidelberg, pp 203–215

Chapter 28
Knowledge Transfer for Conservation and Sustainable Management of Natural Resources: A Case Study from Southern Ecuador

Sven Günter, Baltazar Calvas, Thomas Lotz, Jörg Bendix, and Reinhard Mosandl

28.1 Introduction

Sustainable ecosystem management depends on a diverse knowledge system in which techniques are continuously updated to reflect current understanding and needs (Roux et al. 2006). Since scientific knowledge evolves rapidly, these knowledge systems have to be integrated into adaptive management systems (Cooper et al. 2007; Heinimann 2010; Günter et al. 2011). However, science and practice are often considered to represent rather contrasting worlds than complementary ones. Roux et al. (2006) illustrate exemplarily the mismatch of views between scientists and decision makers (Table 28.1). This raises the question, how scientists can connect interdisciplinary research with societal needs in order to overcome this mismatch (Bradshaw and Bekoff 2001). To meet with the challenges of managing complex socio-ecological systems, scientists must assume the roles of collaborative

S. Günter (✉)
Tropical Agricultural Research and Higher Education Center, Latin-American Chair for Protected Areas and Biological Corridors "Kenton Miller", CATIE 7170, Turrialba 30501, Costa Rica
e-mail: sgunter@catie.ac.cr

B. Calvas
Institute of Forest Management, Center of Life and Food Sciences Weihenstephan, Technische Universitaät München, Hans-Carl-von-Carlowitz-Platz 2, 85354 Freising, Germany

T. Lotz • J. Bendix
Laboratory for Climatology and Remote Sensing, Faculty of Geography, University of Marburg, Deutschhausstraße 10, 35032 Marburg, Germany

R. Mosandl
Institute of Silviculture, Center of Life and Food Sciences Weihenstephan, Technische Universität München, Hans-Carl-von-Carlowitz-Platz 2, 85354 Freising, Germany

J. Bendix et al. (eds.), *Ecosystem Services, Biodiversity and Environmental Change in a Tropical Mountain Ecosystem of South Ecuador*, Ecological Studies 221,
DOI 10.1007/978-3-642-38137-9_28,

Table 28.1 Common reasons for mismatch between scientific culture and management culture (Modified according to Roux et al. 2006)

Managers' views of scientists	Scientists' views of managers
• Science peer-review and reward systems enforce an inward-looking, self-serving culture	• Managers work within a system that rewards organizational and individual interests
• Scientists are arrogant	• Poor understanding of scientific processes
• Production of fragmented information that seldom addresses "real" problems • Little regard for application contexts, and are driven only by intellectual curiosity	• Managers are caught up in day-to-day operations, and spend little time in intellectual reflection and longer term R and D planning
• Scientists do not work at appropriate or useful spatial and temporal scales • Scientists are unable to contribute to the value-based debate that usually governs problem solving in the real world	• Managers do not appreciate ecosystem complexity
• No effective communication to non-scientists	• Managers do not articulate their needs effectively

learners and knowledge managers and actively contribute to a science–management partnership (MacMynowski 2007).

In this chapter we address to the above-mentioned problem by examining the potential contributions and limitations of the working groups of the Research Unit RU816 "Biodiversity and Sustainable Management of a Megadiverse Mountain Ecosystem in South Ecuador" (http://www.tropicalmountainforest.org) for knowledge transfer concerning conservation and sustainable management of natural resources.

28.2 Methods

The Research Unit RU816 is structured into four subprograms according to their thematic focus: "forest ecosystems," "landscape dynamics," "pastoral ecosystems," and "rehabilitation and sustainability." In order to assess their potential contributions to management and conservation of ecosystem services, a Research Unit internal survey has been conducted among principal investigators (PI) of all working groups ($n = 21$). The following questions had to be answered:

- How do you rank the potential of the research unit RU 816 for contributing to knowledge transfer between Ecuador and Germany?
- How do you rank the potential of the scientific methods/results of the whole research unit for implementation into practice?
- What have been the main contributions of your working group to knowledge transfer concerning Education/Capacity building of Ecuadorian staff?

– What have been the main contributions of your working group to management and conservation of ecosystem services?
– What have been the main contributions of your working group to sustainable land use?

Each working group had to rank their potential contributions on a scale from 1 (very low) to 5 (very high). In total, 20 working groups submitted answers to the questionnaire (95 % of total). In Sects. 28.3 and 28.4 we present the answers regarding their potential contributions to knowledge transfer and implementation of the results gained by basic research. Additionally, we will present also the results of semi-structured interviews to four groups of stakeholders in the region of Loja in Sect. 28.5.

28.3 Main Contributions of the Working Groups to Management and Conservation of Ecosystem Services

Ecosystem services (ES) are an important part for the Ecuadorian society and decision makers. Several Ecuadorian policies show direct relevance to ES (Table 28.2). It is notable that conservation of biodiversity, provision of water, renewable energy, social resilience, climate change, and the commitment to sustainability receive a particular higher attention in the listing. These policies are reflecting the four Millennium Ecosystem Assessment categories presented in Chap. 4. The Research Unit RU816 (http://www.tropicalmountainforest.org) with its strong focus on ecosystem services and sustainable land use should thus be able to provide valuable scientific contributions to the Ecuadorian society.

Surprisingly, the overall perception of potential contributions to management and conservation of ecosystem services (ES) between the different subprograms of RU816 is relatively similar. It amounts 3.75 (mean) for "ecosystems," 3.5 for "landscape dynamics," "pastoral land use" 3.9, and "rehabilitation and sustainability" 3.8. It is remarkable that only 25 % (of in total $n = 20$) of the interviewed working groups (WG) classified their potential scientific contributions to ecosystem services according to the MEA categories "provisioning services," "regulating services," "supporting services," and "cultural services" (see Chap. 4). Most scientists used instead a rather technical and specific vocabulary and classified their potential contributions according to their specific field of interest (e.g. "mycorrhiza," "land slide risk," "decomposition," "landscape"), indicating a potential mismatch of definitions and terms used by scientists and decision makers. We tried to classify all scientific parameters mentioned in the questionnaires according to the four categories of the Millennium Ecosystem Assessment, released in 2005 (see Fig. 28.1). The classification is based on the authors' knowledge and resulted in highest scores for supporting services and regulating services, while those services (provisioning and cultural services), which may be of highest priority for stakeholders from developing countries, received lower scores.

Table 28.2 Seven Ecuadorian Policies have direct relevance to environmental services (SENPLADES 2009)

1. To preserve and sustainably manage the natural heritage, land-based and marine biodiversity, which must be regarded as strategic sectors
2. To promote an integral approach to the treatment of Ecuador's hydrographic reality, including the State's strategic use of hydrographic basins, always bearing in mind their socio-cultural and environmental value
3. To diversify the national energy matrix by promoting an efficient and greater participation of sustainable sources of renewal energies
4. To prevent, control, and mitigate environmental damage as a crucial contribution to the improvement of the quality of life
5. To promote the adaptation to—and mitigation of—the variability of weather and climate with an emphasis on the process of climate change
6. To reduce social and environmental vulnerability produced by natural and anthropic processes
7. To incorporate an environmental approach in all social, economic, and cultural public policies

They are part of the national objective for good living. To guarantee rights of nature and promote a healthy and sustainable environment. The politics are directed

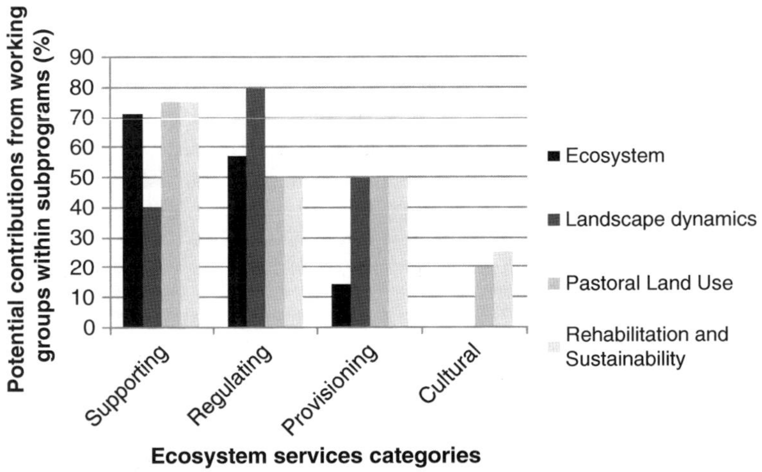

Fig. 28.1 Scores of potential contributions of working groups (WG) to management and conservation of the four categories of ecosystem services (ES) in comparison of the four subprograms. Scores are based on percentage of WG per subprogram with potential contributions to corresponding ES category

28.4 What Are the Main Potential Contributions of the Working Groups to Knowledge Transfer and Implementation of Results?

Figure 28.2 shows that WGs focusing on ecosystem studies perceived lower potential of their work in relation to those focusing on sustainable land use and rehabilitation (mean and standard deviation). While differences between the

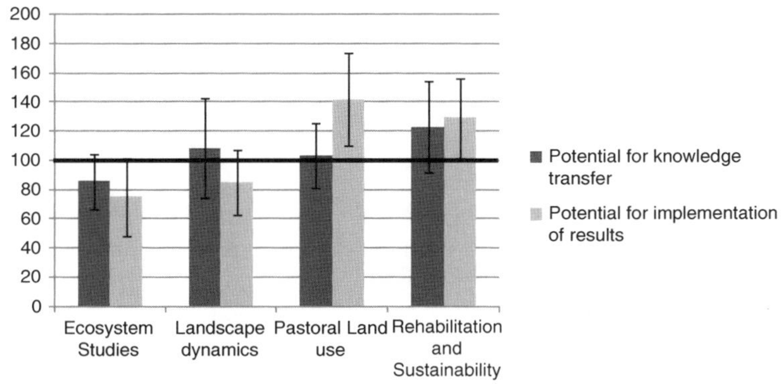

Fig. 28.2 Auto-evaluation of potential contributions from different subprograms to knowledge transfer or implementation of results into practice in relation to the evaluation of the potential of the whole Research Unit (RU816). The mean score for the whole RU816 is 100 (*solid horizontal line*). A perception of lower or higher potential for knowledge transfer or implementation is indicated by lower or higher scores than 100

subprograms are relatively low concerning potential contributions to knowledge transfer, those for implementation of results are relatively pronounced. The subprograms with higher scores in supporting and regulating services (Ecosystem studies and Landscape dynamic) perceived a slightly lower potential of their work for implementation and knowledge transfer.

The comparison with the corresponding scientific output per WG (in number of papers published in peer-reviewed journals from 2007 to 2011) reveals that the perception of implementation potential and scientific output of the corresponding working group is negatively correlated ($r = -0.41$). This could probably indicate that researchers with interest in implementation of their results allocate a higher percentage of time and efforts towards implementation at the cost of lower scientific output. Another reason might be that applied topics cannot be placed at peer-reviewed journals as easy as for basic science papers. Also the number of Ecuadorian PhD students supervised in this period is negatively correlated with the scientific output ($r = -0.34$, see Fig. 28.4). This can certainly be explained to some extent by language barriers and lack of experience in paper writing of the Ecuadorian PhDs, resulting in longer time necessary for publication in international journals, and possibly also in language and cultural barriers between supervisor and students. This is confirmed by the finding that the higher the percentage of local co-authors within a working group, the lower is the scientific output ($r = -0.32$, see Fig. 28.3). It might be hypothesized in this context that this is a potential effect of more time-consuming capacity building for the supervisors. However, no influence of the number of supervised undergraduate students from Ecuador could be revealed ($r = 0.04$). Figure 28.3 illustrates the diversification of the Research Unit RU816 into working groups focused on high scientific output mainly specialized on ecosystem studies, and working groups with higher emphasis on local capacity building and high potential for implementation and sustainable land use. The results

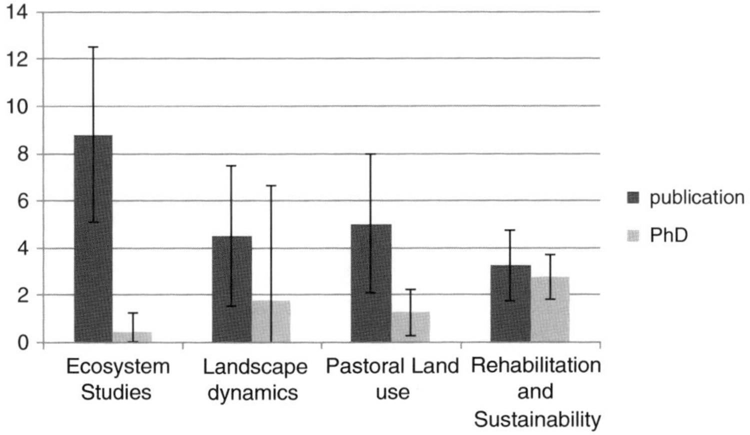

Fig. 28.3 Mean number of publications in peer-reviewed journals and mean number of supervised Ecuadorian PhD students per working group in the period 2007–2011 in comparison of subprograms

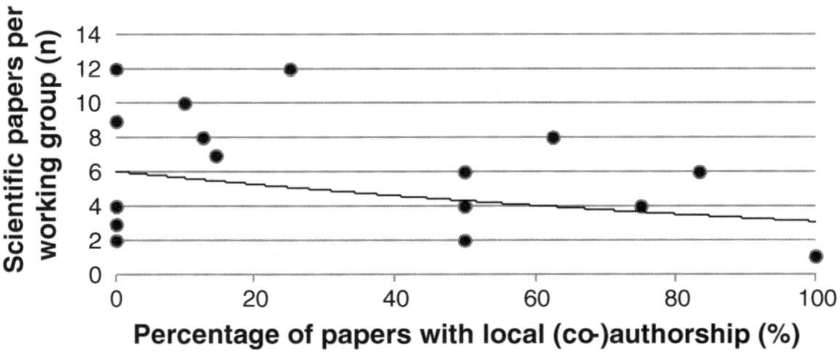

Fig. 28.4 Relationship between proportion of papers with local (co)-authorship and scientific output per working group ($r = -0.32$) in the period from 2007 to 2011

confirm that also for working groups of RU816 it is apparently difficult to combine both high scientific output in international journals and efforts in capacity building or strategies for knowledge transfer to local stakeholders at least on the level of individual research projects and working groups. On the level of the whole Research Unit in turn, there is apparently a balance between both aims.

28.5 What Is the Perception of Local Stakeholders?

Additionally to the internal survey among working groups of RU 816 a survey among counterparts from three stakeholder groups has been conducted: NGOs (e.g. conservation), public sector (e.g. environmental ministry), and academic sector (local universities). In total 23 persons from counterpart institutions with previous working relations with the research group have been interviewed about their perception of own instruments and strategies for knowledge transfer and potential barriers for successful knowledge transfer (Table 28.3). The following questions have been asked:

- Which instruments do you implement for knowledge transfer?
- Where do you see major barriers for knowledge transfer?
- How can knowledge transfer be improved?
- What are the key factors for sustainability in a knowledge transfer process?

Some strategies and instruments are very similar among the different stakeholder groups. All are participating actively in information exchange platforms and round tables as common instruments for knowledge transfer. All of them are developing or using specific planning instruments and participatory approaches for implementing projects. The perceived barriers in contrast are very different. "Missing financial" and "unavailable human resources" were not mentioned by the private sector/NGOs while "lack of institutional reliability and stability" was not mentioned by the public sector. These aspects are most likely attributed to the general development problems of the country and may be typical for the situation in most countries of the tropics. These rather general problems are difficult to be solved by individual stakeholders, but three other mentioned aspects comprise concrete potential for overcoming the barriers for successful knowledge transfer, (1) enhancement of information management (mentioned by all groups), (2) learning from best management practices in other countries (mentioned by public sector), and (3) process-based management instead of project-based management (NGOs/private). The latter aspect indicates a common problem when projects fail to create impact in the society beyond the specific project objectives and outputs. Specific impact evaluation criteria have been developed by the OECD (Organization for Economic Cooperation and Development, OECD 2005).

28.6 Outlook: Towards Science-Based Adaptive Management

The results indicate that research focused on publication in high ranked papers is hardly compatible with short-term priorities of tropical societies, especially those of developing countries. By maintaining a careful balance of projects with different research priorities it might be possible to address both aims on the level of large research groups such as RU816. On the level of individual research projects the PIs probably have to focus either on scientific output or on aims for development and capacity building. Bradshaw & Beckoff (2001) describe this antagonism as two competing models of science: "science apart from society and a science directly engaged with society." A way out of this paradox is seen in leaving traditional

Table 28.3 Strategies and instruments for knowledge transfer of different stakeholder groups in the municipality of Loja and perceived barriers for successful knowledge transfer

	Own strategies and instruments for knowledge transfer	Perceived barriers for successful knowledge transfer
Public sector	– Land-use planning – Integrative, participatory water shed management – Participation in working groups and platforms for exchange of experiences – Prometeo Program (integration of foreign experts) – Experimental stations	Internal – Missing human and financial resources – Missing access to information – Missing political framework for science-based development – Lack of capacity building Internal/external – Missing information management systems – No sufficient learning from best management practices in other countries
Academic sector	– Development of planning instruments – Education/capacity building – Participatory work with communities – Transfer centres in health sector – Participation in working groups and platforms for exchange of experiences	Internal – Missing human and financial resources – Lack of capacity building – Information management not client oriented – Insufficient time management Internal/external – Missing linkage of science to implementation projects – Insufficient linkages to public sector External – Missing institutions for transfer in most thematic areas – Lack of institutional reliability and stability
NGO/ private sector	– Fostering co-management structures – Participatory work with local communities – Participation in working groups and platforms for exchange of experiences	Internal/external – Lack of instruments for information management – Project-based management instead of process-based management as potential long-term approach External – Lack of institutional and political reliability/stability – Missing documentation and limited access to information

pathways by integration of biophysical and social sciences on the one side, and integration of science and practice on the other side. Besides these conceptual pathways, solutions can be found in social responsibility of scientists: "Interdisciplinary collaboration and integrative approaches to science need to extend to education. It is important for senior colleagues to support rather than discourage students in their pursuit to create a career that satisfies science and social criteria" (Bradshaw and Bekoff 2001). Efforts in education and social responsibility consequently have to be acknowledged in scientific review processes.

The transfer-project concept is reflecting the "community of practice" concept (Lave and Wenger 1991; Wenger 1998, 2005). The authors state that two conditions of a community of practice are crucial in the conventionalization of meaning: shared experience over time, and a commitment to shared understanding. A community of practice engages people in mutual sense-making (Eckert 2006). In this context more emphasis is put on creating a common platform of understanding and trust. The common currency of these rather informal networks is knowledge and not the rules of a specific mandate. Membership is based on win–win situations and mutual benefits between the actors of the network (MacMynowski 2007). This kind of relationship can be very successful in societies with low stability of institutional or political structures. While they not necessarily guarantee sustainable management practices or conservation of ES, science and active participation of scientists can be helpful to achieve these goals. A complementary approach, called "citizen science," is characterized by involving citizen participants directly in research activities, such as monitoring and active management, and thus fostering positive and measurable impacts on biodiversity (Cooper et al. 2007). As required from stakeholders of the region Loja, this approach could be very useful as instrument for overcoming project-based management and establishing process-based and adaptive development approaches for the region of Loja.

Community of practice systems and science-based development projects in countries with low institutional and political stability may depend on external support by developing agencies, such as the Deutsche Gesellschaft für Internationale Zusammenarbeit (GIZ), the US Agency for International Development (USAID), or others. There are several potential instruments provided by the development agencies and international institutions which could offer complementary functions, for example the International Model Forest Network (IMFN), the network of CGIAR (Consultative Groups on International Agricultural Research), the CIRAD program from France (Centre de Coopération Internationale en Recherche Agronomique pour le Développement), the integrated expert system CIM from GIZ Germany (Center for Integrated Migration), and several other instruments more. These international platforms could also provide valuable information about best management practice in other countries as requested from local stakeholders in Loja.

In order to reduce the time required for a given scientific finding to assimilate into society scientists have to create transdisciplinary partnerships and aim at joint implementation of pilot projects together with non-scientific counterparts (Bradshaw and Borchers 2000). Also, the role of non-scientific media is of a higher importance than in industrialized countries because English and technical language may build significant barriers for successful transfer of knowledge between developing countries and the global society.

In our query several projects of RU816 have indicated transfer potential regarding different service levels. Two projects have successfully started a knowledge transfer program based on their knowledge gain by previous basic research activities in the RU816. The development of a prototype rain radar network for southern Ecuador is related to supporting / regulating services, while the afforestation program with indigenous trees targets provisioning services.

28.7 From Science to Application: First Examples of Projects for Knowledge Transfer

28.7.1 Transfer Regarding Supporting /Regulating Services: Development of a Rain Radar Network South Ecuador

Water is one of the major resources for Ecuador (hydropower, potable water) (Vergara et al. 2007) and its undisturbed regulation by the ecosystems (regulation services) is essential. Proper natural water regulation is a precondition for other services, ranging from cultural services such as biodiversity until provisions services such as agricultural production. Even if climate projections are highly uncertain, a change in the water cycle due to climate change is most likely (Buytaert et al. 2009; Chaps. 2 and 19). At the same time, land-use change is claimed to threaten the water services of the natural Andean ecosystems considerably (Buytaert et al. 2006; Crespo et al. 2011; Chaps. 2, 9, and 19). The basis to assess water services is a proper observation of the atmospheric input to the system (precipitation) and feedbacks on rainfall formation due to a changing landscape. To date, the climate network in southern Ecuador operated by the national weather service is sparse, not capable to observe the complex spatio-temporal behaviour of rainfall particularly in the remote areas of the Andes and the Amazon. Other technologies such as scanning rain radar networks which are widespread in developed countries (e.g. Junyet and Chandrasekhar 2008) are not available where operating radar networks in mountainous terrain is a specific challenge even in developed countries (see e.g. Andrieu et al. 1997). The lack of a weather radar network in Ecuador is on the one hand due to limited financial resources. On the other hand, national expertise is hitherto not developed to operate this complex technique in a mountainous environment of the Andes. In the scope of the previous basic research activities, a methodology was developed to calibrate and operate straightforward and cost-effective local area weather radar systems (Chap. 1; Rollenbeck and Bendix 2006, 2011). In an ongoing knowledge transfer project funded by the German Research Foundation (Deutsche Forschungsgemeinschaft, DFG) which started recently, a weather radar system for southern Ecuador consisting of three radars (Fig. 28.5) was proposed. The system is intended to act as a prototype for a national network.

First, to overcome national financial constraints, the network is planned on the basis of cost-effective local weather radars. Secondly, several non-university cooperation partners but also university partners (Technical University of Loja, UTPL, University of Cuenca) are involved in the development of the prototype system where the Ecuadorian consortium is led by the provincial government of Loja

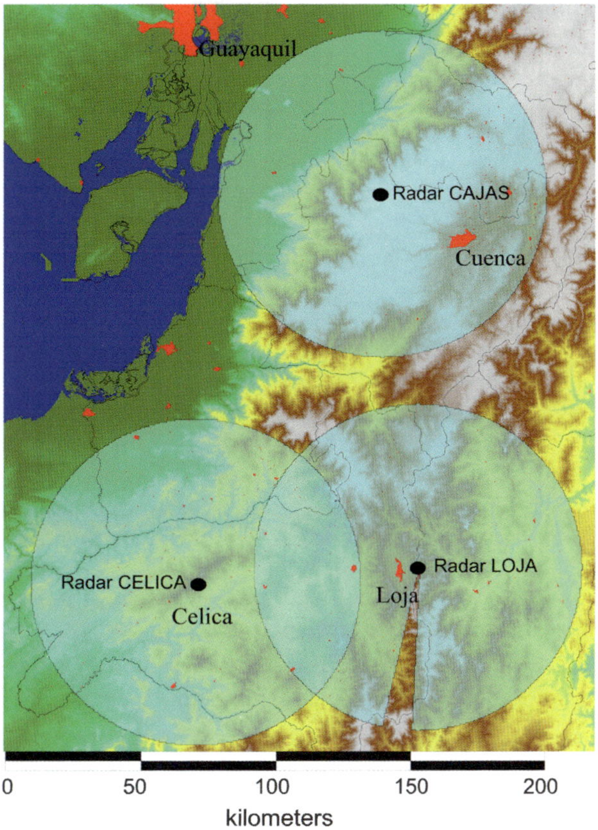

Fig. 28.5 Schematic map of the south of Ecuador depicting the existing Radar installation close to Loja (LOXX) and the location for the two new radars close to Cariamanga (CAXX) and close to Cuenca (CUXX). The *coloured dots* are existing meteorological stations required for calibration, operated by the cooperation partners

(Gobierno Provincial de Loja GPL). The core of the program encompasses several capacity building measures such as e.g. training of local personnel in radar installation, operation and maintenance, data transmission to the head office, calibration with rainfall observations, developing data base and Web applications to disseminate radar data to the public, meteorological interpretation of radar movies for nowcasting, and hazard prevention. The project flow is a good example for a process-based project management approach (Sect. 28.5). The project started with the joint vision to set up and operate a weather radar network serving both public and private stakeholders and the biodiversity research community. The realization is organized by round table and mailing list discussions of all partners where the next steps are planned based on the previous project state and derived priority tasks to be solved in a next step. Currently, the installation is planned where the tasks of consortium were refined as compared to the initial application by a memorandum of understanding developed during a joint discussion, taking up the suggestions of the consortium.

28.7.2 Transfer Regarding Provisioning Services: Transfer Project "New Forests for Ecuador"

Our study region Ecuador suffers the highest deforestation rate in South America (−1.7 %), mainly due to conversion of forests into pastoral land (FAO 2009). However, as the need for subsistence is a major driver for deforestation (Davidar et al. 2007; Byron and Arnold 1999) people will continue to convert native forests into alternative uses, as long as this makes economic sense for them (Tschakert et al. 2007).

Reforestation of deforested land is of particular importance, and could be a promising strategy for reducing the pressure from natural forests and thus, contributing to conservation of biodiversity. While the country faces a yearly loss of approx. 20,000 ha, the yearly reforestation rate amounted only ca. 560 ha between 2000 and 2005 (FAO 2009; Mosandl et al. 2008). In total, only 200,000 ha of plantations have been established so far, and approx. 90 % consisting of introduced species, especially *Pinus* spp. and *Eucalyptus* spp. (FAO 2009). These monocultures are hardly compatible with aims for conservation of biodiversity. Frequently these plantations are accompanied by different ecological problems, e.g. reduced biodiversity, disturbed nutrient cycling, and water imbalance. However they are able to adapt under difficult environmental conditions on poor soils and steep slopes (Günter et al. 2009) and could possibly serve as shelter for economically and ecologically more valuable native species. Our own preliminary results confirm that native species perform better under pine shelter than on adjacent pastures, where light and root competition impede a satisfactory development of valuable tree species with the exception of special soil conditions (Günter et al. 2009; Aguirre et al. 2006). Besides ecological benefits, there is an increasing community emphasizing also the economic benefits of mixed forests (Piotto et al. 2010; Knoke 2008). The shelter effect for the establishment of natural regeneration is a widely accepted and common tool in forestry. It is applied in temperate forests as well as in tropical forests. Of course, the species richness of natural regeneration under exotic species is usually far from that of the native forest, but several authors indicate that exotic plantations can have facilitating effects for regeneration of native species (Cavelier and Tobler 1998; Oberhauser 1997).

Pine species and Eucalyptus species are used worldwide for reforestation because of outstanding good performance and easily available planting material. However, this does not necessarily mean that exotic species always perform better than natives. Our results in Ecuador confirm that *Alnus acuminata* performs even better than *Eucalyptus saligna* under certain conditions (Weber et al. 2008; Günter et al. 2009). *Alnus acuminata* is a widely spread species used for reforestation in the Andes and Central America (Grau 1985; Murcia 1997; Lamprecht 1986). The species is considered to be a pioneer to moderately shade-tolerant species of intermediate economic value (Günter et al. 2009; Mostacedo and Fredericksen 1999). It grows rapidly and improves soil fertility by increasing soil nitrogen, organic matter, and cation-exchange capacity (Grau 1985).

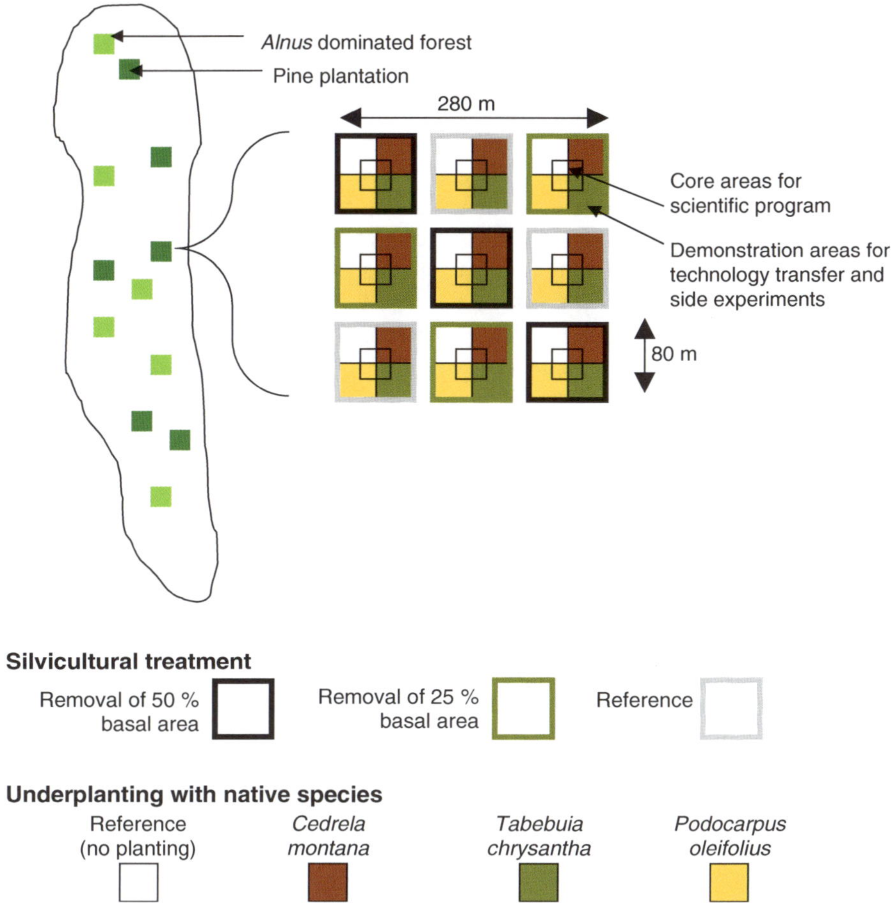

Fig. 28.6 Schematic figure for spatial distribution of experimental blocks (*green squares*) and arrangement of plots within the Andean region of Loja Province. The blocks are completely randomized within the study region. The plots within the blocks are arranged as *latin-square*. In each plot at least three native species plus one reference area will be planted. We expect to achieve results about nutrient dynamics, soil fauna, mycorrhiza, natural regeneration, and socio-economic feasibility. The ecological results are the basis for the development of best management guidelines for Southern Ecuador

Thus, in a pilot project we aim at fostering the establishment of mixed forests with native species by underplanting of pine plantations and *Alnus* stands and by application of silvicultural treatments. By this means monocultures shall be converted into mixed forests with higher ecological and economic stability. Using *Alnus* and Pine stands as shelter for native tree species could be a new instrument for forestry in Ecuador. *Alnus*-dominated stands shall be comparatively underplanted in order to analyse possible positive facilitating effects of this nitrogen fixer to nutrient cycling and biodiversity parameter. Figure 28.6 indicates schematically the implementation in the field.

One major reason for poor acceptance of international experiences and poor implementation in Ecuadorian forestry is the lack of institutions for technology transfer and missing pilot projects, proving the practical feasibility of scientific results. In close cooperation with several counterparts in Ecuador and Germany, we are conducting a pilot project for environmental sciences to overcome these institutional barriers of knowledge transfer and to test the feasibility of scientific results of the research group under realistic practical conditions and a wide range of environmental conditions in the South Ecuadorian Andes. Besides the technical and scientific challenge, how to establish native species in monocultures of exotics, a major challenge of the project is to enhance institutional cooperation. Private land owners, the Municipality of Loja, the government of the Province of Loja, and Nature and Culture International (NCI) will play key roles in the implementation of scientific results, for example to create the fundamentals for a regional seed management program, management guidelines for Pine plantations, and Alder stands, and establishing strategies how to convert monocultures into mixed forests of higher economic and ecological stability. While some funds will be applied for directly from DFG, substantial contributions will also be provided by the multiple Ecuadorian counterparts. This is a new aspect in contrast to "normal" applications to the DFG. For example all project areas (approx. 94 ha) will be loaned for the duration of the project. Additional infrastructural support will be provided by the municipality of Loja and the provincial government. NCI shall be in charge of upscaling and local coordination. The local universities will support with laboratories and release own staff for advanced training and capacity building under supervision of German counterparts. So far, it is the first time for South Ecuador that governmental institutions, NGOs, private land owners, and scientists from Germany and Ecuador join forces for the development of scientific knowledge combined with immediate implementation in the field.

References

Aguirre N, Günter S, Weber M, Stimm B (2006) Enriquecimiento de plantaciones de Pinus patula con especies nativas en el sur del Ecuador. Lyonia 10(1):33–45

Andrieu H, Creutin JD, Delrieu G, Faure D (1997) Use of weather radar for the hydrology of a mountainous area. Part I: radar measurement interpretation. J Hydrol 193:1–25

Bradshaw G, Bekoff M (2001) Ecology and social responsibility: the re-embodiment of science. Trends Ecol Evol 16:460–465

Bradshaw GA, Borchers JG (2000) Narrowing the science–policy gap: uncertainty as information. Conserv Ecol 4(1):1–7

Buytaert W, Célleri R, De Bièvre B, Cisneros F, Wyseure G, Deckers J, Hofstede R (2006) Human impact on the hydrology of the Andean páramos. Earth Sci Rev 79:53–72

Buytaert W, Célleri R, Timbe L (2009) Predicting climate change impacts on water resources in the tropical Andes: effects of GCM uncertainty. Geophys Res Lett 36, L07406. doi:10.1029/2008GL037048

Byron N, Arnold M (1999) What futures for the people of the tropical forests? World Dev 27 (5):789–805

Cavelier J, Tobler A (1998) The effect of abandoned plantations of Pinus patula and Cupressus lusitanica on soils and regeneration of a tropical montane rain forest in Colombia. Biodivers Conserv 7:335–347

Cooper CB, Dickinson J, Phillips T, Bonney R (2007) Citizen science as a tool for conservation in residential ecosystems. Ecol Soc 12(2):1–11

Crespo P, Feyen J, Buytaert W, Bücker A, Breuer L, Frede H-G, Ramirez M (2011) Identifying controls of the rainfall–runoff response of small catchments in the tropical Andes (Ecuador). J Hydrol 407(1–4):164–174. doi:10.1016/j.jhydrol.2011.07.021

Davidar P, Arjunan M, Pratheesh CM, Garrigues JP, Puyravaud JP, Roessingh K (2007) Forest degradation in the Western Ghats biodiversity hotspot: resource collection, livelihood concerns and sustainability. Curr Sci 93(11):1573–1578

Eckert P (2006) Communities of practice http://www.stanford.edu/~eckert/PDF/eckert2006.pdf. Assessed 7 Aug 2012

FAO (2009) State of the world's forests 2009. FAO, Rome

Grau A (1985) La expansión del aliso del cerro (*Alnus acuminata* H.B.K. subsp. acuminata) en el noroeste de Argentina. Lilloa 36:237–247

Günter S, González P, Alvarez G, Aguirre N, Palomeque X, Haubrich F, Weber M (2009) Determinants for successful reforestation of abandoned pastures in the Andes: soil conditions and vegetation cover. For Ecol Manage 258:81–91

Günter S, Weber M, Stimm B, Mosandl R (2011) Five recommendations to improve tropical silviculture. In: Günter S, Weber M, Stimm B, Mosandl R (eds) Silviculture in the tropics, vol 8, Tropical forestry. Springer, Heidelberg, pp 527–546

Heinimann HR (2010) A concept in adaptive ecosystem management: an engineering perspective. For Ecol Manage 259:848–856

Junyet F, Chandrasekhar V (2008) Theory and characterization of weather radar networks. J Atmos Oceanic Technol 26:474–491

Knoke T (2008) Mixed forests and finance-methodological approaches. Ecol Econ 65:590–601

Lamprecht H (1986) Silviculture in the tropics. Parey, Hamburg and Berlin, 317 pp

Lave J, Wenger E (1991) Situated learning: legitimate peripheral participation. Cambridge University Press, Cambridge

MacMynowski DP (2007) Pausing at the brink of interdisciplinarity: power and knowledge at the meeting of social and biophysical science. Ecol Soc 12(1):1–20

Mosandl R, Günter S, Stimm B, Weber M (2008) Ecuador suffers the highest deforestation rate in South America. In: Beck E, Bendix J, Kottke I, Makeschin F, Mosandl R (eds) Gradients in a tropical mountain ecosystem of Ecuador, vol 198, Ecological studies. Springer, Berlin, pp 37–40

Mostacedo CB, Fredericksen T (1999) Regeneration status of important tropical forest tree species in Bolivia: assessment and recommendations. For Ecol Manage 124:263–273

Murcia C (1997) Evaluation of Andean Alder as a catalyst for the recovery of tropical cloud forests in Colombia. For Ecol Manage 99:163–170

Oberhauser U (1997) Secondary forest regeneration beneath pine (*Pinus kesiya*) plantations in the northern Thai highlands: a chronosequence study. Forest Ecol Manage 99:171–183

OECD (2005) http://www.oecd.org/dac/evaluationofdevelopmentprogrammes. Assessed 14 Sept 2012

Piotto D, Craven D, Montagnini F, Alice F (2010) Silvicultural and economic aspects of pure and mixed native tree species plantations on degraded pasturelands in humid Costa Rica. New For 39(3):369–385. doi:10.1007/s11056-009-9177-0

Rollenbeck R, Bendix J (2006) Experimental calibration of a cost-effective X-band weather radar for climate-ecological use in southern Ecuador. Atmos Res 79:296–316

Rollenbeck R, Bendix J (2011) Rainfall distribution in the Andes of southern Ecuador derived from blending weather radar data and meteorological field observations. Atmos Res 99:277–289

Roux DJ, Rogers KH, Biggs HC, Ashton PJ, Sergeant A (2006) Bridging the science–management divide: moving from unidirectional knowledge transfer to knowledge interfacing and sharing. Ecol Soc 11(1):4

SENPLADES (2009) National plan for good living – policies. http://plan2009.senplades.gob.ec/web/en/policies-4. Assessed 20 Sept 2012

Tschakert P, Coomes OT, Potvin C (2007) Indigenous livelihoods, slash-and-burn agriculture, and carbon stocks in eastern Panama. Ecol Econ 60:807–820

Vergara W, Deeb AM, Valencia AM, Bradley RS, Francou B, Zarzar A, Grünwaldt A, Haeussling SM (2007) Economic impacts of rapid glacier retreat in the Andes. Eos Trans AGU 88 (25):261–264. doi:10.1029/2007EO250001

Weber M, Günter S, Aguirre N, Stimm B, Mosandl R (2008) Reforestation of abandoned pastures: silvicultural means to accelerate forest recovery and biodiversity. In: Beck E, Bendix J, Kottke I, Makeschin F, Mosandl R (eds) Gradients in a tropical mountain ecosystem of Ecuador, vol 198, Ecological studies . Springer, Berlin, pp 447–458

Wenger E (1998) Communities of practice: learning, meaning and identity. Cambridge University Press, New York

Wenger E (2005) Knowledge management as a doughnut: shaping your knowledge strategy through communities of practice. Ivey Bus J 68(3):1–8

Chapter 29
The Role of Biodiversity Research for the Local Scientific Community

Perdita Pohle, Maria Fernanda López, Erwin Beck, and Jörg Bendix

29.1 Collaborative German–Ecuadorian Biodiversity and Ecosystem Research in Southern Ecuador

Joint German–Ecuadorian biodiversity research in the study area is looking back on 15 years of fruitful collaboration and benefit sharing (Bendix and Beck 2012, see also "Preface"). Currently, 50 individual projects are working in the area of southern Ecuador. Biodiversity research started in 1997 with six projects funded by the German Research Foundation (Deutsche Forschungsgemeinschaft, DFG) in the Rio San Francisco valley, mainly aiming at biotic and abiotic inventories of this widely unknown area which turned out as a hot spot of biological diversity. Four years later (in 2001), a first collaborative Research Unit (RU402 "Functionality in a Tropical Mountain Rainforest: Diversity, Dynamic Processes and Utilization Potentials under Ecosystem Perspectives") started with a number of 17 DFG-funded projects, increasing to 25 in the second funding phase of that Research Unit (2005–2007). The main focus of the research program was to investigate ecosystem functioning along various environmental gradients (Beck et al. 2008). Aims and

P. Pohle
Institute of Geography, Friedrich-Alexander University Erlangen-Nürnberg, Kochstr. 4/4, 91054 Erlangen, Germany

M.F. López
Escuela de Geografía, Pontificia Universidad Católica del Ecuador (PUCE), Apt. 17-01-2184, Quito, Ecuador

E. Beck
Department of Plant Physiology and Bayreuth Centre of Ecology and Environmental Research, University of Bayreuth, 95440 Bayreuth, Germany

J. Bendix (✉)
Faculty of Geography, Laboratory for Climatology and Remote Sensing, University of Marburg, Deutschhausstraße 10, 35032 Marburg, Germany
e-mail: bendix@staff.uni-marburg.de

J. Bendix et al. (eds.), *Ecosystem Services, Biodiversity and Environmental Change in a Tropical Mountain Ecosystem of South Ecuador*, Ecological Studies 221,
DOI 10.1007/978-3-642-38137-9_29, © Springer-Verlag Berlin Heidelberg 2013

results of the subsequent and current Research Unit (RU816 "Biodiversity and Sustainable Management of a Megadiverse Mountain Ecosystem in South Ecuador" 2007–2013) are presented in detail in this book. In accordance with the statutes and the funding policy of the German Research Foundation, all individual projects contributing to the mentioned programs were and are in the category of basic, non-applied research, which, however, includes capacity building and benefit sharing as documented by the "Supplementary Instructions for Funding Proposals Concerning Research Projects within the Scope of the Convention on Biological Diversity (CBD)". In this context, one main issue was the science-directed development of sustainable land-use systems in the selected research area, the valley of the Rio San Francisco. An important side effect of the programs was the establishment of an effective research infrastructure both in the field and at the cooperating universities. This attracted two further basic research programs collaboratively working in the area: (1) The EU-funded EDIT program (*All Taxa Biodiversity Inventories and Monitoring ATBI + M*) with two additional projects and (2) a DFG-funded project bundle "ABA-Ecuador" (*Accelerated Biodiversity Assessment*).

Regarding benefit sharing activities, two new types of projects arose recently from the basic research programs. First, a so-called Cooperation Project jointly funded by four Ecuadorian universities and the DFG, aims at the autonomous development of research capacity and teaching staff for the South Ecuadorian universities. This is in line with the endeavor of many Latin American universities to change from a merely teaching system to a research-based teaching as also stressed by the World Bank (Thorn and Soo 2006, refer to Sect. 29.5). Furthermore, the success so far achieved by the collaborative research programs has recently been acknowledged by a memorandum of understanding on future joint research promotion by the Ecuadorian National Secretariat for Higher Education, Science and Technology (SENESCYT) and the German Research Foundation (DFG). Second, two knowledge transfer projects co-funded by DFG and Ecuadorian non-university cooperation partners could be launched, aiming on reforestation with indigenous tree species (see Sect. 28.7.2) and the establishment of a rain radar network in South Ecuador (see Sect. 28.7.1).

29.2 Addressing the Claims of the Convention on Biological Diversity: Benefit Sharing as a Major Goal of the German Research Programs

Besides the scientific endeavor, a major goal of the German research programs related to RU816 was benefit sharing from the access to biological resources with the academic, but also with the nonacademic communities in southern Ecuador. This scope is in full agreement with the spirit of the Convention on Biological Diversity (CBD) which in article 12 (Research and Training) underlines the specific needs of developing countries in scientific and technical education of academic staff at local universities (UN 1993). Additionally, article 13 (Public Education and Awareness) highlights the importance of the involvement of the public regarding the conservation and sustainable use of biological diversity which is also a major concern of the German research programs. Particularly, transfer of knowledge from basic science to application, environmental education, and training shall foster public awareness of the importance of biological diversity for ecosystem functioning and in turn livelihood and daily life. To achieve acceptance by the public, learning from traditional knowledge is also an integral part of the German research programs.

Following a mutual understanding and appreciation of a real partnership, besides access, benefit sharing is the focal point for successful and sustainable research in developing countries summarized under the term Access and Benefit Sharing (ABS) of the CBD. Since no commercial intentions exist in the programs of the German Research Units, benefit sharing is purely research-, education-, and knowledge transfer-oriented.

29.3 The Collaborative Approach of Research Unit 816

In accordance with the goals of the CBD and within the scope of ABS, the research activities of the German Research Unit 816 of the German Research Foundation (DFG) are focused on the following four pillars (Bendix et al. 2013):

- Conducting and promoting joint multidisciplinary research to investigate biodiversity, ecosystem functioning, and ecosystem services under environmental change in the hotspot area of the south-eastern Ecuadorian Andes.
- Supporting academic education, academic staff development, and the implementation of relevant research technologies at the Ecuadorian partner universities regarding all interdisciplinary issues of biodiversity research.
- Developing science-directed recommendations for the sustainable management of the extraordinary biodiversity, including strict protection and conservation by adequate use, and supporting respective administrative structures in a participatory approach together with the national authorities, NGOs, and local communities.
- Facilitating transfer of the compiled knowledge to the public to boost awareness at site for the needs and benefits of biodiversity research in order to safeguard ecosystem services and human well-being, and in turn, to attain acceptance of the local population.

The successful operation of the German Research Unit in South Ecuador benefits from a well-developed network of focal actors (Bendix et al. 2013). Researchers from Ecuadorian and German universities, together with those from other countries (e.g., Brazil, Peru, Belgium, Austria, and USA), are collaborating in multidisciplinary research projects. In Ecuador the main academic partners are the two universities of Loja (the Technical UTPL and the National University UNL), the Pontifical Catholic University PUCE in Quito, and more recently also the Universities of Cuenca and Azuay in Cuenca.

Nonacademic partners of the Research Unit are manifold. The network encompasses national institutions such as the Ministry of Environment (MAE) which is supporting the research activities by granting research permissions and, in following the National Strategy of Ecuador on Biodiversity, is benefiting from the results. Additionally, good relations exist with the national planning (SENPLADES) and the research funding (SENESCYT) agencies as well as with the national weather service (INAMHI). On the regional level the provincial governments of Loja and Zamora-Chinchipe are supporting biodiversity research by participating in knowledge transfer projects developed from previous joint studies (Chap. 28). This also holds for the local administrations, e.g., the municipality of Loja and the municipal agency of the city of Cuenca (ETAPA) which is in charge of the Cajas National Park. Private actors particularly related to benefit sharing are NGOs such as Nature and Culture International (NCI). The Ecuadorian branch of NCI is the main partner regarding environmental education and was the driving force for the implementation of the UNESCO Biosphere Reserve Podocarpus—El Condor. Furthermore, NCI has developed community-based participatory approaches for the protection and rehabilitation of important ecosystems (Aguilar 2008). At a local level, private landowners are part of the benefit sharing process who on their side provide land for reforestation experiments (Sect. 28.7.2).

29.4 Environmental Education and Capacity Building

Environmental education and capacity building in Ecuador as major aims of ABS endeavors are supported by the Research Unit 816 and related programs along four lines: (1) involving Ecuadorian scientists at all qualification levels in the research process, (2) supporting the autonomous development of scientific staff at the local universities, (3) promoting biodiversity research and conservation through basic training in research techniques and methodologies for graduate and undergraduate students, academics, and institutional staff, and (4) enhancing public environmental education through public presentations, workshops, and training.

(1) The research program of RU816 was designed and conducted as truly collaborative, employing assistants from the host country at different qualification levels—from the unskilled community worker, the trainee and diploma student, up to the PhD and PostDoc researcher. Whereas the design of the individual research projects of the program was jointly developed by the Ecuadorian and the German principal investigators (PI), the funding of staff (e.g., Ecuadorian and German PhD positions) and instrumentation was mainly provided by the German Research Foundation (DFG). Figure 29.1 shows the noticeable high number of Ecuadorian participants in the research program (in total 134),

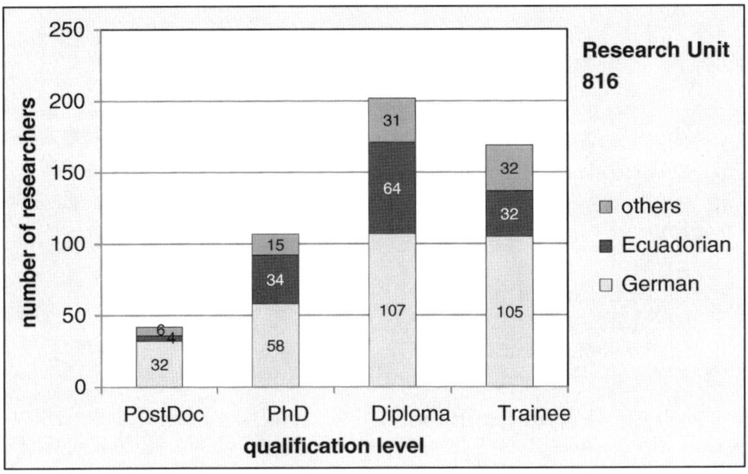

Fig. 29.1 Researchers at different qualification levels working in the Research Unit 816 (2007–2012). *Source*: RU816 Data Warehouse (http://www.tropicalmountainforest.org). Data compilation: T. Lotz, August 2012

particularly in the field of Diploma (Ecuadorian equivalent = licenciatura or ingeniería) students (64).

The capacity building effect of the collaborative research is also expressed in the increasing number of Ecuadorian PhD students which currently even exceeds the number of German PhDs (Fig. 29.2). Many of today's PhD students were also members of the Research Unit during their undergraduate studies. Thus, not only the absolute number but also the academic level of the involved Ecuadorian scientists has significantly increased during the 15 years lifetime of the research. A reasonable part of the Ecuadorian today's staff members have been former undergraduate and doctoral students of the German research programs. This underpins the contribution of the Research Unit to the careers of Ecuadorian scientists from PhD students to leadership positions in universities, NGOs, and public administration. It should also be stressed that students from other Latin American countries such as Brazil and Peru have been and still are attracted by the research program.

The success in staff promotion has also led to a greater internationalization of the research activities of the collaborating Ecuadorian universities (especially UNL and UTPL). This is mainly due to the increased numbers of publications in international peer-reviewed journals (Fig. 29.3). Whereas the relative contributions by Ecuadorian scientists as authors/coauthors were about 25 % at the beginning (2007), it doubled in the course of the RU816, amounting to 51.4 % in 2009. Out of a total of 155 articles published by RU816 members in international peer-reviewed journals between 2007 and September 2012, one-third (54) has Ecuadorian authors/coauthors.

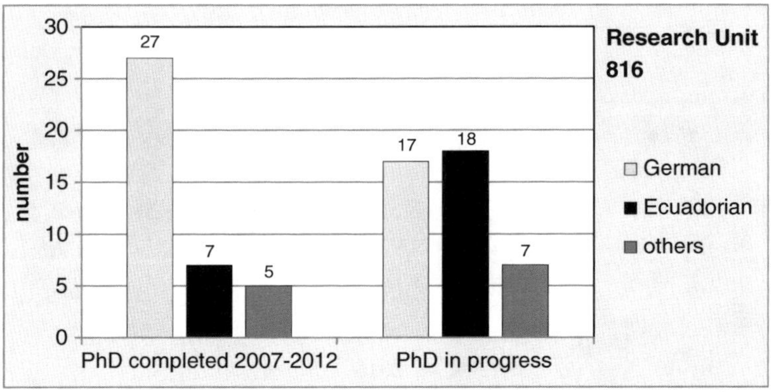

Fig. 29.2 Number of completed PhDs and PhDs in progress with respect to nationality in the Research Unit 816. *Source*: RU816 Data Warehouse (http://www.tropicalmountainforest.org). Data compilation: T. Lotz, June 2012

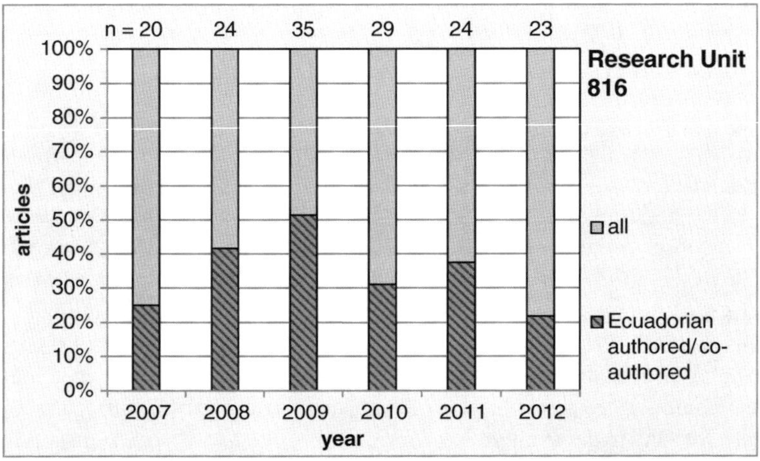

Fig. 29.3 Articles of RU816 members in international peer-reviewed journals (2007–September 2012). *Source*: RU816 Data Warehouse (http://www.tropicalmountainforest.org). Data compilation: T. Lotz, September 2012

(2) The second line, which promotes autochthonous development of university staff, started in 2009 with a specific program of cooperation between the German Research Foundation (DFG) and South Ecuadorian universities. Here, the project was designed by Ecuadorian PIs, who adapted the research goals to those of the program of the German research group to warrant a synergetic use of available resources. The Ecuadorian PhD students and their living costs were provided by the Ecuadorian side whereas the DFG contributed grants for visiting the collaborating German institutes. In addition to scientific staff promotion, technical staff at the Ecuadorian universities is trained by the

German scientists to properly use jointly established research infrastructure (e.g., biochemical and genetic laboratory facilities, field instrumentation, etc.).

(3) The third line aims to increase the scientific skills of graduate/undergraduate students, academics, and institutional staff in biodiversity research through capacity building. Students, academics, and university staff have been involved as research assistants in both the field and laboratory work. Ecuadorian undergraduates have prepared their final theses under the co-supervision of German and Ecuadorian researchers. This has positively influenced their skills in academic work and their understanding of basic procedures of scientific research. Complementary to this, further training activities in relevant topics of biodiversity research (e.g., plant diversity, soil biology, ethnobotany) have been carried out by both Ecuadorian and German researchers in the form of lectures, seminars, workshops, and field and laboratory courses (Table 29.1).

Most activities were conducted in the city of Loja, at universities or public agencies. However, several seminars and workshops have taken place in the field, at the facilities of the Estación Científica San Francisco (ECSF), but also in other locations of the country (Quito, Zamora, El Coca, Santa Elena), or direct in the local communities (Shaime, Laipuna, Sabanilla). During the running period of RU816 (2007–2012) a notable number of 45 capacity building and training activities with a total of 1,515 participants have been conducted by German researchers and Ecuadorian collaborators (cf. documentation in Table 29.1). The highest number of participants (580) and activities (11) was achieved in 2010 (Table 29.2).

Two-thirds of the participants in the capacity building and training activities were academics, but one-third of the activities were either open to the public or specifically addressed to non-academics, e.g., farmers (Table 29.3). The activities (Table 29.1) ranged from 1 day (22) to 1 week (19) and even up to 12 days (4). Additionally, in 2006, an international and interdisciplinary summer school on "Integrative assessment and planning methods for sustainable agroforestry in humid and semiarid regions" (funded by Robert Bosch Foundation, UTPL, DIU,[1] DFG, Organisation: F. Makeschin) was held in Loja with 30 participating students from throughout Latin America.[2]

(4) The fourth line aims to enhance the awareness and acceptance of biodiversity research and conservation in the public by environmental education through public presentations, workshops, and trainings (Table 29.1). A considerable number of lectures and workshops open to the public (15) were held during recent years (cf. Tables 29.1 and 29.3). Also, workshops in the communities with local representatives and farmers have been organized for knowledge transfer. In addition, environmental information on a more popular scientific

[1] Dresden International University.

[2] Lectures from the RU816 by F. Makeschin, (TU Dresden); R. Mosandel, B. Stimm (TU München); M. Richter, A. Gerique (University of Giessen, now University of Erlangen-Nuremberg); E. Beck (University of Bayreuth); J. Bendix, R. Rollenbeck (University of Marburg).

Table 29.1 Capacity building statistics (2007–2012), Research Unit 816, DFG-Ecuador

Project RU816	Type of capacity building and training activity	Title of capacity building and training activity	Teaching staff (with affiliation)	Place	Host institution	Date	Duration of activity	Addressed audience	Number of participants
A5	Public presentation	"Effectos de la perturbación humana a comunidades de epífitas den bosques andinos"	F. Werner (Univ. of Oldenburg)	Casa de la Cultura Ecuatoriana, Loja	RU816	2008, May	1 day	Open to the public, interested people of Loja	20
A5	Field seminar	Curso De Campo Sobre Diversidad De Plantas Vasculares (Árboles Y Epífitas) Y Almacenamiento De Carbono En Bosque Seco Tumbesiano	F. Werner (Univ. of Oldenburg), J. Homeier (Univ. of Göttingen)	ECSF, Loja, field (Reserva Laipuna)	NCI	2010, March	10 days	Academics, university teaching staff, university students	16
A5	Field seminar	Curso De Campo Sobre Diversidad De Plantas Vasculares (Árboles Y Epífitas) Y Almacenamiento De Carbono En Bosque Montano	F. Werner (Univ. of Oldenburg), J. Homeier (Univ. of Göttingen)	ECSF, Loja, field (Reserva Laipuna)	NCI	2012, March/April	11 days	Academics, university teaching staff, university students	13
A7	Meeting	First meeting on Collaborative Doctorate Program (sandwich) and Research Incubators	Organizer: L.M. Romero (UTPL); Presentation: I. Kottke (Univ. of Tübingen)	Loja	UTPL	2008, June	3 days	Rectors and officials of the American Universities	200
B1/D3	Colloquium	"La circulación atmosférica global"	R. Rollenbeck (Univ. of Marburg), T. Peters (Univ. of Erlangen-Nuremberg)	Casa de la Cultura Ecuatoriana, Loja	RU816	2007, Oct.	1 day	Open to the public, interested people of Loja	20
B2	Field excursion	Geographical field excursion to northern Peru	M. Richter, T. Peters (Univ. of Erlangen-Nuremberg)	Peru, dry forest South Ecuador	RU816, UTPL, UNL	2007, Sept./Oct.	5 days	Students, institutional staff	25

B3	Seminar	Methods in soil biology	K. Potthast (TU Dresden)	Loja	UTPL	2007, May	1 day	Students	20
B3	Seminar	"Prueba de fertilización de pasto—impacto en el suelo y consecuencias económicos-sociales y productivas"	H. Lucero (UTPL)	Casa de la Cultura Ecuatoriana, Loja	RU816	2007, Dec.	1 day	Open to the public, interested people of Loja	20
B3	Seminar	Soil mineralogy and soil petrology	F. Haubrich (TU Dresden)	Loja	UTPL	2008, March/April	3 days	Students	20
B3	Seminar	Methods in soil chemistry	K. Potthast (TU Dresden)	Loja	UTPL	2008, June	1 day	Students	20
B3	Seminar	Scientific writing in soil science	E. Bahr (TU Dresden)	Loja	UTPL	2008, April/May	10 days	Students	20
B3	Seminar	Nutrient balances in Ecuadorian soils	E. Bahr (TU Dresden)	Loja	UTPL	2009, Feb.	1 day	Students	10
B3	Workshop	Nutrient balances and methods for soil improvement	E. Bahr (TU Dresden)	Yantzaza, El Tambo, San Lucas	UTPL	2010, March	3 days	Farmers	15
B3	Excursion	Soil properties in coffee plantations	E. Bahr (TU Dresden), D. Chamba (UTPL)	Loja, surroundings	UTPL	2011, March	1 day	Students	10
B3	Seminar	Biophysical characteristics and agroecological indicators for S-Ecuadorian soils	E. Bahr (TU Dresden)	Loja	UTPL	2011, March	1 day	Students	10
B3	Workshop	Preparation of knowledge transfer to farmers	E. Bahr (TU Dresden), L. Izquierdo (UTPL)	Loja	UTPL	2012, Feb.	1 day	Students	10
B3	Workshop	Protection of Ecuadorian soils	E. Bahr (TU Dresden), L. Izquierdo (UTPL)	Yantzaza, El Tambo	UTPL	2012, Feb.	3 days	Farmers	25

(continued)

Table 29.1 (continued)

Project RU816	Type of capacity building and training activity	Title of capacity building and training activity	Teaching staff (with affiliation)	Place	Host institution	Date	Duration of activity	Addressed audience	Number of participants
C1	Course	Professional tree climbing course	A. Wörle (Bayer. Landesanstalt für Wald und Forstwirtschaft); D. Kübler, B. Calvas (TU München)	ECSF	NCI	2012, April	12 days	Institutional staff	16
C2/T1	Public presentation	"Una visión general de las micorrizas y sus roles en los ecosistemas"	J.P. Suárez (UTPL)	Casa de la Cultura Ecuatoriana, Loja	RU816	2008, Feb.	1 day	Open to the public, interested people of Loja	20
C2	Internat. workshop	Mycorrhizas in tropical forest	Organizer: I. Kottke (Univ. of Tübingen) and J.P. Suárez (UTPL); 26 lecturers from 12 countries	Loja	UTPL	2008, Sept.	5 days	Students of Ecuador and RU816	40
C4	Course	Introduction into ArcGis	A. Tutillo Vallejo (Univ. of Erlangen-Nuremberg/ PUCE Quito)	Loja	UNL	2008, Sept.	6 days	PhD students and professionals	12
C4	Course	Introduction into ArcGis	A. Tutillo Vallejo (Univ. of Erlangen-Nuremberg/ PUCE Quito)	Loja	UNL	2008, Dec.	4 days	PhD students and professionals	10
C4	Seminar	"Manejo sustenible del bosque tropical de montaña—El factor humano"	A. Gerique (Univ. of Erlangen-Nuremberg)	Casa de la Cultura Ecuatoriana, Loja	RU816	2007, Sept.	1 day	Open to the public, interested people of Loja	20

C4	Seminar, workshop	Levantamiento y análisis cuantitativo y cualitativo de información etnobotánica con fines comerciales, con práctica de campo	A. Gerique (Univ. of Erlangen-Nuremberg), M.F. López (PUCE, Quito)	Quito	PUCE Escuela de Geografia	2009, Oct.	2 days	Under graduate students	25
C4	Presentation	"Seminario-Taller Nacional de Biodiversidad en Recursos Genéticos Nativos para la Alimentación y la Agricultura": La biodiversidad como recurso: el uso de las plantas, su manejo y alternativas sostenibles en comunidades rurales del sur del Ecuador	A. Gerique (Univ. of Erlangen-Nuremberg)	El Coca, INIAP Research Station	SENACYT/ INIAP	2010, Oct.	2 days	Authorities, teaching staff of universities, students	100
C4	Workshop	Plantas útiles de los Shuar del Alto Nangaritza	A. Gerique (Univ. of Erlangen-Nuremberg)	Shaime, Alto Nangaritza	Centro Shaime	2010, Oct.	1 day	Shuar community members	40
C4	Workshop	El uso sustentable de la fitodiversidad como alternativa a la deforestación	A. Gerique (Univ. of Erlangen-Nuremberg)	Zamora, Consejo Provincial Zamora Chinchipe	Consejo Prov., GTZ	2010, Nov.	1 day	Farmer associations, technicians	20
C4	Colloquium	Forum "La Cordillera del Cóndor: Territorio Pluricultural y Biodiverso" (video-conference): La Biodiversidad de la Cordillera del Cóndor como un recurso económico sustentable	A. Gerique (Univ. of Erlangen-Nuremberg)	Zamora, Consejo Provincial Zamora Chinchipe	Consejo Prov., Zamora Chinchipe	2012, May	1 day	Authorities, experts, general public	100

(continued)

Table 29.1 (continued)

Project RU816	Type of capacity building and training activity	Title of capacity building and training activity	Teaching staff (with affiliation)	Place	Host institution	Date	Duration of activity	Addressed audience	Number of participants
D2	Course	Laboratory course on wood anatomy and dendroecology	A. Bräuning (Univ. of Erlangen-Nuremberg); Ing. H. Maza (UNL)	Loja	UNL	2007, Sept.	6 days	Students, institutional staff	15
D2	Course	Laboratory course on wood anatomy and dendroecology	A. Bräuning (Univ. of Erlangen-Nuremberg); Ing. H. Maza, Ing. O. Ganzhi, Ing. D. Pucha (UNL)	Loja	UNL	2011, Sept./Oct.	6 days	Students, institutional staff	15
D2	Lab.Course	Reconstrucción climática a través de métodos dendrocronológicos	A. Bräuning (Univ. of Erlangen-Nuremberg); Ing. D. Pucha (UNL)	Loja	UNL	2011, Sept./Oct.	6 days	Students	15
D3	Workshop	1. El clima y los elementos del clima	A. Fries (Univ. of Marburg/UTPL)	ECSF	RU816	2010, Feb.	1 day	MAE, Park Ranger	20
D3	Workshop	2. El clima en Ecuador e instrumentos meteorológicas en la ECSF	A. Fries (Univ. of Marburg/UTPL)	ECSF	RU816	2010, Feb.	1 day	MAE, Park Ranger	20
D3	Workshop	3. Interpolación de datos climáticos para el área de la ECSF	A. Fries (Univ. of Marburg/UTPL)	ECSF	RU816	2010, Feb.	1 day	MAE, Park Ranger	20
Z1	Workshop	Data warehouse (data management)	T. Lotz (Data Manager, Univ. of Marburg)	Loja	UTPL	2010, Oct.	1 day	RU816 members, Ecuadorian PhD students	29
Z1	Workshop	Data warehouse (data management)	T. Lotz (Data Manager, Univ. of Marburg)	Loja	UTPL	2011, Oct.	1 day	RU816 members, Ecuadorian PhD students	38
Z2	Course	Practice and theory of GIS	PhD student and Postdocs of RU816	ECSF	NCI	2008	2 days	MAE	20
Z2, others	Seminar	Taller de Capacitación para coordinadores de los ecoclubes de los colegios	PhD student and Postdocs of RU816	ECSF	NCI	2008, Dec.	2 days	Teachers	20

Z2	Workshop	Practice and theory of DGPS and GIS	J. Zeilinger (Project Unit Manager, ECSF); Jorge Cueva, Luis Chalán (NCI)	ECSF	RU816/NCI	2009, Sept.	3 days	MAE	20
Z2	Presentation	"Seminario-Taller Nacional de Biodiversidad en Recursos Genéticos Nativos para la Alimentación y la Agricultura": La Unidad de Investigación RU816 "Biodiversidad y Manejo Sostenible de un Ecosistema de Montaña Megadiverso en el Sur del Ecuador"	J. Zeilinger (Project Unit Manager, ECSF)	El Coca, INIAP Research Station	SENACYT/ INIAP	2010, Oct.	2 days	Authorities, teaching staff of universities, students	100
Z2	Presentation	La Unidad de Investigación RU816 "Biodiversidad y Manejo Sostenible de un Ecosistema de Montaña Megadiverso en el Sur del Ecuador"	J. Zeilinger (Project Unit Manager, ECSF)	Santa Elena	SENACYT	2010, Nov.	3 days	Authorities, teaching staff of universities, students	200
Z2	Presentation	La Unidad de Investigación RU816 "Biodiversidad y Manejo Sostenible de un Ecosistema de Montaña Megadiverso en el Sur del Ecuador"	J. Zeilinger (Project Unit Manager, ECSF)	Quito	FAO/ SENAGUA	2011, Nov.	5 days	Authorities, teaching staff of universities, students	60

(continued)

Table 29.1 (continued)

Project RU816	Type of capacity building and training activity	Title of capacity building and training activity	Teaching staff (with affiliation)	Place	Host institution	Date	Duration of activity	Addressed audience	Number of participants
Z2	Presentation and excursion	La Unidad de Investigación RU816 "Biodiversidad y Manejo Sostenible de un Ecosistema de Montaña Megadiverso en el Sur del Ecuador"	J. Zeilinger (Project Unit Manager, ECSF)	ECSF	RU816	2012, Jan.	1 day	Students of the UNIVERSIDAD ESTATAL AMAZÓNICA	22
Z2	Presentation	Socialization of the LiDAR projekt in the Rio San Francisco valley	J. Zeilinger (Project Unit Manager, ECSF)	ECSF	RU816	2012, Feb.	1 day	Authorities, Institutions of San Francisco, Zamora	12
Z2	Presentation	Socialization of the LiDAR projekt in the Rio San Francisco valley	J. Zeilinger (Project Unit Manager, ECSF)	Sabanilla, Junta Parroquial	RU816	2012, Feb.	1 day	Local communities of San Francisco and surroundings	12

Table 29.2 Number of participants and events in capacity building and training activities per year of the RU816 running period (based on Table 29.1)

Years	2007	2008	2009	2010	2011	2012	Total
Participants	120	402	55	580	148	210	1,515
No. of events	6	11	3	11	6	8	45

Table 29.3 Addressed audience of capacity building and training activities in the course of the running period of RU816 (based on Table 29.1)

Addressed audience	Academics	Non-academics	Academics and non-academics
Participants	1,131	192	192
No. of events	30	9	6

level is necessary for stakeholders and interested locals. The results of international research programs are mainly published in English in scientific journals or books making their access by the local people difficult. Thus, it is necessary to translate the results into laymen's language in order to raise awareness and foster the feeling of responsibility of all citizens for biodiversity and related ecosystem services. One example is the joint publication of a booklet by NCI and the Research Unit (Kiss and Bräuning 2008) summarizing and "translating" the scientific results into Spanish for a wider readership. Additionally, annual symposia with keynote talks in Spanish and monthly research meetings both open to the public were and are still organized in Loja.[3] The lecture hall building of the research station is used to conduct classes and field courses for pupils on biodiversity and the environment, but also specific courses for local administrators (Table 29.1).

29.5 Shared Access to Research Facilities, Technology, and Information

The importance of the research programs for the development of the universities in southern Ecuador is noticeable (Bendix et al. 2013). At the beginning of the research in 1997, UTPL for instance was just a teaching university without any biological or ecological disciplines. Fifteen years later, the situation has completely changed. Recently, UTPL has promoted biological research personnel and infrastructure as well as teaching staff. In the scope of collaboration with the German programs, laboratories for molecular biology, soil analysis, and geographical information systems have been established and equipped. Improving their facilities in this way, the cooperating Ecuadorian university could increase its international attractiveness far beyond the cooperation with the German research group: UTPL

[3] As regularly and obligatory activity of the entire RU816 they are not included in Table 29.1.

hosts more than 600 visiting professors per year and supports more than 300 research visits of its own scientists abroad. The funding of science development by national agencies was increased to 4 million US$ in 2009 (Bendix et al. 2013).

At the UNL, the joint research program similarly led to the establishment and extension of important research infrastructure: (1) Improvement of the soil analysis lab, (2) complementation of the UNL Herbarium "Reinaldo Espinosa" by 3,500 new specimens, (3) improvement of the equipment of UNL laboratories for dendrochronology and (4) plant physiology, and (5) a tree nursery which is indispensable for the long-term reforestation experiments of the research group.

Very important for multidisciplinary biodiversity research in a foreign country is the availability and guaranteed unlimited access to a field station with well-managed experimental and monitoring sites. Above all, the research program benefits from the close cooperation with the local branch of the foundation NCI which provides the well-equipped research station ECSF (Estación Científica San Francisco). It offers accommodation and board, provides basic research infrastructure like soil, water, and IT labs, runs a herbarium, and holds a lecture hall. Furthermore, many parts of the research area are owned by the foundation, e.g., the protected natural mountain forest RBSF (Reserva Biológica San Francisco). Similarly, wide areas where the natural forest has been converted into pastures or exotic tree plantations are also available for research. Access to research areas of the cooperating universities is permitted.

A central element of benefit sharing is the access to and the transfer of knowledge. The knowledge on biodiversity and underlying ecosystem processes/services produced by the Research Unit is compiled in a central data warehouse (Nauß et al. 2007; Lotz et al. 2012) which is open to all contributing scientists and cooperating organizations. To date, the data warehouse is not only keeping more than 5 Mio stored data values, but also offers access to digital publications which is usually not possible for Ecuadorian universities. Furthermore, the quarterly newsletter of the RU816 has continuously gained importance as a fast outlet for science news of the Ecuadorian and German researchers, but also as a medium of communication between all academic and nonacademic partners. Since the hosting institute of the RU816 data base project was accepted as an "official" data center by DataCite (a member of the International DOI Foundation IDF), the newsletter is officially citable by a Digital Object Identifier (DOI) and thus became a persistently available document which helps to boost the international visibility of the joint research, in particular of the Ecuadorian partners.

29.6 Conclusions

Fifteen years of joint collaborative research have significantly changed the scientific scene in southern Ecuador, in particular with regard to interdisciplinary ecological and biodiversity research. The main achievements can be summarized as follows:

- The local universities have established new and internationally compatible curricula related to ecology and biodiversity.
- They have established and implemented respective research infrastructure, particularly labs for geo- and biosciences including the first sequencing capabilities.
- The first generation of Ecuadorian graduates is now in leading positions, thus promoting the development of biodiversity research and teaching in their universities towards an internationally competitive institution.
- With this, the cooperating Ecuadorian universities have started the transition process from teaching universities to international competitive research universities which is underpinned by their excellent positions in the national university ranking (e.g., UTPL is at second place, all contributing universities within the leading 15).
- The younger Ecuadorian graduates and PhD students are now familiar with the processes and demands of scientific work and have gained English and/or German language and publication skills. Once graduated as PhDs, they are capable to push the development of Ecuadorian research universities through their publications and their involvement as lecturers.
- Last but not least, 15 years of ecosystem research have provided a wealth of scientific data covering all aspects of interdisciplinary biodiversity and ecosystem research. Although the research area represents one of the most complex ecosystems of the world, synthesis of the up to now available data (see Chap. 27) allows a level of understanding which appears unparalleled in tropical ecosystem analysis. Ongoing field experiments open new opportunities for further joint research activities.

The Ecuadorian state assigns a primary service function for the public to research subjects investigated by Ecuadorian scientists. Its national strategy on biodiversity, as part of the "Regional Biodiversity Strategy for the Tropical Andean Countries" (2005), puts much emphasis on conservation of biodiversity by habitat conservation. National priority fields are the production of bio-knowledge, including knowledge on biodiversity and conservation, and the assessment of impacts of environmental change (climate and land-use changes) on ecosystems and their services. Of major importance is research on the provisioning service *water* and on the potentials of diverse *land uses,* which can help to protect water resources and biodiversity. With the current local expertise and the baseline data stock the consortium can and will come into a new phase of purely co-funded German–Ecuadorian research which is currently under development. Based on the gained knowledge and the established infrastructure, an extension of the research program to the dry forest in the South-West Ecuador (Laipuna Natural Reserve) and to the páramo (El Cajas National Park) is envisaged. In that respect a German–Ecuadorian consortium has compiled a research plan for a "Platform for Biodiversity and Ecosystem Monitoring and Research in South Ecuador": A subprogram focused on basic research of environmental impacts on biodiversity and ecosystem services and two subprograms on sustainable land use and on global change monitoring will transfer results gained from basic research into application.

This is intended to be realized with knowledge transfer programs as presented in Chap. 28. The development of prototypes for a sustainable management and a functional monitoring system with Ecuadorian university and nonuniversity cooperation partners will further boost the capacity building activities and, thus, will at the same time match the national aims of Ecuador and the CBD objective "...fair and equitable sharing of the benefits arising out of the utilization of genetic resources" (UN 1993).

References

Aguilar Z (ed) (2008) Guía de vida silvestre del área de conservación y desarrollo – La Ceiba. Naturaleza y Cultura Internacional, Quito, Ecuador, 87 pp

Beck E, Bendix J, Kottke I, Makeschin F, Mosandl R (eds) (2008) Gradients in a tropical mountain ecosystem of Ecuador. Ecological studies, vol 198. Springer, Berlin

Bendix J, Beck E (2012) 15 Years of German-Ecuadorian research. In: TMF Newsletter of the DFG Research Unit 816, Issue 17. Laboratory for climatology and remote sensing (LCRS), University of Marburg, Marburg, Germany, pp 4–6. doi:10.5678/lcrs/for816.cit.1132. Accessed 29 Sept 2012

Bendix J, Paladines B, Ribadeneira-Sarmiento M, Romero LM, Valarezo C, Beck E (2013) Benefit sharing by research, education and knowledge transfer – a success story of biodiversity research in southern Ecuador. In: Brooks LA, Arico S (eds) Tracking key trends in biodiversity science and policy. UNESCO, Paris, pp 116–121

Kiss K, Bräuning A (2008) El bosque húmedo de montaña – Investigaciones sobre la diversidad de un ecosistema de montaña en el sur del Ecuador. Unidad de investigación FOR402. DFG, TMF y Naturaleza y Cultura Internacional, Loja – Ecuador, 64 pp (also available in English)

Lotz T, Nieschulze J, Bendix J, Dobbermann M, König-Ries B (2012) Diverse or uniform? Intercomparison of two major German project databases for interdisciplinary collaborative functional biodiversity research. Ecol Inform 8:10–19

Nauß T, Göttlicher D, Dobbermann M, Bendix J (2007) Central data services in multidisciplinary environmental research projects. J Inform Agric 2:1–13

Thorn K, Soo M (2006) Latin American Universities and the third mission—trends, challenges and policy options. World Bank Policy Research Working Paper, 4002. 23 pp

United Nations (1993) Multilateral convention on biological diversity (with annexes). Concluded at Rio de Janeiro on 5 June 1992, vol 1760, I-30619, 83 pp. http://www.cbd.int/convention/convention.shtml. Accessed 28 June 2012

Index

A
Aboveground NPP (ANPP), 118
Acceptance, 413
Adaptive radiation, 73
Afforestation, 388
Agriculture, sustainable, 237
Agrobiodiversity, 232, 242
Agro-pastoral activities, 227
Aichi biodiversity targets, 34–36
Air humidity, 279
Air temperatures, 24, 279
Alnus acuminata, 180, 278, 379, 387, 406
Altitude, 251, 382
AM. *See* Arbuscular mycorrhizal (AM) fungi
Amazon, 287
Ammonium input, 291
Amoebae, 82, 248
Amotape-Huancabamba Depression, 6
Andean bear, 35, 37
Andes uplift, 74
Andes, 3–4, 19–27
  uplift, 377
Angiosperm species richness, 95
ANN. *See* Artificial neural networks (ANN)
ANPP. *See* Aboveground NPP (ANPP)
Approaches
  collaborative, 413–414
  participatory, 401, 413
Arbuscular mycorrhizal (AM) fungi, 81, 83–85, 189, 378
Artificial neural networks (ANN), 277
Atmospheric deposition, 6–8
Atmospheric fertilization, 390
Atractiellales, 191

B
Balloon, 357
Base metals, 144, 166
Basic research, 404
Benefit sharing, 411
Biodiversity, 46, 53, 76, 93, 103, 153, 188, 247–260, 310, 390, 427
  causes, 13–15
  distribution, 377–378
  evolution, 57
  hotspots, 20, 35
  hypotheses, 153
  loss, 295
  origin, 67, 377–378
  persistance, 377–378
  threats, 26, 373
  underlying mechanisms, 196–197
Biodiversity research, 426
  collaborative, 411
Biogeochemical cycles, 6, 299
Biological corridor, 37
Biomass, 122, 125–127, 208–211, 359
  above-ground, 363
  bracken frond, 363
  burning, 26
  fine root, 320, 323
  response to FERPAST, 367
  root, 128
Biomass grass production, 367, 369
Biosphere Reserve Podocarpus-El Cóndor, 174
Bombuscaro, 5
Bowen ratio, 276

J. Bendix et al. (eds.), *Ecosystem Services, Biodiversity and Environmental Change in a Tropical Mountain Ecosystem of South Ecuador*, Ecological Studies 221,
DOI 10.1007/978-3-642-38137-9,

Bracken, 99, 100, 207–209, 212, 214, 215, 332, 335, 355, 359–362, 364, 365, 367, 381, 390
control, 357–358, 363, 388, 391
fronds, 362
germination, 209
nutrient acquisition, 210
rhizome system, 356
sexual propagation by spores, 356
succession experiment location, 7
Brundtland Report, 31
Buffer zone, 37
Burning, 207, 208, 210
experimental, 362

C
Cajanuma, 5
Cajas National Park, 173
Calibrated phylogenies, 69
Cambium, 293
Canopy, 124
Capacity building, 400, 414, 425
Carbon
allocation, 118, 133–135
balance, 117–135
cycle, 118, 119, 121, 315–325, 327, 331–340
emission, 46
fluxes, 118, 128–133
pools, 119, 121, 125, 129, 135
regulation services, 44
sequestration, 379, 380
stocks, 379
storage, 128, 135, 316, 379
transfer, 132
turnover, 390
Carbon relations
climate change, 380–381
food web, 380
forest, 378–380
pasture, 378–380
Carbon stocks, aboveground, 339
Catchment, 108, 110
water balance, 110
Catchment modeling framework (CMF), 278
Cattle grazing, 390
Cattle ranching, 206, 227
CBD. *See* Convention on Biological Diversity (CBD)
Ecosystem, CBD approach, 35, 36
CE. *See* Choice experiment (CE)
Cedrela montana, 63, 175, 179–182, 293–294, 387, 407
Charcoal, 62
analyses, 56
Choice experiment (CE), 45, 47
Climate, 275–285
data, 198
fluctuations, 249
history, 376–377
peculiarities research area, 8–11
regulation services, 44
scenario, 267
trends, 24, 266
Climate changes, 4, 19, 20, 23, 247–260, 265–274, 331–340, 374, 389–390, 427
impact, 268
past, 54
CLM. *See* Community land model (CLM)
Cloud, 292
formation processes, 10
CMF. *See* Catchment modeling framework (CMF)
C:N ratio, 211
Coevolution, 76
Colonization, 221
Commodity, 236
Community composition, 339
Community land model (CLM), 278
Compensation payments, 235, 238, 239, 350, 391
Competitive strength, 339
Conservation, 32, 76, 104, 171–183, 350, 351, 377, 389, 391, 396, 427
payments, 235–242, 348, 352
policies, 350
seed sources, 179
Convention on Biological Diversity (CBD), 32, 37, 412
Conversion, 206
Cooperation partners, 404
CO_2-uptake, 334
Cropland, 348

D
Data warehouse, 426
Decision making, 33, 395
Decomposer communities, 248
Decomposer, 81, 86, 380
Decomposition, 134
Deforestation, 3, 20, 53, 62, 222
Degree distributions, 195
DEM. *See* Digital elevation model (DEM)
Denudation rates, 12

Deposition, 143, 145, 149, 290, 316
athmospheric, 383
scenario, 292
Development, sustainable, 31–38
Digital elevation model (DEM), 156
Digital soil map, 162
Discharge, 111–113, 268, 271, 272
simulated, 269
Disturbances, 15, 83, 93, 96
human, 97
natural, 96
Diversification, 73, 239, 242
Diversity, 3, 6, 378
belowground, 89
biological, 14, 32, 48
epiphytes, 101
functional, 103–104
insects, 4
molecular, 82
mycobionts, 191
mycorrhizal fungi, 199
plants, 82, 93–104, 188, 191, 199
protist, 86
Shannon, 99
vascular plants, 3, 14
Droplets, 290
Drought, 257–259, 389
sensitivity, 324–326
Dry conditions, 147
Dry days, 266
Dry phases, 25
Dynamic pressure maps, 159

E
Easterlies, 9, 376–377
Ecological-Economic Farm Diversification (EEFD), 241, 389
Ecosystem services, regulating, 153–168
Economic valuation, 45
Economy, 386–389
Ecosystem
functioning, 260
service, 135
stability, 197, 391
state, 377
sustainable management, 395
Ecosystem services (ES), 32, 46, 107, 168, 282, 397, 398
categories, 42
climatic related, 285
concept, 41–48
concept MEA, 41
cultural, 43, 200, 375
diversity related, 93–94
hydrologic related, 44, 285
optimized, 238
preserving, 43, 81
provisioning, 45, 113–114, 171–183, 200, 205–215, 219, 240, 273, 282, 355–369, 375, 380, 385–390, 406–408
regulating, 44, 111–113, 141–149, 282, 298, 375, 379–381
RU 816, 43–45
supporting, 44, 108–111, 141–149, 375, 379, 381, 383, 385, 390
threats, 373
water-related, 107, 114, 115, 273, 285
Ecotourism, 174, 386
ECSF. *See* Estación Científica San Francisco (ECSF)
Ecuador, 19–27
Andes, Ecuadorian, 26
EEFD. *See* Ecological-economic farm diversification (EEFD)
El Niño Southern Oscillation (ENSO), 383
El Tibio, 223, 225
Emissions, 288
forest fire, 146
inventories, 289
scenarios, 268, 272, 292, 294, 333, 336
trends, 289
Emisstrajekt, 289
Endemic plant, 46
Endemism, 6
Energy balance, 279
Enrichment planting experiment, 181
ENSO. *See* El Niño Southern Oscillation (ENSO)
Environmental changes, 19–27
Environmental education, 413, 414
Environmental equity, 33
Environmental justice, 33
Eois, 70, 71, 377
Epiphytes, 101, 103, 260
communities, 102
Ericaceae, 189
hemiepiphytic, 189
terrestrial, 189
Erosion, 11
ES. *See* Ecosystem services (ES)
Estación Científica San Francisco (ECSF), 4
Ethnic groups
Mestizos, 172, 177, 206
Saraguros, 172, 177, 206, 215, 220
Shuar, 172, 177

Eucalyptus saligna, 180–181, 406
Eutrophication, 295
Evapotranspiration, 110, 267, 282, 381
Exotic species, 15, 406
Experimental sites location, 7
Experiments, ecological, 6–8

F
Farmers, 220
  risk-averse, 239
FERPAST. *See* Fertilization of Active Pastures (FERPAST)
Fertilization, 213, 240, 287, 299, 302, 303, 308, 310, 316, 356, 365, 383, 384, 390, 391
Fertilization of Active Pastures (FERPAST), 301, 309, 358, 361, 365–368
  location, 7
Field cropping, 227–228
Financial incentives, 47
Financial return, 240, 241
Financial yield, 346
Fine litter production, 323
Fine root
  biomass, 320, 323
  production, 132, 135
Fire, 100, 209, 374
  events, 64
  experiment, 362
Fog, 288
Fondo Regional del Agua (FORAGUA), 174
Food, 227–229, 237, 344, 388
  prices, 238
  production, 220
  sovereignty, 242
  webs, 380
FORAGUA. *See* Fondo Regional del Agua (FORAGUA)
Forcing, 332
Forest management
  experiment, 387
  sustainable, 343
Forestry options, 351
Forests, 279, 297–310, 316, 319, 325, 331, 343, 348, 373, 379, 380, 389
  conversion, 37, 387
  dynamics, 96–97
  evergreen tropical mountain, 53
  loss, 35
  management, 171–183
  natural, 175, 391
  organic layer, 308
  plants used, 178
  protection, 386
  tree species, 189
  vegetation, 98
FORMIND model, 333–334
Fragmentation, 94, 101, 103
Plant, functional traits, 104

G
Gaps, 97
Gardening, 229
GCM. *See* General circulation model (GCM)
Gene-ecological zones, 180
General circulation model (GCM), 266
Geometrid moths, 70, 377
German Research Foundation, 411
Glacial environment, 57–58
Glaciation, 11
Global change, 374–375
Global warming, 248, 339
Glomeromycota, 191, 378
GPP. *See* Gross primary production(GCM)
Gradients, 8
  altitudinal, 95, 110, 135, 146, 292
  land use, 211
  topographical, 96
Grass biomass production, 367, 369
Grassland, 279
Gross photosynthetic production, 380
Gross primary production (GPP), 118, 131
Gynoxys, 376

H
H^+ concentration, 145
Heat shock, 210
Hedyosmum, 58, 64, 254
Herbicides, systemic, 364
Holcus, 212, 213
  H. lanatus, 97, 207
Holocene, 58, 60, 64, 254, 259, 376
Home gardens, 229, 230
Hotspot, 89
Households, 214, 215, 223
  GNP, 215
Huancabamba depression, 376
Huertas, 229
Human needs, 33
Human Settlement History, 60–63
Hydrological fluxes, 281, 283
Hydrological models, 267, 271, 272
Hydrology, 271–273, 275–285, 381–382
Hydro-meteorological fluxes, 278, 281

I

IDH. *See* Intermediate disturbance hypothesis (IDH)
IDRISI Andes land use change modeller, 277
Implementation, 399
Improvement thinning experiment, 176, 387
Inclination, 13
Income portfolio, 215, 388
Information management, 401
Interaction services, 44
Interdisciplinary, 402, 413
Intergovernmental Panel on Climate Change (IPCC), 23, 268, 292, 336, 377, 389
Intergovernmental Science-Policy Platform for Biodiversity and Ecosystem Services (IPBES), 41
Intermediate disturbance hypothesis (IDH), 154
Interviews, 46
Invasive species, 390–392
Inventory, 189
  ethnobotanical, 221
  floristic, 189
  mycobiont, 189
  plants, 229
IPBES. *See* Intergovernmental Science-Policy Platform for Biodiversity and Ecosystem Services (IPBES)
IPCC. *See* Intergovernmental Panel on Climate Change (IPCC)

K

Knowledge transfer, 396, 399, 402, 412

L

Lagunas Natosas Forest, 59
LAI. *See* Leaf area index (LAI)
Land, 236
  management, 47
  ownership, 236
  reforms, 3
  rights, 221
  scarcity, 238
Landform, 11
Landowners, tropical, 345
Landsat, 277
Landscape
  approach, 343
  history, 376–377
  scale, 238
  transformation, 223, 225
  view, 343
Landslides, 11, 13, 96, 113, 154, 167, 333, 338, 339, 385–386
  dynamics, 385
  erosion, 386
  factors controlling occurrence, 154
  factors controlling susceptibility, 155
  frequency, 337
  predictor for probability, 154
  risk, 390
  susceptibility, 161
  triggering factors, 385
Land-use, 211, 214, 219–233, 236
  alternatives, 232, 235, 391
  approach, 344
  changes, 19, 82, 148, 149, 275, 277, 281, 284, 285, 384–385, 390, 427
  concepts, 240, 241
  conflict, 235
  diversification, 345
    risk minimization, 391
  diversified strategy, 241
  dynamics, 20–23
  intensification, 391
  options, 344, 345
  patterns, 236
  plant diversity, 100
  portfolio, 350
  scenarios, 349
  sustainable, 226, 391
Land-use change
  future, 285
  prediction, 280–281
Land use cover change (LUCC), 280–281
Land use/land cover change (LULC), 223–226, 232
Land-use modelling, risk-sensitive, 352
Land-use portfolio, 391
Language barriers, 399
La Niña event, 293
Law of agrarian reform, 221
Leaf area index (LAI), 121, 127, 131, 320, 324
Leaf-bud break, 362
Life-style, 388
Lineage through time (LTT) plot, 70, 73
Litterfall, 160, 164, 300
Litter leachate, 305, 382
Livelihood, 32, 37, 388
  strategies, 227
Livestock density, 214
LLJ. *See* Low level jet (LLJ)
Los Guabos, 223, 224, 230
Lower Mountain Rainforest (LMF), 56
Low-impact management, 345, 351

Low level jet (LLJ), 9
LTT. *See* Lineage through time plot (LTT)
LUCC. *See* Land use cover change (LUCC)
LULC. *See* Land use/land cover change (LULC)
Lysimeters, 144

M

Machine learning, 156
Macronutrients, 367
Maize, 227
Management decisions, 35
Management plan, 36
Mass movements, 11
MAT. *See* Mean annual temperature (MAT)
Matorral, 220
Matrix plots, 121
Matter turnover, 297–310
MEA. *See* Millennium Ecosystem Assessment (MEA)
Mean annual temperature (MAT), 118
Mean transit time, 112
Meat
  prices, 238
  production, 214
Melinis, 207, 208, 212, 355
  biomass production, 207
  M. minutiflora, 97, 206, 207, 390
Mestizos, 220, 231, 387
Microarthropods, 81–89
Microbial activity, 258
Microbial community structure, 309
Micro-catchment, 143
Mid-successional species, 337
Milk
  prices, 238
  production, 214
Millennium Ecosystem Assessment (MEA), 33, 41, 42, 383
Mineralization, 147, 367
MLP. *See* Multi-layer perceptron (MLP)
Model ensemble, 268
Monoculture, 239
Multi-layer perceptron (MLP), 277
Mycobionts, 188
Mycorrhizae, 188, 199
  specialization, 377

N

National forest and reforestation programme, 182
Native species, 406
Near-surface flow, 143
NEP. *See* Net ecosystem production (NEP)
Nestedness, 377
Net ecosystem production (NEP), 118
Net primary production (NPP), 118, 135, 379
Net revenues, 176
Networks
  architecture, 188
  degree distributions, 193
  graphs, 192
  metrics, 194
  mutualistic, 191
  nested distribution of links, 193
  NODF, 193
  theories, 188
NH_4^+ concentrations, 291
Nitrate input, 291
Nitrogen (N), 141, 166, 289, 315
Nitrogen deposition, 19, 20
  anthropogenic, 297
NO_3^-concentrations, 291
NODF. *See* N = paired Overlap, Decreasing Fill (NODF)
NO_3^-–N concentration, 145, 147
N_2O–N losses, 305
Non-timber forest products (NTFPs), 176, 222
N = paired Overlap, Decreasing Fill (NODF), 192
NPP. *See* Net primary production (NPP)
NTFPs. *See* Non-timber forest products (NTFPs)
NUMEX. *See* Nutrient Manipulation Experiment (NUMEX)
Nursery, 177
Nutrient concentrations, foliar, 302
Nutrient Manipulation Experiment (NUMEX), 84, 299, 300, 302, 305, 306, 308, 316, 319, 383
  location, 7
  study design, 317, 318
Nutrients, 361, 365, 367, 383
  addition, 297–310, 315–327
  analysis, 357
  application, 309
  atmospheric deposition, 382
  availability, 142, 166, 168
  cycles, 141–149, 298, 382, 385
  deposition, 26, 288, 290, 294, 325, 374, 390
  input, 287, 297
  land use change, 384–385
  limitations, 299
  manipulation, 316
  services, 44
  short term effects, 383
  tree rings, 383
  use efficiency, 367

O
Operational Taxonomic Units (OTUs), 82, 84, 190
Orchidaceae, 189
Organic layers, 142, 160, 163, 167, 300, 382
Organismic groups, 4
Oribatid mites, 82, 86–88, 380
OTUs. *See* Operational Taxonomic Units (OTUs)

P
Palynological records, 57
Palynological study, 55, 249
Papers, 400
Parameterization, 332
Páramo, 57, 58, 60, 64, 250, 389
Pastos
  P. azules, 207, 212, 355
  P. mieles, 208, 355
  P. naturales, 207
Pasture fertilization experiment, 299
Pastures, 98, 206–209, 211, 212, 215, 240, 297–310, 331, 348, 362, 373, 379, 384
  abandoned, 6, 212, 348, 356, 365, 367, 373
  active, 365, 367
  development, 205–215
  dry matter production, 359
  erosion, 374
  expansion, 223
  farming, 205, 213–215, 369, 387, 388
  fertilization, 307, 358, 384
  grasses, 206, 302, 339, 355
  high resolution satellite image, 214
  intensity of grazing, 360
  management study, 205–215, 240, 355–369
  microbial biomass, 210
  recovered, 365
  regeneration experiment, 364, 366
  rehabilitation experiment, 232, 357–358, 363–365
  revenue, 214
  rotational grazing, 213
  soil, 209
  species composition, 210
  sustainable, 365
  types, 207–209
  use, 205–215
  utilization, 365
  vegetation, 210
  water regulation, 381
Patterns, altitudinal, 89
Payments, 346
Payments for Ecosystem Services (PES), 38, 45
Pennisetum clandestinum, 97, 207, 390
Perception, 400–401
PES. *See* Payments for Ecosystem Services (PES)
PFT. *See* Plant functional types (PFT)
Phospholipid fatty acid (PLFA), 301, 308
Phosphorus (P), 141, 315
Photosynthesis, 124, 129
  C4-type, 355
  temperature dependence, 334, 335
Photosynthetic capacity, 124
Photosynthetic performance, 334
pH rainfall, 146
Phytopathogenic ascomycete
  Cryptomycella pteridis, 360
Pine shelter, 406
Pinus patula, 99, 180, 379, 384, 387
Pioneer
  plant, 96
  species, 337
  trees, 99
Piper, 73
Planning instruments, 401
Plant functional types (PFT), 333
Plants, 83, 248
  family composition, 98
  frequency distributions, 190
  functional types, 213
  resources, 229
  uses, 179, 219–233
  vascular, 97, 99, 101
Pleistocene, 53, 57, 64, 252, 259
Pleistocene refuge hypothesis, 68
PLFA. *See* phospholipid fatty acid (PLFA)
Plots
  forest, 199
  reforestation, 199
PNP. *See* Podocarpus National Park (PNP)
Poaceae, 58
Podocarpus–El Cóndor, 34, 222
Podocarpus–El Condor UNESCO MAB Biosphere Reserve, 373
Podocarpus National Park (PNP), 5, 34, 53, 55, 386
Podocarpus oleifolius, 176, 222, 326, 386, 407
Pollen
  analysis, 56
  compositional change, 255
  data, 253
  records, 254, 259

Polylepis, 58, 64, 256, 376
Portfolio
  crops, 242
  optimization, 242
Potable water, 46
Potential crop trees (PTC), 387
Precipitation, 109, 111, 266, 269, 270, 381
Process-based management, 401
Productivity, 121, 122, 128, 315–327, 380
Project-based management, 401
Protists, 81–89
PTC. *See* Potential crop trees (PTC)
Prumnopytis montana, 386
Pteridium arachnoideum, 4, 15, 99, 209, 332, 335
Public, 417
Publications, number of, 400

Q
Quaternary climatic oscillations, 67

R
Rabadilla de Vaca Mire, 61
Radar, 405
Radiation, 11
  climate, 124
  solar, 359
Radiocarbon ages, 56
Rain
  exclusion, 258, 259
  formation processes, 10
Rainfall, 8, 9, 25, 109, 279
Rainforest
  lower mountain, 56
  upper mountain, 56
Range, altitudinal, 257
Rarefaction analysis, 88
Rare species, 199
RBSF. *See* Reserva Biológica San Francisco (RBSF)
Recultivation, 348
Reforestation, 180–182, 232, 345, 349, 351, 387, 388, 406
  pastures, 171–183
  sites, 84
Regionalization, 157
Regulative idea, 34
Rehabilitation experiment, 356–358
Relative humidity (rH), 11, 15
Remote fertilization, 290–291, 295
  tree growth, 292–294
Repasturization, 345, 350, 352, 355–369, 387, 388
Reproductive material, 177
Research, 34–38
  collaborative, 415, 426
  infrastructure, 7, 414
Research areas, 4
  geomorphologic zonation, 12
  landform distribution, 12
  location, 4, 5
Researchers qualification levels, 415
Reserva Biológica San Francisco (RBSF), 4, 20, 94, 111, 189, 249, 266, 332
Resilience, 97, 102, 356, 374–378, 391
Resorption potential, 367
Respiration, 124, 133, 134
Revenue, 349
rH. *See* Relative humidity (rH)
Rio San Francisco, 108, 111
Rio San Francisco catchment, 108, 110, 111, 280–281, 381
Rio San Francisco valley, 95, 98, 153, 205, 265, 273, 280, 285, 355, 381, 411
Risk, 242
Risk minimization, 389
Road construction, 20, 37
Road cut, 199
Root growth, 379

S
Sahara dust, 383
Saraguros, 231
Scenarios, 23, 268, 271, 284, 291–292, 336, 377, 348–349, 380, 381, 389
Sebacinales, 191
Sediments, 113
  total suspended, 113
Seed
  centres, 182
  harvesting calendars, 180
Setaria, 100, 207–210, 212, 213, 282, 302, 355, 359, 360, 362, 364, 365, 367
  diet, 208
  nutritive value, 208
  oxalate content, 208
  S. sphacelata, 6, 97, 206, 209, 278, 301, 303, 332, 335, 339, 355, 359, 361, 367, 381, 388, 390
Shelter effect, 406
Shelter tree, 387
Shrubland, 356
  succession fallow, 356
Simplified Management Programme (PAFSI), 175
Sistema Nacional de Áreas Protegidas (SNAP), 173
  PANE, 173

Slash-and-burn, 15, 205, 209, 226
Small holders, 369
SNAP. *See* Sistema Nacional de Áreas Protegidas (SNAP)
SoBraCoMo. *See* Southern bracken competition model (SoBraCoMo)
SOC. *See* Soil organic carbon (SOC)
SOM. *See* Soil organic matter (SOM)
Socio-Bosque, 38
Soil, 301, 359, 362, 365, 390
  acidification, 141
  alkalization, 384
  biological diversity, 378
  biota, 81
  buffering capacity, 390
  fertility, 142
  food web, 81, 380
  fungi, 81–89
  macro-nutrients, 360
  matric potentials, 144, 148
  microbial biomass, 383
  microfauna, 248
  microorganisms, 148, 259, 307, 310
  moisture regime, 143, 146–147, 149, 248
  nutrients, 367
  properties, 211, 380
  protists, 82
  regionalization, 156
  respiration, 319, 321–323
  soaking, 376
  solution, 164
  water content, 359
  water logging, 167
Soil organic carbon (SOC), 125, 211
Soil organic matter (SOM), 134, 209
Soil-Vegetation-Atmosphere-Transfer-Model (SVAT), 278
Southern Bracken, 339, 360, 374
Southern Bracken Competition Model (SoBraCoMo), 278, 332, 356, 359
Speciation, 67–76, 377
  adaptive, 377
Species composition, plants, 212
Species numbers, 3–4
Species richness, 67, 95, 191, 377
Stagnic color pattern, 161
Stagnic horizon, 162
Stagnic horizon occurrence probability, 162
Stakeholders, 45, 46
Statistical downscaling method, 267
Statistical landslide models, 155–156, 160–161
Stock density, 227, 367
Stream flow, 143
Study area, 3–15
Subpáramo, 57, 60, 64
Subsidies, 346, 348
Succession, 96, 99, 100, 103, 223, 338
Survey, 45
Survival, trees, 180
Sustainability, 343
Sustainability portfolio, 392
Sustainable agriculture, 235–242
Sustainable development, 32–34
  socio-economic development, 37
Sustainable Forest Management Programme (PAFSU), 175
Sustainable management, 403
  case study, 395–408
Sustainable use, 386–389
  forests, 343–352
SVAT. *See* Soil-Vegetation-Atmosphere-Transfer-Model (SVAT)

T
Tabebuia chrysantha, 175, 180, 222, 387, 407
Taxonomy, 257
Temperature, 266, 268, 270
  altidudinal lapse rate, 376
  increase, 377
Tensiometers, 144
Testate amoebae, 82, 86–87, 251, 257–260, 380
Timber
  extraction, 222
  poaching, 386
  resources, 348
Topographic shelter, 164
Training activities, 417, 425
Transfer project, 403, 406–408
Tree growth, 54, 256, 292–294, 321, 324–326
Treeline, 60, 248, 251, 252, 256, 259, 376, 377
Tree line ecotone, 168
Trees
  isolated, 101, 103
  rings, 63, 288, 293, 295
  species, 256, 257
Tremarctos ornatus, 37
Trends, altitudinal, 133–135
Tropical easterlies, 287
Tropical forests, 19
Tulasnellales, 191

U
UMF. *See* Upper Mountain Rainforest (UMF)
Uncertain portfolio yield, 347

Unsustainability, 374
Upper Mountain Rainforest (UMF), 56

V
Valle Pequeño, 62
Valley-breeze system, 9
Vegetation
  history, 376
  natural, 94, 97
  secondary, 97–99
Vegetation dynamics, 331–340
  past, 251–256

W
Warming, 389–390
  current, 376
Water cycle, 107
Water fluxes, 111
Waterlogging, 163
Weather conditions, 356, 359
Weeds, 4, 205–215, 355, 362, 390
Weibull density function, 159
Weighted model ensembles, 160
Willingness to pay (WTP), 46–48
Wind
  direction, 291
  pressure, 159
  speed, 164, 165, 167
WTP. *See* Willingness to pay (WTP)

Y
Younger Dryas event, 64

Z
Zea mays, 227

Ecological Studies

Further volumes can be found at springer.com.

Volume 204
Biological Invasions in Marine Ecosystems: Ecological, Management, and Geographic Perspectives (2009)
G. Rilov and J.A. Crooks (Eds.)

Volume 205
Coral Bleaching: Patterns, Processes, Causes and Consequences (2009)
M.J.H. van Oppen and J.M. Lough (Eds.)

Volume 206
Marine Hard Bottom Communities: Patterns, Dynamics, Diversity, and Change (2009)
M. Wahl (Ed.)

Volume 207
Old-Growth Forests: Function, Fate and Value (2009)
C. Wirth, G. Gleixner, and M. Heimann (Eds.)

Volume 208
Functioning and Management of European Beech Ecosystems (2009)
R. Brumme and P.K. Khanna (Eds.)

Volume 209
Permafrost Ecosystems: Siberian Larch Forests (2010)
A. Osawa, O.A. Zyryanova, Y. Matsuura, T. Kajimoto, and R.W. Wein (Eds.)

Volume 210
Amazonian Floodplain Forests: Ecophysiology, Biodiversity and Sustainable Management (2010)
W.J. Junk, M.T.F. Piedade, F. Wittmann, J. Schöngart, and P. Parolin (Eds.)

J. Bendix et al. (eds.), *Ecosystem Services, Biodiversity and Environmental Change in a Tropical Mountain Ecosystem of South Ecuador*, Ecological Studies 221,
DOI 10.1007/978-3-642-38137-9, © Springer-Verlag Berlin Heidelberg 2013

Volume 211
Mangrove Dynamics and Management in North Brazil (2010)
U. Saint-Paul and H. Schneider (Eds.)

Volume 212
Forest Management and the Water Cycle: An Ecosystem-Based Approach (2011)
M. Bredemeier, S. Cohen, D.L. Godbold, E. Lode, V. Pichler, and P. Schleppi (Eds.)

Volume 213
The Landscape Ecology of Fire (2011)
D. McKenzie, C.S. Miller, and D.A. Donald (Eds.)

Volume 214
Human Population: Its Influences on Biological Diversity (2011)
R.P. Cincotta and L.J. Gorenflo (Eds.)

Volume 215
Plant Desiccation Tolerance (2011)
U. Lüttge, E. Beck, and D. Bartels (Eds.)

Volume 216
Forest Hydrology and Biogeochemistry: Synthesis of Past Research and Future Directions (2011)
D.F. Levia, D. Carlyle-Moses, and T. Tanaka (Eds.)

Volume 217
Peatlands of the Western Guayana Highlands, Venezuela: Properties and Paleogeographic Significance of Peats (2011)
J.A. Zinck and O. Huber (Eds.)

Volume 218
Aralkum – A Man-Made Desert: The Desiccated Floor of the Aral Sea (Central Asia) (2012)
S.-W. Breckle, W. Wucherer, L.A. Dimeyeva, and N.P. Ogar (Eds.)

Volume 219
Seaweed Biology: Novel Insights into Ecophysiology, Ecology and Utilization (2012)
C. Wiencke and K. Bischof (Eds.)

Volume 220
Growth and Defence in Plants: Resource Allocation at Multiple Scales (2012)
R. Matyssek, H. Schnyder, W. Oßwald, D. Ernst, J.C. Munch, and H. Pretzsch (Eds.)

Volume 221
Ecosystem Services, Biodiversity and Environmental Change in a Tropical Mountain Ecosystem of South Ecuador (2013)
J. Bendix, E. Beck, A. Bräuning, F. Makeschin, R. Mosandl, S. Scheu, and W. Wilcke (Eds.)

Printed by Printforce, the Netherlands